This book is dedicated to our families.

Graph Theory: Modeling, Applications, and Algorithms

Geir Agnarsson

Department of Mathematics
George Mason University

Raymond Greenlaw

School of Computing
Armstrong Atlantic State University

PEARSON
Prentice
Hall

Upper Saddle River, NJ 07458

Library of Congress Cataloging-in-Publication Data on File.

Vice President and Editorial Director: *Marcia J. Horton*
Senionr Editor: *Holly Stark*
Editorial Assistant: *Dolores Mars*
Executive Managing Editor: *Vince O'Brien*
Managing Editor: *David A. George*
Production Editor: *Craig Little*
Director for Creative Services: *Paul Belfanti*
Creative Director: *Juan Lopez*
Cover Designer: *Bruce Kenselaar*
Cover Artist: *Dan Sandin*
Managing Editor, AV Management and Production: *Patricia Burns*
Art Editor: *Gregory Dulles*
Manufacturing Manager: *Alexis Heydt-Long*
Manufacturing Buyer: *Lisa McDowell*
Executive Marketing Manager: *Tim Galligan*

12 2025

ISBN 978 0-13-142384-8

Pearson Education Ltd., *London*
Pearson Education Australia Pty. Ltd., *Sydney*
Pearson Education Singapore, Pte. Ltd.
Pearson Education North Asia Ltd., *Hong Kong*
Pearson Education Canada, Inc., *Toronto*
Pearson Educación de Mexico, S.A. de C.V.
Pearson Education—Japan, *Tokyo*
Pearson Education—Malaysia, Pte. Ltd.
Pearson Education, Inc., *Upper Saddle River, New Jersey*

Contents

Preface

Graph Theory

This introductory textbook gives you up-to-date coverage of the fundamentals in graph theory. Once considered an unimportant branch of topology (some even called it the "slum of topology"), graph theory has long since justified its existence through many important contributions to a wide range of fields. Rich in interesting problems and applications, it is now one of the most studied and fastest-growing areas within discrete mathematics and computer science. Countless problems involving a collection of discrete objects can be phrased and solved by graph-theoretic methods. Many such methods have become standard and are considered a natural part of every curriculum in discrete mathematics and computer science.

Who Should Read This Book?

Anyone interested in learning graph theory, discrete structures, or algorithmic design for graph problems can benefit from reading this book. It is a good starting point for students majoring in or having a major interest in any of the following fields:

- mathematics
- computer science
- information technology
- engineering
- economics
- physics
- biology

All of these areas have problems that can be solved using graph theory. Hence, this book will be useful for students interested in solving scientific problems involving any form of discrete structures.

Organization of the Text

The material is organized for a one-semester college course or for a self-study of graph theory. We present the subject matter in an order appropriate for a junior- or senior-level mathematics, computer science, or engineering major or a beginning-level graduate student. Many topics regarding applications or interesting implications are introduced via the exercises. Among other things, the book covers the following:

- basic concepts, definitions, theorems, and examples from graph theory
- a collection of interesting results from mathematics that involve key concepts and proof techniques
- design and analysis of computer algorithms for solving problems in graph theory
- applications of graph theory to the sciences: physics, electrical engineering, and various other fields.

Each chapter presents material of varying difficulty. We have attempted to organize the material based on its usefulness, importance, and practical impact rather than mere difficulty. The presentation, however, is as "linear" as we could make it, relying at each point only on what has been covered earlier.

The book is written in LaTeX 2_ε, which is the document preparation system of current choice for most mathematical and scientific textbooks and articles. This typesetting system enhances the *logical design* of the entire text.

Teaching from This Book

In a standard one-semester undergraduate course in graph theory, the majority of material in Chapters 1–7 should be covered. This part contains most of the essential definitions and important theorems in graph theory. The first two chapters can be viewed as a "Cook's tour" through the basic setup of graph theory and can therefore be covered quickly.

The later chapters are each self-contained, provided that the suggested Chapters 1–7 have been covered already. The instructor can choose material from the later chapters as desired. Some suggestions:

- Although interesting for all, §4.6 is optional for undergraduate mathematics and computer science majors.
- The last three sections in Chapter 7 (§7.7–§7.9) are optional for undergraduate non-mathematics majors.
- When discussing graph coloring, §8.6 and all of Chapter 9 are optional.
- When discussing covers and matchings in graphs, Chapter 10 is a must, but Chapter 11 is optional.
- Regarding the last two chapters, Chapter 12 is optional for students who are neither mathematics nor computer science majors. Chapter 13 should be covered by computer science and electrical engineering majors.

A preliminary version of this book was used as a text in graph theory for first-year graduate students in mathematics, computer science, statistics, and electrical engineering at George Mason University. For such use, all of the chapters should be covered, possibly supplemented by/with a term paper, a large programming assignment, or an additional special topic in graph theory (of the instructor's choice).

Why Choose This Book?

A wealth of excellent, good, decent, and not-so-good textbooks in graph theory are currently available. Do we need another?

To answer this question, keep in mind that each instructor has his or her own opinion on how to present graph theory. Should graph theory be made a self-contained branch of pure mathematics? Should applications be emphasized? Should graph algorithms be emphasized equally as much as the theory itself? Should the classical results in graph theory be presented in the book in the same order in which they were discovered? What is pedagogically the best way to present nontrivial results in graph theory? No matter how much work is put into a textbook on graph theory, or any other technical subject, it will *never* be the ideal textbook for every instructor in the field.

Most textbooks on graph theory we feel, fall into one of three restrictive categories: (i) books that treat graph theory as a branch of pure mathematics, addressing neither the recent applications of graph theory outside mathematics nor any algorithmic aspects, (ii) books that emphasize only the practical aspects of graph theory, lacking the necessary rigor in either presentations or proofs, and finally (iii) books on graph algorithms and data structures, where a wealth of computer programs on graphs, especially trees and binary trees, are presented but the general treatment of graph theory (both practical and theoretical) is sparse.

Our aim, by careful selection of topics, has been to present a textbook on graph theory that is mathematically rigorous, but also practical, intuitive, and algorithmic.

We have been mathematically rigorous by presenting formal statements such as theorems, definition, lemmas, and such in a precise and unambiguous manner, such that any mathematically inclined person can get a firm grasp of the subject matter right from the beginning. This style is particularly important for a programmers who need to know exactly what is required and how to encode a graph-related program.

We have addressed what is important from a practical point of view. For example, the various notions of trees are covered relatively early in the book, since many researchers in the sciences have been working with trees and binary trees without even knowing the basic graph-theory definitions. Also, when possible, we have included constructive algorithmic proofs of known theorems, rather than nonconstructive ones, even when the nonconstructive ones are shorter, so that the alert reader can extract an algorithm from a given constructive proof. In particular, many inductive proofs have been chosen over other existing proofs. (Of course,we also include some nonconstructive proofs, if they are classical and particularly elegant.)

Finally, we have included numerous algorithms written in a precise pseudocode. Anyone who knows a programming language (like Java, C++, C, and so on) can write a computer program directly from our pseudocode. Both the input and output are listed in a precise manner to indicate what data structures are most suitable for a programmer. All the assignments within the pseudocodes are mathematically precise and accurate. Some commonly used graph algorithms that were not discussed earlier in the text are covered in Chapter 13, which is devoted entirely to graph algorithms. We have included full-fledged computer programs in Java for these graph algorithms. The programs are carefully commented and can be entered, compiled, and executed right away. We chose Java because it is the current programming language of choice at many universities. Java is also a "small language," which encourages clear and elegant computer-code writing.

We have addressed all three aspects of graph theory rather thoroughly on the topics that we have chosen to cover: the pure mathematical one, the practical one, and the computer science aspect. A solid introductory book on graph theory should address all these aspects carefully.

The main features of this textbook are as follows.

Self-containment: The textbook is self-contained. Almost all proofs of some exceptionally technical theorems (such as the Four Color Theorem and a Minor Theorem) are contained in the text. The text does not assume any special mathematical background beyond the standard undergraduate mathematics courses.

Precise presentation: All definitions, theorems, and such are stated in a precise and formal manner, so there should be no ambiguity about the topic being discussed.

Explanatory notes: Numerous notes and remarks are included to explain the "common-sense" point of view, the motivation, and many "hand-waving" arguments, so that the reader gets the best of both worlds—the rigor and the intuition.

Examples: A wealth of examples explain both the idea of technical definitions and theorems and the applications in graph theory itself, computer science, and other sciences.

Exercises: A wealth of exercises are presented at the end of each chapter, all solvable with the material presented in the text. These 500+ exercises vary greatly in difficulty, but they should all be attempted by the diligent and alert reader. We have not included hard research problems or unsolved problems, since we believe that exercises should *exercise*, stimulate, and encourage the reader in mastering the fundamental concepts that are treated in the text itself. As you might already know, problems in graph theory can look innocent, yet have no readily accessible solution.

Hints and clues: Numerous hints and suggestions are provided for many of the more involved exercises, especially those that are important to the development of the field of graph theory itself. Hence, gaps in the overall treatment are reduced to an absolute minimum.

Figures: The book contains approximately 250 figures and diagrams to explain and illuminate graph-theory concepts that are being defined or discussed. The figures are designed to be simple, uncluttered, and "to the point," rather than to impress. They are all drawn in the `xfig` program, which allows the use of *exactly* the same fonts within the figures as within the text itself. Formulas can be displayed around the figures.

Algorithms in pseudocode: A wide range of graph algorithms are presented in a precise pseudocode for easy implementation in any programming language.

Java programs: A collection of graph algorithms written in Java are included, that are ready for compiling and running.

Some uniqueness: Many topics are presented that are, not always covered in modern textbooks including the following:

- An early treatment of trees, binary trees, and rooted trees, both ordered and unordered.

- A fairly comprehensive topological discussion regarding surface graphs.
- Numerous types of graph colorings and related algorithms.
- A brief introduction to chordal graphs and related algorithms.
- A full chapter devoted to certain basic graph algorithms.
- Both theoretical and algorithmic treatment of graph matchings.
- Numerous enumerative aspects, such as the Matrix Tree Theorem, chromatic polynomials, rook polynomials, and matching polynomials.
- An introduction to generating functions, exponential generating functions, and their applications in graph countings.

Historical remarks: Numerous classical results, their brief history, and historical remarks about the authors of various concepts in graph theory are presented. Sometimes interesting biographical tidbits are included, too; all these are indexed.

User-friendly index: A convenient, comprehensive, and above all user-friendly index is presented, for quick look-ups of both concepts in graph theory and well-known graph theorists.

Primary citation: A fairly thorough list of citations of other books, relevant papers, and websites is presented. The list of citations is not too long, since we have made the textbook as self-contained as possible.

Not overwhelming: Graph theory is a large field that is rapidly expanding. Thus, we have made careful choices about what material to include and what to exclude. The text is not overly comprehensive, but it treats the topics covered in a precise and complete manner.

The text does not assume any special knowledge other than a basic college linear algebra course, some calculus in one variable, and a bit of mathematical maturity. A one-semester course in discrete mathematics at the sophomore or junior level and an abstract algebra course will be most helpful, but by no means necessary to understand this text in its entirety. In addition, some experience in computer programming will be helpful for understanding the algorithms presented.

We have avoided the temptation to cover all interesting topics in graph theory, which can be confusing for a newcomer to graph theory.

We have deliberately omitted the interesting topic of matroids, which is best understood *after* one has successfully completed an introductory course in graph theory.

Likewise, we have omitted any probabilistic graph theory, although it is a very interesting and currently a growing field within graph theory. Probabilistic methods in graph theory are useful for proving the existence of graphs with certain properties, but they are nonconstructive and by no means necessary in a first course in graph theory.

We have also completely omitted Ramsey Theory for graphs. Although it is a deep and profoundly interesting subject from a mathematical point of view, Ramsey Theory is a somewhat advanced topic that cannot easily be covered thoroughly in an introductory course.

Other Books in Graph Theory

We kept many books, research articles, and notes at hand when writing this book, both to compare various presentations, and to determine the most common terminology and notation. All such material is cited in the Bibliography.

For the interested reader, we mention some of these references here and offer our own personal opinions of them.

The book by Douglas West [57] is arguably the most comprehensive textbook on general graph theory available as of this writing. It is already a standard reference in other books/articles on graph theory. This graduate-level text is an excellent source of exercises and references. It covers a wide range of topics of current interest in graph theory and is excellent for one who has had some prior experience in graph theory.

The book by Reinhard Diestel [12] is a beautiful graduate-read text, in which graph theory is viewed as a branch of pure mathematics. It emphasizes style, short and often tricky proofs, and existence theorems rather than algorithmic ones. It is highly rigorous mathematically and recommended for those who appreciate the beauty of graph theory, rather than its applications.

The book by Béla Bollobás [5] is a classic graduate-read text, which treats very well the topics that it covers. It emphasizes *extremal graph theory* and how to obtain various existence theorems for graphs. The text is mathematically rigorous, and the topics as well as the exercises are very well chosen. Like Diestel's book it emphasizes elegance rather than direct practical applications (not too surprisingly, since Bollobás was Diestel's thesis adviser!).

The book by Ronald Gould [21] is a well-written textbook, by an author with as much background in computer science as in mathematics. This upper-division undergraduate text/beginning-graduate text is mathematically rigorous and includes numerous graph algorithms.

The book by Robin J. Wilson [59] is an excellent introduction to graph theory. This upper-division undergraduate textbook is short, concise, and written in the "get-right-to-the-point" style common in scholarly texts in Europe. The topics are well chosen, and the book has a good variety of exercises.

The book by Narsingh Deo [11] is one of the first comprehensive textbooks on graph applications in various other disciplines, including computer science, engineering (especially network analysis in electrical engineering), and operations research. This beginning-graduate text emphasizes practical computational applications and includes an early account of the use of graphs in the previously mentioned fields. It is a classic, which has been used as a classroom text throughout the United States and India for decades.

The book by Russell Merris [39] is a well-written and relatively short treatment of general graph theory. This upper-division undergraduate text is not overly comprehensive, but rather contains an excellent choice of topics. It provides a good source of exercises. The writing is friendly and inviting, as well as mathematically rigorous. The book covers some nontrivial topics, which are not too common in the graph-theory literature. This book includes an excellent discussion on chordal graphs, interval graphs, and a special class of interval graphs, called *threshold graphs*.

Finally, we would like to mention the lecture notes of Reynir Axelsson [3], presented in a graph-theory course given at the University of Iceland in the fall of 1989, which one of the authors (Agnarsson) attended. These notes are well written, rigorous, and have influenced many of our choices in precise presentations in our current textbook.

Suggestions and Corrections

The text may (and most likely will!) contain some errors, and certain topics may have been omitted that readers feel are especially relevant for inclusion. In anticipation of possible future printings, we would like to correct any mistakes and incorporate as many suggestions as possible. Please send comments via email to

geir@math.gmu.edu

About the Authors

Dr. Geir Agnarsson was born in Norway and grew up in Iceland. He received the B.S. degree in mathematics from the University of Iceland in 1990 and the Ph.D. degree in mathematics from the University of California at Berkeley in 1996. Geir has won a number of awards including a Fulbright Scholarship. He has been a visiting scholar at Arizona State University and Los Alamos National Laboratory. He has given a number of invited talks throughout the United States. Geir has published a variety of papers on graph theory. He was formerly a Professor with the School of Computing at Armstrong Atlantic State University and is currently a Professor with the Department of Mathematics at George Mason University.

Dr. Raymond Greenlaw grew up in Rhode Island. He received the B.A. degree in mathematics from Pomona College in 1983, the M.S. degree in computer science in 1986 from the University of Washington, and the Ph.D. degree in computer science in 1988 from the University of Washington. Ray has won a number of international awards including three Senior Fulbright Scholarship Research Awards, a Humboldt Fellowship, a Spanish Fellowship for Scientific and Technical Investigations, and a Japan Society for the Promotion of Science Invitation Fellowship. He is also a two-time ARCS scholar. Ray has written 14 books; they have been translated into several languages. His books are in the areas of complexity theory, graph theory, the Internet, operating systems, networks, parallel computation, theoretical computer science, and the World Wide Web. His books have been used in over 100 computer science programs in the United States as well as internationally. Ray's technical papers have appeared in over 35 different journals and conference proceedings. He has received research funding from 10 different agencies and groups including the National Science Foundation. His research has been supported by the governments of Germany, Hong Kong, Iceland, Italy, Japan, Malaysia, Spain, and Thailand, as well as the United States. Ray has given invited talks all over the world. He is currently a Professor of Computer Science and the Dean of the School of Computing at Armstrong Atlantic State University.

Acknowledgments

Thanks to our families for their tremendous support throughout the years and while we were writing this book. We hope that these two time periods appear different to them.

A very special thanks to George Lobell for his encouragement and support during this project. He kept the project alive.

A special thanks to Tamara Burton for her careful reading of the manuscript. Her comments were very helpful to us. Thanks to Greg Geller for his careful reading of portions of the manuscript. His comments also were very helpful to us.

Thanks to Ruslan Hristov for giving generously of his time to this project. Ruslan coded and typeset the Java programs for us. He also helped tremendously with formatting issues, layout, indexing, and style issues.

Thanks to Mirna Morrison for helping with manuscript preparation. She learned a lot more about LaTeX than she really wanted to. Her patience while correcting typos is much appreciated. Thanks to Janice Stanford who helped us with editing many of the latter chapters. Her thorough work is much appreciated. Thanks to Ben Page for creating the chapter openers for us and for working with us on the cover design. Thanks to Chris McCarthy for assisting us with typesetting. Thanks to Jonnie Chandler for assisting with manuscript preparation.

Thanks to Mary Thornton for drawing many of the figures for us in `xfig`. Her efforts are greatly appreciated and helped to improve significantly the readability of the figures in the text. Thanks to Tunishe Timmons for assistance with editing. She helped draw many of the figures for us.

Thanks to the students in the MATH–641 class, Combinatorics and Graph Theory, at George Mason University (GMU) in the fall semesters of both 2003 and 2005 for their many helpful comments and for pointing out some silly and embarrassing mistakes in a preliminary version of this text. Special thanks to Toni Robertson from the 2003 class for her careful reading and pointing out many typos, and to Jill Dunham from the 2005 class for carefully working through many of the exercises and finding additional typos. Thanks also to our GMU colleagues Walter D. Morris Jr. and Igor Griva for helpful discussions regarding linear programs in general.

Big thanks to those who reviewed this text at various stages, for their careful reading and many helpful comments. These folks include John Caughman, Eddie Cheng, Curtis R. Cook, Nancy Eaton, Abdol-Hossein Esfahanian, Fred Galvin, Richard Alan Gillman, Renu Laskar, Paulius Micikevicius, Darren A. Narayan, William J. B. Oldham, Uri N. Peled, Andrzej Proskurowski, and Robert W. Robinson. Special thanks to Professors Uri Peled and Renu Laskar for their pointed comments some faulty proofs, which helped us fix them.

Ray completed part of the work on this book while he enjoyed a Fulbright Scholarship to Chiang Mai University in Thailand. He would like ot thak the Fulbright Foundation's of the United States and Thailand for supporting this project, and the Department of Computer Science at Chiang Mai University, for hosting him. Special thanks go to Sanpawat "Bobby" Kantabutra.

Finally, many thanks to the graph-theory team at the University of Iceland, consisting of Reynir Axelsson, Magnús M. Halldórsson, and Rögnvaldur G. Möller, for many helpful discussions (which they most probably do not remember!). Without them, this book would be far inferior.

Authors' Note

It is our sincere hope that you, dear reader, will experience the power of graph theory as a mathematical tool for problem solving in mathematics, computer science, engineering, or some other chosen field. Good luck, and above all, enjoy!

Authors' Contact Information

GEIR AGNARSSON
Department of Mathematics
George Mason University, MS 3F2
4400 University Drive
Fairfax, Virginia 22030

email address: `geir@math.gmu.edu`
World Wide Web page:
`math.gmu.edu/~geir`

RAYMOND GREENLAW
School of Computing
Armstrong Atlantic State University
11935 Abercorn Street
Savannah, Georgia 31419-1997

email address: `greenlaw@armstrong.edu`
World Wide Web page:
`www.cs.armstrong.edu/greenlaw`

CHAPTER

1

Introduction to Graph Theory

Chapter Goals

Motivate the subject of graph theory.

Present some applications of graph theory.

Describe mathematical preliminaries.

Define and illustrate graphs and simple graphs.

Define a directed graph and provide examples.

Present the Hand-Shaking Theorem.

Discuss the notion of a subgraph and a subdigraph.

1.1 Introduction

This book is about a branch of discrete mathematics called *graph theory*. Discrete mathematics—the study of discrete structures (usually finite collections) and their properties—includes combinatorics (the study of combination and enumeration of objects), algorithms for computing properties of collections of objects, and graph theory (the study of objects and their relations). Many problems in discrete mathematics can be stated and solved using graph theory. Therefore graph theory is considered by many to be one of the most important and vibrant fields within discrete mathematics.

Chapters 1 and 2 contain introductory material and may be covered quickly or perhaps skipped by an advanced reader.

1.2 Why Study Graphs?

The best way to illustrate the utility of graphs is via a "Cook's tour" of several simple problems that can be stated and solved via graph theory. We do this in an intuitive manner prior to presenting formal definitions in §1.4. Graph theory has many practical applications in various disciplines, including, to name a few, biology, computer science, economics, engineering, informatics, linguistics, mathematics, medicine, and social science. As will become evident after reading this chapter, graphs are excellent modeling tools. We now look at several classic problems.

We begin with the Bridges of Kőnigsberg. This problem has a historical significance, as it was the first problem to be stated and then solved using what is now known as graph theory. Leonard Euler fathered graph theory in 1736 when his general solution to such problems was published. Euler not only solved this particular problem, but more importantly introduced the terminology for graph theory.

Problem 1.1 The Bridges of Kőnigsberg

Figure 1.1 shows the layout of the Kőnigsberg bridges. The Pregel River in Kőnigsberg[1] has two banks (labeled B_1 and B_2), and its splitting forms two islands (labeled I_1 and I_2). These islands were connected to each other and to the banks with seven bridges as shown in Figure 1.1. The problem was to make a round trip through downtown Kőnigsberg, traversing each bridge exactly once.

Euler distilled the problem into its simplest form using the diagram (that is, *graph*) shown in Figure 1.2. The black dots (that is, *vertices*) represent the areas of land, and the lines (that is, *edges*) represent the bridges.

Looking at Figure 1.2, we see that the problem now is to start at some black dot, go along each line exactly once, and end up at the starting dot. As we will see in Chapter 5,

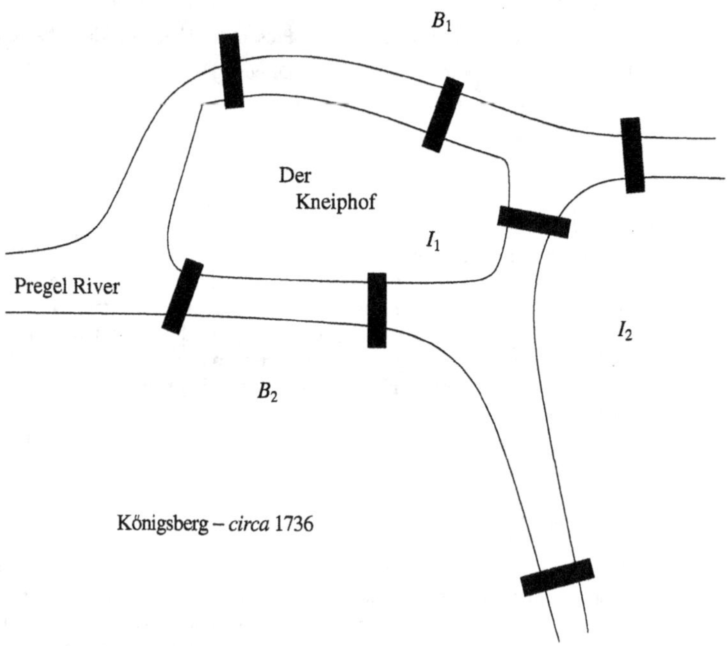

Figure 1.1: The bridges of Kőnigsberg.

[1] Now Kaliningrad.

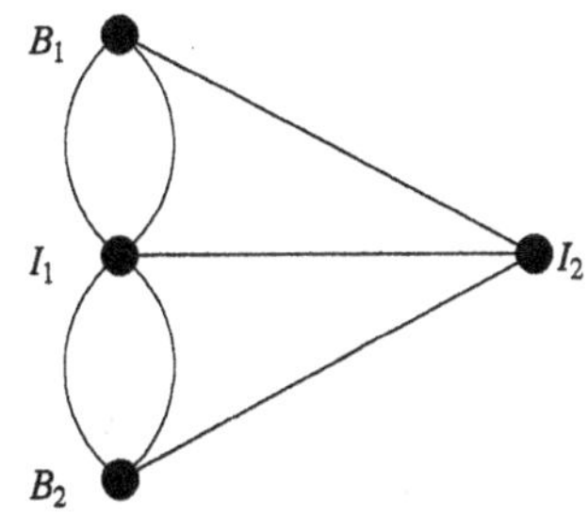

Figure 1.2: The graph of the bridges of Kőnigsberg problem.

Euler gave an exact characterization of when such problems have solutions, in terms of each dot having an even number of line-ends going into it, thereby showing in particular that the problem of the bridges of Kőnigsberg has no solution.

The next problem we discuss concerns discovering communities on the World Wide Web (WWW).

Problem 1.2 World Wide Web Communities

The World Wide Web can be modeled as a graph, where the Web pages are represented by dots or vertices and the hyperlinks between them are represented by lines or edges in the graph. One can discover interesting information by examining this "Web graph." As an example, the graph shown in Figure 1.3 is termed a *Web community*. This name is bestowed because the vertices represent two different classes of objects, and each vertex representing one type of object is connected to every vertex representing the other kind of objects. In graph theory such a graph is called a *complete bipartite graph*.

On the left-hand side of Figure 1.3, the five vertices are displayed with labels travelagencyX.html, where $1 \leq X \leq 5$. Each such label travelagencyX.html represents an HTML file for a given travel agency. On the right-hand side of the figure, the three vertices have labels www.alitalia.com, www.delta.com, and www.united.com. Each of these labels is a Web site of a well-known airline. Note that there is an edge between each HTML file and each airline, but there are no edges between two HTML files, nor are there any edges between two airlines. Such a graph structure is common on the Web between competitors in the same industry. Rival companies do not have hyperlinks on their Web pages to their competitors. For example, Delta has no hyperlink to United Airlines. To be able to offer the best deals for their customers, travel agents often have hyperlinks from their Web pages to all of the airlines, as shown in Figure 1.3. The travel agencies are also competing among themselves, so they too have no hyperlinks from one agency to any other.

Such Web communities can be discovered by finding *complete bipartite subgraphs* in the Web graph. Web community information can be used for marketing purposes or for examining the relationships among companies in a given industry.

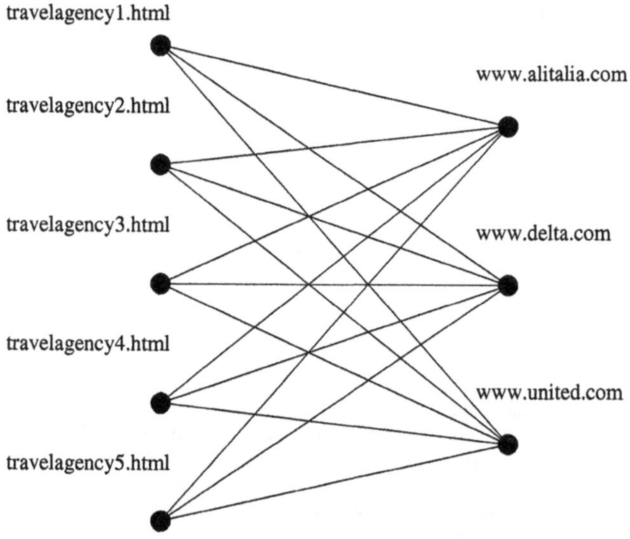

Figure 1.3: World Wide Web community.

Our next problem deals with job assignments.

Problem 1.3 Job Assignments

A certain company has five different jobs labeled by J_i, where $1 \leq i \leq 5$, and wants to hire people for these positions. The seven applicants, labeled by A_j for $1 \leq j \leq 7$, are qualified for some of the positions, but not for others. An edge between an applicant A_j and a job J_i means that applicant A_j is qualified for job J_i. The applicants are qualified for jobs as shown in the graph from Figure 1.4. This graph is an example of a *bipartite graph*, where the vertices represent two kinds of objects: the applicants and the jobs. The question in this problem is: Can the company fill all its open positions with qualified applicants?

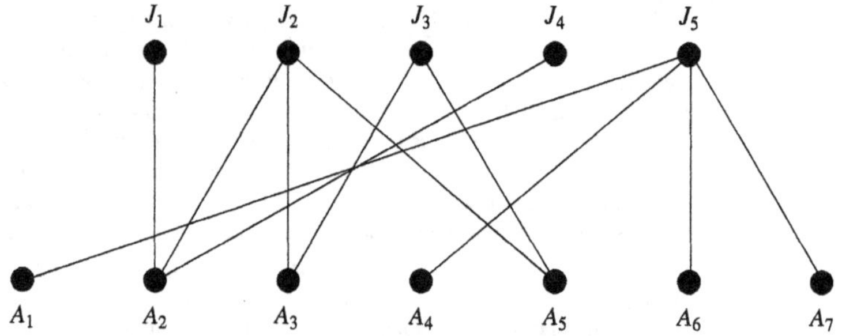

Figure 1.4: Bipartite graph describing job qualifications.

In Chapter 10 we will derive necessary and sufficient conditions for such job-assignment problems to have a positive solution. We will also derive a method to find the best possible hiring scheme when not all the positions can be filled with qualified applicants.

Note that as in Problem 1.2, there are no edges connecting vertices representing the same kind of objects: jobs or applicants. In this problem, however, an applicant need not be qualified for every job, and hence there is not an edge between *every* pair of vertices representing different objects, as was the case in Problem 1.2.

The next problem illustrates the use of graph theory when determining the minimum number of warehouses needed to store volatile chemicals.

Problem 1.4 Storing Volatile Chemicals

A particular chemical factory produces seven different kinds of chemicals, labeled as $C_1, C_2, \ldots, C_7$. For security reasons some of the chemicals should not be stored in the same warehouse. The graph depicted in Figure 1.5 shows the situation of volatility between the chemicals, where an edge between chemicals C_i and C_j indicates a grave danger in storing these chemicals in the same warehouse, whereas lack of an edge indicates it is safe to store the chemicals in the same warehouse. Here the question is: What is the minimum number of warehouses the factory needs in order to store its chemical products safely?

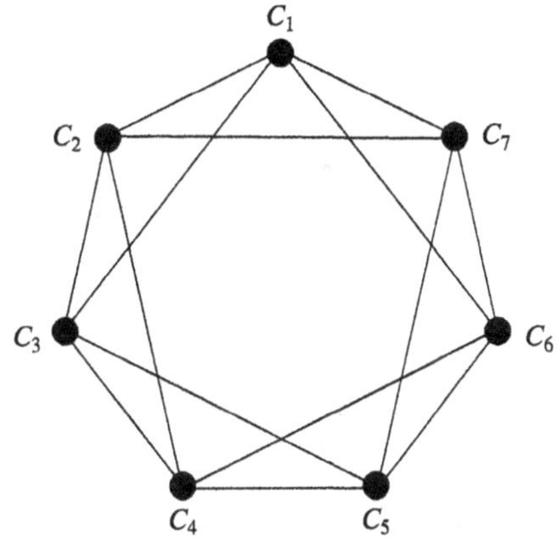

Figure 1.5: Connected chemicals should not be stored together.

If we think of each warehouse as a specific *color* assigned to each vertex, the problem boils down to using the least number of colors to assign to the vertices of the graph such that no two vertices with an edge between them receive the same color. The least number of colors needed to color the vertices of a graph in this manner is called the *chromatic number* of the graph. We are able to assign four colors to the vertices of the graph in Figure 1.5 such that no edge is between two vertices of the same color. We can, for example, color vertex C_1 with color 1, C_2 with color 2, C_3 with color 3, C_4 with color 1, C_5 with color 2, C_6 with color 3, and C_7 with color 4. The graph cannot be vertex colored using three colors, and hence the chromatic number of this graph is four; that is, the minimum number of warehouses needed for our problem is four.

In general, vertex coloring of graphs is a hard problem from an algorithmic point of view. There is no easy solution that can be applied to all such problems. Each problem must be considered individually. In Chapter 8 we will discuss colorings in general and develop a systematic way of finding the chromatic number of any fixed graph. We will also study graph colorings in various other settings. Graph coloring is currently an active and rich area of research among mathematicians and computer scientists.

The next application we present deals with resistance in an electrical network.

Problem 1.5 Electrical Networks

In Figure 1.6 each edge represents a fixed electrical resistance of 1 Ω (here "Ω" is pronounced "ohm" after the German physicist G. S. Ohm). How can we compute the total resistance between the vertices u and v? Those who know elementary physics (which is not something we will assume) realize that this problem is easy to solve for certain kinds of electrical networks, but harder for other networks, where a significant physical insight is needed. In §4.6 in Chapter 4 we will develop methods for solving problems of this kind that involve a general class of electrical networks, where there is no restriction nor special condition for their corresponding underlying graphs.

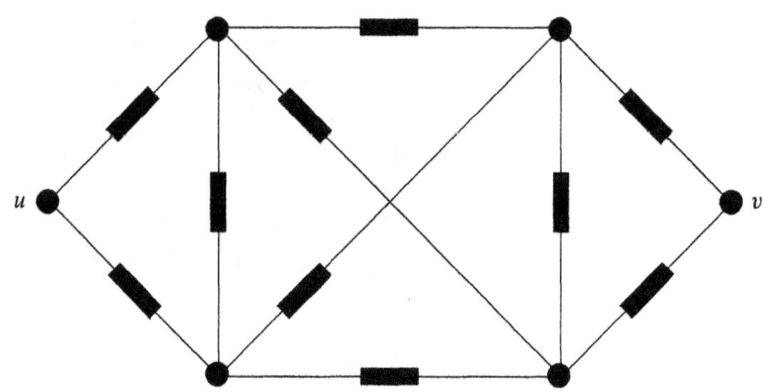

Figure 1.6: Electrical resistance network.

So far, we have presented a handful of simply stated and practical problems from graph theory. You can now probably cook up numerous other problems that can be modeled using graphs.

In order to tackle more challenging and interesting problems and develop a coherent theory about graphs, it is necessary to formalize the notions from these examples in precise definitions. Before we do that, we will review a few mathematical preliminaries.

1.3 Mathematical Preliminaries

In this section we give a brief description of some mathematical concepts that will be used throughout the book. We first specify set-theoretic concepts, describe notation, and present basic facts—some of whose proofs we leave as exercises. Several facts that are considered common knowledge will be stated without their proofs.

A *set* is a collection of objects. An object in a set is called an *element* of the set. The unique set containing no elements is called the *empty set* and is denoted by $\varnothing$. Let A and B be sets. If a is an element contained in a set A, we denote this by $a \in A$. If for every $a \in A$, it is also true that $a \in B$, we say that A is a *subset* of B and denote this by $A \subseteq B$. Two sets A and B are *equal*, denoted by $A = B$, when each element in A is also an element in B and vice versa—that is, we have both $A \subseteq B$ and $B \subseteq A$.

We have the following set operations: The *union* of A and B, denoted by $A \cup B$, is the set of those elements that are either in A or in B. The *intersection* of A and B, denoted by $A \cap B$, is the set consisting of those elements that are contained in both A and B. If $A \cap B = \varnothing$, then we say that A and B are *disjoint*. The *difference* of A and B, denoted by $A \setminus B$, consists of those elements that are contained in A and are not in B. The *symmetric difference* of A and B, denoted by $A \bigtriangleup B$, is equal to $(A \setminus B) \cup (B \setminus A)$. It consists of those elements that are contained in exactly one of the sets A and B. The notion of union, intersection, and symmetric difference can be generalized to a collection of finitely many sets rather than just two; that is, one can show that if $A_1, A_2, \ldots, A_n$ is a finite collection of sets, then the notions of $A_1 \cup \cdots \cup A_n$, $A_1 \cap \cdots \cap A_n$, and $A_1 \bigtriangleup \cdots \bigtriangleup A_n$ are all unambiguous and sensible, since the operations $\cup$, $\cap$, and $\bigtriangleup$ are all associative. The last expression denotes the set of elements that are contained in an odd number of the sets $A_1, \ldots, A_n$. (See Exercises 1.1 and 1.2.)

A set S containing finitely many elements is a *finite set*. The number of elements in the set S is denoted by $|S|$ and is often called the *cardinality of* S. For example, the set $\{0, 1, 2, 3, 4\}$ has five elements and hence $|\{0, 1, 2, 3, 4\}| = 5$. If S is infinite, that is, not finite, then we write $|S| = \infty$. The set of natural numbers, $\mathbb{N} = \{1, 2, 3, \ldots\}$, is an infinite set. Nearly all sets in this book are finite sets. If S is a set, then the set of all subsets of S, denoted by $\mathbf{P}(S)$, is called the *power set* of S. For example, if $S = \{1, 2, 3\}$, then

$$\mathbf{P}(S) = \{\varnothing, \{1\}, \{2\}, \{3\}, \{1, 2\}, \{1, 3\}, \{2, 3\}, \{1, 2, 3\}\}.$$

Let $k \in \mathbb{N}$ and $a_1, a_2, \ldots, a_k$ be any k objects. Then $(a_1, a_2, \ldots, a_k)$ is an *ordered k-tuple* or just *k-tuple* for short. For each $i \in \{1, 2, \ldots, k\}$, we say that a_i is the ith *component* of $(a_1, a_2, \ldots, a_k)$. If $A_1, \ldots, A_k$ are sets, then their *Cartesian product*, denoted by $A_1 \times \cdots \times A_k$, is the set of all the k-tuples $(a_1, \ldots, a_k)$, where $a_i \in A_i$ for each i.

For sets A and B, a *binary relation* from A to B is a subset $R \subseteq A \times B$. For $a \in A$ and $b \in B$ we write $a \sim b$ and say that "a is related to b" if $(a, b) \in R$. When there is no danger of ambiguity, we talk about the binary relation $\sim$ instead of the subset R. We will discuss two important types of binary relations in the next chapter.

For a natural number n we denote by $n!$ the product of all numbers from 1 to n, that is, $n! = 1 \cdot 2 \cdots (n-1) \cdot n$. By convention $0! = 1$. The notation $n!$ is read *n-factorial*, and it describes the number of distinct ways to order the numbers $1, \ldots, n$, one after another in a string; that is, it is the number of *permutations* of the numbers $1, \ldots, n$.

Let k and n be natural numbers with $k \leq n$. The number of k-tuples $(a_1, \ldots, a_k)$ of nonrepeating numbers from $\{1, \ldots, n\}$ is denoted by $P(n, k)$. That is to say,

1. $a_i \in \{1, \ldots, n\}$ for each $i \in \{1, \ldots, k\}$,
2. $a_i \neq a_j$ for $i \neq j$.

The fact that

$$P(n,k) = n(n-1)(n-2)\cdots(n-k+1) = \frac{n!}{(n-k)!}$$

is left to the reader as Exercise 1.3. When all the a_i's are distinct, there are precisely $k!$ ordered k-tuples $(a_1, \ldots, a_k)$ formed from the set $\{a_1, \ldots, a_k\}$. Therefore, if $C(n,k)$ denotes the number of k-element subsets of $\{1, \ldots, n\}$, then we have $k!C(n,k) = P(n,k)$, and so $C(n,k) = P(n,k)/k!$. The number $C(n,k)$ is called a *binomial coefficient*; it is read "n choose k." The most common way of denoting $C(n,k)$ is by $\binom{n}{k}$. To summarize, the number of k-element subsets of any n-element set is given by

$$C(n,k) = \binom{n}{k} = \frac{P(n,k)}{k!} = \frac{n!}{k!(n-k)!}.$$

From this formula we can deduce the following formulas, which themselves give a recursive definition of the binomial coefficients for nonnegative k and n with $k \leq n$.

$$\binom{n}{0} = 1,$$
$$\binom{n}{k} = \binom{n}{n-k},$$
$$\binom{n}{k} = \binom{n-1}{k} + \binom{n-1}{k-1}.$$

Note that $\binom{n}{n} = \binom{n}{0} = 1$ for all n. The previous formulas can also be derived using combinatorial arguments. (see Exercise 1.4.)

In chapters to follow, we will often use *induction* to prove theorems and statements about finite graphs. If $n \in \mathbb{Z}$ is an integer and $P(n)$ is a statement involving n as the only unknown parameter, the truth value of $P(n)$ depends only on n. The general concept of induction states that to prove $P(n)$ is true for all $n \geq a$, where a is a fixed integer (possibly negative), it suffices to prove the following two statements:

1. $P(a)$ is true.
2. For every $n \geq a$, if $P(k)$ is true for each $k \in \{a, a+1, \ldots, n\}$, then so is $P(n+1)$.

In most cases when proving statements about graphs, we will be using a simplified version of induction. Namely, to prove that $P(n)$ is true for each $n \geq 1$, we prove the following two statements:

1. $P(1)$ is true.
2. For every $n \geq 1$, if $P(n)$ is true, then so is $P(n+1)$.

There are many variations of proofs by induction (see Exercises 1.5 and 1.6), but the two just mentioned are the most common forms.

We will frequently use several other sets of numbers. The set of *integers* is $\mathbb{Z} = \{\ldots, -2, -1, 0, 1, 2, \ldots\}$. The set of *rational numbers* is $\mathbb{Q} = \{a/b : a, b \in \mathbb{Z} \text{ and } b \neq 0\}$. The set of *real numbers* is denoted by $\mathbb{R}$. The notation $[a; b]$ denotes the set of real numbers x such that $a \leq x \leq b$. The notation $[a; b[$ denotes $[a; b] \setminus \{b\}$, $]a; b]$ denotes $[a; b] \setminus \{a\}$, and $]a; b[$ denotes $[a; b] \setminus \{a, b\}$. The set of *complex numbers* is $\mathbb{C} = \{x + yi : x, y \in \mathbb{R}\}$, where

$i = \sqrt{-1}$ denotes the imaginary number satisfying $i^2 = -1$. The arithmetic operations of addition and multiplication of the complex numbers are given by

$$\begin{aligned}(x + yi) + (x' + y'i) &= (x + x') + (y + y')i,\\ (x + yi) \cdot (x' + y'i) &= (xx' - yy') + (xy' + x'y)i.\end{aligned}$$

From this we see that when $y = y' = 0$, then we have the usual addition and multiplication of $\mathbb{R}$. Note that the definition of the addition is natural, and the multiplication is a direct consequence of the usual distribution law and the fact that $i^2 = -1$:

$$\begin{aligned}(x + yi) \cdot (x' + y'i) &= x \cdot x' + x \cdot y'i + yi \cdot x' + yi \cdot y'i\\ &= (xx') + (xy')i + (x'y)i + (yy')(i^2)\\ &= (xx' - yy') + (xy' + x'y)i.\end{aligned}$$

Subtraction and division are then derived directly from the definitions of addition and multiplication.

For a function (also called a mapping) $f : X \to Y$, the set X is called the *domain* of f and the set Y is called the *codomain* of f. The set $f(X) = \{f(x) : x \in X\} \subseteq Y$ is called the *range* of f. The function $f : X \to Y$ is *injective* or *one-to-one*, if $f(x_1) = f(x_2)$ always implies that $x_1 = x_2$. Also, f is *surjective* or *onto* if the range equals the codomain $f(X) = Y$—that is, if for every $y \in Y$, there is an $x \in X$ such that $y = f(x)$. A map that is both injective and surjective is called *bijective*. Such functions are also called *bijections* or *one-to-one correspondences*.

Example 1.6 Consider the function $f : \mathbb{N} \to \mathbb{N}$ defined by $f(n) = n^2$. Suppose $f(n) = f(m)$. This means $n^2 = m^2$ and hence $n = \pm m$. Since $n, m \in \mathbb{N}$, and hence both n and m are nonnegative, this implies $n = m$ must hold. Therefore, f is an injective mapping. Notice that f is not onto $\mathbb{N}$, since there is no $n \in \mathbb{N}$ such that $n^2 = 2$. This fact means f is not a one-to-one correspondence. It is worth noting that if the domain of f had been $\mathbb{Z}$ instead of $\mathbb{N}$, then f would not be injective.

The map from X to X that has a value of x for each $x \in X$ is called the *identity map*. The identity map on X is denoted by $\iota_X : X \to X$. The function ι_X is bijective for any set X. Let $f : X \to Y$ and $g : Y \mapsto Z$ be any two maps. The *composition* of g with f denoted by $g \circ f : X \to Z$ is the mapping such that $(g \circ f)(x) = g(f(x))$ for each $x \in X$. If $f : X \to Y$ is a bijective map, then there is a unique map, denoted by $f^{-1} : Y \to X$, such that $f^{-1} \circ f = \iota_X$ and $f \circ f^{-1} = \iota_Y$. The map f^{-1} is called the *inverse* of the map f. Finally, if $f : X \to Y$ is a map and $X' \subseteq X$, then the *restriction* of f to X' is the map $f_{X'} : X' \to Y$ defined by $f_{X'}(x) = f(x)$ for each $x \in X'$.

Example 1.7 Let $f : \mathbb{N} \to \mathbb{R}$ be defined by $f(n) = 2n$. Let $g : \mathbb{R} \to \mathbb{R}$ be defined by $g(x) = 1 - 3x$. Then $g \circ f : \mathbb{N} \to \mathbb{R}$ is defined. In fact,

$$(g \circ f)(n) = g(f(n)) = g(2n) = 1 - 6n.$$

Note that in this case $f \circ g$ is not defined. It is easy to see that g is a bijection of $\mathbb{R}$, so the inverse of g exists. In fact, if $h : \mathbb{R} \to \mathbb{R}$ is given by $h(x) = 1 - \frac{x}{3}$, then $h = g^{-1}$, since

$$\begin{aligned}(g \circ h)(x) &= g(h(x)) = g(\tfrac{1-x}{3}) = 1 - (1 - x) = x,\\ (h \circ g)(x) &= h(g(x)) = h(1 - 3x) = \tfrac{1-(1-3x)}{3} = x,\end{aligned}$$

for each $x \in \mathbb{R}$, and hence $g \circ h = h \circ g = \iota_{\mathbb{R}}$. The restriction of g to $\mathbb{N}$ is given by $g_{\mathbb{N}}(x) = 1 - 3x$ for each $x \in \mathbb{N}$.

For a function $f : A \to B$, where both A and B are finite, we have the following:

1. If f is injective, then $|A| \leq |B|$.
2. If f is surjective, then $|A| \geq |B|$.
3. If $|A| = |B|$, then f is injective if, and only if, f is surjective.

Note the contrapositive of the first statement: If $|A| > |B|$, then f is *not* injective. This form is a powerful principle called the *Pigeon-Hole Principle*. It can be stated in the following form: "If m pigeons are put into n holes, where $m > n$, then one hole must contain at least two pigeons."

This statement can be generalized to: "If m pigeons are put into n holes, then one hole must contain at least $\lceil m/n \rceil$ pigeons." Here $\lceil x \rceil$ denotes the smallest integer $m \geq x$ ($\lfloor x \rfloor$ denotes the largest integer $n \leq x$).

Having discussed some preliminaries, we are ready to begin looking at graphs more closely.

1.4 The Definition of a Graph

We have discussed graphs informally to provide a feeling as to what they are like. An intuitive feeling is always good, but not sufficient when, for example, a computer scientist wants to implement a graph in a computer program. Informally, a graph is simply a collection of vertices and edges together with a rule about how the vertices are connected to one another with the edges. In §1.2 we saw a number of pictorial representations of graphs. To be precise, the next definition captures the essentials of the intuitive description of graphs.

The following defines the most general form of what we will call a graph throughout this book. We will shortly discuss some more restricted versions of this definition.

Definition 1.8 A *graph* or a *general graph* is an ordered triple $G = (V, E, \phi)$, where

1. $V \neq \varnothing$,
2. $V \cap E = \varnothing$,
3. $\phi : E \to \mathbf{P}(V)$ is a map such that $|\phi(e)| \in \{1, 2\}$ for each $e \in E$.

The elements of V are the *vertices* of G, and the elements of E are the *edges* of G. The map ϕ is called an *edgemap*, and the vertices in $\phi(e)$ are called the *endvertices* of the edge e. When the ordered triple defining a graph G is not given, then the set $V(G)$ will always denote the set of vertices of G, and $E(G)$ will always denote the set of edges of G.

We look at an example to illustrate this definition.

Example 1.9 Graphs are in most cases represented by diagrams consisting of dots, which represent the vertices, and curves drawn between the dots representing the endvertices, which represent the edges between the vertices. When there is no danger of ambiguity, the diagram itself is called a graph, since one can read all the information from the diagram and record the abstract definition from the diagram in a unique way.

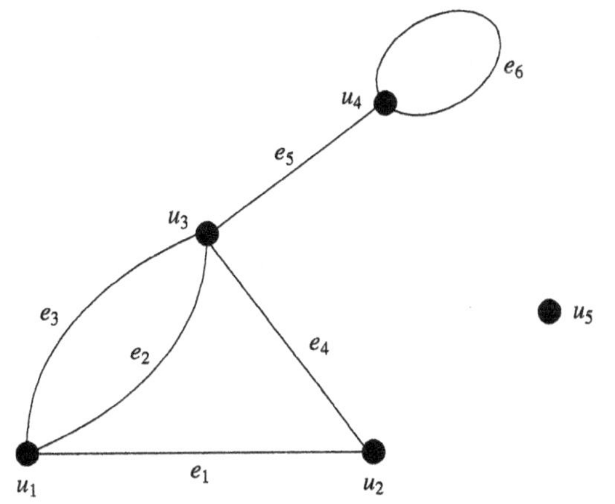

Figure 1.7: A graph with five vertices and six edges.

Consider the diagram shown in Figure 1.7. Here we see that $G = (V, E, \phi)$, where $V = \{u_1, u_2, u_3, u_4, u_5\}$, $E = \{e_1, e_2, e_3, e_4, e_5, e_6\}$, and the edgemap ϕ is given by

$$\begin{aligned}\phi(e_1) &= \{u_1, u_2\},\\ \phi(e_2) = \phi(e_3) &= \{u_1, u_3\},\\ \phi(e_4) &= \{u_2, u_3\},\\ \phi(e_5) &= \{u_3, u_4\},\\ \phi(e_6) &= \{u_4\}.\end{aligned}$$

The next definition presents the most basic terminology on graphs.

Definition 1.10 Let $G = (V, E, \phi)$ be a graph.

1. Vertices u and v in V are *adjacent* or *neighbors*, if they are the endvertices of some edge $e \in E$. That is, they are adjacent if there is an $e \in E$ such that $\phi(e) = \{u, v\}$.
2. Two distinct edges e and f are *adjacent* to each other, if they have a common endvertex. That is, $\phi(e) \cap \phi(f) \neq \varnothing$.
3. A vertex u and an edge e are *incident* if $u \in \phi(e)$—that is, if u is an endvertex of e.
4. A *loop* is an edge e, whose endvertices are equal. That is, $|\phi(e)| = 1$.
5. We say that $E' \subseteq E$ is a set of *multiple edges* or *parallel edges*, if $|E'| \geq 2$ and all $e' \in E'$ have the same set of endvertices. That is, $\phi(e') = \phi(f')$ for all $e', f' \in E'$.
6. A vertex u is called *isolated* if it is not an endvertex of any edge. That is, $u \notin \phi(e)$ for all $e \in E$.

Example 1.11 Referring to the graph given in Figure 1.7, we see the following: The endvertices of e_2 are u_1 and u_3. The vertices u_1 and u_2 are adjacent, but u_1 and u_4 are not; so u_1 and u_2 are neighbors, but u_2 and u_4 are not. The vertex u_4 is adjacent to itself. The edges e_1 and e_2 are adjacent, but e_1 and e_5 are not. The edge e_6 is adjacent both to itself and to e_5. The edge e_6 is the only loop in this graph. The edges e_2 and e_3 are parallel. The vertex u_5 is isolated.

In Example 1.9 both $V(G)$ and $E(G)$ are finite sets. By Definition 1.10 this is not necessary, although most often graphs that we consider have both a finite vertex set and a finite edge set. A graph G with both $V(G)$ and $E(G)$ as finite sets is called a *finite graph*. If either of the sets $V(G)$ or $E(G)$ is infinite, then G is called an *infinite graph*. In Figure 1.8 we have three infinite graphs G_1, G_2, and G_3. There $V(G_1)$ is infinite and $E(G_1)$ is finite; $V(G_2)$ is finite and $E(G_2)$ is infinite; finally, both $V(G_3)$ and $E(G_3)$ are infinite.

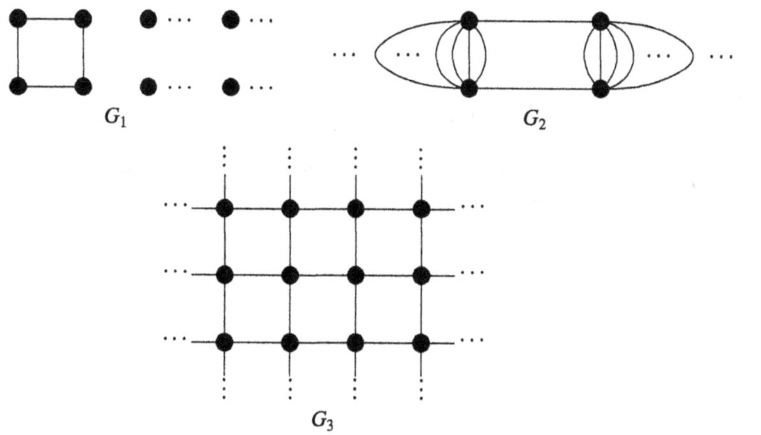

Figure 1.8: Three infinite graphs G_1, G_2, and G_3.

Although the study of infinite graphs is an interesting topic, we will focus solely on finite graphs.

Note 1.12 Unless otherwise stated, the term "graph" will always mean finite graph.

When tackling a problem that can be phrased in graph-theoretic terms, often it can be reduced to a problem involving a graph having no multiple edges or any loops. Such graphs constitute an important class called *simple graphs*. Since there is at most one edge between a pair of vertices in a simple graph, the edges are in one-to-one correspondence with their distinct endvertices. Therefore, a simple graph can be defined without the edgemap ϕ from Definition 1.8. Because of the importance of simple graphs, we state their formal definition separately.

Definition 1.13 A *simple graph* is an ordered pair $G = (V, E)$, where V is a nonempty set of vertices and E is a set of 2-element subsets of V, that is,

$$\begin{aligned} E &\subseteq \{X : X \subseteq V, |X| = 2\} \\ &= \{\{u, v\} : u, v \in V, u \neq v\}. \end{aligned}$$

1.5 Examples of Common Graphs

We next provide examples of some common simple graphs. In the examples that follow, $m, n \in \mathbb{N}$.

Example 1.14 The *null graph* on n vertices shown in Figure 1.9 is the simple graph $G = N_n$, where

$$\begin{aligned} V(N_n) &= \{u_1, u_2, \ldots, u_n\}, \\ E(N_n) &= \varnothing. \end{aligned}$$

Figure 1.9: Null graph N_n.

Example 1.15 The *path graph* on $n \geq 2$ vertices shown in Figure 1.10 is the simple graph $G = P_n$, where

$$\begin{aligned} V(P_n) &= \{u_1, u_2, \ldots, u_n\}, \\ E(P_n) &= \{\{u_1, u_2\}, \{u_2, u_3\}, \ldots, \{u_{n-1}, u_n\}\}. \end{aligned}$$

By convention we set $P_1 = N_1$.

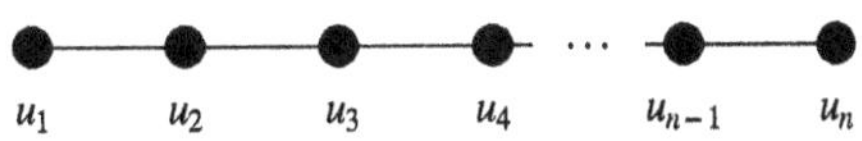

Figure 1.10: Path graph P_n.

Example 1.16 The *cycle graph* or simply *cycle* on $n \geq 3$ vertices shown in Figure 1.11 is the simple graph $G = C_n$, where

$$\begin{aligned} V(C_n) &= \{u_1, \ldots, u_n\}, \\ E(C_n) &= \{\{u_1, u_2\}, \{u_2, u_3\}, \ldots, \{u_{n-1}, u_n\}, \{u_n, u_1\}\}. \end{aligned}$$

Note that C_n is obtained by connecting u_1 and u_n in P_n.

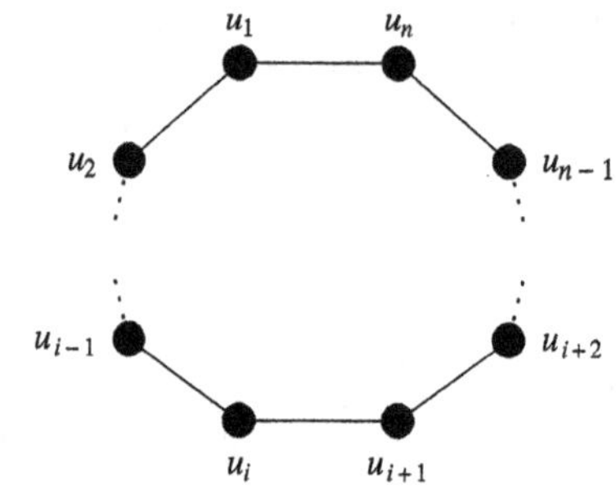

Figure 1.11: Cycle graph C_n.

For $n = 1$, we let C_1 be defined by $(V(C_1), E(C_1), \phi)$, where $V(C_1) = \{u_1\}$, $E(C_1) = \{e_1\}$, and $\phi(e_1) = \{u_1\}$. Likewise, for $n = 2$, we let C_2 be defined by $(V(C_2), E(C_2), \phi)$, where $V(C_2) = \{u_1, u_2\}$, $E(C_1) = \{e_1, e_2\}$, and $\phi(e_1) = \phi(e_2) = \{u_1, u_2\}$. Note that C_n is simple if, and only if, $n \geq 3$.

Example 1.17 The *complete graph* on n vertices shown in Figure 1.12 is the simple graph $G = K_n$, where

$$V(K_n) = \{u_1, u_2, \ldots, u_n\},$$
$$E(K_n) = \{\{u_i, u_j\} : 1 \leq i < j \leq n\}.$$

That is, every pair of distinct vertices is connected by an edge.

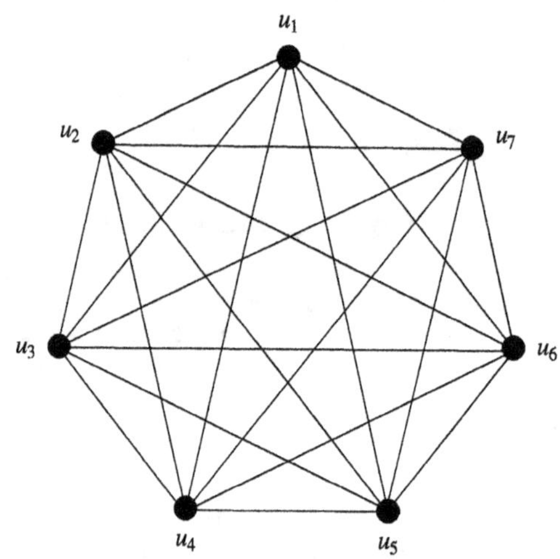

Figure 1.12: Complete graph K_7.

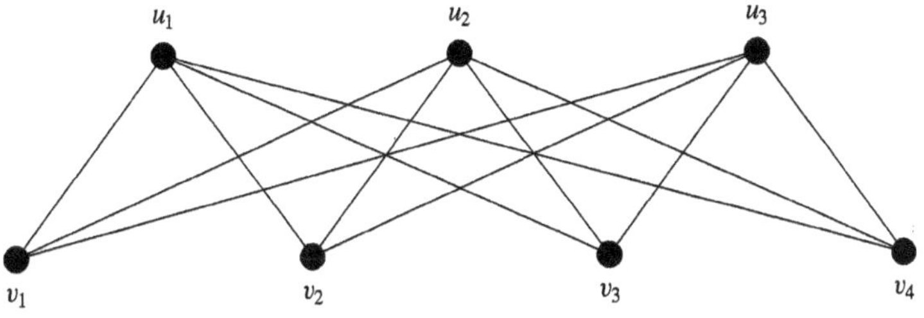

Figure 1.13: Complete bipartite graph $K_{3,4}$.

Example 1.18 The *complete bipartite* (m, n)*-graph* on $m + n$ vertices shown in Figure 1.13 is the simple graph $G = K_{m,n}$, where

$$V(K_{m,n}) = \{u_1, u_2, \ldots, u_m\} \cup \{v_1, v_2, \ldots, v_n\},$$
$$E(K_{m,n}) = \{\{u_i, v_j\} : 1 \leq i \leq m, 1 \leq j \leq n\},$$

that is, the vertices are of two types, where every pair of vertices of different types are adjacent, but no two vertices of the same type are adjacent.

Note 1.19 In this book, graphs are finite with possibly some loops and multiple edges. In many places in the literature simple graphs are called graphs, and graphs that are possibly nonsimple are called *multigraphs*, referring to their multiple edges.

When we present a graph by means of a diagram in a figure, what matters from a graph-theoretical point of view is whether the vertices (the dots) are connected by edges (the curves) or not. Whether the dots representing the vertices are small, fat, circular, or egg shaped is irrelevant. Likewise, whether the curves connecting the dots are straight, curved, or zigzag-like does not matter either. As an example, each diagram shown in Figure 1.14 represents the same graph from Example 1.9. Looking at the four diagrams in Figure 1.14, it is evident that some are more complicated or "messy" than others, and some have more edges crossing than others. When the vertices are represented by dots that are not much thicker than the points on the curves representing the edges, it becomes difficult to see, when two edges cross, whether they are meeting at a vertex or are simply crossing. It is common practice when drawing a graph to have a minimum number of crossings in its diagram and to draw the edges as straight as possible.

1.6 Degrees and Regular Graphs

Many properties of graphs must be stated in terms of numerical values associated with the graph or some of its components. The first such attribute we define is the *degree* of a vertex in a graph.

Definition 1.20 Let $G = (V, E, \phi)$ be a graph, and $u \in V$ a vertex of G. The *degree* of u, denoted $d_G(u)$, or $d(u)$ when there is no danger of ambiguity, is defined by

$$d_G(u) = |\{e \in E : u \in \phi(e), |\phi(e)| = 2\}| + 2|\{e \in E : u \in \phi(e), |\phi(e)| = 1\}|.$$

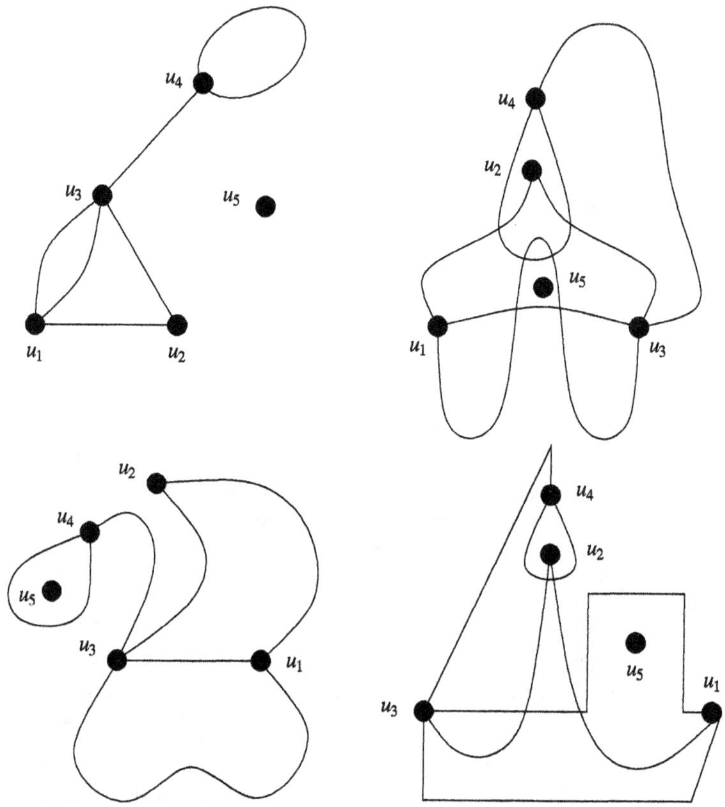

Figure 1.14: The same graph drawn in four different ways.

Informally speaking, $d_G(u)$ just tallies the number of "edge ends" connected to u. Thus, it makes intuitive sense to count the loops twice, since both of a loop's ends go into its endvertex.

Remark 1.21 For a graph G and a vertex $u \in V(G)$ we have:

- If G is a simple graph, then $d_G(u)$ is the number of neighbors of u in G.
- If G is a loopless graph, then $d_G(u)$ is the number of edges that have u as an endvertex.
- If G is a general graph, then $d_G(u)$ is the number of edges having u as an endvertex, where the loops are counted twice.

Definition 1.22 Let $G = (V, E, \phi)$ be a graph and $u \in V$ a vertex of G. The *open neighborhood* of u or just the *neighborhood* of u, denoted by $N_G(u)$ or just $N(u)$, is the set of all of the neighbors of u in G. Likewise, the *closed neighborhood* of u, denoted by $N_G[u]$ or just $N[u]$, is the set of neighbors of u together with u itself—that is,

$$\begin{aligned} N_G(u) &= \{v \in V(G) : \phi(e) = \{u, v\} \text{ for some } e \in E(G)\}, \\ N_G[u] &= N_G(u) \cup \{u\}. \end{aligned}$$

In general, for $S \subseteq V$ we define the following.

$$\begin{aligned} N_G(S) &= \{v : v \in N_G(s) \text{ for some } s \in S\}, \\ N_G[S] &= N_G(S) \cup S. \end{aligned}$$

When G is understood, we write $N(S)$ and $N[S]$, respectively.

Lemma 1.23 provides some interesting properties of $N_G(S)$ and $N_G[S]$. We ask for its proof in Exercise 1.18.

Lemma 1.23 The following properties hold:

$$\begin{aligned} N_G(S_1 \cup S_2) &= N_G(S_1) \cup N_G(S_2), \\ N_G[S_1 \cup S_2] &= N_G[S_1] \cup N_G[S_2], \\ N_G(S_1 \cap S_2) &\subseteq N_G(S_1) \cap N_G(S_2), \\ N_G[S_1 \cap S_2] &\subseteq N_G[S_1] \cap N_G[S_2]. \end{aligned}$$

Note that the number of neighbors of a vertex u in a graph G is never more than $d_G(u)$. However, if G is simple, then $d_G(u)$ is precisely the number of neighbors of u in G. We summarize in the following.

Observation 1.24 For a graph G we have

$$|N_G(u)| \leq d_G(u) \tag{1.1}$$

for every vertex $u \in V(G)$. Moreover, G is a simple graph if, and only if, equality holds in (1.1) for every vertex $u \in V(G)$.

Example 1.25 For the graph shown in Figure 1.15, we have

$$\begin{aligned} d(u_1) &= 3, \\ d(u_2) &= 2, \\ d(u_3) &= 4, \\ d(u_4) &= 3. \end{aligned}$$

Also, we have

$$\begin{aligned} N(u_1) &= \{u_2, u_3\}, \\ N[u_1] &= \{u_1, u_2, u_3\}, \\ N(u_4) &= N[u_4] = \{u_3, u_4\}. \end{aligned}$$

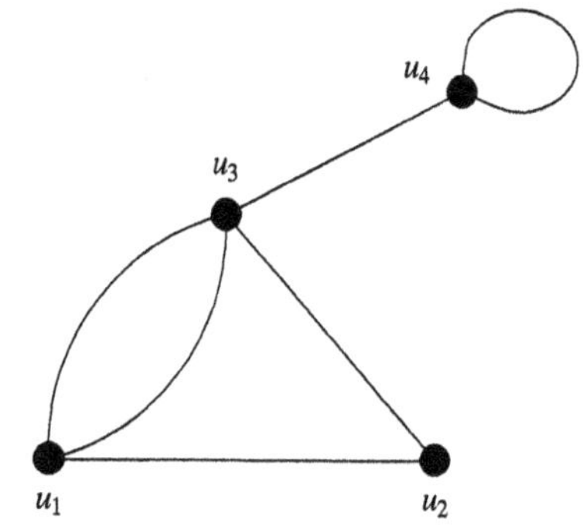

Figure 1.15: Various degrees.

We now state our first theorem. It relates degrees of the vertices to the number of edges of a given graph.

Theorem 1.26 **Hand-Shaking**

For a graph G we have

$$\sum_{u \in V(G)} d_G(u) = 2|E(G)|.$$

The name "Hand-Shaking Theorem" comes from the following party analogy: Consider a collection of guests at a party. Suppose some guests shook hands with some other guests. If we asked everyone at the party how many guests they shook hands with and added those numbers all up, this sum would be equal to twice the number of total hand shakes. In graph-theory terms, each vertex represents a guest, and an edge between two guests represents a hand shake between them.

Proof (Theorem 1.26) Each edge in $E(G)$ has endvertices, say u and v (which could be identical, if the edge is a loop), and hence contributes two to the sum on the left—one to $d_G(u)$ and one to $d_G(v)$ (or two to $d_G(u)$, if the edge is a loop). ■

As a consequence of the Hand-Shaking Theorem, we obtain the following corollary.

Corollary 1.27 In any graph there is always an even number of vertices having odd degree.

Proof By the Hand-Shaking Theorem, the value $\sum_{u \in V(G)} d_G(u)$ is always even, and hence the number of odd $d_G(u)$-terms is even. This means precisely that the number of vertices of odd degree is even. ■

Graphs with a uniform property often play an important role in applications. The following notion is an important uniformity property.

Definition 1.28 A graph G is called *k-regular* if $d_G(u) = k$ for all $u \in V(G)$. A graph G is *regular* if it is k-regular for some nonnegative integer k.

In the following example we mention a few regular graphs.

Example 1.29 Let $m, n \in \mathbb{N}$.

- The null graph N_n is a 0-regular graph.
- The cycle C_n is a 2-regular graph, since $d_{C_n}(u) = 2$ for all $u \in V(C_n)$.
- The complete graph K_n is an $(n-1)$-regular graph.
- The complete (m, n)-bipartite graph $K_{m,n}$ is a regular graph if, and only if, $m = n$, in which case it is m-regular.

By the Hand-Shaking Theorem, we obtain the following result.

Corollary 1.30 Every k-regular graph on n vertices has $kn/2$ edges. In particular, the complete graph K_n has $(n-1)n/2$ edges.

1.7 Subgraphs

When studying a specific graph, many times our attention is focused solely on a special part of the graph, perhaps on a smaller graph lying inside the larger graph. This motivates the following definition of a *subgraph*.

Definition 1.31 For graphs $G' = (V', E', \phi')$ and $G = (V, E, \phi)$, we say that G' is a *subgraph* of G if

1. $V' \subseteq V$,
2. $E' \subseteq E$,
3. $\phi'(e) = \phi(e)$ for all $e \in E'$.

We denote the subgraph relation on graphs using the standard subset notation $\subseteq$. That is, if G' is a subgraph of G, then we write $G' \subseteq G$. In particular, for simple graphs $G' = (V', E')$ and $G = (V, E)$, we have that G' is a subgraph of G if $V' \subseteq V$ and $E' \subseteq E$.

Subgraphs are often obtained from a specific set of vertices or edges, as described in the following definition.

Definition 1.32 For a nonempty subset $W \subseteq V(G)$ of vertices, the subgraph of G *induced by* W, denoted $G[W]$, is the graph with vertex set W whose edge set consists of all the edges of G having their endvertices in W.

For a nonempty subset $F \subseteq E(G)$ of edges, the subgraph of G *induced by*, or simply *formed by*, F, denoted $G[F]$, is the graph with edge set F whose vertex set consists of all the endvertices of edges of F.

When the subset $W \subseteq V(G)$ is given by a specific list, say $W = \{w_1, \ldots, w_m\}$, then we often write $G[w_1, \ldots, w_m]$ instead of $G[\{w_1, \ldots, w_m\}]$ for the subgraph induced by W. We also use similar convention for the edges. The following example illustrates Definitions 1.31 and 1.32.

Example 1.33 Consider the graph G given in Figure 1.16(a). Figure 1.16(b) shows a subgraph G' of G. Figure 1.16(c) shows the subgraph $G[u_1, u_2, u_3]$, the subgraph induced by the vertex set $\{u_1, u_2, u_3\}$. Figure 1.16(d) shows the subgraph $G[e_1, e_2, e_3, e_4]$ induced by (or formed by) the edges in $\{e_1, e_2, e_3, e_4\}$ together with their endvertices.

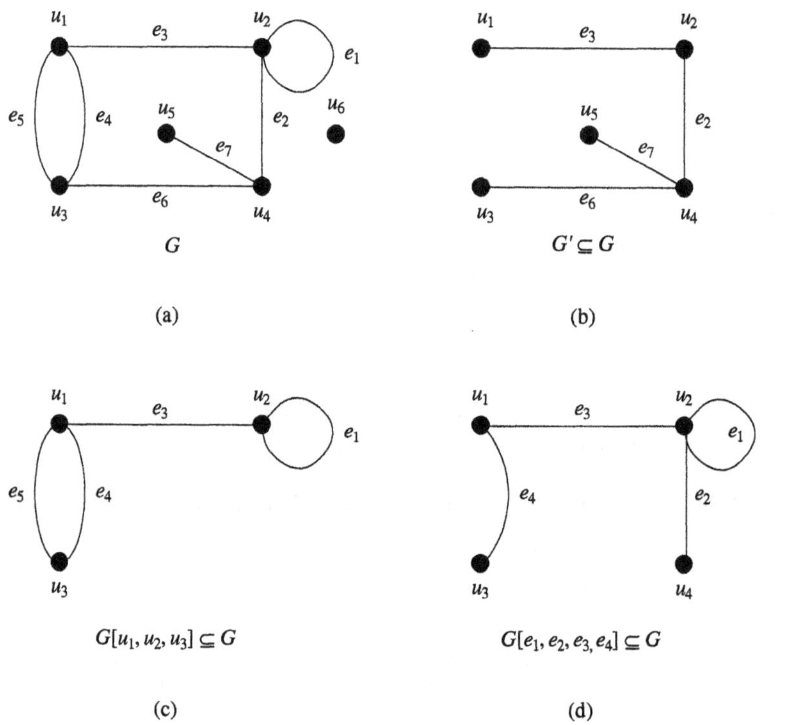

Figure 1.16: Various subgraphs.

The next definition provides us with a mechanism for relating various graphs to one another.

Definition 1.34 Let S be a set. A binary relation $\preceq$ between the elements of S satisfying the following properties is called a *partial order* on S:

1. $s \preceq s$ holds for all $s \in S$ (*reflexive* condition),
2. $s \preceq t$ and $t \preceq s$ implies $s = t$ (*antisymmetric* condition),
3. $s \preceq t$ and $t \preceq u$ implies $s \preceq u$ (*transitive* condition).

The set S together with a partial order $\preceq$ is called a *partially ordered set* or a *poset*, for short.

Notice that the following properties hold for graphs and the subgraph relation:

1. $G \subseteq G$, every graph is a subgraph of itself.
2. If $G \subseteq G'$ and $G' \subseteq G$, then $G = G'$.
3. If $G' \subseteq G''$ and $G'' \subseteq G$, then $G' \subseteq G$.

These properties mean that we have a partial order on any collection of subgraphs of a given graph. Suppose G' and G'' are both subgraphs of a graph G. It is possible that $G' \subseteq G''$ or $G'' \subseteq G'$, but these inclusions are not necessary. The graphs may be incomparable using the subgraph relation. The word "partial" is used to denote this type of ordering.

Observation 1.35 Any collection of subgraphs of a given graph forms a poset with the relation $\subseteq$ ("is a subgraph of").

In the next section we define another important type of graph.

1.8 The Definition of a Directed Graph

So far, we have defined the two most important concepts in graph theory, a (general) graph and a simple graph, and briefly discussed many of their basic properties. In some situations we also need to keep track of directions between vertices, not unlike knowing which streets in a big city are one-way streets and which ones are not.

Definition 1.36 A *directed graph* or *digraph* is an ordered triple $\vec{G} = (V, E, \eta)$, where

1. $V \neq \varnothing$,
2. $V \cap E = \varnothing$,
3. $\eta : E \mapsto V \times V$ is a map.

The set V is the set of vertices, and the set E the *directed* edges, or just *edges*. If $\eta(e) = (u, v)$, then u is called the *tail* of e and v the *head* of e. Also, v is called a *successor* of u, and u is called a *predecessor* of v.

If $\eta(e) = (u, u)$, then e is called a *directed loop*. If $\vec{G}$ is a digraph, then $V(\vec{G})$ and $E(\vec{G})$ will always denote the set of vertices and directed edges of $\vec{G}$, respectively.

Two directed edges e and e' are said to be *parallel edges* if $\eta(e) = \eta(e')$. That is, the edges are mapped onto the same ordered pair of vertices.

Remark 1.37 Although Definition 1.36 does not insist upon it, we will assume that both V and E are finite sets, unless stated otherwise.

A digraph is represented similarly to a graph by a diagram, where the directed edges are not line segments but arrows from one dot (representing the tail) to another dot (representing the head). The next example illustrates Definition 1.36.

Example 1.38 Consider the digraph $\vec{G} = (V, E, \eta)$ with five vertices and six directed edges as shown in Figure 1.17. Here we see that $\vec{G} = (V, E, \eta)$, where $V = \{u_1, u_2, u_3, u_4, u_5\}$, $E = \{e_1, e_2, e_3, e_4, e_5, e_6\}$, and the edgemap η is given by

$$\begin{aligned}\eta(e_1) &= (u_1, u_2),\\ \eta(e_2) &= (u_3, u_1),\\ \eta(e_3) &= (u_1, u_3),\\ \eta(e_4) &= (u_2, u_3),\\ \eta(e_5) &= (u_3, u_4),\\ \eta(e_6) &= (u_4, u_4).\end{aligned}$$

In particular, the directed edge e_6 is a directed loop.

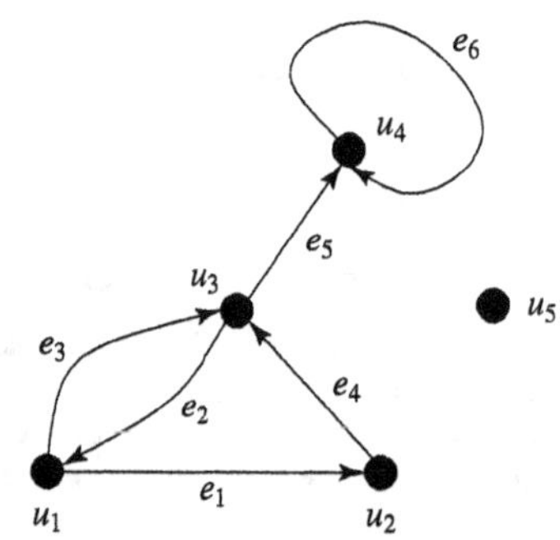

Figure 1.17: A digraph with five vertices and six directed edges.

If $\vec{G} = (V, E, \eta)$ is a directed graph, then in a natural way we can form the corresponding *underlying graph* $G = (V, E, \phi)$, where

$$\eta(e) = (u, v) \Rightarrow \phi(e) = \{u, v\}.$$

Conversely, forming a directed graph from a given undirected one cannot be done in a unique natural way. In fact, if a graph G has exactly m nonloop edges, then there are precisely 2^m digraphs having G as an underlying graph. Hence, there are 2^m choices for forming a directed graph $\vec{G}$ from G, none of which is more natural than another. In Figure 1.18 we see a digraph $\vec{G}$ and its underlying graph G.

Remark 1.39 If $\vec{\mathbf{G}}$ is the class of all directed graphs, and $\mathbf{G}$ is the class of all undirected graphs, then we can define a formation, or what is sometimes called a *functor*, $\mathbf{F} : \vec{\mathbf{G}} \to \mathbf{G}$, by

$$\mathbf{F}(\vec{G}) = G,$$

where G denotes the underlying graph of the digraph $\vec{G}$. Such a natural formation is called a *forgetful functor*, since it simply disregards or "forgets" the directions of the edges.

However, since we do have to make choices as to which direction should be given to each edge, there is no one specific digraph that naturally associates to each given undirected

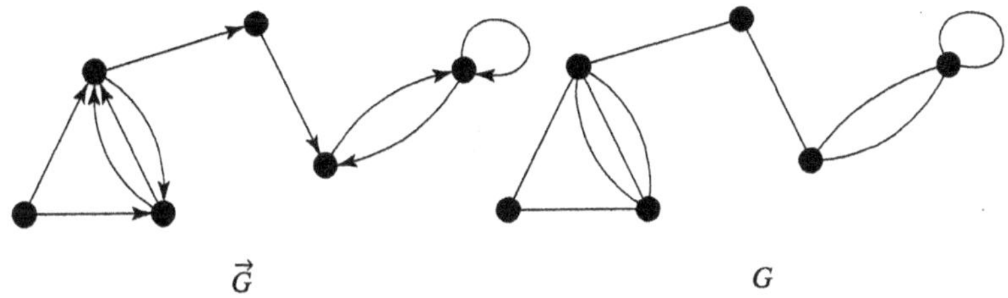

Figure 1.18: A digraph and its underlying undirected graph.

graph G. In the words of category theory, there is no functor $\mathbf{F}' : \mathbf{G} \to \tilde{\mathbf{G}}$ from the class of all undirected graphs to the class of all digraphs.

In the next several examples we illustrate some important digraphs.

Example 1.40 The *path digraph* on $n \geq 2$ vertices, which is shown in Figure 1.19, is the digraph $\vec{P}_n = (V(\vec{P}_n), E(\vec{P}_n), \eta)$, where

$$V(\vec{P}_n) = \{u_1, u_2, \ldots, u_n\},$$
$$E(\vec{P}_n) = \{e_1, e_2, \ldots, e_{n-1}\},$$

where η is given by

$$\eta(e_i) = (u_i, u_{i+1}) \qquad \text{for all } i \in \{1, \ldots, n-1\}. \tag{1.2}$$

Clearly, the underlying graph of $\vec{P}_n$ is the path graph P_n on n vertices.

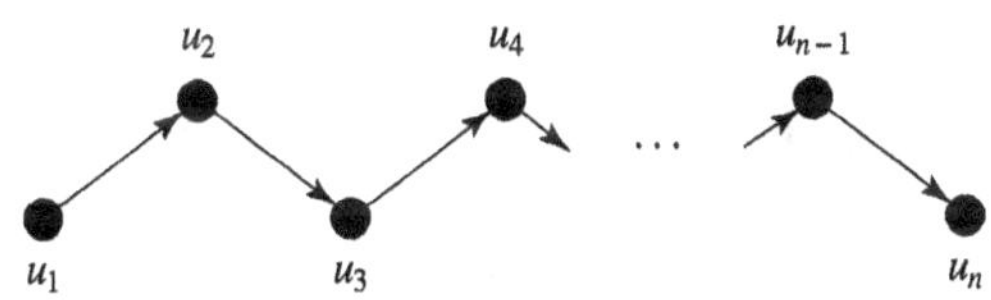

Figure 1.19: A directed path.

Example 1.41 The *cycle digraph* on $n \geq 1$ vertices, which is shown in Figure 1.20, is the digraph $\vec{C}_n = (V(\vec{C}_n), E(\vec{C}_n), \eta)$, where

$$V(\vec{C}_n) = \{u_1, u_2, \ldots, u_n\},$$
$$E(\vec{C}_n) = \{e_1, e_2, \ldots, e_{n-1}, e_n\}.$$

Here η is given by (1.2) together with $\eta(e_n) = (u_n, u_1)$. Clearly the underlying undirected graph of $\vec{C}_n$ is the cycle graph C_n on n vertices. In a diagram representing a directed cycle, the edges point either in a clockwise or a counterclockwise direction.

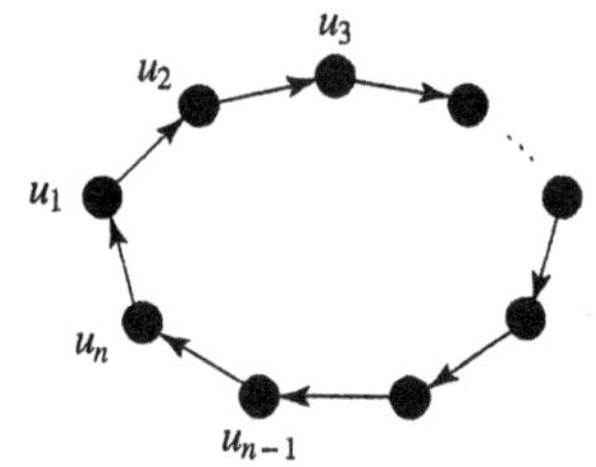

Figure 1.20: A directed cycle.

As we saw in the previous two examples, the edgemap η is really not necessary, since both in the path digraph and the cycle digraph there is at most one directed edge from one vertex to another. Such digraphs are called *simple digraphs*.

Definition 1.42 A digraph having no directed loops and no parallel directed edges is called a *simple digraph.*

Any simple digraph $\vec{G}$ can be presented as an ordered pair $\vec{G} = (V, E)$, where V is a set of vertices and $E \subseteq V \times V$. Note that there is a slight difference between a simple digraph and having the edges represented by a subset of $V \times V$:

- A digraph $\vec{G}$ has a representation $\vec{G} = (V, E)$, where $E \subseteq V \times V$ if, and only if, $\vec{G}$ has no parallel directed edges.

- A digraph is simple if, and only if, $\vec{G}$ has a representation $\vec{G} = (V, E)$, where $E \subseteq V \times V$ *and* $\vec{G}$ has no directed loops.

1.9 Indegrees and Outdegrees in a Digraph

Vertex degree is one of the most basic parameters obtained from a graph. We have similar parameters for digraphs.

Definition 1.43 Let $\vec{G} = (V, E, \eta)$ be a digraph, and $u \in V$ a vertex in $\vec{G}$.

1. The *indegree* of u, denoted by $d^-_{\vec{G}}(u)$ or $d^-(u)$, is the number of directed edges having u as head, that is,

$$d^-(u) = |\{e \in E : \eta(e) = (x, u) \text{ for some } x \in V\}|.$$

2. The *outdegree* of u, denoted by $d^+_{\vec{G}}(u)$ or $d^+(u)$, is the number of directed edges having u as tail, that is,

$$d^+(u) = |\{e \in E : \eta(e) = (u, y) \text{ for some } y \in V\}|.$$

For a given vertex $u \in V(\vec{G})$, the *inneighbors*, denoted by $N^-_{\vec{G}}(u)$ or just $N^-(u)$, and the *outneighbors*, denoted by $N^+_{\vec{G}}(u)$ or just $N^+(u)$, of u are given by the following:

$$\begin{aligned} N^-(u) &= \{x \in V(\vec{G}) : \eta(e) = (x, u) \text{ for some } e \in E(\vec{G})\}, \\ N^+(u) &= \{y \in V(\vec{G}) : \eta(e) = (u, y) \text{ for some } e \in E(\vec{G})\}, \end{aligned}$$

respectively.

Note that each directed edge counts exactly once toward an indegree and once toward an outdegree. Therefore, analogous to the Hand-Shaking Theorem 1.26, we have the following result.

Theorem 1.44 For a digraph $\vec{G}$, we have

$$\sum_{u \in V(\vec{G})} d^-_{\vec{G}}(u) = \sum_{u \in V(\vec{G})} d^+_{\vec{G}}(u) = |E(\vec{G})|.$$

We now illustrate Theorem 1.44.

Example 1.45 Consider the digraph $\vec{G}$ shown in Figure 1.21. It has five vertices $V(\vec{G}) = \{u_1, u_2, u_3, u_4, u_5\}$ and six directed edges $E(\vec{G}) = \{e_1, e_2, e_3, e_4, e_5, e_6\}$. For $\vec{G}$ we have the following degrees:

$$\begin{aligned} d^-(u_1) = 1, \quad d^+(u_1) = 2, \\ d^-(u_2) = 1, \quad d^+(u_2) = 1, \\ d^-(u_3) = 2, \quad d^+(u_3) = 2, \\ d^-(u_4) = 2, \quad d^+(u_4) = 1, \\ d^-(u_5) = 0, \quad d^+(u_5) = 0. \end{aligned}$$

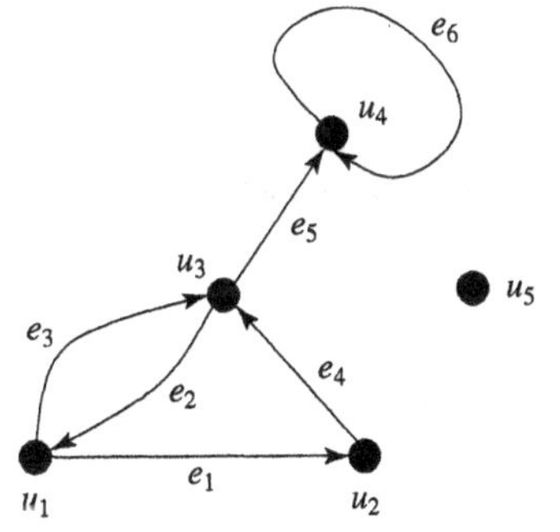

Figure 1.21: A digraph on five vertices and six directed edges.

This listing is consistent with the following:

$$\sum_{i=1}^{5} d^-(u_i) = \sum_{i=1}^{5} d^+(u_i) = 6 = |E(\vec{G})|,$$

the number of directed edges in $\vec{G}$. Also, in this case we have

$$N^-(u_3) = \{u_1, u_2\},$$
$$N^+(u_3) = \{u_1, u_4\}.$$

The notion of regularity for digraphs can be defined similarly as for graphs.

Definition 1.46 A digraph $\vec{G}$ is called *balanced* if for every vertex $u \in V(\vec{G})$ the indegree equals the outdegree, that is, $d^-_{\vec{G}}(u) = d^+_{\vec{G}}(u)$.

A balanced digraph $\vec{G}$ is called *regular* if $d^-_{\vec{G}}(u) = d^-_{\vec{G}}(v)$, or equivalently $d^+_{\vec{G}}(u) = d^+_{\vec{G}}(v)$ for every pair of vertices $u, v \in V(\vec{G})$.

Note that the underlying graph of a balanced digraph has an even degree at every vertex. Also note that the underlying graph of a regular digraph is a regular graph, where each vertex has the same (even) degree.

Example 1.47 For $n \geq 3$, the directed cycle $\vec{C}_n$ is a regular digraph. Indeed

$$d^-_{\vec{C}_n}(u_i) = d^+_{\vec{G}}(u_i) = 1,$$

for each vertex u_i of $\vec{C}_n$.

Just as for subgraphs, one can talk about a subdigraph $\vec{G}'$ of a given digraph $\vec{G}$. Definition 1.31 carries over directly to digraphs with η replacing ϕ. For simple digraphs the definition is the same. $\vec{G}'$ is a *subdigraph* of $\vec{G}$, if $V(\vec{G}') \subseteq V(\vec{G})$ and $E(\vec{G}') \subseteq E(\vec{G})$, as a subset of $V(\vec{G}) \times V(\vec{G})$.

Also, for a given digraph $\vec{G}$, and subsets $V' \subseteq V(\vec{G})$ and $E' \subseteq E(\vec{G})$, the subgraphs $\vec{G}[V']$ and $\vec{G}[E']$ denote the subdigraph induced by V', and the subdigraph induced by or formed by E' together with all the corresponding heads and tails, as in Definition 1.32.

Finally, it should be evident that a given collection of subdigraphs of a given digraph forms a poset with the relation $\subseteq$ ("is a subdigraph of"), just as in Observation 1.35.

1.10 Exercises

1.1. Show that for sets A, B, and C, we have the following:

$$(A \triangle B) \triangle C = A \triangle (B \triangle C).$$

This result shows that the symmetric difference operator is an associative operation; hence, it makes sense to talk about $A \triangle B \triangle C$.

1.2. Use Exercise 1.1 to show that if $A_1, \ldots, A_n$ is a collection of sets, then the expression $A_1 \triangle \cdots \triangle A_n$ makes mathematical sense. Show further that this set contains precisely all the elements which are contained in an odd number of the sets $A_1, \ldots, A_n$.

1.3. Show that the number of k-tuples $(a_1, \ldots, a_k)$ from the set $\{1, \ldots, n\}$, where each number appears at most once in the k-tuple, is given by the following:

$$P(n,k) = n(n-1)\cdots(n-k+1) = \frac{n!}{(n-k)!}.$$

1.4. Use combinatorial arguments to prove the following formulas:

$$\binom{n}{0} = 1,$$
$$\binom{n}{k} = \binom{n}{n-k},$$
$$\binom{n}{k} = \binom{n-1}{k} + \binom{n-1}{k-1}.$$

[*Hint:* (i) Count the number of ways to choose a k-element subset from $\{1, \ldots, n\}$ and its complement set containing $n-k$ elements. (ii) Notice that every k-element subset of $\{1, \ldots, n\}$ either contains the element n or does not.]

1.5. It is known that a certain statement $P(n)$ satisfies the following two conditions:

(a) $P(1)$ and $P(2)$ are both true.
(b) For $n \geq 1$, we know that if $P(n)$ is true, then so is $P(n+3)$.

Can you conclude from these facts alone that $P(n)$ is true for all $n \in \mathbb{N}$? What if, in addition, we know $P(3)$ is true?

1.6. It is known that a certain statement $Q(n)$ satisfies the following conditions:

(a) $Q(1001)$ is true.
(b) If $n \geq 4$ and $Q(n)$ is true, then so is $Q(n-3)$.
(c) If $Q(n)$ is true, then so is $Q(2n)$.

For what $n \in \mathbb{N}$ can you conclude that $Q(n)$ is true?

1.7. Draw a portion of your academic department's Web page as a graph by modeling Web pages as vertices and hyperlinks as edges. Include at least ten different Web pages and draw edges for all of the existing hyperlinks between the Web pages you select. Label the vertices with their uniform resource locators.

1.8. Refer to the job assignments problem shown in Figure 1.4. Is the company able to meet its hiring needs? If so, provide a possible set of hires that meet their needs.

1.9. Write out the explicit 3-tuples (V, E, ϕ) that define the graphs shown in Figures 1.4 and 1.5.

1.10. For what values of $n < 20$ can you draw a 3-regular graph having n vertices? What if you insist upon the 3-regular graph being simple?

1.11. For $n \in \{1, 2, 3, 4, 5\}$, draw the following graphs: N_n, P_n, C_n, K_n, and $K_{n,n}$. Which of the graphs are the same?

1.12. Construct a table whose entries are formulas for the number of vertices and edges for the graphs N_n, P_n, C_n, K_n, and $K_{m,n}$ for arbitrary values of n and m.

1.13. Express both the maximum and minimum degree of vertices in the graphs N_n, P_n, C_n, K_n, and $K_{m,n}$ in terms of n and m.

1.14. How are N_n, P_n, C_n, and K_n related via the subgraph relationship?

1.15. For what values of n is K_n a subgraph of $K_{\ell,m}$ for given values of ℓ and m?

1.16. Specify a minimum value of n in terms of ℓ and m such that $K_{\ell,m}$ is a subgraph of K_n.

1.17. For what kind of vertices u in a graph G do we have $N_G(u) = N_G[u]$?

1.18. Prove Lemma 1.23.

1.19. For every k-regular graph is there a $(k + 1)$-regular graph that contains the k-regular graph as a subgraph?

1.20. Let $m_{30}(n)$ be the maximum number of edges that a simple graph on n vertices can have, if each vertex of the graph is connected to at most 30% of the other vertices. Determine a quadratic polynomial $p(x) = Ax^2 + Bx + C$ such that

(a) $m_{30}(n) \leq p(n)$ for every $n \in \mathbb{N}$.
(b) $m_{30}(n) = p(n)$ holds for infinitely many $n \in \mathbb{N}$.

Is $p(n)$ uniquely determined by the two conditions above?

1.21. Is there a simple graph on n vertices such that the vertices all have distinct degrees? Is there a (general) graph with this property?

1.22. Let G be the simple graph, where the vertices correspond to each of the squares of an 8×8 chess board, and where two squares are adjacent if, and only if, a knight can go from one square to the other in one move. What is/are the possible degree(s) of a vertex in G? How many vertices have each degree? How many edges does G have?

1.23. Prove that the usual less-than relation $(<)$ is a partial order on the set of *integers* $\mathbb{Z} = \{\ldots, -2, -1, 0, 1, 2, \ldots\}$.

1.24. Define a relation on graphs (other than "is a subgraph of") that is a partial order, and include a proof that your relation is indeed a partial order.

1.25. When do we have $N^-(u) \cap N^+(u) \neq \varnothing$ for a vertex u in a digraph $\vec{G}$?

1.26. For $n \geq 3$, let $\mathcal{C}(n)$ be the collection of digraphs $\vec{G}$ with $V(\vec{G}) = \{u_1, \ldots, u_n\}$ that satisfy the following two conditions:

(a) The underlying graph of each digraph in $\mathcal{C}(n)$ is the cycle C_n.

(b) No digraph in $\mathcal{C}(n)$ is a directed cycle.

What is the cardinality of $\mathcal{C}(n)$?

1.27. Let $\mathcal{C}(n)$ be as in Exercise 1.26. Let $k \leq n$ and $\vec{G} \in \mathcal{C}(n)$ be such that when $\vec{G}$ is "turned k steps counterclockwise" we obtain a graph identical to the digraph $\vec{G}$—that is to say, the digraph $\vec{G}_k$ on the same set of vertices as $\vec{G}$ with the edges given by

$$(u_{i+k}, u_{i+k+1}) \in E(\vec{G}_k) \Leftrightarrow (u_i, u_{i+1}) \in E(\vec{G}).$$

(Note that here we compute modulo n. That is, u_{n+1} is identified with u_1, and so on.) Show that if $\vec{G}_k = \vec{G}$, then $G_{\gcd(k,n)} = \vec{G}$ also.

1.28. This problem is a continuation of Exercise 1.27. Show that when n is a prime number $n = p$, then for any $\vec{G}$, all the digraphs $\vec{G}_0, \vec{G}_1, \ldots, \vec{G}_{p-1}$ are distinct digraphs of $\mathcal{C}(p)$. (Here $\vec{G}_0 = \vec{G}$.)

1.29. Use Exercise 1.28 to prove that for any prime number $p \geq 2$, the quantity $2^p - 2$ is always divisible by p. (This result provides a graph-theoretical proof of a special case of Fermat's Little Theorem.)

1.30. For a prime number $p \geq 3$, let $\tilde{a} = (a_1, \ldots, a_p)$ be an ordered p-tuple of elements from $\{1, \ldots, n\}$. Let C_n be a cycle with vertices $V(C_n) = \{u_1, \ldots, u_n\}$. Form a graph $G_{\tilde{a}}$ by replacing each edge (u_i, u_{i+1}) in C_n by a_i identical edges, making it a_i-fold. (Note that here we compute cyclically modulo n. That is, u_{n+1} is identified with u_1.) How many graphs $G_{\tilde{a}}$ are there, where not all adjacent pairs of vertices are connected by the same number of edges?

1.31. Continuing from Exercise 1.30, mimic the arguments from Exercises 1.26, 1.27, and 1.28, and give a graph-theoretical proof of the following version of Fermat's Little Theorem:

Theorem 1.48 **Fermat's Little**

If $n \in \mathbb{N}$ and p is a prime number, then $n^p - n$ is divisible by p.

1.32. Let S be a finite poset with $\preceq$ as its partial order. A *maximal element* $x_{\max} \in S$ is an element such that

$$x_{\max} \preceq x \Rightarrow x = x_{\max}.$$

Likewise, a *minimal element* $x_{\min} \in S$ is an element such that

$$x \preceq x_{\min} \Rightarrow x = x_{\min}.$$

Show that if T is a finite poset, then T has at least one maximal element and at least one minimal element.

1.33. Let S be a finite poset with $\preceq$ as its partial order. Construct a simple digraph $\vec{G}_{\preceq}$ by

$$\begin{aligned} V(\vec{G}_{\preceq}) &= S, \\ V(\vec{G}_{\preceq}) &= \{(u, v) \in S \times S : u \preceq v,\ u \neq v\}. \end{aligned}$$

Show that $\vec{G}_{\preceq}$ is a simple digraph, which has no cycle digraphs as subdigraphs. (Such digraph are called *acyclic digraphs.*)

1.34. Does the opposite of Exercise 1.33 hold? That is to say, can we associate a poset to any finite simple acyclic digraph $\vec{G}$? [*Hint:* What happens when you define $u \preceq v$ as "One can walk from v to u along the directed edges of $\vec{G}$"?]

1.35. Construct a time line that includes 25 of the most important events in graph theory. [*Hint:* Use the World Wide Web to conduct your research.]

CHAPTER

2

Basic Concepts in Graph Theory

Chapter Goals

Define and study paths and cycles of both graphs and digraphs.

Define and study homomorphisms and isomorphisms of both graphs and digraphs.

Discuss connectivity of graphs and digraphs.

Present some common operations on graphs.

Define contractions and minors of graphs.

State a Minor Theorem by Robertson and Seymour.

2.1 Paths and Cycles

By the mere definition of a graph, from previous Chapter 1, the essence of a graph is whether two given vertices in it are connected or not. In many situations, though, we do not need a *direct* connection with an edge, but rather a route to be able to "go" from one vertex to another in some number of steps. Although two vertices in a given graph may not be adjacent, they might be adjacent to a common neighbor, or more generally they might be connected by a sequence of edges. The following definition captures this idea and some varieties of it.

Definition 2.1

1. A *walk* in a graph $G = (V, E, \phi)$ is an alternating sequence

$$(u_0, e_1, u_1, e_2, \ldots, e_k, u_k)$$

of vertices and edges that begins and ends with a vertex. For each $i \in \{1, 2, \ldots, k\}$, the endvertices of e_i are u_{i-1} and u_i. That is,

$$\phi(e_i) = \{u_{i-1}, u_i\}.$$

The vertex u_0 is the *initial vertex* of the walk. The vertex u_k is the *final vertex* of the walk. The initial or final vertex of a walk is also called an *endvertex* of the walk. The natural number k is the *length* of the walk.

2. A *trail* in G is a walk with all of its edges $e_1, e_2, \ldots, e_k$ distinct.
3. A *path* in G is a walk with all of its vertices $u_0, u_1, \ldots, u_k$ distinct.
4. For vertices u and v in G, a u, v-*walk* (u, v-*trail* or u, v-*path*) is a walk (respectively, trail or path) with initial vertex u and final vertex v.
5. A walk or trail of length at least one is *closed* if the initial vertex and the final vertex are the same. A closed trail is also called a *circuit*.
6. A *cycle* is a closed walk with distinct vertices except for the initial and final vertex, which are the same.

By convention, we view a single vertex as a walk, trail, or path of length zero. However, cycles always have positive length, having at least one edge. The only cycles of length one are loops. Also, the set of vertices and edges constituting a given walk, trail, path, or cycle in a graph G clearly forms a subgraph of G.

Assuming G is a simple graph, then a walk, trail, path, or cycle can be specified by a sequence of vertices $(u_0, u_1, \ldots, u_k)$ instead of $(u_0, e_1, u_1, e_2, \ldots, e_k, u_k)$. This is because a pair of adjacent vertices u_{i-1} and u_i completely determine the only edge $e_i = \{u_{i-1}, u_i\}$ between them.

Example 2.2 We illustrate Definition 2.1 using the graph shown in Figure 2.1.

- $w = (u_1, e_2, u_3, e_3, u_1, e_1, u_2, e_4, u_3, e_4, u_2, e_8, u_5, e_7, u_4)$

  Here w is a walk of length seven. Note that w is not a trail, since the edge e_4 appears twice. A walk that is not a trail is clearly not a path. Hence, w is not a path.

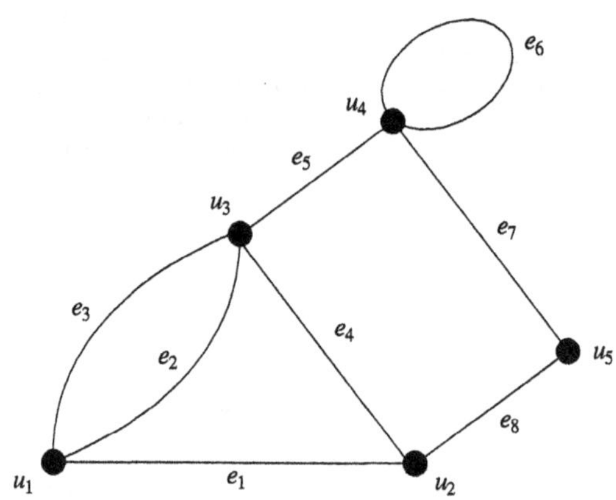

Figure 2.1: Graph used to illustrate walks, trails, paths, and cycles.

- $t = (u_1, e_2, u_3, e_3, u_1, e_1, u_2, e_8, u_5, e_7, u_4)$
  The walk t is a trail of length five. Note that t is not a path, since the vertex u_1 appears twice.
- $p = (u_1, e_2, u_3, e_4, u_2, e_8, u_5, e_7, u_4)$
  The walk p is a path of length four.
- $c_1 = (u_3, e_2, u_1, e_2, u_3, e_4, u_2, e_8, u_5, e_7, u_4, e_5, u_3)$
  The walk c_1 is a closed walk of length six.
- $c_2 = (u_3, e_2, u_1, e_3, u_3, e_4, u_2, e_8, u_5, e_7, u_4, e_5, u_3)$
  The walk c_2 is a closed trail of length six.
- $c_3 = (u_3, e_4, u_2, e_8, u_5, e_7, u_4, e_5, u_3)$
  The walk c_3 is a cycle of length four.

In many situations we need to combine walks, trails, or paths. Clearly, if we know how to go from point A to point B and from point B to point C, then we can get from point A to point C. Likewise, we can go back to point A from point B. This everyday observation inspires the following definition.

Definition 2.3 Let p and q be two walks given by

$$p = (u_0, e_1, u_1, \ldots, e_k, u_k) \text{ and}$$
$$q = (v_0, f_1, v_1, \ldots, f_l, v_l).$$

If $u_k = v_0$, then we can form the *composite* walk

$$pq = (u_0, e_1, u_1, \ldots, e_k, u_k, f_1, v_1, \ldots, f_l, v_l)$$

by first traversing p and then along q.

We can also form the *inverse* walk of p

$$p^{-1} = (u_k, e_k, \ldots, u_1, e_1, u_0).$$

We observe that although the intuitive concepts of a path and a cycle are simple, their formal definitions seem quite involved, since we can only use the terminology that we have already precisely defined.

Note the "left-right" difference between compositions of walks and the composition of two maps g and f. The composition of maps $(g \circ f)(x) = g(f(x))$ means that we first *apply* f to x on its right, and then g to $f(x)$ on its right. Walks are not applied to anything on their right. Instead they are simply concatenated in the same order we read, from left to right.

Figure 2.2 explains the relationship of the items from Definition 2.1 in any given graph G. An arrow from one oval down to another means that the first (upper) oval contains the items in the second (lower) oval, since the items in the first are more general than the items in the second oval. So, for example, walks are more general than trails. Going down the diagram in Figure 2.2 means that we are restricting the structure more and more. The diagram in Figure 2.2 is actually an example of a directed graph, and, in fact, a poset.

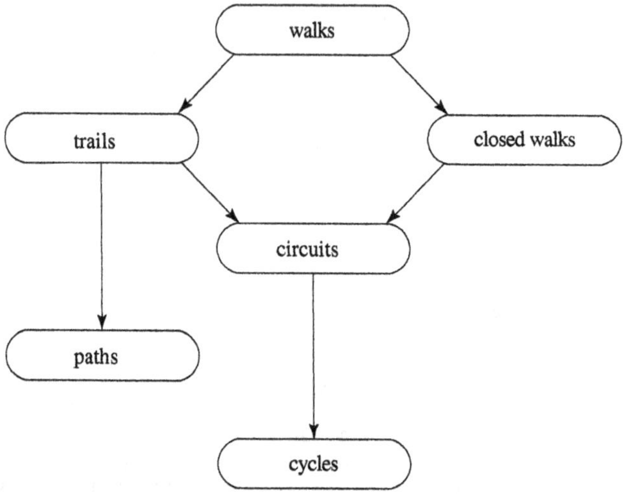

Figure 2.2: Relationships between various subgraphs and walks.

The next result shows that if there is a walk between two vertices in a graph, then there must also be a path between them. Intuitively, the idea is simply to go directly between the two vertices without taking any "detours."

Theorem 2.4 Let G be a graph having distinct vertices u and v. If G contains a u, v-walk, then G contains a u, v-path.

Proof If there is a u, v-walk in G, then there is one of shortest length among all of them. Let p be one such u, v-walk of shortest length k:

$$p = (u_0, e_1, u_1, e_2, \ldots, e_k, u_k),$$

where $u = u_0$ and $v = u_k$.

We now show that p is actually a path. Assume the contrary, that p is not a path. Then there must be a repeated vertex in p—that is, there must be a pair of indices $i, j \in \{0, 1, 2, \ldots, k\}$ with $i < j$ and $u_i = u_j$. In this case we can form a new u, v-walk

$$p' = (u_0, e_1, u_1, \ldots, e_i, u_i, e_{j+1}, u_{j+1}, \ldots, e_k, u_k)$$

of length $k - (j - i) < k$. Figure 2.3 depicts the situation.

This means that p is not the shortest u, v-walk, which contradicts our choice of p. Our assumption that p is not a path is therefore wrong. This proves the theorem. ■

It should be clear from context when we talk about a walk, trail, path, or cycle whether we mean the actual sequence of vertices and edges it is defined as, or simply the underlying subgraph it produces. The phrase "G contains a path of length five" means that the graph G contains a subgraph with five edges and their endvertices such that this very subgraph makes up a path of length five in an unambiguous and clear fashion.

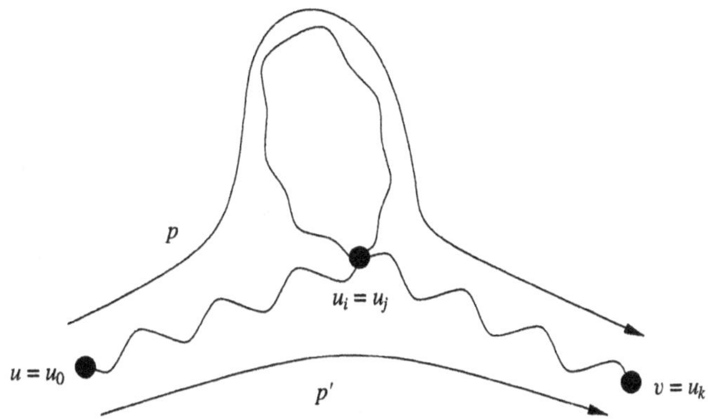

Figure 2.3: Walk p' is *shorter* than walk p.

2.2 Connectivity

It is intuitively clear what we mean by a graph being connected, namely, that we can start at any vertex and walk to any other vertex along the edges of the graph. This leads to the following definition.

Definition 2.5 A graph G is *connected* if for every pair of distinct vertices $u, v \in V(G)$, the graph G has a u, v-path. Otherwise, we say the graph is *disconnected.*

Example 2.6 Consider the graphs G and G' depicted in Figure 2.4. G is a connected graph, but G' is not, and hence disconnected.

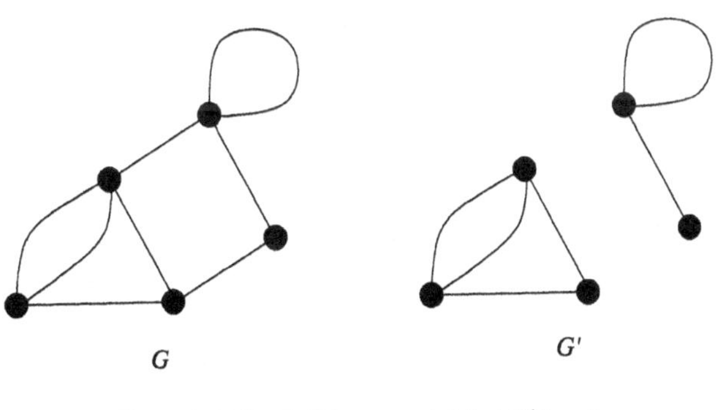

Figure 2.4: Graph G is connected, but G' is not.

As we see in Figure 2.4, the graph G' is made up of two connected parts. Each segment is a subgraph of G' that is itself connected. Such a connected part of a graph is called a *component* or sometimes a *connected component* of the graph. More formally, we have the following definition.

Definition 2.7 Let G be a graph. Let $H_1, \ldots, H_k$ be connected subgraphs of G whose vertex sets and edge sets are pairwise disjoint and such that they *cover* all the vertices and edges of G. That is,

$$V(G) = V(H_1) \cup \cdots \cup V(H_k),$$
$$E(G) = E(H_1) \cup \cdots \cup E(H_k),$$

where $V(H_i) \cap V(H_j) = E(H_i) \cap E(H_j) = \varnothing$ for each distinct i and j.

Each of the subgraphs H_i is called a *component* or a *connected component* of G.

The next result states the intuitive idea that every graph is made up of its components and its components alone.

Theorem 2.8 Every graph G has a unique collection of connected components $H_1, \ldots, H_k$. In particular, the number k of connected components of G is uniquely determined by G.

It is evident that a graph G is connected if, and only if, $k = 1$. Before proving Theorem 2.8, we define a very convenient mathematical tool, which has far-reaching applications.

Definition 2.9 Let S be a set and $\sim$ a binary relation on S. Assume that $\sim$ satisfies the following:

1. $s \sim s$ holds for all $s \in S$ (*reflexive* condition),
2. $s \sim t$ implies $t \sim s$ (*symmetric* condition),
3. $s \sim t$ and $t \sim u$ imply $s \sim u$ (*transitive* condition).

In this case $\sim$ is called an *equivalence relation* on S.

The following result shows that an equivalence relation on a finite nonempty set S is really the same as a *partition* of that set; that is, we can write $S = S_1 \cup \cdots \cup S_k$ for some $k \in \mathbb{N}$, where

1. $S_i \neq \varnothing$ for $1 \leq i \leq k$,
2. $S_i \cap S_j = \varnothing$ for $1 \leq i < j \leq k$.

Theorem 2.10 Let S be a finite set and $\sim$ an equivalence relation on S. Then for some k we have a partition $S = S_1 \cup \cdots \cup S_k$ of S, which is uniquely determined by $\sim$. The partition satisfies

$$S_i = \{t \in S : t \sim s_i\} \text{ for each } s_i \in S_i. \tag{2.1}$$

Each of the subsets S_i is called an *equivalence class* corresponding to $\sim$.

Conversely, the equivalence relation $\sim$ on S yielding this partition is unique, and can be retrieved by

$$s \sim t \Leftrightarrow s, t \in S_i \text{ for some } i \text{ with } 1 \leq i \leq k. \tag{2.2}$$

By the same token, an *arbitrary* partition on S yields an equivalence relation on S given by (2.2).

Remark 2.11 For a given set S, the one-to-one correspondence between equivalence relations on S and partitions of S can be stated precisely in terms of bijective maps as follows:

Let $\mathcal{P}(S)$ and $\mathcal{E}(S)$ be the sets of partitions of S and equivalence relations in S, respectively. Define the map $F : \mathcal{E}(S) \to \mathcal{P}(S)$ by $F :\sim \mapsto \mathbf{P}_\sim$, where the partition is obtained by (2.1) in Theorem 2.10. Likewise, define the map $G : \mathcal{P}(S) \to \mathcal{E}(S)$ by $G : \mathbf{P} \mapsto \sim_\mathbf{P}$, where the $\sim_\mathbf{P}$ is obtained by (2.2). In this way we have $G \circ F = \iota_{\mathcal{E}(S)}$ and $F \circ G = \iota_{\mathcal{P}(S)}$, the identity maps—that is to say,

$$G \circ F :\sim \mapsto \mathbf{P}_\sim \mapsto \sim,$$
$$F \circ G : \mathbf{P} \mapsto \sim_\mathbf{P} \mapsto \mathbf{P}.$$

Proof (Sketch, Theorem 2.10) Choose $s_1 \in S$ and let $S_1 = \{t \in S : t \sim s_1\}$. If S_1 equals S, then we stop. Otherwise, we pick $s_2 \in S \setminus S_1$ and let $S_2 = \{t \in S : t \sim s_2\}$. Note that because $\sim$ is an equivalence relation, we see that $S_1 \cap S_2 = \varnothing$, since otherwise, if $t \in S_1 \cap S_2$, then we have $s_1 \sim t$ and $s_2 \sim t$. Hence, by the symmetric and transitive conditions, we get $s_1 \sim s_2$. But then $s_2 \in S_1$, which is not the case.

Again, if $S = S_1 \cup S_2$, then we stop. Otherwise, we continue in this manner and pick $s_3 \in S \setminus (S_1 \cup S_2)$ and let $S_3 = \{t \in S : t \sim s_3\}$. As before, we have $S_3 \cap S_i = \varnothing$ for $i = 1, 2$.

The procedure outlined here must stop because S is a finite set. At the stopping point, we have that $S = S_1 \cup \cdots \cup S_k$, where $s_1, \ldots, s_k \in S$ and $S_i \cap S_j = \varnothing$ for $1 \le i < j \le k$. Since $s_i \in S_i$ for $1 \le i \le k$, we see that S_i is nonempty.

Finally, let $i \in \{1, \ldots, k\}$ and let $s_i' \in S_i$. Since $s_i' \sim s_i$, we have that if $t \sim s_i$ then $t \sim s_i'$, and the other way, that if $t \sim s_i'$, then $t \sim s_i$. Therefore,

$$S_i = \{t \in S : t \sim s_i\} = \{t \in S : t \sim s_i'\}.$$

This holds for each $s_i' \in S_i$. We leave what remains to the reader as Exercise 2.10. ■

Figure 2.5 depicts the partition of the set $S = \{1, 2, \ldots, 10\}$ corresponding to the relation of being congruent modulo 3. Notice that S is partitioned into three equivalence classes, each class containing the numbers which are congruent modulo 3.

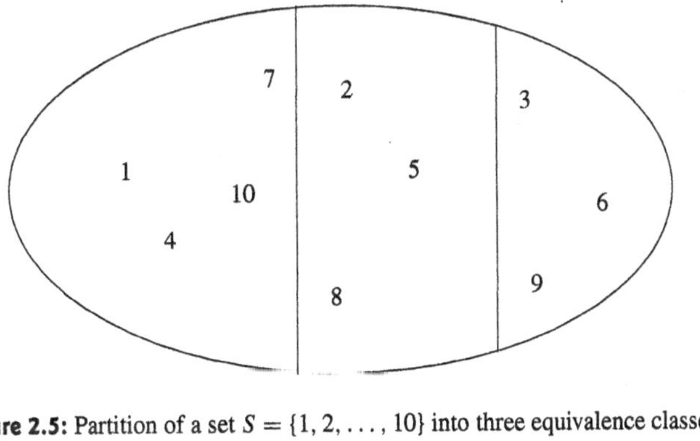

Figure 2.5: Partition of a set $S = \{1, 2, \ldots, 10\}$ into three equivalence classes.

Theorem 2.12 For a graph G let $\sim$ be the *path relation* on $V(G)$ given by

$$u \sim v \Leftrightarrow G \text{ has a } u, v\text{-path.}$$

Then $\sim$ is an equivalence relation on $V(G)$.

Proof For any vertex u the path of length zero is in G and hence $u \sim u$.

Suppose $u \sim v$ and that p is a u, v-path in G. In this case the inverse path p^{-1} is a v, u-path in G, and hence $v \sim u$.

Finally, if p is a u, v-path in G and q is a v, w-path in G, then pq is a u, w-walk in G. By Theorem 2.4 there is a u, w-path in G, which completes the proof. ■

We are now ready to present the proof of Theorem 2.8.

Proof (Theorem 2.8, page 36) By Theorem 2.12 using the path relation $\sim$, we get a partition of the vertices of G

$$V(G) = V_1 \cup \cdots \cup V_k,$$

where $V_i \cap V_j = \varnothing$ for $1 \leq i < j \leq k$. Note that every pair of vertices u and v in each V_i is connected by a path in G.

Let $H_i = G[V_i]$ be the subgraph of G induced by V_i. Let E_i be the edge set of H_i for $1 \leq i \leq k$. Since no two of the H_i's have a vertex in common, they have no edges in common either, implying that $E_i \cap E_j = \varnothing$ for $1 \leq i < j \leq k$.

Finally, if $e \in E$ is an edge of G, then the endvertices of e, say u and v, are connected by the path (u, e, v). So $u, v \in V_i$ for some i, and therefore $e \in E_i$. Thus we have $E(G) = E_1 \cup \cdots \cup E_k$, which completes the proof of the theorem. ■

In Exercise 2.9 we ask for an alternative way to define the components of a graph. We conclude this section with the following theorem.

Theorem 2.13 For a simple graph G with n vertices and k components we have

$$|E(G)| \leq \frac{(n-k)(n-k+1)}{2}.$$

Before we present the proof, we will present and prove the following lemma on a collection of real numbers adding up to a given number.

Lemma 2.14 If $x_1, x_2, \ldots, x_k \geq 1$ are real numbers and $x = x_1 + \cdots + x_k$, then

$$\sum_{i=1}^{k} x_i^2 \leq x^2 - (k-1)(2x-k).$$

Proof For each i let $x_i = y_i + 1$, where $y_i \geq 0$. Let $y = y_1 + \cdots + y_k$. By expressing the right-hand side of the inequality in Lemma 2.14 in terms of y, we have

$$\begin{aligned} x^2 - (k-1)(2x-k) &= (y+k)^2 - (k-1)(2y+k) \\ &= y^2 + 2y + k \\ &= \left(\sum_{i=1}^{k} y_i\right)^2 + 2\left(\sum_{i=1}^{k} y_i\right) + k \\ &\geq \sum_{i=1}^{k} y_i^2 + 2\left(\sum_{i=1}^{k} y_i\right) + k \\ &= \sum_{i=1}^{k} (y_i^2 + 2y_i + 1) \\ &= \sum_{i=1}^{k} x_i^2, \end{aligned}$$

which proves the lemma. ∎

Proof (Theorem 2.13) Let $H_1, \ldots, H_k$ be the components of G. For $1 \leq i \leq k$, denote by n_i the number of vertices of H_i. Note that $n = n_1 + \cdots + n_k$. Observe that each H_i is a simple graph. From Corollary 1.30 on page 19, it follows that H_i can have at most $n_i(n_i - 1)/2$ edges. Hence, using Lemma 2.14, the maximum number of edges of G is at most

$$\begin{aligned} \sum_{i=1}^{k} \frac{n_i(n_i-1)}{2} &= \frac{1}{2}\left(\sum_{i=1}^{k} n_i^2\right) - \frac{n}{2} \\ &\leq \frac{1}{2}\left[n^2 - (k-1)(2n-k)\right] - \frac{n}{2} \\ &= \frac{(n-k)(n-k+1)}{2}, \end{aligned}$$

which proves the theorem. ∎

2.3 Homomorphisms and Isomorphisms of Graphs

In this section we discuss the concept of similarity between graphs, which is considered by many to be a fundamental concept in graph theory. Just as in many branches of mathematics and computer science, various properties of objects can be obtained by studying maps between them that satisfy some special conditions, and the same holds true for graphs. These maps are called *homomorphisms*. A homomorphism from a graph G to another graph G' is a function that maps the vertices of G to the vertices of G', and the edges of G to the edges of G' in the following way: If an edge e in G has endvertices u and v, then the corresponding image edge e' of G' has endvertices u' and v', which are the respective images of u and v under the mapping. To state this precisely and to be in coherence with Definition 1.8 on page 10, we make the following definition.

Definition 2.15 Let $G = (V, E, \phi)$ and $G' = (V', E', \phi')$ be two graphs. A *homomorphism*

$$f : G \to G'$$

from G to G' is an ordered pair $f = (f_1, f_2)$ of maps $f_1 : V \to V'$ and $f_2 : E \to E'$ satisfying the following condition:

$$\phi(e) = \{u, v\} \Rightarrow \phi'(f_2(e)) = \{f_1(u), f_1(v)\}. \tag{2.3}$$

That is, if u and v are the endvertices of e in G, then $f_1(u)$ and $f_1(v)$ are the endvertices of $f_2(e)$ in G'.

If both maps f_1 and f_2 are bijective, then f is called an *isomorphism*. In this case we say that the graph G is *isomorphic* to the graph G', or that G and G' are *isomorphic* and denote that by $G \cong G'$. An isomorphism from G to itself is called an *automorphism* of G.

Before discussing the implication of this definition further, we consider two examples. The first example illustrates the concept of homomorphism.

Example 2.16 Consider the two graphs G and G' shown in Figure 2.6. Define the maps f_1 and f_2 as follows, first f_1 by

$$\begin{aligned} &f_1(u_1) = f_1(u_4) = f_1(u_5) = f_1(u_6) = u'_4, \\ &f_1(u_2) = u'_2, \\ &f_1(u_3) = u'_3, \end{aligned}$$

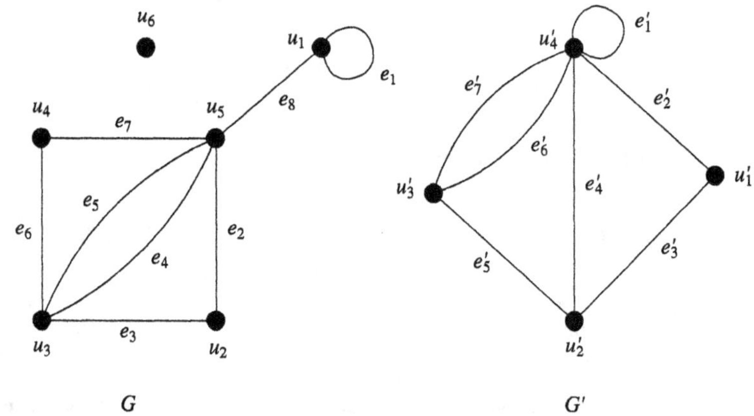

Figure 2.6: Graphs G and G' used in Example 2.16 to illustrate a homomorphism.

and then f_2 for the edges by

$$\begin{aligned}
&f_2(e_1) = f_2(e_7) = f_2(e_8) = e'_1,\\
&f_2(e_2) = e'_4,\\
&f_2(e_3) = e'_5,\\
&f_2(e_4) = f_2(e_5) = e'_6,\\
&f_2(e_6) = e'_7.
\end{aligned}$$

Let $f : G \to G'$ be defined by $f = (f_1, f_2)$. Having defined the maps f_1 and f_2, one needs to consider each edge in G and check whether its endvertices are mapped to the corresponding endvertices of the mapped edge in G'. So we "bite the bullet," so to speak, and check each of edges $e_1, \ldots, e_8$ of G:

- We have $\phi(e_1) = \{u_1\}$ and $\phi'(f_2(e_1)) = \phi'(e'_1) = \{u'_4\} = \{f_1(u_1)\}$.
- We have $\phi(e_2) = \{u_2, u_5\}$ and $\phi'(f_2(e_2)) = \phi'(e'_4) = \{u'_2, u'_4\} = \{f_1(u_2), f_1(u_5)\}$.

  $\vdots$

- We have $\phi(e_8) = \{u_1, u_5\}$ and $\phi'(f_2(e_8)) = \phi'(e'_1) = \{u'_4\} = \{u'_4, u'_4\} = \{f_1(u_1), f_1(u_5)\}$.

We do urge the reader to check all the remaining cases for $\phi(e_i)$. Hence, $f : G \to G'$ is indeed a homomorphism. Note that f is not an isomorphism.

By the definition of homomorphism, any pair of adjacent vertices are mapped to adjacent vertices or to the same vertex. Moreover, any pair of adjacent edges are mapped to adjacent edges. Also a nonloop edge can be mapped to a nonloop edge or to a loop. However, a loop can be mapped only to a loop.

The following example illustrates the notion of isomorphism.

Example 2.17 Consider the two graphs G and G' shown in Figure 2.7. Here both G and G' are simple graphs, and hence each edge in these graphs can be expressed as a two-element set of vertices. This implies that both the corresponding edgemaps ϕ and ϕ' are identity maps. Therefore, if $f_1 : V \to V'$ is the map defined by $f_1(u_i) = v_i$ for each $i \in \{1, 2, \ldots, 10\}$, then the condition in (2.3) that $f_2 : E \to E'$ must satisfy becomes

$$f_2(\{u_i, u_j\}) = \{v_i, v_j\} \tag{2.4}$$

for every edge $\{u_i, u_j\} \in E$. The condition in (2.4), which completely determines f_2, is sufficient for $f = (f_1, f_2)$ to be a homomorphism. The reader should verify that f is indeed a homomorphism.

Note that f_1 is a bijective map. The map f_2 is an injective map from a set E of 15 edges to another set E' of 15 edges, and is therefore also a bijection. Thus, f is an isomorphism. What this means is the graphs in Figure 2.7 are "equivalent" in the sense that every graph-theoretic property fulfilled by one of the graphs is also fulfilled by the other.

In this example it is not hard to see that geometrically G' is a simple topological deformation of G. To see this, simply take G and deform it by stretching to the right, as illustrated in Figure 2.8. Next reflect the vertices u_2 and u_7 about the edge $\{u_6, u_8\}$ and let

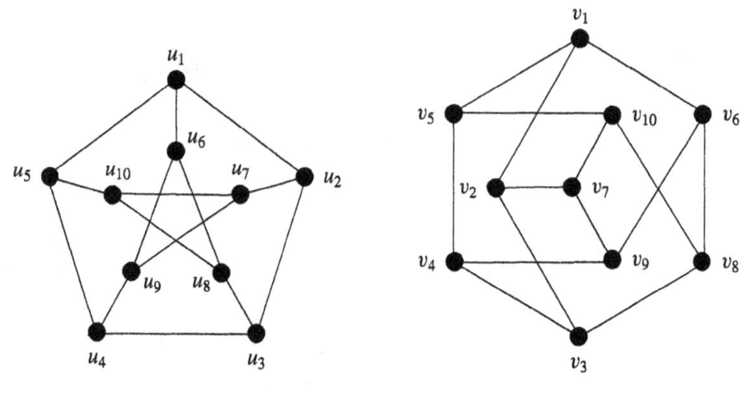

Figure 2.7: Are graphs G and G' isomorphic?

the edges with endvertices u_2 and u_7 follow. This is shown in Figure 2.9. We now see that G and G' "are the same."

The following theorem describes one of the most important equivalence relations among graphs, induced by isomorphisms between graphs.

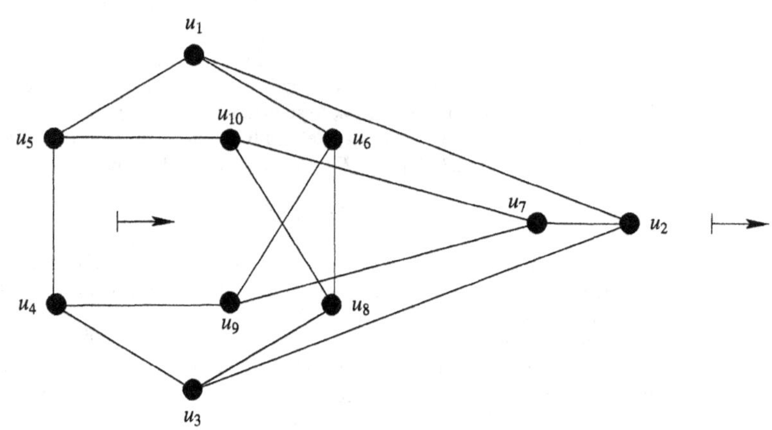

Figure 2.8: In a step to deform G into G' we stretch G to the right, as indicated by the arrows.

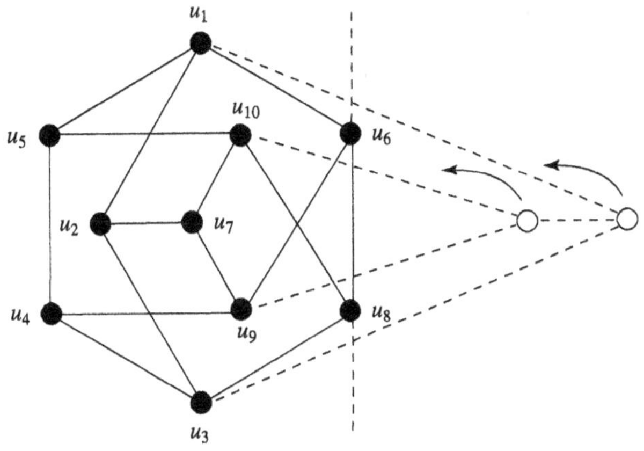

Figure 2.9: Reflection about the edge $\{u_6, u_8\}$ to obtain G' from G.

Theorem 2.18 The isomorphism relation between graphs in a fixed collection of graphs is an equivalence relation. That is,

1. For any graph $G = (V, E, \phi)$, we have $G \cong G$. In fact,
$$\iota_G = (\iota_V, \iota_E)$$
is always an automorphism of G.
2. If $G \cong G'$ and $G' \cong G''$, then $G \cong G''$. In fact, if $f = (f_1, f_2)$ is an isomorphism from $G = (V, E, \phi)$ to $G' = (V', E', \phi')$ and $f' = (f_1', f_2')$ is an isomorphism from $G' = (V', E', \phi')$ to $G'' = (V'', E'', \phi'')$, then the ordered pair
$$f' \circ f = (f_1' \circ f_1, f_2' \circ f_2)$$
is an isomorphism from G to G''.
3. If $G \cong G'$ then $G' \cong G$. In fact, if $f = (f_1, f_2)$ is an isomorphism from $G = (V, E, \phi)$ to $G' = (V', E', \phi')$, then the ordered pair
$$f^{-1} = (f_1^{-1}, f_2^{-1})$$
is an isomorphism from G' to G.

Proof Since the proofs of each of the three statements are very similar, we leave the proofs of the first two enumerated statements as Exercise 2.12.

To prove the third statement, we assume that $f = (f_1, f_2)$ is an isomorphism from G to G'. Since both f_1 and f_2 are bijective, then so are their inverses f_1^{-1} and f_2^{-1}. To show $f^{-1} = (f_1^{-1}, f_2^{-1})$ is an isomorphism from G' to G, we need that the condition
$$\phi'(e') = \{u', v'\} \Rightarrow \phi(f_2^{-1}(e')) = \{f_1^{-1}(u'), f_1^{-1}(v')\}$$
holds for every $e' \in E'$ and its endvertices $u', v' \in V'$.

Assume that $\phi'(e') = \{u', v'\}$ and $\phi(f_2^{-1}(e')) = \{x, y\}$. We want to show that $\{x, y\} = \{f_1^{-1}(u'), f_1^{-1}(v')\}$. Set $f_2^{-1}(e') = e$, so we have that $f_2(e) = e'$ and $\phi(e) = \{x, y\}$. Since $f = (f_1, f_2)$ is an isomorphism, we have

$$\{u', v'\} = \phi'(e') = \phi'(f_2(e)) = \{f_1(x), f_1(y)\}.$$

Therefore, either $u' = f_1(x)$ or $u' = f_1(y)$. For reasons of symmetry we can assume the former. We then have that $u' = f_1(x)$ and $v' = f_1(y)$ (note that this holds regardless of whether u' and v' are distinct or not). From this we get $f_1^{-1}(u') = x$ and $f_1^{-1}(v') = y$. Therefore, $\{x, y\} = \{f_1^{-1}(u'), f_1^{-1}(v')\}$. This proves that $f^{-1} = (f_1^{-1}, f_2^{-1})$ is an isomorphism. ■

Theorem 2.18 implies that when describing a graph-theoretic property of all graphs that are isomorphic to each other (that is, equivalent by $\cong$), it suffices to consider one representative among them and work with it. Once we have deduced the graph-theoretic property we desire, then we can conclude that all the graphs isomorphic to the one we worked with also have that property. Such graph-theoretic properties include "being connected," "containing a vertex of degree less than or equal to five," "containing a simple path with more than half the number of vertices of the entire graph," and so on. This is true because we can transform back and forth between the graphs in the equivalence class defined via the isomorphisms.

This allows us to talk about "the graph N_n" (as we have done already, without justification, in Example 1.14 on page 13, and the examples following that one!) as just one representative of the countless graphs isomorphic to N_n. This is because all graphs isomorphic to N_n have exactly the same graph-theoretic properties as N_n. Therefore, they can be viewed to be the same graph as the one represented in Example 1.14.

As we saw in Example 2.17 on page 41, the graphs G and G' represented by the diagrams in Figure 2.7 were isomorphic. Hence, we can say that both G and G' are copies of the graph, which is called the *Petersen graph*. It is usually presented as the diagram of G on the left of Figure 2.7.

What happens when we have two graphs that look similar, but we seem unable to find an isomorphism between them? In this case either the isomorphism itself is well hidden somehow, or the two graphs are simply not isomorphic.

Example 2.19 Consider the two graphs that are shown in Figure 2.10. Both graphs have five vertices, and both have eight edges, both have one loop, and both have one double edge, so a priori it looks as if these graphs could be isomorphic. However, on closer inspection, we see that the two vertices u and v of the graph on the left of degree four are adjacent, but the two vertices u' and v' of degree four in the graph on the right are not adjacent. This is a graph-theoretic property that distinguishes the two graphs. From this alone, we can conclude that these graphs are *not* isomorphic, since an isomorphism from the graph on the left to the one on the right would simply carry the property from left to right.

To make this even clearer, consider another property that distinguishes the graphs. Both graphs have exactly one vertex of degree two, w in the left graph and w' in the right graph. If there were an isomorphism $f = (f_1, f_2)$ from the graph on the left to the one on the right, then we must have $f_1(w) = w'$. Here we see that the neighbors of w have degrees of three and four, and both neighbors of w' have the same degree, namely four. Hence, f

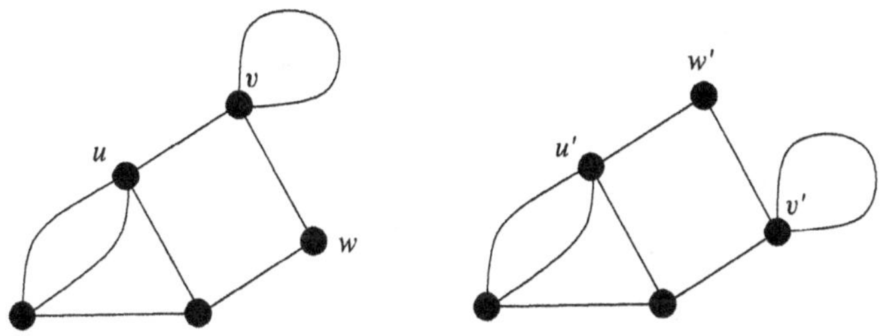

Figure 2.10: Graphs used in Example 2.19. Are the graphs isomorphic?

cannot be an isomorphism. The property "*G has precisely one vertex of degree two, and both its neighbors have the same degree*" is satisfied by the graph on the left, but not by the one on the right.

2.4 More on Isomorphisms on Simple Graphs

In Example 2.17, the isomorphism from G to G' was completely determined by the vertex map f_1. This is indeed the case for all pairs of simple graphs. Since many graph-theoretic problems can actually be reduced to the study of simple graphs, we can simplify our general Definition 2.15 on page 40 of homomorphism and isomorphism between simple graphs. Recall that by Definition 1.13 on page 13, an edge in a simple graph is a two-element subset of the set of its vertices.

Definition 2.20 Let $G = (V, E)$ and $G' = (V', E')$ be two simple graphs. A *homomorphism*

$$f : G \to G'$$

from G to G' is an ordered pair $f = (f_1, f_2)$ of maps, $f_1 : V \to V'$ and $f_2 : E \to E'$, satisfying the compatibility condition

$$f_2(\{u, v\}) = \{f_1(u), f_1(v)\} \tag{2.5}$$

for every edge $\{u, v\} \in E$.

The homomorphism f is an *isomorphism* if both f_1 and f_2 are bijective.

Note that this definition is completely compatible with the more general Definition 2.15 on page 40 restricted to simple graphs, since in the simple case both ϕ and ϕ' are identity maps, as we saw in Example 2.17. Hence, the isomorphism relation (2.5) defines an equivalence relation between any collection of simple graphs as well.

Moreover, for any homomorphism $f = (f_1, f_2)$ from G to G', the vertex map $f_1 : V \to V'$ completely determines the edge map $f_2 : E \to E'$ by the compatibility condition (2.5). Hence, in checking whether or not the graphs G and G' are isomorphic we "only" need to check all the injective maps $f_1 : V(G) \to V(G')$. Note that when both G and G' are complete, then *any* such f_1 will induce an isomorphism. Hence we have the following observation.

Observation 2.21 To check whether two graphs G and G' on n vertices are isomorphic, it suffices to consider the possible homomorphisms $G \to G'$ induced by the $n!$ bijective vertex maps $V(G) \to V(G')$.

Needless to say, $n!$ grows very fast when n grows, so such a strategy to test isomorphism is by no means a practical one for large graphs. There are some simplifications we can make, but they also do not amount to much for worst-case scenarios.

Any subgraph of a given graph G that is isomorphic to a complete graph K_h for some $h \in \mathbb{N}$ is called an *h-clique* in G. When the number of vertices in the subgraph is irrelevant, we simply call it a *clique*. Since two adjacent vertices in a simple graph must be mapped to *different* adjacent vertices, we get the following.

Observation 2.22 In any homomorphism between simple graphs, an h-clique is always mapped to an h-clique, for every $h \in \mathbb{N}$.

Determining if two simple graphs are isomorphic is a well-known, but algorithmically difficult, open problem in graph algorithms. Given two graphs, the problem boils down to determining whether all graph-theoretic properties satisfied by one graph are also satisfied by the other. If we can spot a property that one graph satisfies but the other one does not, then we can rest assured that the graphs are not isomorphic. But even if two graphs are not isomorphic, finding a particular property satisfied by one and not the other can be quite a tricky task, even though we know for a fact that such a property exists.

There are times when we can tell right away that two graphs are not isomorphic. We note some properties that can easily be checked. Such properties are called *invariants* of a graph. If two graphs are isomorphic, they must satisfy the following:

1. They must have the same number of vertices.
2. They must have the same number of edges.
3. They must have the same number of components.
4. They must have the same number of vertices of degree i, for each nonnegative i.
5. They must have the same number of h-cliques, for each $h \in \mathbb{N}$.

Let G and G' be two simple graphs on n vertices, where for each nonnegative i, the number of vertices in both G and G' of degree i is given by n_i. Assume further that the maximum degree of a vertex in both graphs is k, and hence $n = n_0 + n_1 + \cdots + n_k$. We can now state a slight improvement of Observation 2.21, the proof of which we leave as Exercise 2.19.

Observation 2.23 Let G and G' be two simple graphs both with maximum degree of k and with n_i vertices of degree i for $1 \le i \le k$.

To check whether G and G' are isomorphic, it suffices to consider the possible homomorphisms $G \to G'$ induced by at most $n_1!n_2! \cdots n_k!$ bijective vertex maps $V(G) \to V(G')$.

It becomes harder to determine whether two simple graphs are isomorphic when the number of edges is increased. Recall that a simple graph with n vertices can have at most $n(n-1)/2$ edges. We will see that when the simple graphs both have either very few edges or close to the maximum of $n(n-1)/2$ edges, it is relatively quick to check for isomorphisms. Before presenting that result, we need to introduce the notion of the *complement* of a graph.

Definition 2.24 For a simple graph $G = (V, E)$, the *complement* of G is defined as

$$\overline{G} = (V, \overline{E}),$$

where $\overline{E} = \{\{u, v\} : \{u, v\} \notin E\}$. In other words, $\overline{G}$ is a simple graph with the same set of vertices as G and where two vertices are adjacent if, and only if, they are not adjacent in G.

Note that $\overline{N_n} = K_n$ and $\overline{K_n} = N_n$. In general, we have $\overline{\overline{G}} = G$ for any simple graph G. With Definition 2.24 in mind, we obtain the following theorem, whose proof we leave as Exercise 2.20.

Theorem 2.25 For simple graphs G and G', we have

$$G \cong G' \Leftrightarrow \overline{G} \cong \overline{G'}.$$

That is, two graphs are isomorphic if, and only if, their complements are isomorphic.

For a graph G on n vertices, we have

$$|E(G)| + |E(\overline{G})| = \frac{n(n-1)}{2},$$

and hence either G or $\overline{G}$ has at most $n(n-1)/4$ edges. Therefore, by Theorem 2.25 we see that in order to check whether two simple graphs on n vertices and on the same number of edges are isomorphic, we may assume they both have at most $\lfloor n(n-1)/4 \rfloor$ edges.

Example 2.26 Consider the two graphs illustrated in Figure 2.11. Both graphs are 3-regular simple graphs on six vertices. We note that the graph G on the left has a cycle of length three, but G' on the right has no 3-cycles. Therefore, these graphs are not isomorphic.

Another argument can be used by examining the complements. The complements of the graphs are shown in Figure 2.12, from which we see that the complement $\overline{G}$ is the cycle C_6 on six vertices and in particular connected. However, the complement $\overline{G'}$ is the disjoint union of two copies of the cycle C_3 on three vertices, which has two components. So $\overline{G}$ is not isomorphic to $\overline{G'}$. Therefore, by Theorem 2.25 the graphs G and G' are not isomorphic either.

It is not hard to see that the graph G' on the right side of Figure 2.11 is isomorphic to $K_{3,3}$. The graph $K_{3,3}$ is *nonplanar*, in the sense that it cannot be drawn in the plane without edges crossing each other, something we will discuss in more detail in Chapter 7. Clearly, the graph G on the left side of Figure 2.11 is *planar*; there are no edge crossings. Hence, we have yet another way to see that G and G' are not isomorphic.

We could continue to list graph-theoretic properties satisfied by one graph and not by the other. However, just one such property suffices for the two graphs not to be isomorphic.

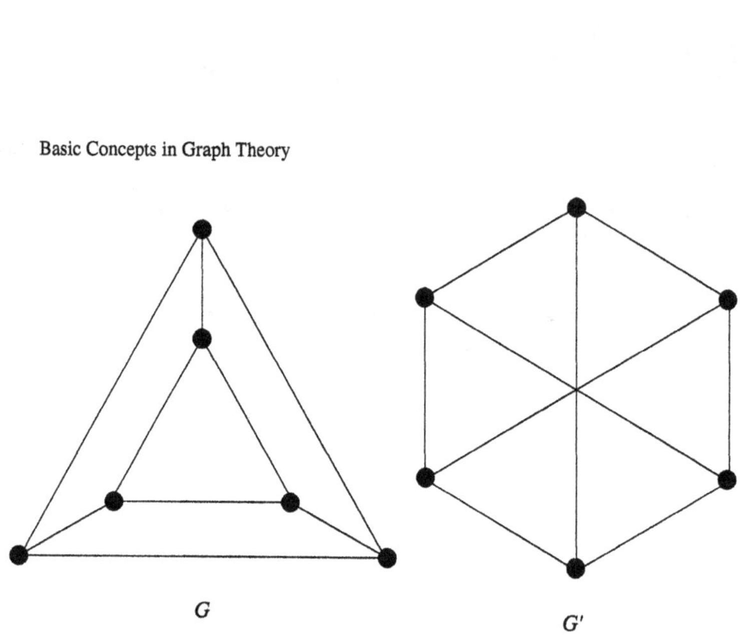

Figure 2.11: Graphs used in Example 2.26. Are the graphs isomorphic?

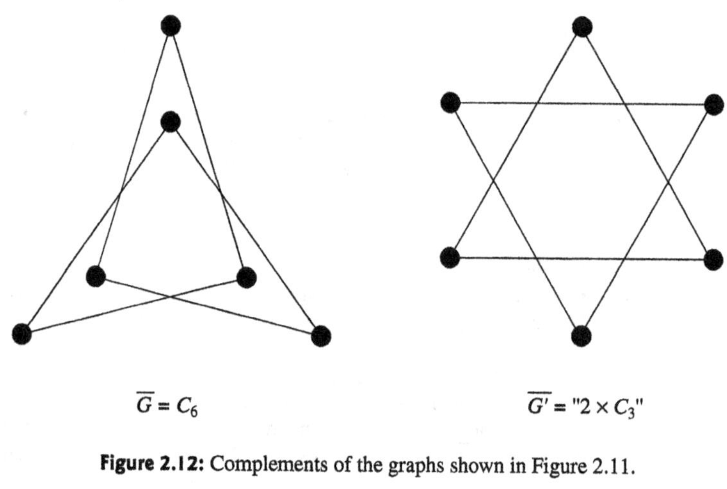

Figure 2.12: Complements of the graphs shown in Figure 2.11.

2.5 Formations and Minors of Graphs

There are several natural ways of forming graphs from a collection of other graphs. As for sets, one can form unions, intersections, and symmetric differences of graphs and obtain new graphs that are better suited for one's purpose. This is particularly straightforward for simple graphs, since every simple graph is by definition an ordered pair of sets.

Definition 2.27 Let $G_1 = (V_1, E_1)$ and $G_2 = (V_2, E_2)$ be simple graphs. The *union* of G_1 and G_2 is the simple graph

$$G_1 \cup G_2 = (V_1 \cup V_2, E_1 \cup E_2).$$

If $V_1 \cap V_2 \neq \varnothing$, then the *intersection* of G_1 and G_2 is the simple graph

$$G_1 \cap G_2 = (V_1 \cap V_2, E_1 \cap E_2).$$

The *symmetric difference* of G_1 and G_2 is the simple graph

$$G_1 \bigtriangleup G_2 = (V_1 \cup V_2, E_1 \bigtriangleup E_2).$$

For general graphs the same ideas hold. However, we need to be more careful in our definitions. Figure 2.13 demonstrates how the union, intersection, and symmetric difference of two (general) graphs are defined. As illustrated in the figure, if an edge is in both G_1 and G_2, then the edge's endvertices in G_1 must be the same as in G_2, since otherwise, these three operations cannot be sensibly defined. To be precise, we make the following definition that generalizes Definition 2.27.

Definition 2.28 Let $G_1 = (V_1, E_1, \phi_1)$ and $G_2 = (V_2, E_2, \phi_2)$ be subgraphs of a universal graph $G = (V, E, \phi)$.

The *union* of G_1 and G_2 is the graph

$$G_1 \cup G_2 = (V_1 \cup V_2, E_1 \cup E_2, \phi_{1\cup 2}),$$

where $\phi_{1\cup 2}$ is the restriction of ϕ to $E_1 \cup E_2 \subseteq E$.

If $V_1 \cap V_2 \neq \varnothing$, then the *intersection* of G_1 and G_2 is the graph

$$G_1 \cap G_2 = (V_1 \cap V_2, E_1 \cap E_2, \phi_{1\cap 2}),$$

where $\phi_{1\cap 2}$ is the restriction of ϕ to $E_1 \cap E_2 \subseteq E$.

The *symmetric difference* of G_1 and G_2 is the graph

$$G_1 \bigtriangleup G_2 = (V_1 \cup V_2, E_1 \bigtriangleup E_2, \phi_{1\bigtriangleup 2}),$$

where $\phi_{1\bigtriangleup 2}$ is the restriction of ϕ to $E_1 \bigtriangleup E_2 \subseteq E$.

Note 2.29 Our assumption stating that both G_1 and G_2 are subgraphs of a universal graph G implies that both ϕ_1 and ϕ_2 agree on $E_1 \cap E_2$; that is to say, for every $e \in E_1 \cap E_2$ we have

$$\phi_1(e) = \phi_2(e). \tag{2.6}$$

If we had not made the assumption that G_1 and G_2 were subgraphs of G, then we would have had to insist that (2.6) hold for all $e \in E_1 \cap E_2$, which would be an equivalent definition.

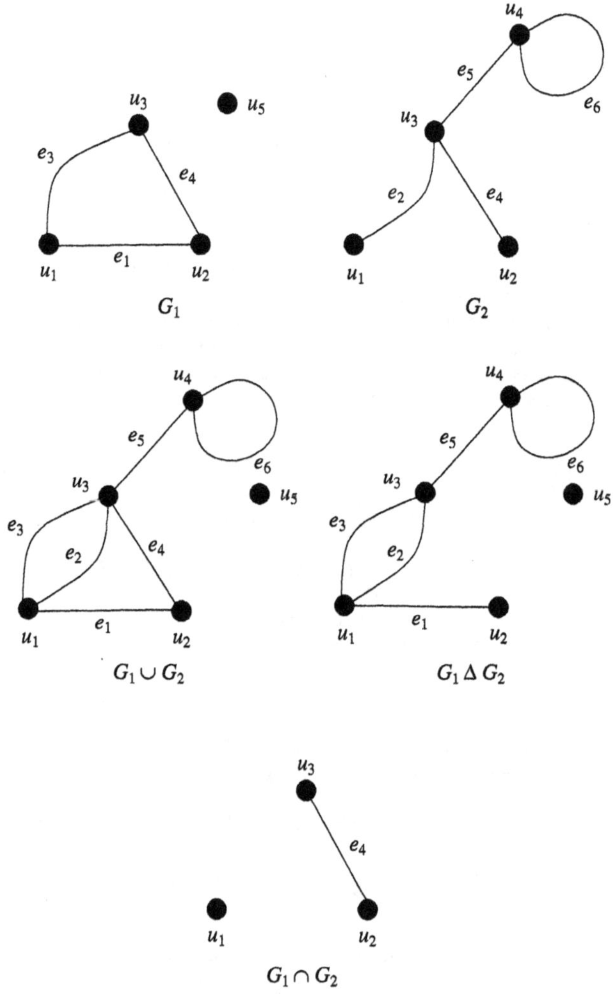

Figure 2.13: Union, intersection, and symmetric difference of two graphs.

It is obvious from their definitions that union, intersection, and symmetric difference are commutative operations, as they are for sets:

$$G_1 \cup G_2 = G_2 \cup G_1,$$
$$G_1 \cap G_2 = G_2 \cap G_1,$$
$$G_1 \bigtriangleup G_2 = G_2 \bigtriangleup G_1.$$

One can also show that these operations are associative, as for sets:

$$(G_1 \cup G_2) \cup G_3 = G_1 \cup (G_2 \cup G_3),$$
$$(G_1 \cap G_2) \cap G_3 = G_1 \cap (G_2 \cap G_3),$$
$$(G_1 \bigtriangleup G_2) \bigtriangleup G_3 = G_1 \bigtriangleup (G_2 \bigtriangleup G_3).$$

Because of these two algebraic properties—being commutative and associative—it should be clear that these operations can be extended to any k graphs. By associativity it makes sense to drop the parentheses that indicate the order in which operations are performed, since it does not matter. By commutativity the order of the listing of the graphs themselves is also irrelevant. For example, if $\sigma : \{1, \ldots, k\} \to \{1, \ldots, k\}$ is any permutation of indices (i.e., a bijective map), then $G_{\sigma(1)} \bigtriangleup G_{\sigma(2)} \bigtriangleup \cdots \bigtriangleup G_{\sigma(n)}$ is the graph on $V_1 \cup \cdots \cup V_k$ that contains precisely all those edges included in an odd number of the graphs G_i, where $1 \leq i \leq k$. (See the first two exercises of Chapter 1.)

In addition to forming unions, intersections, and symmetric differences, we often need to discuss the graph we obtain by deleting a vertex or an edge from it.

Definition 2.30 Let $G = (V, E, \phi)$ be a graph. Suppose $v \in V$.

$$G - v = G[V \setminus \{v\}]$$

is the subgraph of G resulting from eliminating v from V.

Suppose $e \in E$.

$$G - e = (V, E \setminus \{e\}, \phi')$$

is the subgraph of G resulting from eliminating e from E, where ϕ' is the restriction of ϕ to $E \setminus \{e\}$.

Figure 2.14 provides an example of deleting a vertex from a graph and an example of deleting an edge from that same graph.

Note 2.31 For any graph the following conditions hold:

- If a vertex is deleted, then so are all the edges having that vertex as an endvertex.
- If an edge is deleted, then its endvertices are not deleted.

In a natural way, we can generalize Definition 2.30 to the deletion of an arbitrary subset of vertices or edges. It does not matter whether one thinks of such deletions as a repeated deletion of one vertex or edge at the time, or simply all at once.

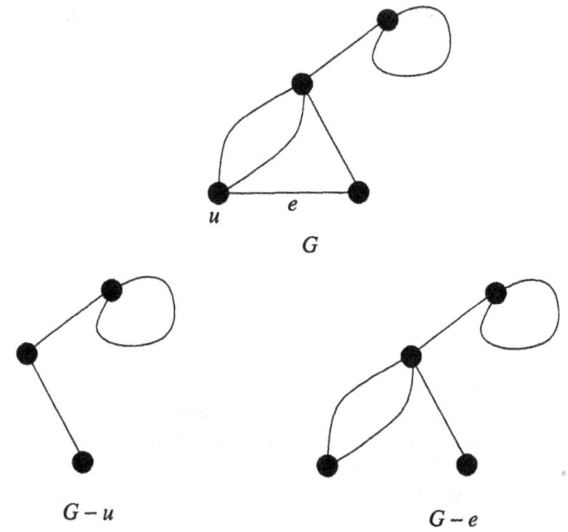

Figure 2.14: Vertex deletion and edge deletion.

1. If $V \subseteq V(G)$, then by $G - V$ we mean the resulting graph when every vertex from V has been deleted from G.
2. If $E \subseteq E(G)$, then by $G - E$ we mean the resulting graph when every edge from E has been deleted from G.

Note 2.32 When we view an edge $e \in E(G)$ as a subgraph of G, together with its endvertices, we have

$$G - e = G \bigtriangleup e.$$

Also, by removing a vertex or edge from a graph G we could increase the number of components of G.

Definition 2.33 A vertex v is a *cut-vertex* if $G - v$ has more components than G. An edge e is a *cut-edge* or a *bridge* if $G - e$ has more components than G.

Next we define yet another way of forming a new graph from a given graph, obtained by contracting an edge of the given graph into its endvertices.

Definition 2.34 Let $G = (V, E, \phi)$ be a graph and e an edge of E having distinct endvertices u and v. Let w be a new vertex such that $w \notin V$. Let α be the mapping defined by $\alpha : V \to (V \backslash \{u, v\}) \cup \{w\}$,

$$\alpha(x) = \begin{cases} x & \text{if } x \notin \{u, v\}, \\ w & \text{if } x \in \{u, v\}. \end{cases}$$

The *contraction* of G by e is the graph

$$G \cdot e = (\alpha(V), E \setminus \{e\}, \alpha' \circ \phi),$$

where $\alpha'(\{x, y\}) = \{\alpha(x), \alpha(y)\}$ for any vertices x and y.

In Definition 2.34, the graph $G \cdot e$ is formed from G by replacing the vertices u and v by the single new vertex w. Except for the edge e, every edge that was incident to u, v, or both is now incident to the new vertex w. Figure 2.15 provides an example of a graph before and after an edge has been contracted.

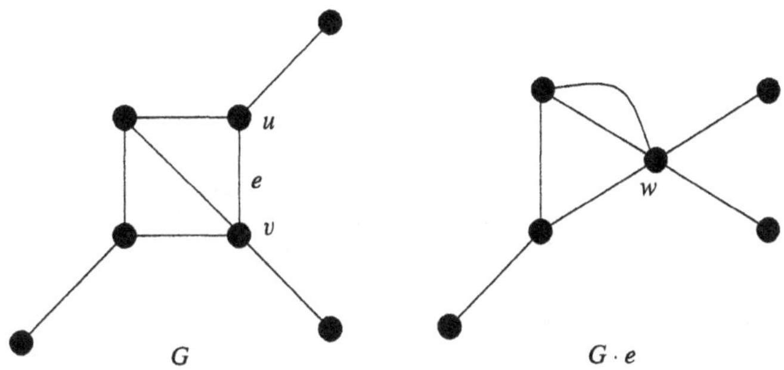

Figure 2.15: Contraction by the edge $e = \{u, v\}$.

Note 2.35 Observe that $\alpha' \circ \phi$ is applied only to $E \setminus \{e\}$ but not all of E. Although $(\alpha' \circ \phi)(e) = \{w\}$ is well defined, this loop is not contained in $G \cdot e$. Hence, we see that a contraction by an edge reduces the number of edges by one and the number of vertices by one.

Figure 2.15 illustrates that contraction is easy to understand geometrically. Although formal definitions can be somewhat involved in graph theory, one should always keep the geometrical pictures in mind. If G is a simple graph and $e \in E(G)$, note that the contracted graph $G \cdot e$ is not necessarily simple, since some edges might become twofold. We can, however, define a related *simple contraction* by an edge e that will produce a simple graph by simply discarding the multiple edges in the contracted graph $G \cdot e$.

Definition 2.36 Let $G = (V, E)$ be a simple graph and e an edge in E having endvertices u and v. Let w be a new vertex such that $w \notin V$. Let α be as in Definition 2.34, that is, $\alpha : V \to (V \setminus \{u, v\}) \cup \{w\}$,

$$\alpha(x) = \begin{cases} x & \text{if } x \notin \{u, v\}, \\ w & \text{if } x \in \{u, v\}. \end{cases}$$

The *simple contraction* of G by e is the graph

$$G/e = (\alpha(V), \alpha'(E \setminus \{e\})),$$

where $\alpha'(E \setminus \{e\}) = \{\{\alpha(x), \alpha(y)\} : \{x, y\} \in E \setminus \{e\}\}$.

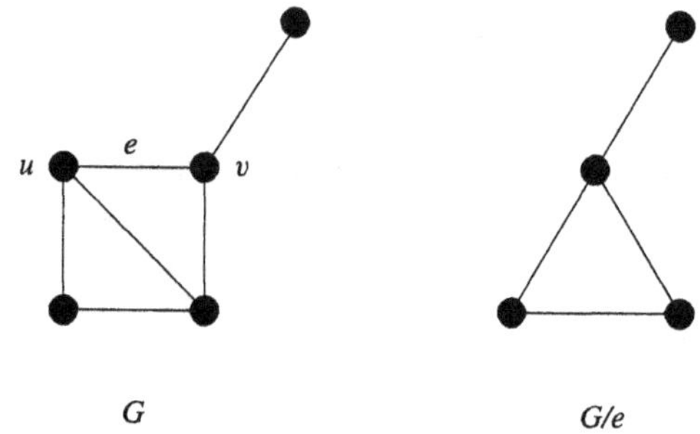

Figure 2.16: Simple contraction by the edge $e = \{u, v\}$.

Figure 2.16 provides an example of the simple contraction of an edge.

Another important way to form graphs from a given graph can now be presented. The notion of a contraction graph and a *minor* graph defined next is of great importance to understanding hidden geometrical properties of graphs.

Definition 2.37 For a connected simple graph G, we say that the simple graph G' is a *contraction* of G, if G' can be obtained from G by a sequence of simple contractions—that is, if there is a finite sequence $G_0, G_1, \ldots, G_k$ of simple graphs such that

1. $G = G_0$ and $G' = G_k$,
2. $G_{i+1} = G_i/e_i$, for some edge e_i in G_i for each $i \in \{1, \ldots, k-1\}$.

If G' is a contraction of a subgraph of G, then we say that G' is a *minor* of G.

The following example provides a useful way of viewing contractions and minors of a given graph.

Example 2.38 Given a simple and connected graph G, we can think of any contraction G' of G as follows: The vertices $V(G') = \{u'_1, \ldots, u'_m\}$ correspond to a partition of $V(G)$ into disjoint subsets $X_1, \ldots, X_m$ such that

- Each X_i induces a connected subgraph $G[X_i]$ of G, and
- The vertices u'_i and u'_j are connected by an edge in G' if and only if there is some edge in G with one endvertex in X_i and the other in X_j.

On the left-hand side of Figure 2.17 is a graph G together with a partition of its vertices. The right-hand side of the figure shows a contraction of G. Notice the correspondence between the $G[X_i]$'s and the u'_i's.

We can define a relation $\preceq_m$ on a given set of graphs by

$$G' \preceq_m G \Leftrightarrow G' \text{ is a minor of } G. \tag{2.7}$$

With this definition we have the following, the proof of which we leave as Exercise 2.29.

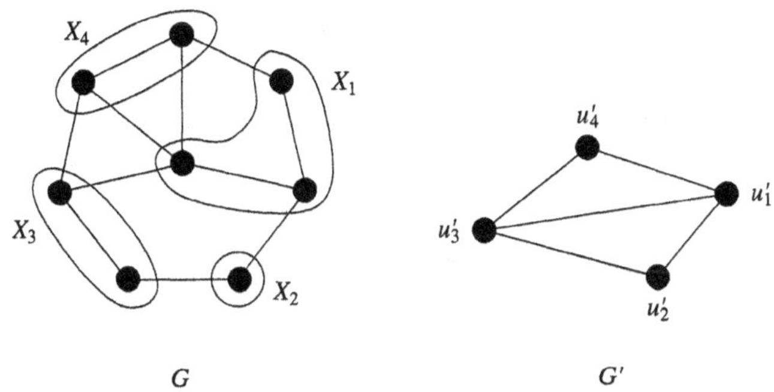

Figure 2.17: The graph G' is a contraction of G.

Theorem 2.39 A set of simple graphs, together with the minor relation $\preceq_m$, forms a poset.

We conclude this section by stating a deep yet easily understood theorem in graph theory, a *Minor Theorem* by Neil Robertson and Paul Seymour. This Minor Theorem is one of many such theorems that Robertson and Seymour have published in a series of more than twenty papers since 1983, labeled Graph Minors I, Graph Minors II, Graph Minors III, ..., Graph Minors XXIII and counting. The proof takes more than 500 pages of technical mathematical reasoning, and hence will be omitted!

Theorem 2.40 **Minor**

In every infinite set of simple graphs, there are two graphs such that one is a minor of the other.

This theorem has had a wide range of consequences, both within graph theory and in other branches of mathematics.

2.6 Homomorphisms and Isomorphisms for Digraphs

A lot of what we have discussed so far can be done in a natural way for digraphs as well. The notions of union, intersection, and symmetric difference of a universal digraph all are defined in exactly the same manner as for graphs.

There are, however, some items one must discuss with a bit more care for digraphs than for graphs. The definition of a homomorphism and isomorphism from one digraph to another is almost exactly the same as Definition 2.15 for graphs.

Definition 2.41 Let $\vec{G} = (V, E, \eta)$ and $\vec{G}' = (V', E', \eta')$ be two digraphs. A *homomorphism*

$$f : \vec{G} \to \vec{G}'$$

from $\vec{G}$ to $\vec{G}'$ is an ordered pair $f = (f_1, f_2)$ of maps $f_1 : V \to V'$ and $f_2 : E \to E'$ satisfying the following condition:

$$\eta(e) = (u, v) \Rightarrow \eta'(f_2(e)) = (f_1(u), f_1(v)).$$

That is, if u and v are the tail and head of e in $\vec{G}$, respectively, then $f_1(u)$ and $f_1(v)$ are the tail and head of $f_2(e)$ in $\vec{G}'$, respectively.

If both maps f_1 and f_2 are bijective, then f is called an *isomorphism*. In this case we say that the digraph $\vec{G}$ is *isomorphic* to the digraph $\vec{G}'$, or that $\vec{G}$ and $\vec{G}'$ are *isomorphic* as digraphs, denoted by $\vec{G} \cong \vec{G}'$. An isomorphism from $\vec{G}$ to itself is called an *automorphism* of the digraph $\vec{G}$.

We do not have as much freedom in constructing a homomorphism for digraphs as we had for graphs. For example, in Figure 2.18 we see the images of two vertices u and v under the vertex map f_1. If $\eta(e) = (u, v)$, then we see that the possibilities for $f_2(e)$ are either e_1 or e_3, but not e_2.

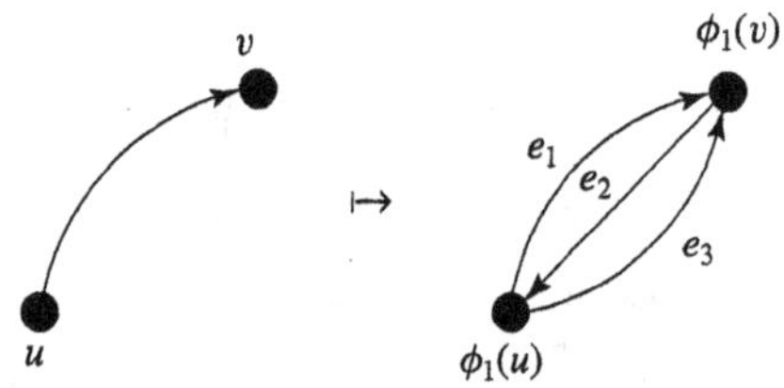

Figure 2.18: A homomorphic image of a directed edge.

Example 2.42 Consider the two digraphs $\vec{G} = (V, E, \eta)$ and $\vec{G}' = (V', E', \eta')$ in Figure 2.19. We define the maps f_1 and f_2 by the following assignments.

$$\begin{aligned} &f_1(v_1) = f_1(v_2) = f_1(v_5) = v_1', \\ &f_1(v_3) = v_2', \\ &f_1(v_4) = v_3', \end{aligned}$$

and then f_2 by

$$\begin{aligned} &f_2(e_1) = f_2(e_2) = f_2(e_3) = e_1', \\ &f_2(e_4) = e_2', \\ &f_2(e_5) = f_2(e_6) = e_3', \\ &f_2(e_7) = f_2(e_8) = f_2(e_9) = e_4', \\ &f_2(e_{10}) = e_6'. \end{aligned}$$

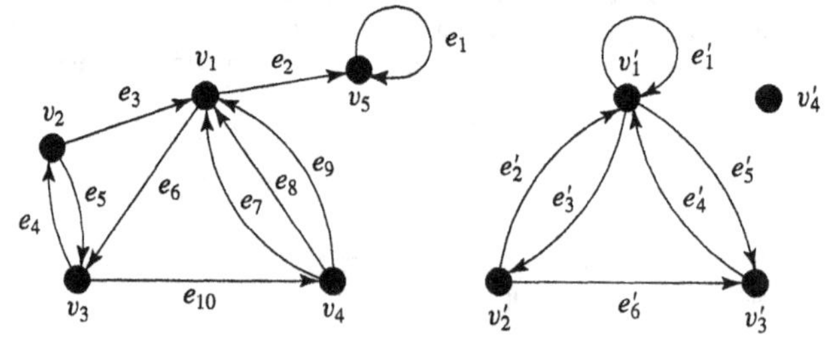

Figure 2.19: A digraph and a homomorphic image of it.

Now, if we let $f : \vec{G} \to \vec{G}'$ be given by $f = (f_1, f_2)$, then we can easily verify that f is indeed a homomorphism. We need to check each of the edges $e_1, \ldots, e_{10}$:

- We have $\eta(e_1) = (v_5, v_5)$, and $\eta'(f_2(e_1)) = \eta'(e_1') = (v_1', v_1') = (f_1(v_5), f_1(v_5))$.
- We have $\eta(e_2) = (v_1, v_5)$, and $\eta'(f_2(e_2)) = \eta'(e_1') = (v_1', v_1') = (f_1(v_1), f_1(v_5))$.

$\vdots$

- We have $\eta(e_{10}) = (v_3, v_4)$, and $\eta'(f_2(e_{10})) = \eta'(e_6') = (v_2', v_3') = (f_1(v_3), f_1(v_4))$.

We urge the reader to check the "unchecked" cases of $\eta(e_i)$. Hence, $f : \vec{G} \to \vec{G}'$ is indeed a homomorphism of digraphs. Note that f is not an isomorphism of digraphs.

We note that if $f : \vec{G} \to \vec{G}'$ is a homomorphism between digraphs, then in a natural manner f yields a homomorphism $f_o : G \to G'$ of the corresponding underlying graphs, in the following way: If $f = (f_1, f_2)$, then $f_o = (f_{o;1}, f_{o;2})$, where

$$\begin{aligned} f_{o;1} &= f_1, \\ f_{o;2} &= s \circ f_2, \end{aligned}$$

where $s : E(\vec{G}) \to E(G)$ is the set-map given by $s(u, v) = \{u, v\}$, which simply disregards the order of the vertices. Clearly, if f is an isomorphism, then f_o is an isomorphism between the corresponding underlying graphs. It is quite natural to insist that if two digraphs $\vec{G}$ and $\vec{G}'$ are isomorphic as digraphs, then, in particular, their underlying graphs G and G' should be isomorphic as graphs. This is a *necessary* condition but not sufficient, as we discuss in the following example.

Example 2.43 Consider the digraphs shown in Figure 2.20. Clearly all three have isomorphic underlying graphs, namely C_4. However, viewing them as digraphs, we notice that the leftmost digraph has all of its vertices of indegree and outdegree one, but both the middle digraph and the rightmost digraph have two vertices each of indegree and outdegree one, one vertex of indegree two, and one of outdegree two. Hence, the leftmost digraph is not isomorphic to either of the other ones. Further note that the two vertices of indegree and outdegree of one in the middle graph are adjacent in the underlying graph C_4, but that is not the case for the rightmost digraph. Hence, these two digraphs are not isomorphic either. So we have three mutually nonisomorphic digraphs.

Note that in this example, each property distinguishing one digraph from the other can *only* be phrased using the directed edges, since if there were a property distinguishing them that did not mention the directions of the edges, then their underlying graphs could not be isomorphic either.

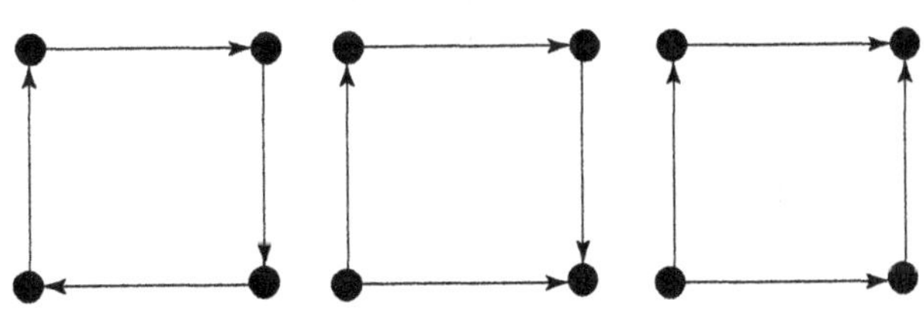

Figure 2.20: Three nonisomorphic digraphs.

2.7 Digraph Connectivity

The other concept we need to discuss with more care for digraphs than for graphs is the concept of connectivity. Basically, there are two natural ways to view connectivity in a digraph. One is simply to adapt directly the definition from graphs, and say that a digraph $\vec{G}$ is connected if, and only if, its underlying graph G is connected. In that case we say that $\vec{G}$ is *weakly connected.* By a *component* of $\vec{G}$ we simply mean the subdigraph induced by the vertices of the corresponding component of the underlying graph G.

There is, however, a more useful and common way of describing connectivity in digraphs. For that we need the following definition, analogous to Definition 2.1.

Definition 2.44 A *directed walk* $\vec{w}$ in a digraph $\vec{G}$ is an alternating sequence

$$\vec{w} = (u_0, e_1, u_1, e_2, \ldots, e_k, u_k)$$

of vertices and directed edges, where for each $i \in \{1, 2, \ldots, k\}$ the tail and head of e_i are u_{i-1} and u_i, respectively. That is,

$$\eta(e_i) = (u_{i-1}, u_i).$$

The notions of an *initial vertex,* a *final vertex,* the *length,* a *directed trail,* a *directed path,* and a *directed* u, v*-walk, trail, path* are the same as in Definition 2.1, but here we add the word "directed" before each of the words "walk," "trail," and "path." Likewise, the notion of *closed directed* trail and path are called *directed circuit* and *directed cycle,* respectively.

Remark 2.45 Since each directed edge has a unique tail and head, there is no ambiguity in writing a directed walk as

$$\vec{w} = (e_1, e_2, \ldots, e_k),$$

where it is understood that the initial vertex of $\vec{w}$ is the tail of e_1 and the final vertex is the head of e_k.

Likewise, if our digraph $\vec{G}$ has no parallel directed edges, and hence the set of directed edges is given by a subset $E \subseteq V \times V$, then we can write a directed walk as a sequence of vertices

$$\vec{w} = (u_0, u_1, \ldots, u_k).$$

Example 2.46 Consider the digraph shown in Figure 1.17 from Example 1.38. There $\vec{p}$ given by

$$\vec{p} = (u_1, e_1, u_2, e_4, u_3, e_5, u_4)$$

is a directed path of length three from u_1 to u_4. Note that $\vec{p}$ can also be represented as $\vec{p} = (e_1, e_4, e_5)$. It is interesting to note that the sequence $(u_1, e_2, u_3, e_4, u_2)$ does not define a directed walk, but does define a path in the underlying graph of the digraph.

Likewise, $\vec{c}$ given by

$$\vec{c} = (e_1, e_4, e_2)$$

is a directed cycle, whereas the tuple (e_1, e_4, e_3) does not define a directed cycle but does define a cycle in the underlying graph.

We can now state the more common definition for connectivity in digraphs, called *strong* connectivity, to emphasize its difference from weak connectivity.

Figure 2.21: The condition for strong connectivity in a digraph.

Definition 2.47 We say that a digraph $\vec{G}$ is *strongly connected* if for every pair $u, v \in V(\vec{G})$ of distinct vertices there is a directed walk from u to v in $\vec{G}$.

The *strong components* of $\vec{G}$ are the maximal strongly connected subdigraphs of $\vec{G}$.

Note 2.48 The Definition 2.47 implies that for every pair $u, v \in V(\vec{G})$ of vertices there is a directed path from u to v *and* a directed path from v to u as well (by reversing the role of u and v.) Also note that these two directed paths, one from u to v and the other from v to u, are not necessarily edge disjoint!

As we see in Figure 2.21, the condition of a digraph $\vec{G}$ being strongly connected implies that for every pair $u, v \in \vec{G}$ of vertices there is a directed walk from u to v in $\vec{G}$, as well as a directed walk from v back to u in $\vec{G}$. Note that any connected simple graph G corresponds naturally to a strongly connected simple digraph $\vec{G}$, simply by replacing every edge $\{u, v\}$ by two directed edges (u, v) and (v, u), as shown in Figure 2.22. Such digraphs are sometimes called *symmetric*.

Example 2.49 The two digraphs in Figure 2.23 both have the same connected underlying graph, and hence are both weakly connected. The one on the right is also strongly connected, since there is a directed walk from every vertex to every other. However, the one digraph on the left is not strongly connected, since it contains a vertex of outdegree zero. Any nontrivial digraph containing a vertex of either outdegree zero or indegree zero is not strongly connected. (See Exercise 2.38.)

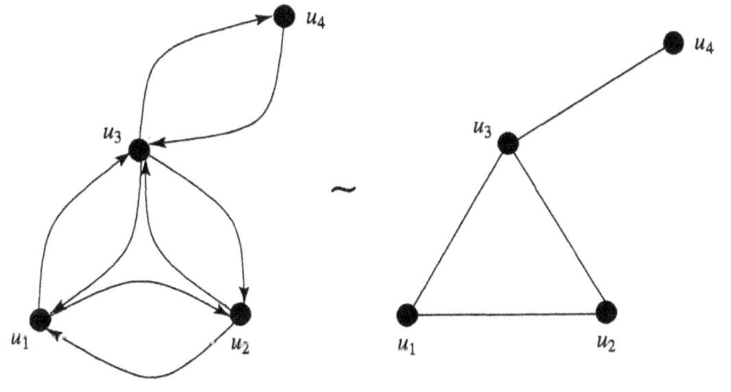

Figure 2.22: A simple symmetric digraph.

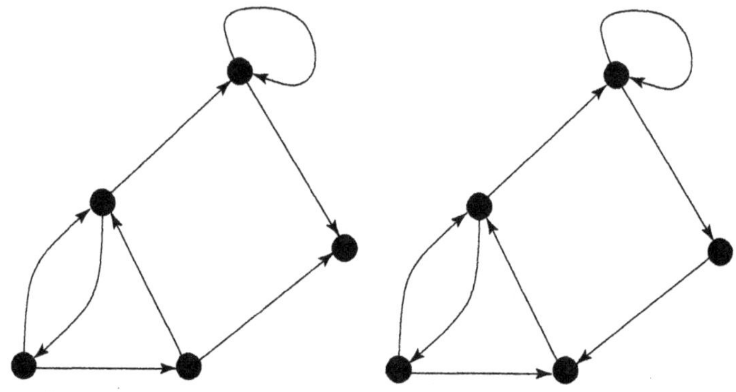

Figure 2.23: A weakly connected and a strongly connected digraph.

One type of digraph where the difference between digraphs and their underlying graphs is most evident is digraphs with complete graphs as underlying graphs. Such digraphs on n vertices with the complete graph K_n as an underlying graph possess a rich structure, which has fueled a lot of research.

Example 2.50 A simple digraph having exactly one directed edge between every pair of distinct vertices is called a *complete antisymmetric digraph* or a *tournament digraph*. A tournament digraph on n vertices has $n(n-1)/2$ directed edges. Hence, if the vertices are labeled, then there are precisely $2^{n(n-1)/2}$ tournament digraphs on those labeled vertices. In Figure 2.24 we see one out of the $2^{10} = 1{,}024$ tournament digraphs on five labeled vertices.

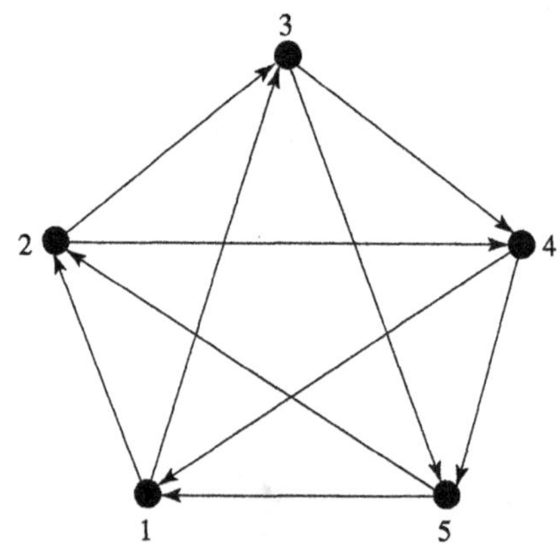

Figure 2.24: A tournament digraph on five vertices.

2.8 Exercises

2.1. Is it possible to draw a graph that has a trail of length seven but no path of length seven? If so, draw such an example. Is it possible to draw a graph that has a path of length seven but no trail of length seven? If so, draw such an example.

2.2. When considered separately, for what values of n do N_n, P_n, C_n, K_n, and $K_{n,n}$ have

- **(a)** trails of length n?
- **(b)** walks of length n?
- **(c)** cycles of length n?

2.3. Show that if a graph G contains a closed trail of length seven, then G contains C_3, C_5, or C_7 as a subgraph. Can something similar be stated if G contains a closed trail of length eight?

2.4. Prove that if a graph G has exactly two vertices u and v of odd degree, then G has a u, v-path. [*Hint:* Assume the contrary, then use Corollary 1.27 on page 18.]

2.5. What is the minimum number of edges a simple graph on n vertices must have, so we can rest assured it is connected? [*Hint:* What is the maximum number of edges a disconnected graph on n vertices can have? Use Theorem 2.13.]

2.6. For each natural number $n \geq 3$, what is the maximum number of edges a simple graph with n vertices can have and contain three components? For $n = 7$ draw a graph that realizes this maximum. What are the degrees of its vertices?

2.7. Suppose G and G' are two graphs having n vertices. For what values of n is it possible for G to have more components and edges than G'?

2.8. For $a, b \in \mathbb{N}$ with $a \leq b$, show that K_a and K_b have collectively fewer edges than K_{a-1} and K_{b+1}; that is, show that

$$\frac{a(a-1)}{2} + \frac{b(b-1)}{2} < \frac{(a-1)(a-2)}{2} + \frac{(b+1)b}{2}.$$

Use this to show that a simple graph on n vertices with k components has the maximum number of edges when one component is K_{n-k+1} and the remaining $k-1$ components are each a single vertex. Use this to give an alternative proof of Theorem 2.13.

2.9. Show that for a graph G, the components of G are precisely the maximal connected subgraphs of G. (*Note!* This gives an alternative way to define the components of G.)

2.10. Complete the proof of Theorem 2.10; that is, show that the relation $\sim$ given by (2.2) is indeed an equivalence relation, which, in addition, is identical with the one yielding the partition stated in Theorem 2.10.

2.11. Define a relation on graphs such that G and G' are related if, and only if, the maximum degree of G is less than or equal to the maximum degree of G'. Does this relationship satisfy the reflexive, transitive, and antisymmetric properties? Is it a partial order? Does it satisfy the symmetric property? Is the relation an equivalence relation?

2.12. Complete the proof of Theorem 2.18; that is, prove (i) that $\iota_G = (\iota_V, \iota_E)$ is always an automorphism of G, and (ii) that if both $f = (f_1, f_2)$ and $f' = (f'_1, f'_2)$ are isomorphisms, then so is $f' \circ f = (f'_1 \circ f_1, f'_2 \circ f_2)$.

2.13. Show that the graphs shown in Figure 2.25 are isomorphic.

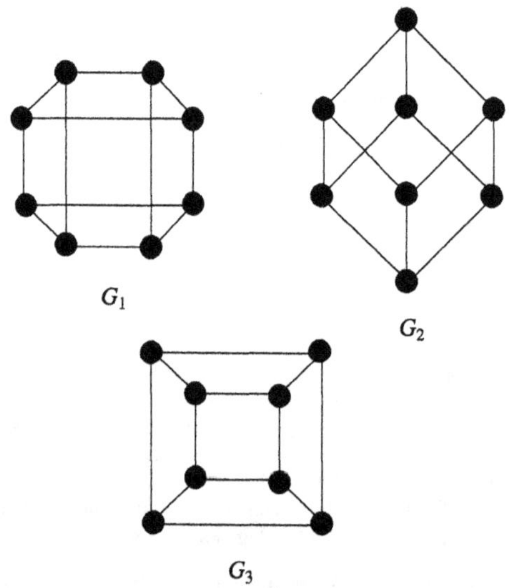

Figure 2.25: Isomorphic graphs used in Exercise 2.13.

2.14. Which, if any, of the graphs G_1, G_2, G_3, and G_4 in Figure 2.26 are isomorphic? For those that are isomorphic, construct an isomorphism between them; for those that are not, state a graph-theoretic property satisfied by one and not the other.

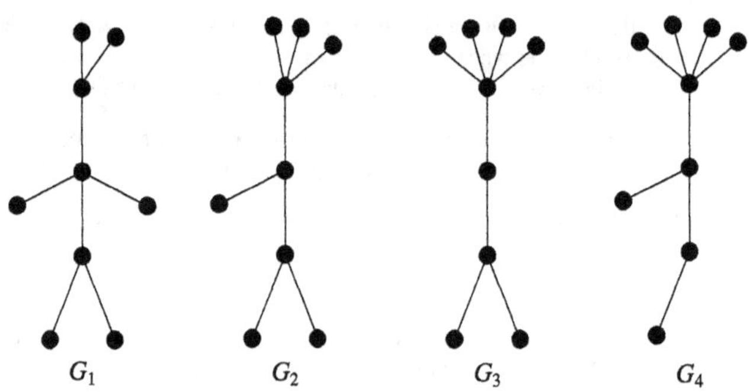

Figure 2.26: Graphs used in Exercise 2.14. Are any of them isomorphic?

2.15. Show that for $n \in \mathbb{N}$, the graph $K_{n,n}$ has a subgraph isomorphic to C_k for any even $k \in \{4, \ldots, 2n\}$.

2.16. Which, if any, of the graphs G_1, G_2, and G_3 in Figure 2.27 are isomorphic? For those that are isomorphic, construct an isomorphism between them; for those that are not, state a graph-theoretic property satisfied by one and not the other.

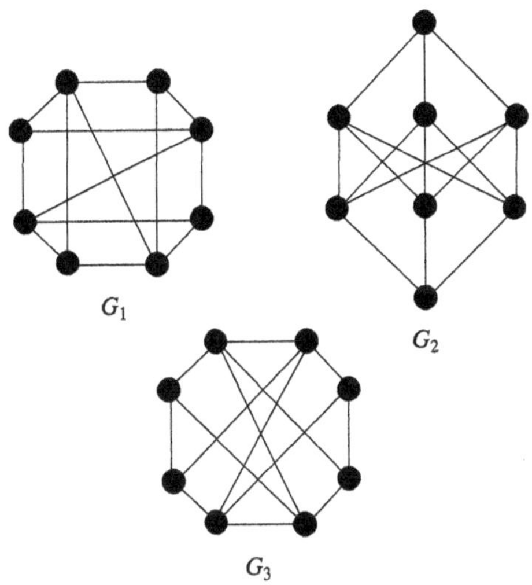

Figure 2.27: Graphs used in Exercise 2.16. Are these graphs isomorphic?

2.17. Enter a query of "Graph Isomorphism" to your favorite search engine. Read a number of Web pages, selecting from the hyperlinks returned by your query. Summarize your findings in a one-page paper.

2.18. An editor from the *New York Times* has contacted you to write a one-page article for the *Times* that makes the concepts of connectivity and isomorphism understandable to the masses. Develop such an article, being sure to include lots of intuition about these concepts.

2.19. Prove Observation 2.23. [*Hint:* Note that a degree-i vertex must be mapped to another degree-i vertex.]

2.20. Prove Theorem 2.25. [*Hint:* Use one isomorphism to establish the other isomorphism, and use the fact that $\overline{\overline{G}} = G$.]

2.21. Are any of the graphs N_n, P_n, C_n, K_n, and $K_{n,n}$ complements of each other?

2.22. Prove whether or not the complement of every regular graph is regular.

2.23. Show that if a simple graph G is isomorphic to its complement $\overline{G}$, then G has either $4k$ or $4k + 1$ vertices for some natural number k. Find all simple graphs on four and five vertices that are isomorphic to their complements.

2.24. Let G be a simple graph. Show that either G or its complement $\overline{G}$ is connected.

2.25. Assuming that $n_i = m$ for each i, and hence $n = km$, show that Observation 2.23 is more than $(k!)^m$ times faster than Observation 2.21 to check for isomorphisms between two graphs, both on n vertices.

2.26. Let G be a graph, $u, v \in V(G)$ two distinct vertices, and p_1 and p_2 two different u, v-paths. Viewing these graphs as subgraphs of G, show that $p_1 \triangle p_2$ constitutes a disjoint union of cycles in G. Does the same hold for digraphs?

2.27. Draw $K_{2,4}$ in the usual manner and label the edges $1, \ldots, 8$ from left to right. Draw the graphs resulting from the contraction of the edges one at a time in increasing numerical order. Repeat the process by deleting edges in the order 2, 4, 6, 8, 1, 3, 5, and 7. Are any of the resulting graphs the same? Explain your answer.

2.28. Let $n \in \mathbb{N}$ be given. Which of the graphs N_n, P_n, C_n, K_n, and $K_{n,n}$ contain a cut-vertex?

2.29. Prove Theorem 2.39; that is, prove that the minor relation $\preceq_m$ is reflexive, antisymmetric, and transitive.

2.30. Show that K_n is always a contraction of K_{n+1}. Is $K_{n,n}$ always a contraction of $K_{n+1,n+1}$?

2.31. Show that for each $n \in \mathbb{N}$ the complete graph K_n is a contraction of $K_{n,n}$.

2.32. For $n \in \mathbb{N}$, can K_n be a contraction of $K_{m,n}$ if $m < n$?

2.33. Let G be a simple graph and G' a contraction of G. Is G' a *homomorphic image* of G? That is, is there a homomorphism $f : G \to G'$ such that both f_1 and f_2 are surjective? What if G' is a minor of G? [*Hint:* What is $f_2(e)$, where e is the contracted edge of G?]

2.34. Let $N \in \mathbb{N}$ be a fixed number. Consider the set $\mathcal{G}_N$ of graphs given by

$$\mathcal{G}_N = \{K_{m,n} : m, n \in \mathbb{N},\ m + n = N\}.$$

Show that for any two distinct graphs in $\mathcal{G}_N$, one is not a minor of the other. Does this contradict the Minor Theorem 2.40?

2.35. Let $\mathcal{G}$ be the set of graphs given by

$$\mathcal{G} = \{K_{m,n} : m, n \in \mathbb{N},\ m + n \geq 1\}.$$

Show that $\mathcal{G}$ contains a subset of arbitrary cardinality in which, for any two graphs, one is not a minor of the other. [*Hint:* Consider the sets of graphs given by $\mathcal{G}_N$ from Exercise 2.34.]

2.36. Show that in an infinite collection of simple graphs, there is no bound on how many graphs one needs to consider in order to obtain a pair of graphs with one being a minor of the other.

2.37. Let $\vec{G}$ be a digraph and $u, v \in V(\vec{G})$ be two distinct vertices. Show that if $\vec{G}$ has a directed u, v-walk, then it has a directed u, v-path as well.

2.38. Show that any digraph $\vec{G}$ containing a vertex u such that $d^-(u) = 0$ or $d^+(u) = 0$ is not strongly connected.

2.39. How many tournament digraphs on n labeled vertices contain two vertices, one of indegree zero and the other of outdegree zero?

2.40. How many possible outcomes are there in a boxing match among n boxers, where there is no draw and every boxer wins at least one match and loses at least one match? [*Hint:* Solve Exercise 2.39 first.]

CHAPTER

3

Trees and Forests

Chapter Goals

Present the concepts of trees and forests.

Give a complete characterization of trees.

Demonstrate inductive proof techniques for trees.

Prove a theorem by Erdős and Szekeres and discuss the corresponding data tree.

Discuss the concept of centers for graphs, in particular for trees.

Present the concept of rooted and binary trees.

Present some counting problems for rooted trees and define the Catalan number.

Discuss levels of a rooted tree and the minimum average level of the leaves.

3.1 Trees and Some of Their Basic Properties

In many scientific disciplines, *trees* and *forests* are the most frequently occurring graphs, especially in computer science when organizing certain data structures or developing search algorithms. Trees are important in computer science largely because they minimally connect the vertices which represent some form of data (to be discussed in detail shortly). For example, trees are useful in the following applications:

- The hierarchy in a business.
- The directories in a UNIX operating system.
- An alphabetical sorting procedure for words. (Such a tree is called a *decision tree* or *sorting tree*.)
- In laying out a computer network or even the hardware of a parallel computer.

Trees and forests are also one of the simplest, yet nontrivial, types of graphs in graph theory. Despite this, there is still a lot of research being done on trees and forests, even within the field of graph theory.

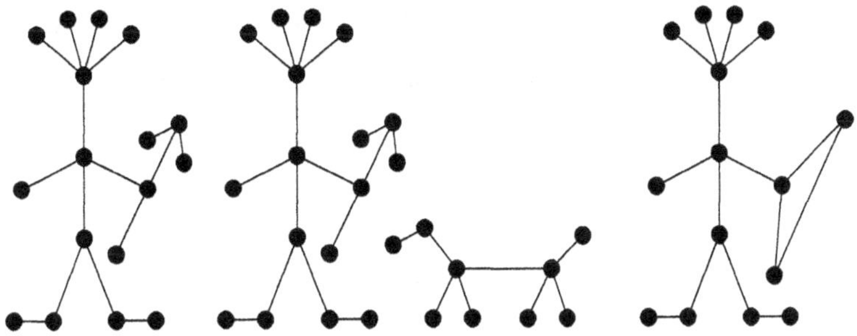

Figure 3.1: (a) A tree. (b) A forest. (c) A graph that is neither a tree nor a forest.

Definition 3.1 A *tree* is a connected graph that has no cycle as a subgraph. A *forest* is a graph in which every component is a tree.

The shapes of "trees" and "forests" can resemble those of ordinary trees and bushes, with their roots, branches, and leaves. However, the resemblance depends on the actual plane drawing. We are concerned about the graph-theoretic *properties* of trees and forests, regardless of their looks; nevertheless, a particular plane drawing can help us understand the properties we are seeking. For example, Figure 3.1 shows three graphs: (a) a tree on 17 vertices, (b) a forest on 26 vertices, and (c) a simple graph on 15 vertices, which is neither a tree nor a forest, since it contains a cycle of length three (namely the "bow").

- By Definition 1.8 on page 10, any graph, including a tree, must have at least one vertex.
- Since a loop together with its endvertex is a cycle of length one, a tree or a forest cannot have any loops.
- Since a double edge together with its two distinct endvertices is a cycle of length two, a tree or a forest cannot have any multiple edges.

Hence, we see that each tree or forest is a *simple* graph. For a graph to be a tree is a strong condition. Consider the set of all possible graphs on n vertices. Relatively few of those will be trees. In Figure 3.2 we see all possible shapes for a tree on six vertices. On the far left of the figure is a path, and on the far right is a *star*.

Example 3.2 A digital dictionary keeps all the words linked in a tree similar to the one shown in Figure 3.3. (Such a tree is generally called a *rooted tree* and will be discussed further shortly.) When you type in a word, say "*BABY*," the initial letter is one of the 26 letters of the alphabet $A, B, C, \ldots, Y, Z$, in this case B, so the information pertaining to this word is contained below the vertex labeled B. The next letter is A, so the information is contained below the vertex labeled A, and so on. In this manner, one can quickly obtain the information on the word "*BABY*." Note that we stop after only four steps, since the word "*BABY*" has only four letters!

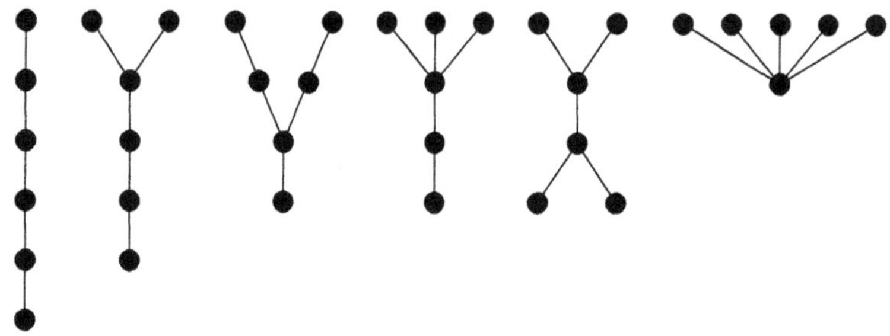

Figure 3.2: All possible trees on six vertices.

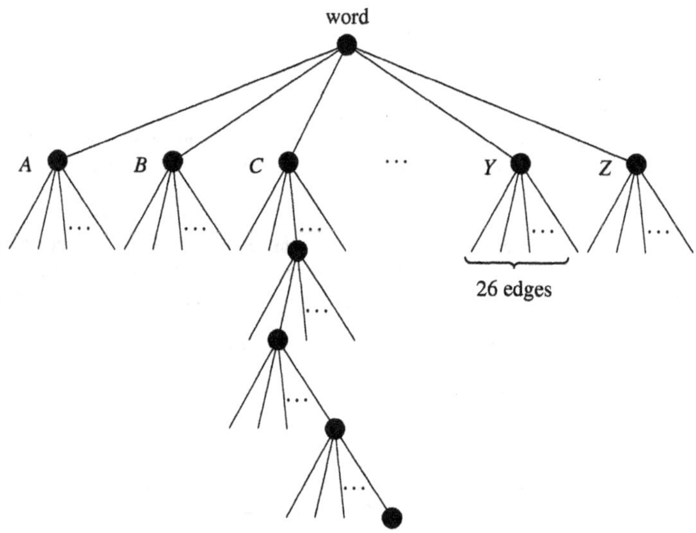

Figure 3.3: A decision tree for finding a word from a collection of words ordered alphabetically.

Practically all searches on the World Wide Web are based on this underlying principle, although many complex modifications are necessary to make the searches even more efficient.

Many trees like the one shown in Figure 3.3 have precisely two choices, instead of the 26 in this case. So the choices for searching at each vertex are to go down either left or right, that is, 0 or 1. We take a closer look at such trees in §3.7. The final vertex we arrive at in a search of this nature is a vertex where we cannot go down any further, so we stop. Such a vertex always has a degree of one (unless, of course, the tree contains just one vertex).

Definition 3.3 A vertex u of a simple graph G is called a *leaf* if $d_G(u) = 1$. A vertex that is not a leaf is called an *internal vertex.*

Many fundamental properties about trees can be derived from the following theorem.

Theorem 3.4 Every tree with at least two vertices has at least two leaves.

Proof Consider a tree T having two or more vertices. Let p be the longest path in T. Let u and v be the initial and endvertex of p, respectively. We show that u and v are leaves—that is, $d(u) = d(v) = 1$. It suffices to consider just one of the vertices, say u, since the argument would be similar for the other vertex v.

If $d(u) \geq 2$, then there is an edge e, not in the path p, having endvertices $u, w \in V(T)$. We now have two cases to consider.

CASE ONE: If w is not a vertex in the path p, then the composite path $p' = (w, e, u)p$ is a path in G of length one more than that of p, which contradicts the fact that p was a path of longest length in T.

CASE TWO: Suppose the vertex w is a vertex in the path p. If p'' is the path from u to w along the path p, then we get a cycle $c = (w, e, u)p''$ of length three or more in T. This contradicts the fact that T is a tree.

Hence, $d(u) = 1$, thereby completing the argument. ■

Note 3.5 Theorem 3.4 provides a useful tool in proving many properties for trees. In many cases, the fact that every tree on two or more vertices has at least *one* leaf is all we need in order to demonstrate a certain property of a tree. Let T be a tree with a leaf u, as shown in Figure 3.4. Notice that the subgraph $T' = T - u$ has no cycles, since T has none. Also, T' is connected, and hence T' is a tree and it has one fewer vertices than T. Many proofs about T can be done inductively by assuming the argument is true for T'.

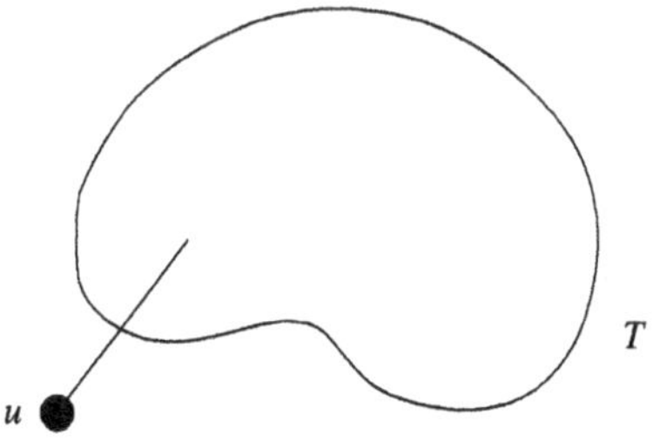

Figure 3.4: Every tree has at least one leaf.

Another useful fact that can be used on trees is the following lemma.

Lemma 3.6 A simple graph G on n vertices with k components has at least $n - k$ edges.

Proof We use induction on m, the number of edges of G. The statement is clearly true for $m = 0$, since G is then the null graph on n vertices.

We continue by induction and assume the assertion is true for all natural numbers less than m, where $m \in \mathbb{N}$ is given. If $e \in E(G)$, then the simple graph $G - e$ has n vertices, $m - 1$ edges, and $k + \delta$ components, where $\delta \in \{0, 1\}$. By the induction hypothesis, we have that $n - (k + \delta) \leq m - 1$, and hence, $n - k \leq m - (1 - \delta) \leq m$, completing the inductive argument, and this proves the lemma. ■

It is instructive to compare Lemma 3.6 with Theorem 2.13 on page 38, which provides the upper bound on the number of edges in a simple graph on n vertices and k components. Lemma 3.6 provides the lower bound.

3.2 Characterizations of Trees

Theorem 3.7 contains some of the most frequently used characterizations of trees, and we will prove some of these statements using the technique described in Note 3.5.

Theorem 3.7 If T is a simple graph on n vertices, then the following statements are equivalent:

1. T is a tree.
2. T has $n - 1$ edges and no cycles.
3. T has $n - 1$ edges and is connected.
4. T is connected and each edge is a bridge.
5. Between every pair of distinct vertices in T there is exactly one path.
6. T has no cycles, but add an edge to T between a pair of nonadjacent vertices and exactly one simple cycle is formed.

Proof We prove the theorem by showing the following "cycle" of implications:

$$1 \Rightarrow 2 \Rightarrow 3 \Rightarrow 4 \Rightarrow 5 \Rightarrow 6 \Rightarrow 1.$$

$1 \Rightarrow 2$: We argue by induction on the number of vertices n. We need to show that T has $n - 1$ edges. The statement is clearly true if $n = 1$. Let T be a tree on $n \geq 2$ vertices. By Theorem 3.4, T has a leaf u. Hence, $T' = T - u$ is a tree on $n - 1$ vertices. Therefore, by the induction hypothesis T' has $n - 2$ edges. Since u is a leaf, T' has one fewer edges than T. So T has $n - 1$ edges.

$2 \Rightarrow 3$: We need to show that T is connected. By definition we have that T is a forest. Let $k \geq 1$ be the number of components in T. Every component is a tree and therefore has one fewer edges than vertices, as we just saw in proving $1 \Rightarrow 2$. Hence, the number of edges in T is $n - k$. Thus $n - k = n - 1$, and hence $k = 1$, so T is connected.

$3 \Rightarrow 4$: We must show that each edge is a bridge. By Lemma 3.6 every connected graph *must* have at least $n - 1$ edges. Therefore, the removal of any edge from T will disconnect it. So each edge is a bridge.

$4 \Rightarrow 5$: We need to show that there is a unique path between each pair of vertices in T. By connectivity of T there is at least one path between any two vertices of T. If there are two vertices u and v having two distinct u, v-paths, then we can assume we have two p_1 and p_2, where p_1 is one of shortest possible length. We now argue that p_2 must contain an edge which is not contained in p_1.

Since p_1 is of shortest possible length, the sequence of edges constituting the path p_2 cannot be a rearrangement of those in p_1, since if that were the case, we could use these common edges of p_1 and p_2 to form a u, v-path which is strictly shorter than p_1, by starting at u, say, and always traversing the edge which is closer to v on either p_1 or p_2. Therefore, there must be an edge e in p_2 which is not contained in p_1. Since there is a walk in $T - e$ connecting the endvertices of e, formed by the edges of p_1 and the remaining edges of p_2, the edge e is not a bridge of T, which is a contradiction.

$5 \Rightarrow 6$: Clearly, T has no cycles, since that would imply two distinct paths between a pair of vertices. We must show that adding any edge to T adds exactly one simple cycle. Let u and v be two vertices and p a u, v-path in T. If we add an edge e between u and v, then $c = \gamma(v, e, u)$ is a simple cycle in T. If there is another such cycle c', then c' would have to contain e, and by removing e from c', one would get yet another u, v-path.

$6 \Rightarrow 1$: We must show that T is indeed connected. Let u and v be vertices in T. By adding an edge e between u and v, we obtain a single cycle. This cycle must contain the edge e. The removal of e yields a simple path from u to v. Hence, T is connected and therefore a tree. ■

Depending on the situation, one of the characterizations given in Theorem 3.7 may be preferable to the others.

3.3 Inductive Proofs on Trees

Inductive proofs on trees arise in a natural way because of Theorem 3.7: Every edge is a bridge. In an n-vertex tree T with an edge e, the removal of e gives us exactly two trees T_1 and T_2 with $|V(T_1)| + |V(T_2)| = |V(T)| = n$, as illustrated in Figure 3.5. Regardless of which edge e is chosen, note that both $|V(T_1)|$ and $|V(T_2)|$ are strictly less than n. Thus induction is a natural proof method for trees. We illustrate this inductive proof technique in the following example.

Example 3.8 Let $n > 1$, and consider all trees on n or fewer vertices, where all of the vertices have degree three or less. Among these are very important class of trees called *regular binary trees*, which will be discussed further in §3.7.

Let $\mathbf{d}_3(n)$ denote the maximum number of vertices of degree three that such a tree can have. Suppose that we want to show that

$$\mathbf{d}_3(n) \leq \left\lfloor \frac{n}{2} \right\rfloor - 1. \tag{3.1}$$

We will use induction on n to show that $\mathbf{d}_3(n) \leq n/2 - 1$, from which (3.1) follows directly. This statement is clearly true for $n = 3$, where there is no vertex of degree three.

Assume now that $n \geq 4$ and that $\mathbf{d}_3(k) \leq k/2 - 1$ holds for all trees on $k \leq n - 1$ vertices. Let T be a tree on n vertices and let $D_3(T)$ be the set of vertices of T having degree three. We can assume that T has the maximum number of vertices of degree three—that is,

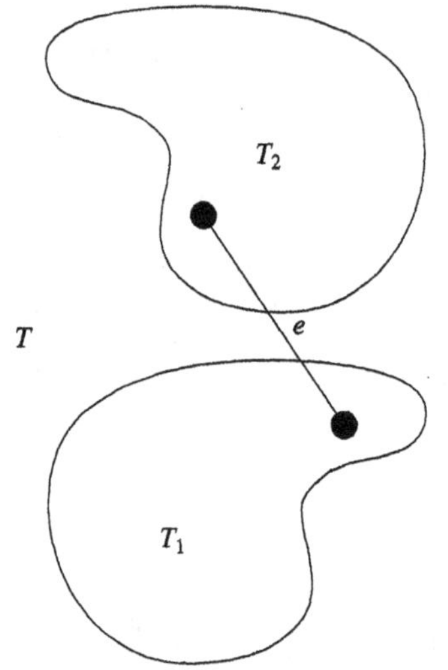

Figure 3.5: The removal of any edge of a tree T yields two trees T_1 and T_2.

$|D_3(T)| = \mathbf{d}_3(n)$. We pick $u \in D_3(T)$ and let its neighbors be u_1, u_2, and u_3. Since every edge is a bridge, by removing the edges $\{u, u_1\}$, $\{u, u_2\}$, and $\{u, u_3\}$ from T, we form three new trees. For each $i \in \{1, 2, 3\}$, we connect a new vertex u_i' to u_i by the edge $\{u_i', u_i\}$ and obtain in this fashion three new trees T_1, T_2, and T_3, where T_i is the tree containing u_i and u_i'. Figure 3.6 illustrates the situation. If $n_i = |V(T_i)|$ for each $i \in \{1, 2, 3\}$, then

$$n_1 + n_2 + n_3 = n + 2. \tag{3.2}$$

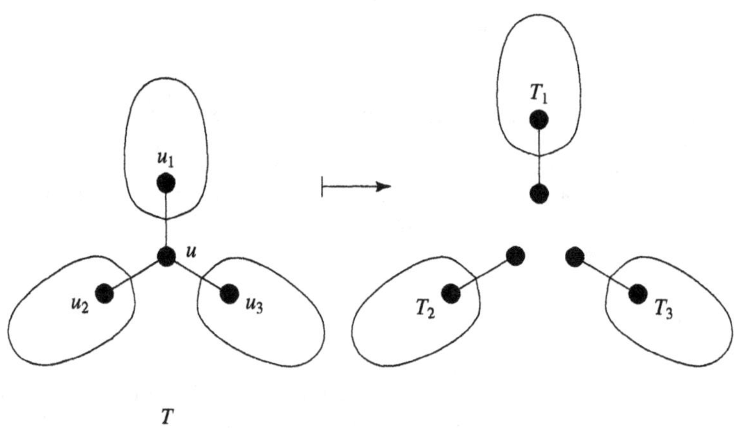

Figure 3.6: From the tree T we obtain three smaller trees T_1, T_2, and T_3.

Since each T_i contains at least two vertices, u_i and u_i', we have that $n_i \leq n - 2$ for each i. Hence, by the induction hypothesis and from (3.2), we have

$$\begin{aligned}
\mathbf{d}_3(n) &= |D_3(T)| \\
&= |D_3(T_1)| + |D_3(T_2)| + |D_3(T_3)| + 1 \\
&\leq \left(\frac{n_1}{2} - 1\right) + \left(\frac{n_2}{2} - 1\right) + \left(\frac{n_3}{2} - 1\right) + 1 \\
&= \frac{n}{2} - 1.
\end{aligned}$$

So by induction we have that $\mathbf{d}_3(n) \leq n/2 - 1$ for every $n \geq 3$. Since $\mathbf{d}_3(n)$ is an integer, we have (3.1) for all $n \geq 3$. It is instructive to compare this result with Theorem 3.31 for a regular binary tree on $n = 2k + 1$ vertices.

Interestingly, the upper bound of $\mathbf{d}_3(n)$ in Example 3.8 can be reached for an infinite number of natural numbers n. We can form a sequence of trees $T_1, T_2, \ldots, T_k, \ldots$ by initially letting T_1 be a single vertex graph. Next we let T_2 be the graph obtained from T_1 by connecting three leaves to the vertex of T_1. In general, for $k \geq 1$, we obtain T_{k+1} from T_k by connecting two new leaves to each leaf of T_k as shown in Figure 3.7. By viewing each T_{k-1} as a subgraph of T_k, we notice that every vertex of T_{k-1} has degree three in T_k. If we let t_k be the number of vertices of T_k, we see that T_k has t_{k-1} vertices of degree three and $t_k - t_{k-1}$ leaves, a total of t_k vertices. By construction, we add two leaves to each leaf of T_k to obtain T_{k+1}. That is, $t_{k+1} = t_k + 2(t_k - t_{k-1})$. Hence, the sequence $(t_k)_{t \geq 1}$ satisfies the following *recurrence relation*:

$$\begin{aligned}
t_1 &= 1, \\
t_2 &= 4, \\
t_{k+1} &= 3t_k - 2t_{k-1} \qquad \text{for } k \geq 2.
\end{aligned}$$

One can now prove by induction that

$$t_k = 3 \cdot 2^{k-1} - 2 \qquad \text{for all } k \geq 1.$$

(To see how to derive an explicit formula from one that is given recursively, see Example 12.18 in Chapter 12, or see Chapter 5 of [47].)

So, we have for each $k \geq 2$ that

$$|D_3(T_k)| = t_{k-1} = \left\lfloor \frac{t_k}{2} \right\rfloor - 1.$$

This shows that the upper bound of (3.1) can be reached for infinitely many n, namely $n = t_k$ for $k \in \mathbb{N}$.

In Example 3.8 we do not have complete freedom to choose which edge we can remove to use our induction hypothesis to prove the upper bound given in (3.1). Since we are not guaranteed the existence of a leaf with a neighbor having degree three, we cannot remove a leaf and directly use the induction hypothesis on the resulting tree, but rather must choose our vertex with more care.

Using induction to prove various statements about trees and a natural number n, as in Example 3.8, does not always yield the simplest or the "slickest" proof, but often it is the

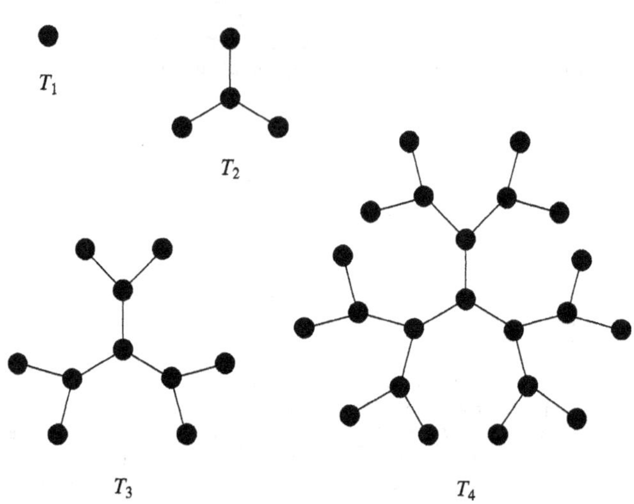

Figure 3.7: The first four trees in the sequence $T_1, T_2, \ldots, T_k, \ldots$.

most natural way of attacking the problem. In Exercise 3.5 a shorter way to prove (3.1) is given.

For $\Delta \geq 3$ one could consider all trees on n or fewer vertices where each vertex has degree Δ or less. The same argument as used in Example 3.8 applies in this more general case for any $\Delta \geq 3$, as we ask in Exercises 3.6 and 3.7.

3.4 Erdős-Szekeres Theorem on Sequences

A problem that occurs often in practical computer science is to compute a strictly increasing or decreasing subsequence, of maximum length, from a given sequence of arbitrarily ordered real numbers. Needless to say, this applies not only to real numbers, but also to elements in a linearly ordered set, such as a collection of words in a nonalphabetic order. Note that the length of the maximum subsequence depends on the length of the original sequence, so a natural question is: For a given sequence of a fixed length, how long a subsequence of this type can we expect to obtain?

The next classical Erdős-Szekeres Theorem, due to Paul Erdős and George Szekeres, answers the basic existence question about the maximum length of such a subsequence. The proof we provide is from [49] and due to Seidenberg.

Theorem 3.9 **Erdős-Szekeres**

If $(x_1, x_2, \ldots, x_{hk+1})$ is a sequence of distinct real numbers, then there is either an increasing subsequence of length $h+1$, or a decreasing subsequence of length $k+1$.

Proof Assume there is no increasing subsequence of length $h+1$. For each $i \in \{1, 2, \ldots, hk+1\}$, let ℓ_i be the length of the longest increasing subsequence with x_i as its first term. Clearly, each ℓ_i falls between 1 and h inclusively. Therefore, by the Generalized Pigeon-Hole Principle there exist $x_{i_1}, x_{i_2}, \ldots, x_{i_{k+1}}$ with $i_1 < i_2 < \cdots < i_{k+1}$ such that their corresponding lengths are all the same—that is, $\ell_{i_1} = \ell_{i_2} = \cdots = \ell_{i_{k+1}} = \ell$. By the definition of the ℓ_i, the sequence $(x_{i_1}, x_{i_2}, \ldots, x_{i_{k+1}})$ of length $k+1$ must be decreasing. ∎

As a direct consequence of the Erdős-Szekeres Theorem, we obtain that every sequence of distinct real numbers of length $k^2 + 1$ contains a subsequence of length $k + 1$ that is either strictly decreasing or strictly increasing. Therefore, so does every sequence of length n, where n satisfies $k^2 + 1 \leq n \leq (k + 1)^2$. Hence we have the following corollary.

Corollary 3.10 Every sequence of distinct real numbers of length n contains a subsequence of length $\lceil \sqrt{n} \rceil$ that is either strictly decreasing or strictly increasing.

Note that this is optimal, since there are sequences of length k^2 having no strictly decreasing or strictly increasing subsequence of length $k + 1$.

Example 3.11 For $k \in \mathbb{N}$, consider the sequence of length k^2 given by

$$s = (s_1, s_2, \ldots, s_k),$$

where each s_i is the strictly decreasing sequence of length k given by $s_i = (ik, ik - 1, ik - 2, \ldots, (i - 1)k + 1)$. Since each subsequence s_i has exactly k terms, then any subsequence s' of length $k + 1$ in s must contain two terms x_i and x_j from distinct sequences s_i and s_j, respectively, where $i < j$. By construction $x_i < x_j$, and hence s' cannot be strictly decreasing. Likewise, by the Pigeon-Hole Principle, the subsequence s' must also contain two terms x_p and x_q with $p < q$ from the same subsequence s_i. Again, by construction $x_p > x_q$, and hence s' cannot be strictly increasing. Since this holds for *any* subsequence s' of length $k + 1$, we see that the bounds in Corollary 3.10 are optimal when the length n of the sequence has the form $n = k^2 + 1$.

Given a sequence of distinct real numbers of a certain length, the Erdős-Szekeres Theorem tells us exactly the maximum length of a strictly increasing subsequence we are *guaranteed* to find if we look long enough. But how do we find all of these subsequences, of maximum length, in a systematic way?

Example 3.12 Consider the sequence s given by

$$s = (8, 3, 7, 5, 1, 6, 2, 9, 4).$$

Here the length of s is $n = 9$, and hence, by the Erdős-Szekeres Theorem, it is guaranteed that there is either a strictly decreasing or a strictly increasing subsequence of length $\lceil \sqrt{9} \rceil = 3$. If we want to find the longest strictly increasing subsequence of s, then we can form a tree, where we start at an initial vertex and connect that to vertices representing every term of the sequence s. To each vertex representing such a term in s, we connect vertices representing all the later terms of s having a greater value. Continuing in this manner until the tree cannot be extended further, we obtain the tree shown in Figure 3.8. Note that every path from the initial vertex down to a leaf in the tree in Figure 3.8 represents a *maximal* subsequence of s that is strictly increasing. In particular, we see that we have seven maximal subsequences of length three (which was ensured by Corollary 3.10!). They are the sequences (3, 7, 9), (3, 5, 9),

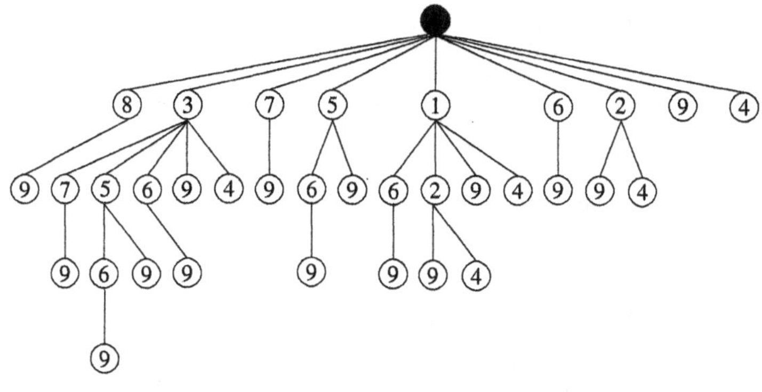

Figure 3.8: The data tree for determining the longest monotonically increasing sequence in the sequence $(8, 3, 7, 5, 1, 6, 2, 9, 4)$.

$(3, 6, 9)$, $(5, 6, 9)$, $(1, 6, 9)$, $(1, 2, 9)$, and $(1, 2, 4)$. In addition, we see that there is a unique strictly increasing subsequence of *maximum* length four as well (which is particular to this sequence s and not something we could expect a priori!). This subsequence is $(3, 5, 6, 9)$ and corresponds to the longest path in the tree that starts at the initial vertex. It was only found because we formed the tree in Figure 3.8. This type of tree is called a *data tree*.

In this sequence s, all the terms were distinct. However, this may not always be the case. We can still apply the technique described here to an input where the values are not distinct. Duplicate values will never occur on the same path based on the way our data tree is constructed. Given a sequence, it should be clear how to construct its data tree and how to determine the sequence's strictly increasing subsequence of maximum length.

As we saw in Example 3.12, by constructing such a data tree we were able to obtain all the maximal (including the maximum!) strictly increasing subsequences for s. Hence, the data tree contains all the information regarding this particular sequence s, regarding strictly increasing subsequences. Needless to say, the longer the sequence is, the larger the tree will be.

Observation 3.13 For a sequence $s = (x_1, \ldots, x_n)$ of distinct real numbers, the data tree T_s for s on the strictly increasing subsequences of s has $\mathbf{n}(T_s)$ vertices, $\mathbf{l}(T_s)$ of them leaves, where

$$n + 1 \leq \mathbf{n}(T_s) \leq 2^n,$$
$$n \leq \mathbf{l}(T_s) \leq 2^{n-1}.$$

Proof Clearly, each data tree must contain the initial vertex, together with one vertex representing each term of s. Hence, we have the lower bound for $\mathbf{n}(T_s)$. Since a tree containing a vertex of degree n must contain at least n leaves (Exercise 3.18), we have the lower bound for $\mathbf{l}(T_s)$ as well.

The upper bound we obtain by induction on n. For $n = 2$, bounds for both $\mathbf{n}(T_s)$ and $\mathbf{l}(T_s)$ are clear. Let $s = (x_1, \ldots, x_n)$ be a sequence of distinct real numbers, and assume that both upper bounds hold for every sequence of length $n-1$, in particular for $s' = (x_1, \ldots, x_{n-1})$. Since x_n is larger than at most every term of s', we can connect a leaf to at most every vertex in the data tree $T_{s'}$ to obtain the data tree T_s. By the induction hypothesis we therefore have

$$\mathbf{l}(T_s) \leq \mathbf{n}(T_{s'}) \leq 2^{n-1}.$$

By the same token we must also have by the induction hypothesis that

$$\mathbf{n}(T_s) \leq \mathbf{n}(T_{s'}) + \mathbf{l}(T_s) \leq 2^n,$$

thereby proving the observation. ■

In particular, we see from Observation 3.13 that a sequence of n distinct real numbers can have at most 2^{n-1} maximal strictly increasing subsequences.

3.5 Centers in Trees

In every tree, we will see that either one or two vertices are "innermost" in the tree. We call these vertices *centers* of the tree. Centers play an important role in developing recursive algorithms for problems on trees. Additionally, centers serve an important role in networking applications, where nodes might communicate with each other through the centers.

Consider the two trees in Figure 3.9. If we were to talk about a center of the first tree, a natural candidate would be the vertex of degree five. For the other tree there is not such an obvious choice for a center, and either of the internal vertices could serve as a center.

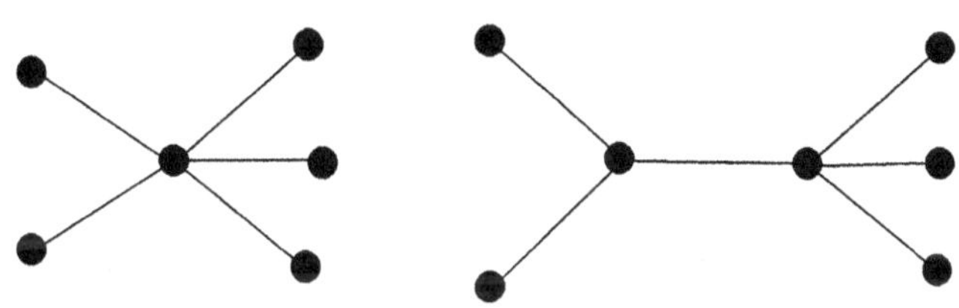

Figure 3.9: Two trees; the first has an obvious center, the second does not.

In order to define and discuss the centers of a tree, we must first discuss the notion of *distance* between vertices in a graph in general. To comprehend distances we first define a *metric*[1] on a given set.

Definition 3.14 Let S be a set. A function $\partial : S \times S \to \mathbb{R}$ is called a *metric* on S if it satisfies the following conditions:

1. $\partial(s, s) = 0$ for all $s \in S$.
2. $\partial(s, t) > 0$ for all distinct $s, t \in S$.

[1] A la the *metric* unit system developed in France during the French Revolution to unify all units!

3. $\partial(s,t) = \partial(t,s)$ for all $s, t \in S$.
4. $\partial(s,u) \le \partial(s,t) + \partial(t,u)$ for all $s, t, u \in S$.

The ordered pair (S, ∂) is called a *metric space*.

The function ∂ in Definition 3.14 can be viewed as a measure of distance. All of the conditions then make intuitive sense, as we now explain.

1. The distance from any point to itself is zero.
2. The distance between any two points is always positive.
3. The distance from one point to a second is the same as the distance from the second point to the first. Hence, we can talk about the distance *between* points as we do in everyday conversation, instead of the distance *from* one point *to* another.
4. The last condition defining a metric space is called the *triangle inequality*. This comes from the special case of the usual distances in the Euclidean plane: For three points A, B, and C in the plane, the length of the line segment AB is always shorter than the sum of the lengths of the line segments of AC and CB.

Let us examine how Definition 3.14 applies to graph theory and in particular to trees.

Definition 3.15 Let G be a graph. For $u, v \in V(G)$, the *distance* between u and v, denoted $\partial_G(u, v)$—or $\partial(u, v)$ when there is no danger of ambiguity—is the length of the shortest u, v-path in G. If there is no path between u and v in G, then we set $\partial_G(u, v) = \infty$.

Note that if G' is a subgraph of G, then for any vertices u and v in $V(G')$ we have that

$$\partial_G(u, v) \le \partial_{G'}(u, v).$$

In fact, if G is connected and we have equality in the above inequality, for all induced connected subgraphs G' of G and for all $u, v \in V(G')$, then G is called a *distance hereditary graph*. Such graphs allow many computationally hard problems to be solved in an efficient manner.

When implementing distances in a computer program, the notion of ∞ must be replaced by some "large" concrete number, which sometimes is chosen to be n, since the distance between any two vertices in a connected graph on n vertices is at most $n - 1$. Hence, whenever the distance of n or greater pops up, we know the vertices are not in the same component.

Observation 3.16 For a graph G, the distance function $\partial_G : V(G) \times V(G) \to \mathbb{N} \cup \{\infty\} \subseteq \mathbb{R} \cup \{\infty\}$ is a metric.

Needless to say, for a large graph on many vertices, the task of finding a shortest path between two given vertices by hand can be a daunting task. We could consider all the possible paths and take one (of possibly many!) of the shortest length.

Example 3.17 Consider the graph shown in Figure 3.10. Suppose we want to compute $\partial(u_1, u_4)$ for this graph. We need to consider only paths and not walks, since for every u_1, u_4-walk, there is a u_1, u_4-path of shorter length. Since u_1 and u_4 are not neighbors, we have $\partial(u_1, u_4) \ge 2$.

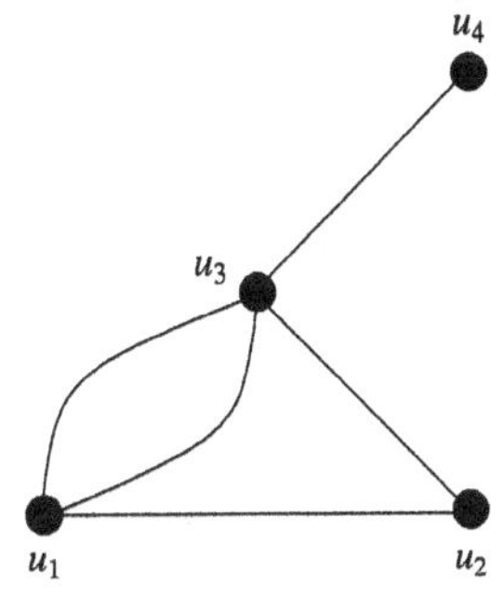

Figure 3.10: The distance between u_1 and u_4, $\partial(u_1, u_4) = 2$.

Also, since both u_1 and u_4 have a common neighbor, namely u_3, there is a path (in fact two!) from u_1 to u_4 via u_3, and hence we have $\partial(u_1, u_4) \leq 2$. This shows us that $\partial(u_1, u_4) = 2$.

The graph in Figure 3.10 is not complicated, and therefore we could "get away" with simple reasoning, to determine quickly the distance between u_1 and u_4. But for a large graph there is no easy way to simply "spot" shortest paths, although there are some efficient algorithms to do this. The algorithm most commonly implemented, called DIJKSTRA'S SHORTEST PATH ALGORITHM, will be discussed in more detail in Chapter 13.

Now, by Theorem 3.7, there is only one path between any two vertices in a tree. Thus, the determination of distance between vertices in trees is much easier. We simply need to find *the* path between two vertices and determine its length.

Using the distance function, we are now able to define centers for any graph.

Definition 3.18 Let G be a graph and $u \in V(G)$.

1. The *eccentricity* $\epsilon_G(u)$ of u in G is the distance from u to the vertex farthest from u in G. That is,

$$\epsilon_G(u) = \max(\{\partial_G(u, v) \mid v \in V(G)\}).$$

2. A *center* of G is a vertex having minimum eccentricity.

When there is no danger of ambiguity, we simply write $\epsilon(u)$ for $\epsilon_G(u)$.

To determine the centers of a "large" graph can be a hard task, but for trees this is particularly simple, as the following example reveals.

Example 3.19 Considering the eccentricities of the vertices of a tree, it is evident that the vertices having the largest eccentricities are among the leaves, since each path of maximal length connects two leaves. In particular, a leaf is never a center of a tree on three or more vertices, and centers must therefore be among the internal vertices. In Figure 3.11, we see that the center

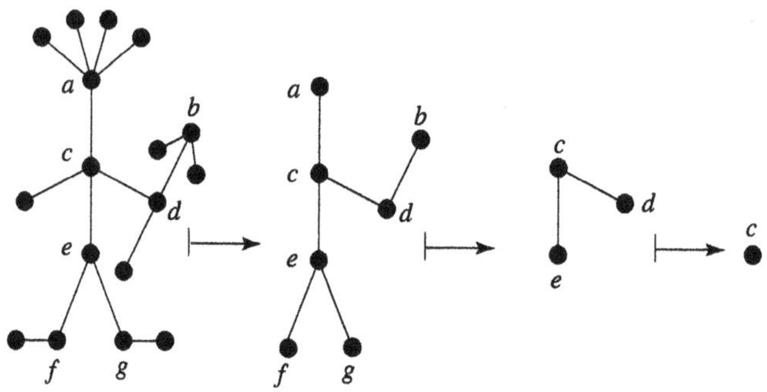

Figure 3.11: Finding the center(s) of a tree.

must be among the vertices labeled a, b, c, d, e, f, or g. By *removing* all the leaves from the original tree, the eccentricity of each internal vertex of a tree will be reduced by one, and hence the center of the original tree is the vertex of the pruned tree with minimum eccentricity. In Figure 3.11 we see that the center of the original tree must therefore be among the nonleaves of the pruned tree, namely among c, d, or e.

This process can be continued until we are left with either a single vertex or two adjacent vertices. In Figure 3.11 we are indeed left with just a single vertex c, which is therefore the unique center of this tree.

The argument in Example 3.19 can be used to prove the following theorem.

Theorem 3.20 There are one or two centers in every tree; in the latter case they are adjacent.

The next theorem explains when there are two centers and when there is just one in a tree. Note that this could have been used to define the center(s) for a tree, but since we defined the centers more generally for graphs, as the vertices with minimum eccentricities, the following theorem needs a proof, which we leave as Exercises 3.26, 3.27, and 3.28.

Theorem 3.21 Let T be a tree on more than one vertex.

1. If a longest path of T has an even length, then T has exactly one center, which is the mid-vertex of each longest path.
2. If a longest path of T has an odd length, then T has exactly two adjacent centers, which are the two midmost vertices of each longest path.

The length of the longest path in a tree (or the longest distance between two vertices in any graph for that matter) is sometimes called the *diameter* of the tree. Because of this, the eccentricity of a center is sometimes referred to as the *radius* of the tree. These two concepts are not always compatible in the sense that the diameter is not always twice the radius (see Exercise 3.29).

3.6 Rooted Trees

As we saw in Example 3.12, the initial vertex in the data tree there is unique and plays a special role of being the starting point of some process. Situations like this are very common in algorithms in computer science and many other applications. This unique vertex is the called the *root* of the tree, and trees with such a root are *rooted trees*.

Definition 3.22 Let T be a tree and $r \in V(T)$. A *rooted tree* is the ordered pair (T, r). The vertex r is called the *root* of T.

The tree in Figure 3.3 has a vertex labeled as "word." This vertex plays a special role in the tree representing the dictionary, since it does not represent any alphabetical letter, but rather the initial starting point of the process of finding a given word in the dictionary.

Example 3.23 Consider the two rooted trees in Figure 3.12. We see that when we disregard the root r, these two trees are isomorphic as graphs. However, they are distinct as rooted trees, since the root of the tree on the left has degree three, but the root of the tree on the right has degree two. Also, in the tree on the left, there is precisely one path of length two from the root r, but there are two paths of length two from the root r in the tree on the right.

Figure 3.12: Two rooted trees, isomorphic as trees but not as rooted trees.

For a rooted tree (T, r), we usually omit this notation and simply talk about the "rooted tree T" when it is clear which vertex r is the root. It is customary to represent a rooted tree in a diagram by drawing the root uppermost and then its neighbors below the root, their neighbors below them, and so on, as seen in Figure 3.12.

Once we have a rooted tree, we can define some new terminology, most of which comes from *family trees* showing the genealogy of a group of related people.

Definition 3.24 Let (T, r) be a rooted tree with at least two vertices, $u \in V(T)$ be a vertex that is not the root r, and p_u be the unique path between r and u in T.

1. The *parent* of u is the neighbor of u on p_u and is denoted by $\text{par}(u)$.
2. Any neighbor of u other than $\text{par}(u)$ is called a *child* of u. The set of children of u is denoted by $\text{Chi}(u)$.
3. Two vertices are called *siblings* if they have the same parent.
4. Any vertex on p_u other than u is called an *ancestor* of u. The set of ancestors of u is denoted by $\text{Anc}(u)$.
5. A vertex v that has u as an ancestor, that is $u \in \text{Anc}(v)$, is called a *descendant* of u. The set of all of the descendants of a vertex u are denoted by $\text{Des}(u)$.

Our reason for having par(∗) as the only lower-case abbreviation is that it is the only one of these concepts that represents a *unique* vertex. The other notions are *sets* that usually contain more than one element and are therefore denoted with capital letters, Chi(∗), Anc(∗), and Des(∗).

The terminology in Definition 3.24 is directly related to family trees, where the root represents the oldest person and an edge is always present between a parent and a child. This is why we usually draw the figures of rooted trees from the top downward.

Example 3.25 Figure 3.13 depicts a rooted tree (T, r). For the vertex $u \in V(T)$, we have the following:

$$\begin{aligned} \text{par}(u) &= a, \\ \text{Chi}(u) &= \{b, c, d\}, \\ \text{Anc}(u) &= \{a, r\}, \\ \text{Des}(u) &= \{b, c, d, e, f, g, h\}. \end{aligned}$$

The vertices e and f are siblings. The vertices c and f are not siblings. Note that every vertex, except the root, in T is a descendant of r. That is, every vertex of T except the root has r as an ancestor. There are no descendants of the leaves of T.

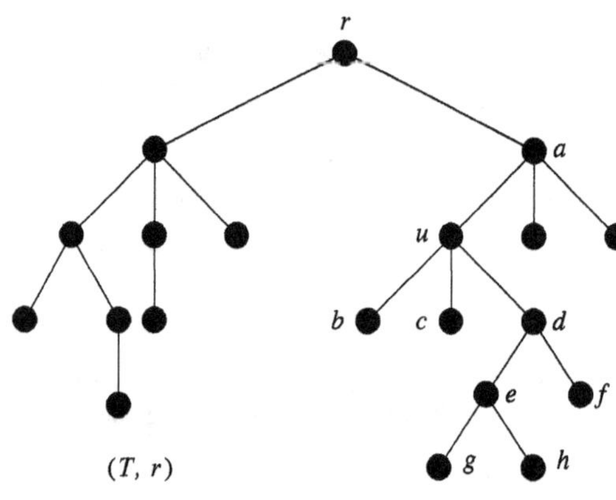

Figure 3.13: A rooted tree and relationships between its vertices.

Often in rooted trees a vertex's position in the representation of the tree is important when reading from left to right, particularly in *binary trees*. This is highlighted in the following definition.

Definition 3.26 A rooted tree (T, r) in which the left/right order of every set of siblings is specified is called an *ordered rooted tree*.

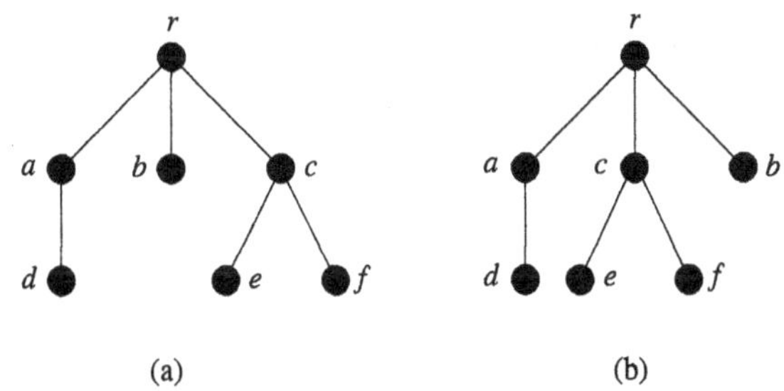

Figure 3.14: Two distinct ordered rooted trees that are identical when viewed as rooted trees.

Example 3.27 Figure 3.14 shows two ordered rooted trees. They are the same when considered just as trees or even as rooted trees. They are not the same ordered rooted tree, because the order of the siblings $\{a, b, c\}$ differs. In the first tree the order is (a, b, c) but in the second the order is (a, c, b).

In the next section we focus on an important class of ordered rooted trees.

3.7 Binary Trees

Binary trees form a very special class of ordered rooted trees in computer science and combinatorics. Countless data structures and search tools are based on binary trees. The word *binary* means formed by or consisting of two parts. In computer science the parts are the two binary numbers 0 and 1. In the following definition "left" can be interpreted as "0," and "right" as "1."

Definition 3.28 A *binary tree* is an ordered rooted tree in which each vertex has at most two children. Each child of a vertex is called either the *left* child or the *right* child. A subtree rooted at the left (right) child of a vertex u is known as u's *left (right) subtree*. By convention, the single vertex tree is considered a "trivial" binary tree.

Figure 3.15 depicts a sample binary tree. In a binary tree drawing, the position of a vertex in the picture determines whether it is a left or right child. For example, in Figure 3.15 vertex a is a left child of r and vertex b is a right child of r.

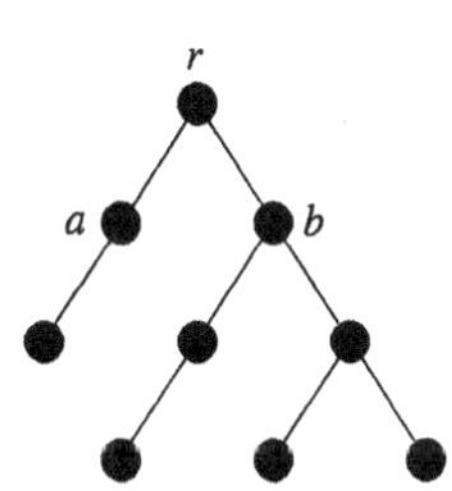

Figure 3.15: A sample binary tree.

In many applications of binary trees it is required that each nonleaf vertex have *exactly* two children. For example, various algorithms in computer science require such a tree as input. Usually we refer to this particular class of trees as the *regular binary trees*. We define them in what follows.

Definition 3.29 A *regular* or a *full binary tree* is a binary tree that either is trivial or meets the following conditions:

1. There is exactly one vertex of degree two, namely the root.
2. All vertices other than the root have degree one or three.

Remark 3.30 The reason that we consider the single-vertex graph as both a "trivial" binary tree and an "trivial" regular binary tree is similar to the reason that by convention we set $n^0 = 1$ and $0! = 1$, which must hold when we are naturally expanding algebraic rules beyond the natural numbers $\mathbb{N}$. The conventions of including "trivial" trees allow more natural approaches in inductive proofs, as we shall see later.

Figure 3.16 depicts a regular binary tree. By Definition 3.28, the vertex of degree two is distinct from all other vertices in a regular binary tree, so this vertex can serve naturally as the root of the tree. Adopting this viewpoint, every regular binary tree can be thought of as a restricted rooted tree.

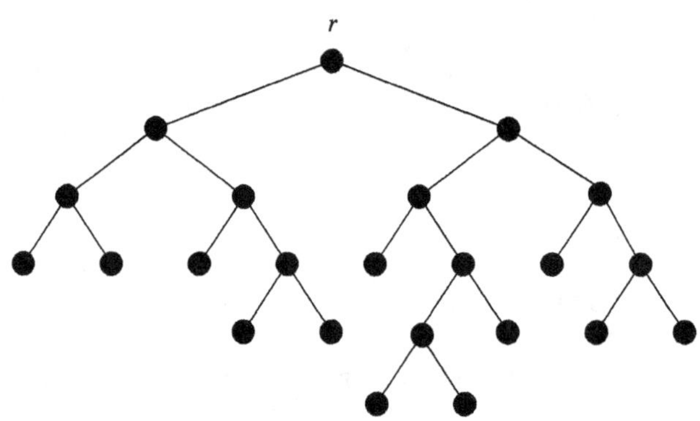

Figure 3.16: An example of a regular binary tree with r as a root.

The next property of regular binary trees follows directly from the definition. It is sometimes referred to as the *Full Binary Tree Theorem.*

Theorem 3.31 **Full Binary Tree**

A regular binary tree has $n = 2k + 1$ vertices, where k of them are internal and $k + 1$ of them are leaves.

Proof Let T be a binary tree on n vertices rooted at r. The statement is clear for $n = 1$, if we view a single vertex as a leaf.

Assume $n \geq 2$, so we proceed by induction on n. If T_ℓ and T_r are the left subtree and right subtree of the root r of T, they contain at least one vertex each, so $n \geq 3$. Since both these subtrees have strictly fewer vertices than T, we have by the induction hypothesis that T_ℓ has $n_\ell = 2k_\ell + 1$ vertices, k_ℓ of them internal and $k_\ell + 1$ of them leaves. Similarly, T_r has $n_r = 2k_r + 1$ vertices, k_r of them internal and $k_r + 1$ of them leaves. The internal vertices of T are the internal vertices of T_ℓ and T_r combined together with the root r, a total of $k = k_\ell + k_r + 1$. The leaves are those of T_ℓ and T_r combined, a total of $k_\ell + k_2 + 2 = k + 1$, which completes our inductive argument. ■

From Theorem 3.31 we obtain as a corollary the following.

Corollary 3.32 For any binary tree T on n vertices, there is a unique regular binary tree T^* with T as a subtree satisfying the following:

1. Every vertex in T is internal in T^*.
2. T^* is a regular binary tree.
3. T^* has $2n + 1$ vertices, n of them in T.

Proof We form T^* by adding leaves to T, one leaf to each vertex with exactly one child in T, and two leaves to every leaf of T. In this way T^* is a regular binary tree with its internal vertices precisely the vertices of T. By Theorem 3.31 the number of internal vertices is n and hence the leaves, which were added to obtain T^*, total precisely $n + 1$. ■

As we have seen, many inductive proofs for trees stem from the fact that every tree has a leaf. Another way to obtain properties about binary trees is to use the root and their left and right subtrees, as in the proof of Theorem 3.31.

Example 3.33 As we saw in Figure 3.12, two isomorphic trees can be different when viewed as rooted trees. Considering regular binary trees, which in particular are rooted, how many different regular binary trees are there on $2n + 1$ vertices?

Let us denote the number of regular binary trees on $2n+1$ vertices by C_n. In Figure 3.17 we see that there is exactly one regular binary tree on one vertex, and one on three vertices, respectively, so $C_0 = C_1 = 1$. Further, we see that there are exactly two regular binary trees on five vertices, so $C_2 = 2$. Finally, we see all of the five possible regular binary trees on seven vertices, so $C_3 = 5$. Evidently, it gets harder to obtain C_n by listing all possible regular binary trees, so we will see what can be said inductively. For $n \geq 1$, the root r of a regular binary tree T on $2n + 1$ vertices has one left subtree T_ℓ and one right subtree T_r, both of which are regular binary trees. Note that T is completely determined by T_ℓ and T_r. The number of vertices of both T_ℓ and T_r are among $1, 3, 5, \ldots, 2n - 1$, since they are both regular; further, they must add up to $2n$, since the total number of vertices in T is $2n + 1$.

If T_ℓ is the trivial tree, then T_r is a regular binary tree on $2n - 1$ vertices. Hence, there are altogether $1 \cdot C_{n-1} = C_0C_{n-1}$ possibilities for T in this case.

If $T_\ell l$ has three vertices, then T_r has $2n - 3$ vertices. Hence, there are C_1C_{n-2} possibilities for T in this case.

Continuing in this manner, we find that the total number of possible shapes of T is given by

$$C_n = C_0C_{n-1} + C_1C_{n-2} + C_2C_{n-3} + \cdots + C_{n-2}C_1 + C_{n-1}C_0.$$

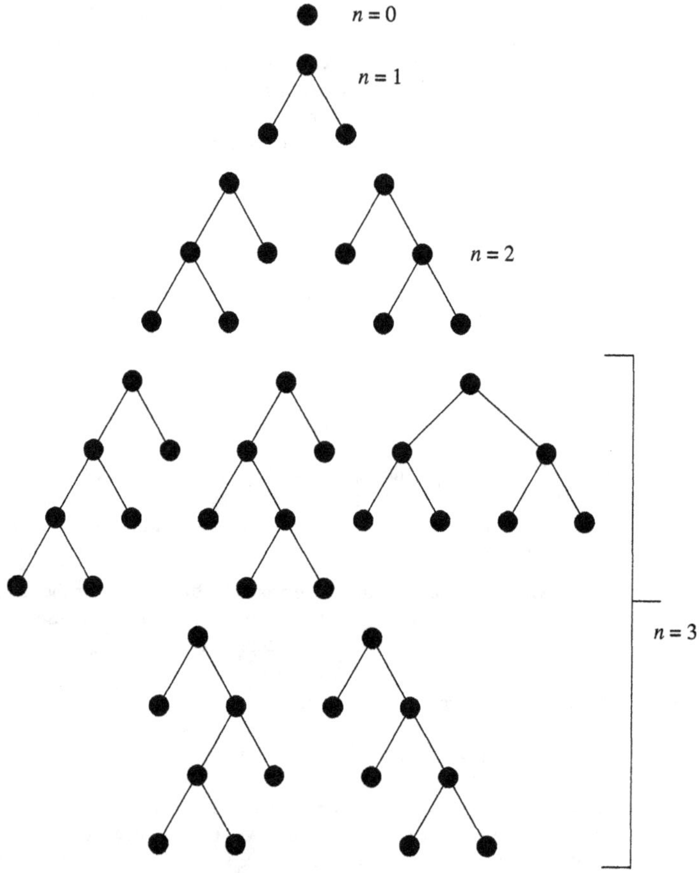

Figure 3.17: Small regular binary trees.

This gives us a way of computing C_n in terms of the previous values. For example, we have in particular

$$\begin{aligned} C_4 &= C_0C_3 + C_1C_2 + C_2C_1 + C_3C_0 \\ &= 1 \cdot 5 + 1 \cdot 2 + 2 \cdot 1 + 5 \cdot 1 \\ &= 14, \end{aligned}$$

which then can be used to compute C_5 and so on.

Note that we did not derive a simple expression for C_n, although we know mechanically how to obtain it from the previous values. This is yet another example of a *recursively* defined number sequence, $(C_n)_{n=1}^{\infty}$.

Definition 3.34 The numbers $\{C_n : n \geq 0\} = \{C_0, C_1, C_2, \dots\}$ defined recursively by

$$C_0 = 1,$$
$$C_1 = 1,$$
$$C_n = \sum_{i=0}^{n-1} C_i C_{n-i-1} \qquad \text{for } n \geq 2,$$

are called the *Catalan numbers*. Here C_n is called the nth *Catalan number*.

As we will see in Example 12.19 in Chapter 12, there is a formula for the Catalan numbers given by

$$C_n = \frac{1}{n+1}\binom{2n}{n}. \tag{3.3}$$

The Catalan numbers often appear in various enumeration problems in graph theory and combinatorics. (See, for example, Exercise 3.36.) For an excellent account of numerous ways to combinatorially interpret the Catalan numbers (66 ways to be exact!), we refer to Stanley [51, p. 219].

Example 3.35 Suppose we would like to count the number r_n of distinct ordered rooted trees on $n+1$ vertices. We clearly have $r_0 = r_1 = 1$. Unlike in Example 3.33, the root of an ordered rooted tree can have an *arbitrary* number of children. For $n \geq 1$, however, we can rest assured that the root of an ordered rooted tree T on $n+1$ vertices has at least one child, and therefore a unique leftmost subtree T_ℓ. Denote the subtree of T, when T_ℓ has been removed, by T'. Note that the root of T is also the root of T', as shown in Figure 3.18. The vertex-disjoint subtrees T_ℓ and T' cover all the vertices of T. In particular, both T_ℓ and T' are trivial trees on one vertex each if, and only if, T is the unique rooted tree on two vertices. For each $i \in \{1, \dots, n\}$, if T_ℓ has i vertices, then T' has $n+1-i$ vertices. Hence

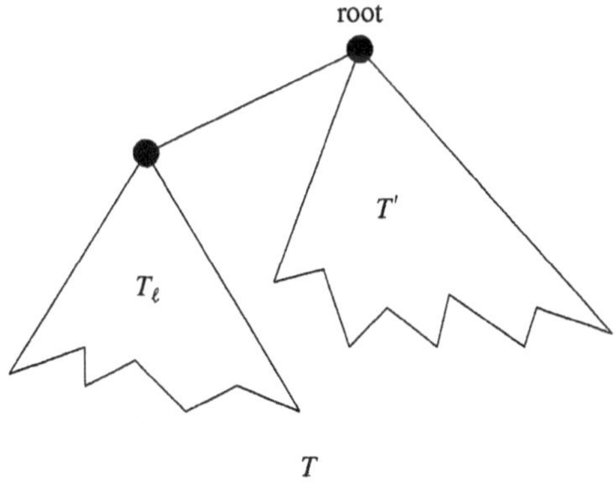

Figure 3.18: Every rooted tree on two or more vertices has a leftmost subtree.

there are r_{i-1} possible T_ℓ's and r_{n-i} possible T''s. From this we see that the total number of possibilities for T is

$$r_n = r_0 r_{n-1} + r_1 r_{n-2} + \cdots + r_{n-1} r_0.$$

Since r_n satisfies the same recursive formula as the Catalan number C_n in Definition 3.34, and has the same initial values for $n = 0, 1$, we have that $r_n = C_n$, the nth Catalan number, for all nonnegative n.

Note that in both Examples 3.33 and 3.35 two isomorphic nonrooted trees could give rise to many different rooted or binary trees. However, we did not assume the vertices were labeled with distinct numbers from $\mathbb{N}$. Every rooted tree on n vertices naturally gives rise to $n!$ labeled rooted trees on that given fixed rooted tree. We summarize in the following theorem.

Theorem 3.36 In what follows, C_n denotes the nth Catalan number.

1. The number of ordered rooted trees on $n + 1$ vertices is C_n.
2. The number of labeled ordered rooted trees on the $n + 1$ vertices $u_1, \ldots, u_{n+1}$ is $(n + 1)!C_n$.
3. The number of regular binary trees on $2n + 1$ vertices is C_n.
4. The number of labeled regular binary trees on the $2n + 1$ vertices $u_1, \ldots, u_{2n+1}$ is $(2n + 1)!C_n$.

3.8 Levels in Rooted and Binary Trees

For many applications of trees, the leaves represent a result, a final step of an algorithm, or a search in a certain data structure. It is therefore essential to choose the root in such a way that the distance from it to as many leaves as possible is small. Optimally, one would like to choose a center of the tree as the root, but that is not always possible. When we do have a choice or some freedom with our tree, we like the distances to the root to be as small as possible. Such considerations lead to the following.

Definition 3.37 Let (T, r) be a rooted tree. For a vertex $u \in V(T)$, the distance $\partial_T(u, r)$ is called the *level* of the vertex u in the rooted tree. The level of u is denoted $\ell_T(u)$ or simply by $\ell(u)$. The maximum level in (T, r) is called the *height* of (T, r) and is denoted $\text{ht}(T)$. A rooted tree of height k is called a $(k + 1)$-*level tree.*

Figure 3.19 depicts a regular binary tree that has 23 vertices and 6 levels. The tree height is five. Note that the root is always on level zero.

A rooted binary tree T with $\text{ht}(T) = k$ has a single vertex on level zero, namely the root. There are no more than two vertices on level one, possibly the left and right child of the root. There are no more than four vertices on level two, possibly left and right children of both of the root's children. In general, there are at most 2^i vertices of T on level i for $0 \le i \le k$. So the total n of vertices that a binary tree T of height k can have satisfies

$$n \le 2^0 + 2^1 + \cdots + 2^k = 2^{k+1} - 1.$$

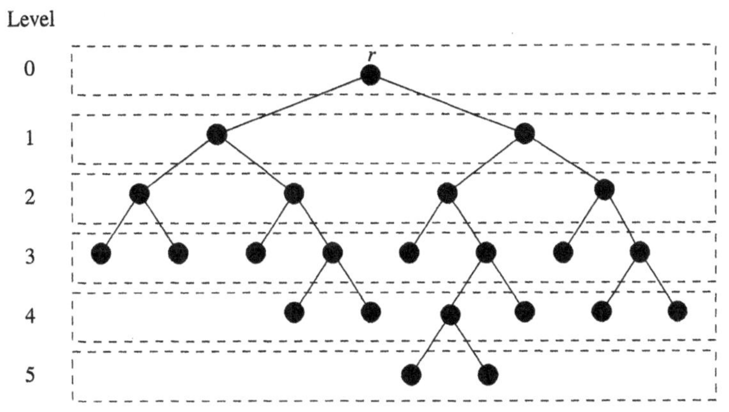

Figure 3.19: The six levels of a regular binary tree on 23 vertices.

In other words the height of T satisfies $2^{\text{ht}(T)+1} - 1 \geq n$. Solving for the height in terms of the number n of vertices, we obtain a lower bound for the height, given by

$$\text{ht}(T) \geq \lceil \lg(n+1) - 1 \rceil,$$

where lg is the logarithm with base number 2. Since every vertex of a regular binary tree, except the leaves, has exactly two children, each level (except level zero) must contain at least two vertices. From this we see that the number of vertices of a tree of height k satisfies

$$n \geq 1 + 2k,$$

from which we deduce the following observation.

Observation 3.38 For any regular binary tree T on n vertices, the height $\text{ht}(T)$ satisfies the following:

$$\lceil \lg(n+1) - 1 \rceil \leq \text{ht}(T) \leq \frac{n-1}{2}.$$

The next example illustrates trees meeting the bounds of Observation 3.38. In particular, the lower bound can be reached for a large group of regular binary trees.

Example 3.39 The regular binary trees in Figure 3.20 both have a height of 3; the one on the left has 9 vertices and the one on the right has 15 vertices. Note that both of these trees have minimum heights, since

$$\lceil \lg(9+1) - 1 \rceil = \lceil \lg(15+1) - 1 \rceil = 3.$$

Another noteworthy point is that the right tree has "almost" twice the number of vertices as the left one.

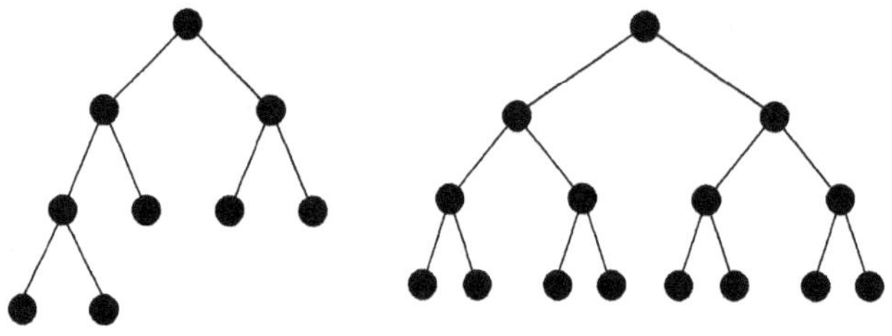

Figure 3.20: Two regular binary trees with minimum heights of three. One is on 9 vertices, the other 15.

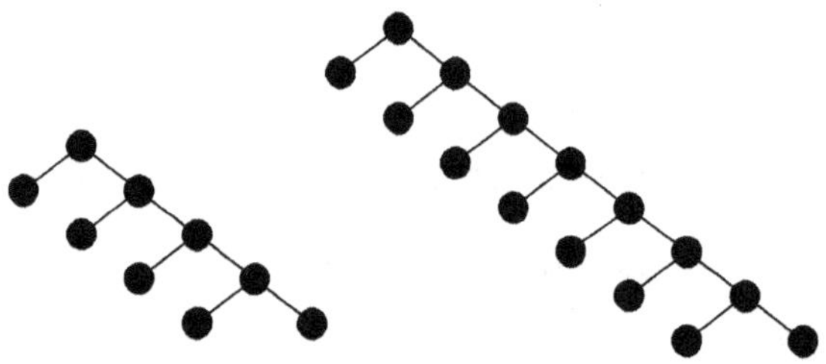

Figure 3.21: Two regular binary trees with maximum heights of 4 and 7, respectively.

In Figure 3.21 we see two regular binary trees of maximum heights; the one on the left has 9 vertices and a height of 4, whereas the one on the right has 15 vertices and a height of 7. Note that every regular binary tree of maximum height must have *exactly* two vertices on each nonzero level, both children of a common parent on the previous level.

As mentioned before, in many applications leaves in binary trees play a special role as "end-of-search" vertices, where the search ends and the result is printed to the user. Therefore, the execution time for a search in a regular binary tree is directly proportional to the level of the leaf containing the information we searched for. Hence, it is of value to ask about the *average* level of all the leaves in a given regular binary tree.

For a rooted tree T, let $L(T)$ denote its leaves. With this we define the *average* level of the leaves by

$$\text{Ave}(T) = \frac{\sum_{u \in L(T)} \ell_T(u)}{|L(T)|}.$$

Example 3.40 Consider the trees in Figure 3.20. The left one has an average level of the leaves given by

$$\text{Ave}(T) = \frac{3+3+2+2+2}{5} = 2.4.$$

The one on the right has an average level of the leaves given by

$$\text{Ave}(T) = \frac{3+3+3+3+3+3+3+3}{8} = \frac{8 \cdot 3}{8} = 3,$$

which is no surprise, since all the leaves are on third level. Both these trees have an average level at most 3, which is their least possible height.

Consider the graph on the right in Figure 3.21 on 15 vertices, 8 leaves and 7 internal vertices. It has an average level of leaves given by

$$\text{Ave}(T) = \frac{1+2+3+4+5+6+7+7}{8} = \frac{35}{8} = 4.375,$$

which is approximately (slightly less than) one plus half the number of internal vertices $(1 + 7/2 = 4.5)$. It holds in general for a regular binary tree on $2n + 1$ vertices, with a maximum height of n, that the average level of its leaves is given by

$$\text{Ave}(T) = \frac{1+\cdots+n+n}{n+1} = 1 + \frac{n}{2} - \frac{1}{n+1}.$$

Before presenting our next theorem, we make the following definition.

Definition 3.41 A binary tree T of height k is called *complete* if all levels $i \in \{0, 1, \ldots, k-1\}$ of T have precisely 2^i vertices, in which case they are called *full levels.*

A complete binary tree of height k has n vertices, where

$$2^0 + 2^1 + \cdots + 2^{k-1} < n \le 2^0 + 2^1 + \cdots + 2^k,$$

and hence $k = \lceil \lg(n+1) - 1 \rceil$, the minimum possible height for a binary tree on n vertices. Note also that all of the leaves are on levels $k - 1$ and k in this case.

We now present a theorem for the average level of leaves among regular binary trees.

Theorem 3.42 Among all regular binary trees on n vertices, the ones with the minimum average level for the leaves are precisely the complete regular binary trees.

Proof Let T be a regular binary tree on n vertices and $\text{ht}(T) = k$ its height. Note that there is at least one leaf on level k. Since T is a regular binary tree, the leaf has a sibling, hence there are at least two leaves at level k.

Assume now that T is a regular binary tree with a leaf on level $i \le k-2$. By removing two leaves at level k from T and connecting them to the leaf on level i, we get another binary tree T'. This tree has three different leaves from T, namely the two leaves moved from level k to level $i + 1$, and the parent of these leaves that became a leaf in T' at level $k - 1$. Note that the leaf at level i in T, where we connected the two removed leaves, is not a leaf in T'. We now get

$$\begin{aligned}
n \cdot \text{Ave}(T') &= \sum_{u \in L(T')} \ell_{T'}(u) \\
&= \left(\sum_{u \in L(T)} \ell_T(u) \right) - (2k - 2(i+1)) + ((k-1) - i) \\
&\le \left(\sum_{u \in L(T)} \ell_T(u) \right) - 1 \\
&= n \cdot \text{Ave}(T) - 1.
\end{aligned}$$

After this procedure, we have one less leaf on level $i \leq k-2$. This procedure can be repeated until there are no leaves on any levels other than k and $k-1$, at which point all the levels $i \in \{0, 1, \ldots, n-1\}$ are full levels, and hence $\text{ht}(T) = k = \lceil \lg(n+1) - 1 \rceil$, and we have achieved the minimum average level of the leaves of T. ■

The next example illustrates one step of the procedure outlined in the proof of Theorem 3.42 for reducing the average level of the leaves in a regular binary tree.

Example 3.43 In Figure 3.22 the average level of the leaves in T is given by

$$\text{Ave}(T) = \frac{3+3+3+3+1}{5} = \frac{13}{5}.$$

We form a new tree T' by carrying out one step of the procedure described in the proof of Theorem 3.42 and obtain an average level of leaves given by

$$\text{Ave}(T') = \frac{3+3+2+2+2}{5} = \frac{12}{5} < \text{Ave}(T).$$

Notice that $\text{Ave}(T')$ cannot be reduced any further, since all of its leaves are on the two highest levels.

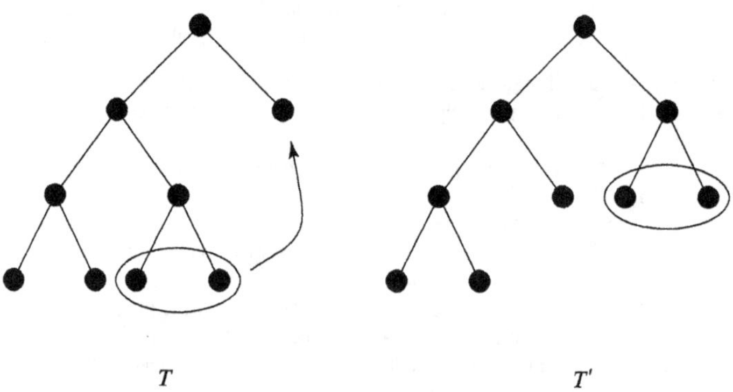

Figure 3.22: The average level of leaves in T is greater than that of T'.

As a corollary from Theorem 3.42, we obtain the minimum average level of the leaves in a regular binary tree on n vertices. Recall that n is necessarily an odd number.

Corollary 3.44 For a regular binary tree T on n vertices, the average level $\text{Ave}(T)$ satisfies

$$\text{Ave}(T) \geq \lceil \lg(n+1) \rceil - \frac{2^{\lceil \lg(n+1) \rceil}}{n+1},$$

where for each $n \in \mathbb{N}$, this lower bound can be obtained.

Proof By Theorem 3.42 the minimum average level of a regular binary tree on n vertices is achieved by a complete regular binary tree on n vertices of height $k = \lceil \lg(n+1) - 1 \rceil$, where all the leaves are either height $k-1$ or k. Note that the number of leaves ℓ_k on level k is given by

$$\ell_k = n - (2^0 + 2^1 + \cdots + 2^{k-1}) = (n+1) - 2^k.$$

Therefore, the number of leaves ℓ_{k-1} on level $k-1$ is given by

$$\ell_{k-1} = 2^{k-1} - l_k/2.$$

Since $\text{Ave}(T) = (\ell_k \cdot k + \ell_{k-1} \cdot (k-1))/(\ell_k + \ell_{k-1})$, where $\ell_k + \ell_{k-1} = (n+1)/2$, the corollary follows. ∎

In many practical applications, there are *weights* attached to each level, and hence in order to obtain the most efficient tree structure, one must obtain the *weighted average* of the levels. An example of an application of such weighted average levels was developed by Huffman [29] for compressing text files. The idea is that some letters appear more frequently than others in language, and hence they should be assigned shorter binary code words. For example, the letter "e" occurs much more frequently in the English language than does the letter "x." Therefore, it makes sense to give the letter "e" a smaller code word than the letter "x" when building a good coding. Such a coding is called a *Huffman coding*, and the corresponding binary tree with the right weights attached to each leaf is called a *Huffman tree*.

For more information on various optimal average level regular binary trees and their applications, the reader is referred to Knuth [32].

3.9 Exercises

3.1. Sketch a tree for sorting people based on nine-digit Social Security numbers. How many vertices does such a tree have? How many leaves?

3.2. Imagine a tree for sorting cars based on license-plate numbers. Suppose each license plate has exactly eight symbols and each symbol can be any capital letter or a digit. How many vertices does such a tree have? How many leaves?

3.3. It can be shown that there are exactly 11 unlabeled trees on seven vertices. (Figure 3.2 does this for trees on six vertices.) Draw these eleven trees, making sure that no two are isomorphic.

3.4. State necessary and sufficient conditions on an ordered n-tuple of positive integers $(d_1, \ldots, d_n)$ with $d_1 \leq \cdots \leq d_n$, in order that there be a tree on vertices $u_1, \ldots, u_n$ with $d_T(u_i) = d_i$ for each $i \in \{1, \ldots, n\}$. (*Note!* Such a sequence is often refered to as a *degree sequence* of a graph.)

3.5. Let T be a tree on n vertices with each vertex of degree three or less. Let x, y, and z denote the number of vertices in T of degree 1, 2, and 3, respectively. Show that $x+y+z = n$ and $x + 2y + 3z = 2n - 2$. Use this to show that $z \leq \lfloor \frac{n-2}{2} \rfloor$, thereby giving another proof of (3.1).

3.6. For $\Delta \geq 3$, consider all trees on n vertices where each vertex has degree Δ or less. Let $\mathbf{d}_\Delta(n)$ denote the maximum number of vertices of degree Δ that such a tree can have. Use induction to show that

$$\mathbf{d}_\Delta(n) \leq \left\lfloor \frac{n-2}{\Delta - 1} \right\rfloor. \tag{3.4}$$

[*Hint:* This is a generalization of Example 3.8.]

3.7. Show that the bound in Exercise 3.6 is tight for infinitely many positive natural numbers n. Do this explicitly or recursively by constructing a sequence $(t_i(\Delta))_{i\geq 1}$ such that for each $n = t_i(\Delta)$ equality holds in (3.4).

3.8. For $\Delta \geq 3$, consider all trees where each vertex has degree Δ or less, and where the longest path is of length k or less. Show that the maximum number $\mathbf{l}_{\max}$ of leaves among such trees satisfies the following:

$$\mathbf{l}_{\max} \leq \Delta(\Delta - 1)^{k/2-1}.$$

3.9. Use a data tree to determine all of the strictly decreasing subsequences of the sequence $(8, 9, 10, 11, 5, 6, 7, 1, 2, 3, 4)$.

3.10. Use a data tree to determine all of the strictly increasing subsequences of the sequence $(5, 7, 3, 8, 6, 7, 5, 3, 0, 9, 5)$.

3.11. Given a sequence of 100 distinct integers, what is the length of the longest strictly increasing or strictly decreasing subsequence you can rest assured it contains? How about a sequence of 1,088, 1,089, or 1,090 distinct integers?

3.12. Let $k \in \mathbb{N}$ and consider the sequence s on the k^2 first natural numbers given by

$$s = (s_k, s_{k-1}, \ldots, s_1),$$

where each $s_i = ((i-1)k+1, (i-1)k+2, \ldots, ik)$. As in Example 3.11, show that this sequence also has no strictly decreasing or strictly increasing subsequence of length $k+1$.

3.13. How many distinct strictly decreasing subsequences of length k does the sequence s from Example 3.11 have? How many strictly increasing subsequences of length k does it have?

3.14. How many distinct strictly decreasing subsequences of length k does the sequence s from Exercise 3.12 have? How many strictly increasing subsequences of length k does it have?

3.15. Construct the data tree for the sequence s given in Example 3.11. How many vertices does it have? How many leaves? Does your answer contradict Observation 3.13? Compare with Exercise 3.13.

3.16. Draw the data tree corresponding to the strictly *decreasing* subsequences of the sequence s in Example 3.12, and list all the subsequences of maximal length. How many are of maximum length?

3.17. Prove that the upper and lower bounds given in Observation 3.13 also hold for data trees on sequences of real distinct numbers with respect to their strictly *decreasing* subsequences.

3.18. Show that every tree containing a vertex of degree k contains at least k leaves.

3.19. Is usual addition (+) a metric on $\mathbb{N}$? Which properties hold?

3.20. Show that $|m - n|$ defines a metric on $\mathbb{N}$. What fails when we omit the absolute-value signs?

3.21. For the graph shown in Figure 3.23 label the vertices and specify the distances between all pairs of them.

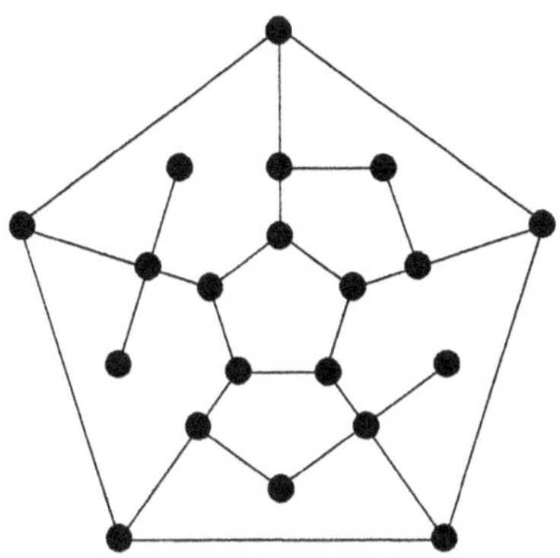

Figure 3.23: Compute the distance between each pair of vertices as requested in Exercise 3.21.

3.22. What are the eccentricities of the vertices of the graph shown in Figure 3.24?

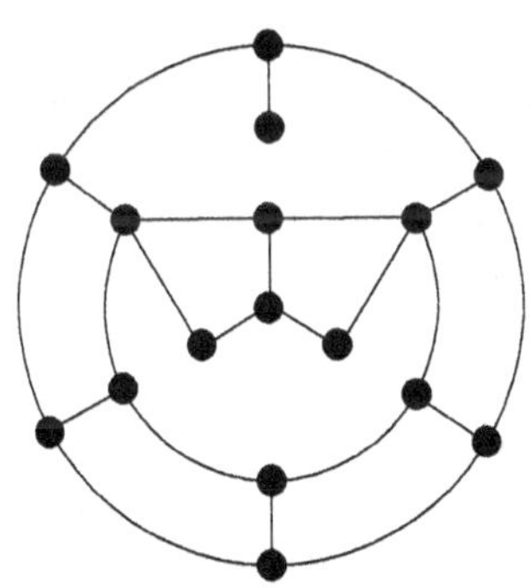

Figure 3.24: Compute the eccentricity as requested in Exercise 3.22.

3.23. For all $n \in \mathbb{N}$, what is the eccentricity of each vertex in K_n? How many centers does K_n have?

3.24. Suppose you delete a center from a tree, thereby separating the tree into several components. Can you prove any general statements about the maximum size of these components in terms of the original size of the tree?

3.25. As was done for trees at the end of §3.5, we can define the *diameter* of a graph G by

$$\operatorname{diam}(G) = \max(\{\partial_G(u, v) : u, v \in V(G)\})$$

Let $d = \operatorname{diam}(G)$ and $\Delta \geq 3$ be the maximum degree G. If G is simple and connected, show that

$$|V(G)| \leq \frac{\Delta(\Delta - 1)^d - 2}{\Delta - 2}.$$

3.26. Show that if a tree T has a longest path of even length, then the mid-vertex of one such longest path is the mid-vertex of every longest path of T. [*Hint:* Show that two longest paths must have a vertex in common, which then must be their common mid-vertex, since otherwise there is a longer path.]

3.27. Show that if a tree T has a longest path of even length, then the two midmost vertices of one such longest path are the two midmost vertices of every longest path of T. [*Hint:* Show first that two such longest paths must have an edge in common, namely their common mid-edge.]

3.28. Use Exercises 3.26 and 3.27 to prove Theorem 3.21.

3.29. Give an example of a tree where the diameter is not twice the radius. Under what circumstances are the diameter and radius the same?

3.30. Can a tree ever have more parents than children? Explain your answer.

3.31. Consider a rooted tree where all the internal vertices have exactly $\Delta \geq 2$ children. Such a tree is called a *full Δ-ary tree*. Given that the height of a full Δ-ary tree is k, find a formula for the maximum number of leaves of such a tree and a formula for the maximum number of internal vertices. What happens when Δ is fixed and k grows large?

3.32. Is it possible to have a tree with more than five vertices such that every vertex except the root and leaves has exactly the same number of ancestors and descendants? Explain.

3.33. Draw the regular binary trees on nine vertices.

3.34. Write a definition for regular binary trees that is equivalent to the one given in Definition 3.29, but that is expressed using the concept of internal vertices.

3.35. Use the Hand-Shaking Theorem 1.26 together with Theorem 3.7 to give an alternative proof of the Theorem 3.31.

3.36. Assume that $*$ is a nonassociative operation of graphs. Show that the number of ways to interpret $G * G * \cdots * G$ (n times) is given by the Catalan number C_n. For example, there are five ways of interpreting $G * G * G * G$, namely as

$$((G * G) * G) * G, \quad (G * (G * G)) * G, \quad (G * G) * (G * G),$$
$$G * ((G * G) * G), \quad G * (G * (G * G)).$$

[*Hint:* Consider first the two pairs of outermost parentheses in each such expression $(\cdots * G * \cdots) * (\cdots * G * \cdots)$, and then use induction on each of these two parts.]

3.37. A *regular* or *full ternary tree* is an ordered rooted tree where each vertex, except the leaves, has exactly three children—a *left* child, a *middle* child, and a *right* child. Hence, all of the internal vertices have degree four, except the root that has degree three. Prove the *Full Ternary Tree Theorem*, which states that a regular ternary tree has $n = 3k + 1$ vertices, k of them internal and $2k + 1$ of them leaves.

3.38. Let t_n be the number of regular ternary trees on $3n + 1$ vertices. Prove that t_n satisfies the "Catalan"-like recursive condition

$$\begin{aligned} t_0 &= 1, \\ t_1 &= 1, \\ t_n &= \sum_{i+j+k=n-1} t_i t_j t_k, \end{aligned}$$

where the sum is taken over nonnegative integers i, j, and k. Use this to compute t_n for $n \in \{2, 3, 4\}$.

3.39. Let b_n be the number of binary trees on n vertices. Show that b_n satisfies the "Catalan"-like recursive condition

$$\begin{aligned} b_1 &= 1, \\ b_2 &= 1, \\ b_n &= b_{n-1} + (b_1 b_{n-2} + b_2 b_{n-3} + \cdots + b_{n-2} b_1). \end{aligned}$$

[*Hint:* Note that if b'_n is the number of binary trees where the root has only one child, and b''_n is the number of binary trees where the root has exactly two children, then $b_n = b'_n + b''_n$ and further $b'_n = b_{n-1}$.]

3.40. Show that the number of regular binary trees on $2n + 1$ vertices of height n is 2^{n-1}.

3.41. Let $b_{\lg}(n)$ be the number of complete binary trees on $2n + 1$ vertices of the optimal low height of $\lceil \lg(n + 1) \rceil$. Show that

$$b_{\lg}(n) = \binom{2^{\lceil \lg(n+1) \rceil - 1}}{2^{\lceil \lg(n+1) \rceil} - (n + 1)}.$$

[*Hint:* Note that all levels but the last have the maximum number of vertices, so the number depends only on the location of leaves on the highest level $2^{\lceil \lg(n+1) \rceil}$.]

3.42. Draw a tree T on 11 vertices with $\text{Ave}(T) = 8/3$. Can you find all such trees on 11 vertices?

3.43. What are the possible values for $\text{Ave}(T)$ when T has 11 vertices?

3.44. Show directly that the lower bound b of $\text{Ave}(T)$ given in Corollary 3.44 satisfies $k - 1 < b \leq k$, where $k = \lceil \lg(n + 1) \rceil$.

3.45. Show that the lower bound b of $\text{Ave}(T)$ given in Corollary 3.44 satisfies $b \neq k + 1/2$; that is, it can never take the value of $k + 1/2$, where $k = \lceil \lg(n + 1) \rceil$.

CHAPTER

4

Spanning Trees

Chapter Goals

Define a spanning tree for a connected graph and a spanning forest for a graph.

Present a recursive method for computing the number of spanning forests in a graph.

Present Cayley's Theorem and a proof using Prüfer codes.

Discuss the adjacency matrix of a graph and its properties.

Discuss the incidence matrix of a graph and its properties.

Present and prove the Matrix-Tree Theorem.

Discuss applications of spanning trees in electrical network analysis.

Present two algorithms, by Kruskal and Prim, for computing the minimum cost spanning tree in a weighted graph.

4.1 Spanning Trees and Forests

As we saw in Theorem 3.7 in Chapter 3, trees are minimally connected in the sense that each edge is a bridge; the removal of any one of them will disconnect the graph. They are the graphs that connect a given collection of vertices $V = \{u_1, \dots, u_n\}$ with the fewest possible edges, namely $n - 1$ of them. If now G is a connected graph with $V(G) = V$ as its set of vertices, an obvious question is whether we can find a subgraph of G that connects all the vertices of G with the fewest possible edges. As we mentioned, such a subgraph *must* by Lemma 3.6 have at least $n - 1$ edges, since otherwise it is not connected. But is it always possible to find such a subgraph of G with exactly $n - 1$ edges that connects all the vertices of G? We will see that this is always possible. Moreover, we give a systematic method to compute the number of all such subgraphs. This result is the main theme of the chapter.

Suppose G has n vertices; then a connected subgraph containing $n - 1$ edges and all n vertices of G is, by Theorem 3.7, necessarily a subtree of G.

Definition 4.1 Let G be a graph.

1. A subtree T of G is called a *spanning tree* of G if $V(T) = V(G)$.
2. A subforest F of G is called a *spanning forest* of G if for each component H of G, the subgraph $F \cap H$ is a spanning tree of H.
3. For a fixed labeling of the vertices of G, the number of distinct spanning forests in G is denoted by $\tau(G)$.

In Figure 4.1 we depict a graph on five vertices. A spanning tree of the graph is shown by thick edges. The set of edges in the spanning tree is given by e_2, e_4, e_5, and e_8.

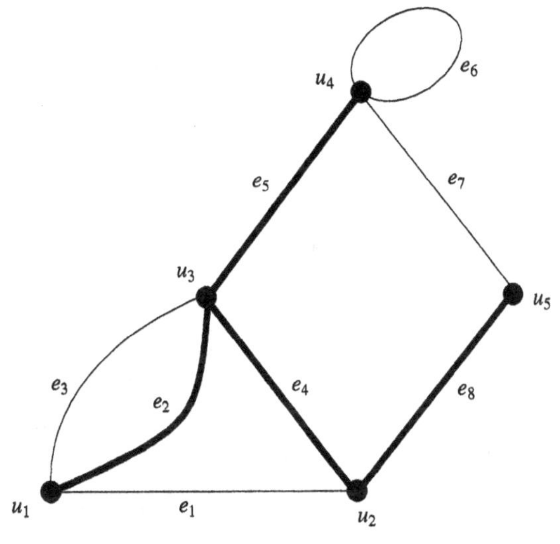

Figure 4.1: A spanning tree of a connected graph.

Remark 4.2

- If a graph G is connected, then $\tau(G)$ is the number of spanning trees in G.
- If G' is the graph we get from G by removing all the loops from G, then $\tau(G') = \tau(G)$.

The following observation is easy to prove, and we leave it as Exercise 4.5.

Observation 4.3 Let G be a graph.

1. If G is a tree, then $\tau(G) = 1$.
2. If $G_1, \ldots, G_k$ are the components of G, then

$$\tau(G) = \tau(G_1) \cdots \tau(G_k) = \prod_{i=1}^{k} \tau(G_i).$$

Before we continue with our discussion of the spanning trees of a graph G, it is helpful to recall Definition 2.34, in which we defined the contraction of G by an edge $e \in E(G)$. This contraction is denoted $G \cdot e$. With this definition in mind, we can prove the following theorem, which is useful for counting the number of spanning trees of a graph.

Theorem 4.4 Let G be a graph and $e \in E(G)$, where e is not a loop. We have that

$$\tau(G) = \tau(G \cdot e) + \tau(G - e).$$

Note that when e is a bridge, then $\tau(G - e) = 0$, since $G - e$ is not connected, so in this case $\tau(G) = \tau(G \cdot e)$.

Proof (Theorem 4.4) By Observation 4.3 we can assume that G is connected. Let $\mathcal{T}(G)$ denote the set of all spanning trees of G, so $|\mathcal{T}(G)| = \tau(G)$. Now let

$$\begin{aligned} \mathcal{T}_1 &= \{T \in \mathcal{T}(G) : e \in E(T)\}, \\ \mathcal{T}_2 &= \{T \in \mathcal{T}(G) : e \notin E(T)\}. \end{aligned}$$

This forms a partition $\mathcal{T}(G) = \mathcal{T}_1 \cup \mathcal{T}_2$.

If e is not a bridge, then $\mathcal{T}_2 = \mathcal{T}(G - e)$. If, on the other hand, e is a bridge, then $\mathcal{T}_2 = \varnothing$, by definition of a bridge. The map $\psi : \mathcal{T}_1 \to \mathcal{T}(G \cdot e)$ defined by $\psi(T) = T \cdot e$ is clearly bijective. Therefore, $|\mathcal{T}_1| = |\mathcal{T}(G \cdot e)|$, and since $\tau(G) = |\mathcal{T}_1| + |\mathcal{T}_2|$, the theorem follows. ■

Note that $G - e$ has one less edge than G but the same number of vertices. Hence, $|V(G-e)| + |E(G-e)| = |V(G)| + |E(G)| - 1$. Also, $G \cdot e$ has one edge and one vertex less than G, so $|V(G \cdot e)| + |E(G \cdot e)| = |V(G)| + |E(G)| - 2$. This suggests a way to use Theorem 4.4 to prove various properties of $\tau(G)$ by induction on $|V(G)| + |E(G)|$. We also can use Theorem 4.4 to calculate the number of spanning trees of any graph G recursively, as shown in Example 4.5. A different, and many times more efficient, way to compute $\tau(G)$ for a given connected graph G is given by the Matrix-Tree Theorem 4.28, which we discuss later in this chapter.

Example 4.5 We illustrate how Theorem 4.4 can be applied to compute the number of spanning trees of the graph G shown in Figure 4.2. Noting that e is not a bridge and applying Theorem 4.4, we get the following:

$$\begin{aligned} \tau(G) &= \tau(G \cdot e) + \tau(G - e) \\ &= \tau(G \cdot e) + 1, \end{aligned}$$

since $G - e$ is a tree. Continuing along these lines, we get

$$\begin{aligned} \tau(G \cdot e) &= \tau((G \cdot e) \cdot e') + \tau((G \cdot e) - e'), \\ &= \tau((G \cdot e)) \cdot e') + 1, \end{aligned}$$

again since $(G \cdot e) - e'$ is a path of length two, and hence a tree. By Remark 4.2 we have by removing the loop that $\tau((G \cdot e) \cdot e') = 1$, hence we have that $\tau(G) = 3$.

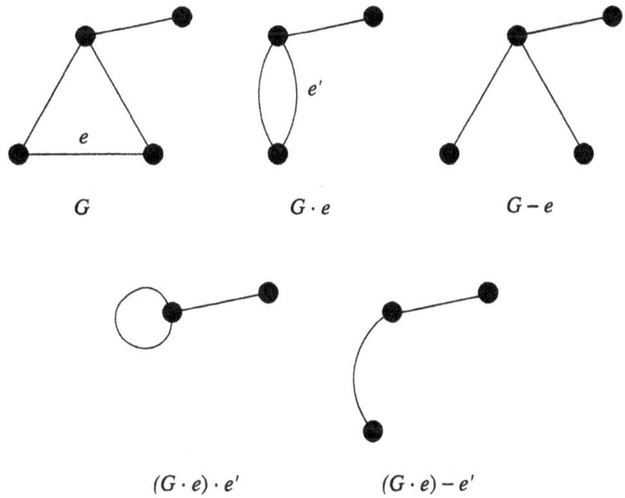

Figure 4.2: A recursive way to calculate $\tau(G)$ for any graph G.

Calculations like those in Example 4.5 become cumbersome for large graphs, since we do have to repeat the calculations for two new graphs, $G - e$ and $G \cdot e$, then at the next step we do have to repeat the calculations for four new graphs, and so on. Without keeping track of possible isomorphisms between graphs, this technique does not provide an efficient method to compute $\tau(G)$, although it is a systematic procedure. However, there is an interesting corollary we get for any graph G by induction on $|V(G)| + |E(G)|$.

Corollary 4.6 For any graph G, we have $\tau(G) \geq 1$. In other words, any graph has a spanning forest. Moreover, any connected graph has a spanning tree.

4.2 Spanning Trees of the Complete Graph

Cayley proved Theorem 4.7 for counting trees in the complete graph K_n [7]. Although his formulation was slightly different, he wanted to compute the number of ways to connect n given labeled vertices $\{u_1, \ldots, u_n\}$ minimally by a tree. This is the same as computing the number of spanning trees of K_n, where the vertices have been labeled by $1, \ldots, n$. There are many different proofs of his theorem, and the one we give here, which is probably the neatest one, is due to Heinz Prüfer.

Theorem 4.7 **Cayley's**

For $n \in \mathbb{N}$ we have $\tau(K_n) = n^{n-2}$.

Proof We sketch the proof using Prüfer's idea. His proof gives a one-to-one correspondence between the set of all labeled spanning trees of K_n and the set of all ordered $(n-2)$-tuples of the numbers $1, \ldots, n$. The encoding of each labeled tree into an ordered $(n-2)$-tuple is called a *Prüfer code*. Once this bijection has been established, we have that the number of spanning trees of K_n is equal to the number of all possible $(n-2)$-tuples of the natural numbers $1, \ldots, n$. Hence, there are n^{n-2} labeled trees on n vertices. The theorem is clearly true for $n = 1, 2$.

Assume that we have a tree T with $n \geq 3$ vertices $u_1, \ldots, u_n$. We want to encode T into an ordered tuple $(a_1, \ldots, a_{n-2})$ of natural numbers $1, \ldots, n$. This procedure is called PRÜFER CODE:

- Let $T' = T$.
- For $i = 1$ to $n-2$ do the following steps:
  1. Let u be the smallest numbered leaf in T'.
  2. Let a_i be the index of u's neighbor in T'.
  3. Let $T' = T' - \{u\}$.

Assume now we have a given ordered $(n-2)$-tuple $(a_1, \ldots, a_{n-2})$ of natural numbers $1, \ldots, n$. We construct a tree on vertices numbered $1, \ldots, n$ by the procedure called TREE TO PRÜFER CODE:

- Let $A = (a_1, \ldots, a_{n-2})$.
- For $i = 1$ to $n-2$ do the following steps:
  1. Let b_i be the least number among $\{1, \ldots, n\}$ that does not occur in A.
  2. Connect the vertices numbered a_i and b_i with an edge.
  3. Remove the now initial term a_i from A and add b_i to the end of A, to get the new $(n-2)$-tuple $(a_{i+1}, \ldots, a_{n-2}, b_1, \ldots, b_i)$.
- Connect the two vertices that are not indexed by any of $b_1, \ldots, b_{n-2}$.

We notice that the procedure TREE TO PRÜFER CODE adds edges in the same order that procedure PRÜFER CODE removes them. We also see that, starting with an $(n-2)$-tuple, applying TREE TO PRÜFER CODE, and then applying PRÜFER CODE, we end up with the original $(n-2)$-tuple. Hence, these procedures give a bijection between the spanning trees of K_n and the ordered $(n-2)$-tuples of the natural numbers $1, \ldots, n$. The theorem follows. ∎

The following example shows how both the procedures PRÜFER CODE and TREE TO PRÜFER CODE work for a specific tree.

Example 4.8 Consider the tree T given in Figure 4.3. Its eight vertices are labeled with $1, 2, \ldots, 8$. Note that $n = 8$ and hence $n - 2 = 6$, so in PRÜFER CODE we need to do six iterations. Here is how PRÜFER CODE works for T in this case:

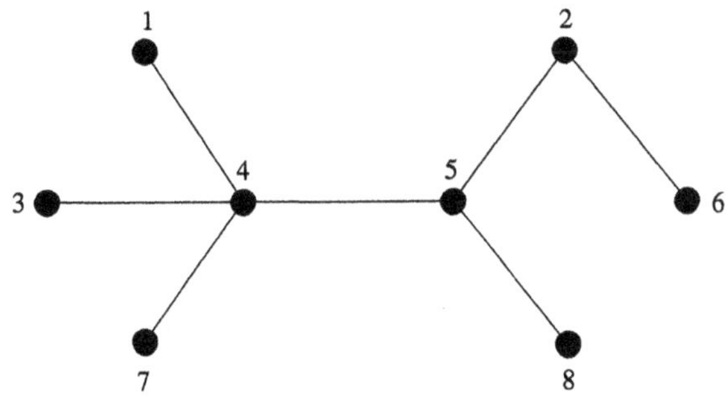

Figure 4.3: The labeled tree T has Prüfer code (4, 4, 2, 5, 4, 5).

$i = 1$. The smallest leaf in T is labeled 1, and its neighbor is labeled by 4, so $a_1 = 4$ and therefore $(a_1) = (4)$.

$i = 2$. Removing leaf labeled 1 from T, the smallest label of a leaf is 3, and its neighbor is labeled 4, so $a_2 = 4$ and therefore $(a_1, a_2) = (4, 4)$.

$i = 3$. Removing leaf labeled 3 from T, the smallest label of a leaf is 6, and its neighbor is labeled 2, so $a_3 = 2$ and therefore $(a_1, a_2, a_3) = (4, 4, 2)$.

$i = 4$. Removing leaf labeled 6 from T, the smallest label of a leaf is 2, and its neighbor is labeled 5, so $a_4 = 5$ and therefore $(a_1, a_2, a_3, a_4) = (4, 4, 2, 5)$.

$i = 5$. Removing leaf labeled 2 from T, the smallest label of a leaf is 7, and its neighbor is labeled 4, so $a_5 = 4$ and therefore $(a_1, a_2, a_3, a_4, a_5) = (4, 4, 2, 5, 4)$.

$i = 6$. Finally, removing leaf labeled 7 from T, the smallest label of a leaf is 4, and its neighbor is labeled 5, so $a_5 = 5$ and therefore $(a_1, a_2, a_3, a_4, a_5, a_6) = (4, 4, 2, 5, 4, 5)$.

This six-tuple $\tilde{a} = (4, 4, 2, 5, 4, 5)$ is the Prüfer code for T.

Now we examine how to construct T from its Prüfer code, $\tilde{a} = (4, 4, 2, 5, 4, 5)$, using TREE TO PRÜFER CODE. In the following display, the lines are numbered 1, 2, 3, 4, 5, 6, and 7, where the first line is the Prüfer code itself, before any changes.

```
( 4 4 2 5 4 5 )
  ( 4 2 5 4 5 1 )
    ( 2 5 4 5 1 3 )
      ( 5 4 5 1 3 6 )
        ( 4 5 1 3 6 2 )
          ( 5 1 3 6 2 7 )
            ( 1 3 6 2 7 4 )
```

From line $i = 1$, we note that the least label not included in $\tilde{a}$ is 1, so we connect $a_1 = 4$ to 1, remove a_1 and append 1 to the back.

From line $i = 2$, we note that the least label not included in $\tilde{a}$ is 3, so we connect $a_1 = 4$ to 3, remove a_1 and append 3 to the back.

From line $i = 3$, we note that the least label not included in $\tilde{a}$ is 6, so we connect $a_1 = 2$ to 6, remove a_1 and append 6 to the back.

From line $i = 4$, we note that the least label not included in $\tilde{a}$ is 2, so we connect $a_1 = 5$ to 2, remove a_1 and append 2 to the back.

From line $i = 5$, we note that the least label not included in $\tilde{a}$ is 7, so we connect $a_1 = 4$ to 7, remove a_1 and append 7 to the back.

Finally, from line $i = 6$, we note that the least label not included in $\tilde{a}$ is 4, so we connect $a_1 = 5$ to 4, remove a_1 and append 4 to the back.

At this point, we are done iterating. Looking at the line $i = 7$, the only two indices not included in $\tilde{a}$ at this point are 5 and 8, which we connect, and obtain our tree T back.

We now see that starting with a tree T, then obtaining the Prüfer code, and then constructing a tree again from its Prüfer code, we obtain the original tree again. This composite procedure shows that there is precisely one code for each tree—in other words, that the number of codes is at least that of the labeled trees.

We could have started with an arbitrary six-tuple of numbers $1, 2, \ldots, 8$, obtained a tree, and then a six-tuple from that tree, which would have been the same six-tuple we started with. We leave this as Exercise 4.14, which shows that the number of trees is at least that of the number of codes.

There are many different proofs of Cayley's Theorem 4.7 with various approaches. An excellent summary of ten such proofs is given by Moon [40]. Other proofs are given in Example 4.29 and in Chapter 12.

So far, we have seen one recursive method, given in Theorem 4.4, for computing $\tau(G)$ for a graph G. Note that we did *not* use this theorem to prove Cayley's Theorem 4.7, but rather a slick trick due to Prüfer, which works only for the specific graphs K_n where $n \in \mathbb{N}$. So, is there a systematic way to compute $\tau(G)$ for a given simple graph G, which works for arbitrary simple graphs G, and is not as cumbersome as the recursive method?

There is such a method, which is given by the Matrix-Tree Theorem 4.28. This method depends on representing a graph in terms of matrices, and we discuss such representations in the next sections, before stating and proving the Matrix-Tree Theorem 4.28.

4.3 The Adjacency Matrix of a Graph

In the three sections to come, we assume some basic knowledge of matrices and their manipulation over the real numbers $\mathbb{R}$. Any course in elementary linear algebra should suffice as a prerequisite. Throughout, an *$n \times m$ matrix* $\mathsf{X} = [x_{ij}]$ denotes the matrix with n rows and m columns, where x_{ij} is the entry in row i and column j. Recall that the for an $n \times m$ matrix $\mathsf{X} = [x_{ij}]$ and an $m \times k$ matrix $\mathsf{Y} = [x_{ij}]$, their product $\mathsf{XY} = [z_{ij}]$ is an $n \times k$ matrix defined by

$$z_{ij} = x_{i1}y_{1j} + x_{i2}y_{2j} + \cdots + x_{im}y_{mj} = \sum_{\ell=1}^{m} x_{i\ell}y_{\ell j},$$

for each $i \in \{1, \ldots, n\}$ and $j \in \{1, \ldots, k\}$.

The *adjacency matrix* is probably the most frequently used matrix representation of a graph. Since it is always a square matrix, it can be manipulated in many more ways than a nonsquare matrix.

Definition 4.9 Let $G = (V, E, \phi)$ be a graph on n vertices labeled by $V(G) = \{u_1, \ldots, u_n\}$. For each $i, j \in \{1, \ldots, n\}$ define the entry a_{ij} by

$$a_{ij} = |\{e \in E(G) : \phi(e) = \{u_i, u_j\}\}|.$$

That is, a_{ij} is the number of edges connecting u_i and u_j. The *adjacency matrix* of G with respect to the labeling of $V(G)$ is defined as the $n \times n$ matrix

$$\mathsf{A}(G) = [a_{ij}]_{i,j\in\{1,\ldots,n\}}.$$

Example 4.10 Consider the graph G shown in Figure 4.4, where the vertices are labeled $V(G) = \{u_1, u_2, u_3, u_4, u_5\}$. The corresponding adjacency matrix $\mathsf{A}(G)$ is given by

$$\mathsf{A}(G) = \begin{bmatrix} 0 & 1 & 2 & 0 & 0 \\ 1 & 0 & 1 & 0 & 0 \\ 2 & 1 & 0 & 1 & 0 \\ 0 & 0 & 1 & 1 & 0 \\ 0 & 0 & 0 & 0 & 0 \end{bmatrix}$$

Remark 4.11

- Note that for any graph G and any listing of its vertices, we have $a_{ij} = a_{ji}$ for all i and j, by definition. Equivalently, $\mathsf{A}(G)$ is a symmetric matrix; that is, it is equal to its transpose $\mathsf{A}(G)^t = \mathsf{A}(G)$. Moreover, any symmetric $n \times n$ matrix A with a_{ij} a nonnegative integer is an adjacency matrix for some graph G on n vertices.

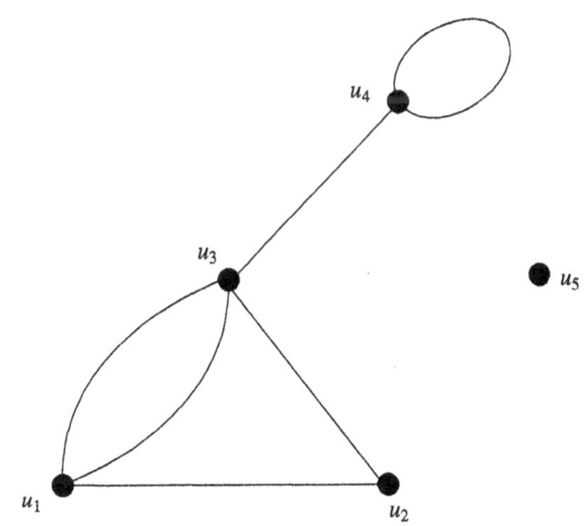

Figure 4.4: A graph G on five vertices has a 5×5 adjacency matrix.

- If G is a simple graph, then $\mathsf{A}(G)$ is a symmetric *binary matrix*—that is, symmetric where $a_{ij} \in \{0, 1\}$ for all i and j. In addition, $a_{ii} = 0$ for each i, since G has no loops.
- Note that each entry a_{ij} in $\mathsf{A}(G)$ counts how many edges are between u_i and u_j. Hence, we have

$$|E(G)| = \sum_{i \geq j} a_{ij} = \sum_{i \leq j} a_{ij}.$$

- For each i the degree $d_G(u_i)$ is the sum of all entries in row (or column) number i in $\mathsf{A}(G)$, where we count the entry a_{ii} on the diagonal twice. That is,

$$d_G(u_i) = \sum_{\ell=1}^{n} a_{i\ell} + a_{ii} = \sum_{\ell=1}^{n} a_{\ell i} + a_{ii}.$$

The notation $\mathsf{A}(G)$ for the adjacency matrix of G with respect to a given labeling $V(G) = \{u_1, \ldots, u_n\}$ of the vertices suggests that the matrix depends only on the graph itself. This is not quite true, since we need a fixed labeling of the vertices. Suppose we have another labeling $V(G) = \{u'_1, \ldots, u'_n\}$, where $u'_i = u_{\sigma(i)}$ for some fixed permutation

$$\sigma : \{1, \ldots, n\} \to \{1, \ldots, n\}.$$

If $\mathsf{A}(G)' = [a'_{ij}]_{i,j \in \{1,\ldots,n\}}$ is the adjacency matrix with respect to this labeling, then clearly

$$a'_{ij} = a_{\sigma(i)\sigma(j)}. \tag{4.1}$$

Let $\widetilde{\sigma}(i)$ denote the column vector with a 1 in row $\sigma(i)$ and zero everywhere else. We can define the square *permutation matrix*

$$\mathsf{P}_\sigma = [\widetilde{\sigma}(1) \mid \cdots \mid \widetilde{\sigma}(n)].$$

Here the vertical bars are used to separate columns. This matrix is *orthogonal*, and its inverse is given by its transpose—that is, $\mathsf{P}_\sigma^{-1} = \mathsf{P}_\sigma^t$. Moreover, if σ and ρ are two permutations on $\{1, \ldots, n\}$, then

$$\mathsf{P}_\sigma \mathsf{P}_\rho = \mathsf{P}_{\sigma\rho},$$

where $\sigma\rho$ denotes the usual composition of permutations (see Exercise 4.17). By (4.1) we have

$$\mathsf{A}(G)' = \mathsf{P}_\sigma^t \mathsf{A}(G) \mathsf{P}_\sigma = \mathsf{P}_\sigma^{-1} \mathsf{A}(G) \mathsf{P}_\sigma.$$

Definition 4.12 We say that two $n \times n$ matrices X and Y are *orthogonally equivalent* or simply *similar* if there is a permutation matrix P such that $\mathsf{Y} = \mathsf{P}^{-1}\mathsf{X}\mathsf{P}$.

The proof of the following observation is left as Exercise 4.18.

Observation 4.13 Orthogonal equivalence among $n \times n$ matrices is an equivalence relation.

If two graphs G and G' are isomorphic, then there is a bijective vertex map $f_1 : V(G) \to V(G')$, where there are equally many edges between u and u' in G as there are between $f_1(u)$ and $f_1(u')$ in G'. If $V(G) = \{u_1, \ldots, u_n\}$ and $V(G') = \{u'_1, \ldots, u'_n\}$, then there must be a permutation σ such that $f_1(u_i) = u'_{\sigma(i)}$ for each $i \in \{1, \ldots, n\}$. Hence, we get the next result.

Observation 4.14 The following statements are equivalent:

1. The graphs G and G' are isomorphic.
2. The adjacency matrices $\mathsf{A}(G)$ and $\mathsf{A}(G')$ are orthogonally equivalent with respect to any labeling of their vertices.

We present an example to illustrate how various matrix manipulations relate to graph-theoretic properties.

Example 4.15 Consider the graph G with its labeling as given in Example 4.10. By squaring the adjacency matrix $\mathsf{A}(G)$, we get the symmetric matrix

$$\mathsf{A}(G)^2 = \begin{bmatrix} 5 & 2 & 1 & 2 & 0 \\ 2 & 2 & 2 & 1 & 0 \\ 1 & 2 & 6 & 1 & 0 \\ 2 & 1 & 1 & 2 & 0 \\ 0 & 0 & 0 & 0 & 0 \end{bmatrix}.$$

We now interpret the entries of this resulting square matrix. For each i and j denote the entry in row i and column j in $\mathsf{A}(G)^2$ by $a_{ij}^{(2)}$. In general, $a_{ij}^{(k)}$ denotes the entry in row i and column j in $\mathsf{A}(G)^k$. By the rules of matrix multiplication, we have

$$a_{ij}^{(2)} = \sum_{\ell=1}^{n} a_{i\ell}a_{\ell j} = a_{i1}a_{1j} + a_{i2}a_{2j} + \cdots + a_{in}a_{nj}.$$

Observe that $a_{i\ell}a_{\ell j} \geq 1$ only if u_i is connected to u_ℓ and u_ℓ is connected to u_j in G. Otherwise, $a_{i\ell}a_{\ell j} = 0$. Therefore, $a_{ij}^{(2)}$ counts the number of walks of length two between u_i and u_j in G.

Similarly, we can consider the third power $\mathsf{A}(G)^3$ of the adjacency matrix of G.

$$\mathsf{A}(G)^3 = \begin{bmatrix} 4 & 6 & 14 & 3 & 0 \\ 6 & 4 & 7 & 3 & 0 \\ 14 & 7 & 5 & 7 & 0 \\ 3 & 3 & 7 & 3 & 0 \\ 0 & 0 & 0 & 0 & 0 \end{bmatrix}.$$

For any square matrix X, we have $\mathsf{X}^3 = \mathsf{X} \cdot \mathsf{X}^2$. So, in particular for $\mathsf{A}(G)$,

$$a_{ij}^{(3)} = \sum_{\ell=1}^{n} a_{i\ell}u_{\ell j}^{(2)} = a_{i1}a_{1j}^{(?)} + a_{i2}a_{2j}^{(2)} + \cdots + a_{in}a_{nj}^{(2)}.$$

We note that $a_{i\ell} \geq 1$ if, and only if, the vertex u_ℓ is a neighbor of u_i. Otherwise, $a_{i\ell} = 0$. For each neighbor u_ℓ of u_i, the entry $a_{\ell j}^{(2)}$ in $\mathsf{A}(G)^2$ counts the number of walks of length two between u_ℓ and u_j. Hence, $a_{ij}^{(3)}$ counts the number of walks from u_i to u_j of length three.

For example, $a_{12}^{(3)} = a_{21}^{(3)} = 6$ indicates six possible walks between u_1 and u_2 of length three in G. The reader is encouraged to look at Figure 4.4 and list all the possible six walks from u_1 to u_2 of length three.

Generalizing Example 4.15, we obtain the following theorem, the proof of which we leave as Exercise 4.20.

Theorem 4.16 Let G be a graph on the n vertices labeled as $V(G) = \{u_1, \ldots, u_n\}$. The entry $a_{ij}^{(k)}$ in $\mathsf{A}(G)^k$ is the number of distinct walks from u_i to u_j of length k in G.

4.4 The Incidence Matrix of a Graph

In this section we consider another important matrix representation of a graph G, the *incidence matrix* of G. Unlike the adjacency matrix, the incidence matrix is in general not a square matrix. In addition to a vertex labeling, the incidence matrix also needs an edge labeling of G. Strictly speaking, the incidence matrix contains more information than the adjacency matrix, since it distinguishes between multiple edges between two given vertices. The adjacency matrix does not.

Definition 4.17 Let $G = (V, E, \phi)$ be a graph with vertex labeling $V(G) = \{u_1, \ldots, u_n\}$ and edge labeling $E(G) = \{e_1, \ldots, e_m\}$. For each $i \in \{1, \ldots, n\}$ and $j \in \{1, \ldots, m\}$ define b_{ij} by

$$b_{ij} = |\{u_i\} \cap \phi(e_j)|.$$

That is, $b_{ij} = 1$ if u_i is an endvertex of e_j, and 0 otherwise. The *incidence matrix* of G with respect to these labelings of $V(G)$ and $E(G)$ is defined as the $n \times m$ binary matrix

$$\mathsf{B}(G) = [b_{ij}]_{i\in\{1,\ldots,n\},j\in\{1,\ldots,m\}}.$$

Example 4.18 Consider the graph G shown in Figure 4.5. Assume the vertices are labeled $V(G) = \{u_1, u_2, u_3, u_4, u_5\}$ and the edges $E(G) = \{e_1, e_2, e_3, e_4, e_5, e_6\}$. According to this labeling, we have that the incidence matrix is given by

$$\mathsf{B}(G) = \begin{bmatrix} 1&1&1&0&0&0 \\ 1&0&0&1&0&0 \\ 0&1&1&1&1&0 \\ 0&0&0&0&1&1 \\ 0&0&0&0&0&0 \end{bmatrix}.$$

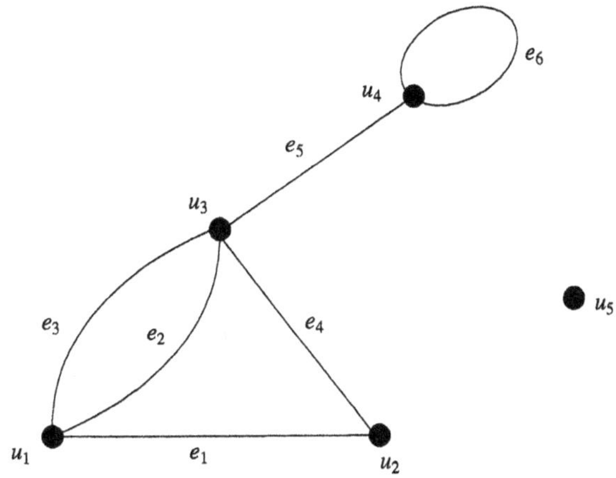

Figure 4.5: A graph G with five vertices and six edges has a 5×6 incidence matrix.

Remark 4.19

- Each column has either one or two 1's. Column j has exactly two 1's if the edge e_j has two distinct endvertices, and one 1 if e_j is a loop.
- The number of 1's in row i, where the 1's corresponding to a loop are counted twice, is equal to $d_G(u_i)$. In particular, if G has no loops, the number of 1's in row i is precisely $d_G(u_i)$.
- Any $n \times m$ binary matrix in which each column has one or two 1's is an incidence matrix for a graph on n vertices and m edges.

As with the adjacency matrix, the incidence matrix $\mathsf{B}(G)$ depends on the labelings of the vertices and the edges. Hence, two incidence matrices for a graph G differ only by a permutation of the rows and columns. To be precise, let G and G' be graphs with

$$\begin{aligned} V(G) &= \{u_1, \ldots, u_n\}, \\ E(G) &= \{e_1, \ldots, e_m\}, \\ V(G') &= \{u'_1, \ldots, u'_n\}, \\ E(G') &= \{e'_1, \ldots, e'_m\}. \end{aligned}$$

Assume $f = (f_1, f_2) : G \to G'$ is an isomorphism with

$$\begin{aligned} f_1(u_i) &= u'_{\sigma(i)}, \\ f_2(e_j) &= e'_{\rho(j)}, \end{aligned}$$

where σ and ρ are permutations on $\{1, \ldots, n\}$ and $\{1, \ldots, m\}$, respectively. Suppose P_σ and P_ρ are the corresponding $n \times n$ and $m \times m$ permutation matrices of σ and ρ, respectively.

The incidence matrices satisfy

$$\mathsf{B}(G') = \mathsf{P}_\sigma^t \mathsf{B}(G) \mathsf{P}_\rho.$$

We summarize in the following observation.

Observation 4.20 Two graphs G and G' on n vertices and m edges are isomorphic if, and only if, there are $n \times n$ and $m \times m$ permutation matrices P and Q, respectively, such that

$$\mathsf{B}(G') = \mathsf{P}^t \mathsf{B}(G) \mathsf{Q}.$$

As for adjacency matrices, this defines an equivalence relation among $n \times m$ matrices (see Exercise 4.27).

Let G be a loopless graph and let $\mathsf{B}(G)$ be the incidence matrix of G with respect to a fixed labeling of the n vertices and m edges of G. Since G has no loops, there are exactly two 1's in each column of $\mathsf{B}(G)$, one for each endvertex of the edge corresponding to that column. Let $\mathsf{B}_{-1}(G)$ be a *digraph incidence matrix* derived from $\mathsf{B}(G)$ by replacing one of the 1's in each column with a -1. In this way we have that each column of $\mathsf{B}_{-1}(G)$ has one 1 and one -1, and the rest 0's. Note that in this way, $\mathsf{B}_{-1}(G)$ is really an incidence matrix of the digraph $\vec{G}$ obtained by directing the edges of G in such a way that -1 indicates the head and $+1$ indicates the tail of the directed edge.

Recall that the *rank* of an $n \times m$ matrix X with real entries, denoted by $\text{rank}(\mathsf{X})$, is the dimension of its row space over the real numbers, or of its column space, since they are equal. We note a few things about the digraph incidence matrix and its rank.

- If P and Q are $n \times n$ and $m \times m$ permutation matrices, respectively, then $\mathsf{P}^t \mathsf{B}_{-1}(G) \mathsf{Q}$ has exactly one 1 and one -1 in each column, and is hence also a digraph incidence matrix for G, where the vertices and edges have been labeled in a different way.

- If P and Q are $n \times n$ and $m \times m$ permutation matrices, respectively, then for any $n \times m$ matrix X (Exercise 4.28)

$$\text{rank}(\mathsf{P}^t \mathsf{X} \mathsf{Q}) = \text{rank}(\mathsf{X}).$$

- Since the sum of the entries in each column of $\mathsf{B}_{-1}(G)$ is zero, the sum of all the rows is the zero vector. Therefore, the row space has dimension at most $n - 1$, that is,

$$\text{rank}(\mathsf{B}_{-1}(G)) \leq n - 1.$$

- If G is a disconnected graph, say $G = G_1 \cup G_2$, then, by labeling first the vertices and edges of G_1 and then the vertices and edges of G_2, we have that $\mathsf{B}_{-1}(G)$ has the following form:

$$\mathsf{B}_{-1}(G) = \begin{bmatrix} \mathsf{B}_{-1}(G_1) & 0 \\ 0 & \mathsf{B}_{-1}(G_2) \end{bmatrix}.$$

  Since the rank of each of $\mathsf{B}_{-1}(G_1)$ and $\mathsf{B}_{-1}(G_2)$ is at most one less than the number of the vertices of G_1 and G_2, respectively, we see that $\text{rank}(\mathsf{B}_{-1}(G)) \leq n - 2$ in this case, when G is disconnected.

- Assume that G has a cycle c. By suitable permutations of the vertices and edges, we may assume that $c = (u_1, e_1, u_2, e_2, \ldots, e_k, u_{k+1})$, where $k \geq 2$ and $u_{k+1} = u_1$. Denoting the ith column by C_i, we have in any case that $C_1 \pm C_2 \pm \cdots \pm C_k = 0$. Hence, the rank of any $n \times k'$ submatrix of $\mathsf{B}_{-1}(G)$ is at most $k' - 1$, if the corresponding edges contain a cycle.

From the previous comments, we have that the rank of *any* submatrix of $\mathsf{B}_{-1}(G)$ is at most $n - 1$. Further, in order for an $n \times (n-1)$ submatrix of $\mathsf{B}_{-1}(G)$ to have rank $n - 1$, the corresponding $n - 1$ edges must form a connected graph and also form no cycles; in other words, they must form a spanning subtree of G. So, a natural question at this point is: Does an $n \times (n-1)$ submatrix of $\mathsf{B}_{-1}(G)$, where the edges form a spanning tree of G, have rank $n - 1$?

Theorem 4.21 Let G be a loopless graph and let $\mathsf{B}_{-1}(G)$ be the digraph incidence matrix of G with respect to a fixed labeling of the n vertices and m edges of G. For any $n \times (n-1)$ submatrix B' of $\mathsf{B}_{-1}(G)$ the following are equivalent:

1. $\text{rank}(\mathsf{B}') = n - 1$.
2. The $n - 1$ edges corresponding to the columns of B' form a spanning tree of G.

Proof By the previous discussion, it suffices to show that if the corresponding $n-1$ edges form a spanning tree of G, then $\text{rank}(\mathsf{B}') = n - 1$.

We proceed by induction on $n \geq 1$. Assume that $e_1, \ldots, e_{n-1}$ form a spanning tree T of G. By a suitable labeling of the vertices and edges of G, we can assume that u_1 is a leaf of T that is connected to u_2 by the edge e_1. In this case B' has a ± 1 in the $(1, 1)$-th entry, and 0 elsewhere in the first row. Since the sum of all the rows in B' is zero, it suffices to consider the $(n-1) \times (n-1)$ submatrix B'' of B', where the nth row has been removed from B' (in fact, we can remove *any* ith row, where $2 \leq i \leq n$). It suffices to show that $\det(\mathsf{B}'') \neq 0$. Expanding the determinant along the first row, we obtain

$$\det(\mathsf{B}'') = \pm \det(\mathsf{B}''_{1,1}),$$

where $\mathsf{B}''_{1,1}$ is the $(n-2) \times (n-2)$ matrix obtained from B'' by removing the first row and column. Since now $\mathsf{B}''_{1,1}$ corresponds to the tree on $e_2, \ldots, e_{n-1}$ (except for the last row), it has by the induction hypothesis the rank of $n - 2$ and hence a nonzero determinant. By induction we have the theorem. ∎

From the proof of Theorem 4.21 we obtain a slightly stronger statement, which we state as a corollary.

Corollary 4.22 For an $(n-1) \times (n-1)$ submatrix B'' of $\mathsf{B}_{-1}(G)$, let E'' be the set of $n - 1$ edges of G corresponding to the columns of B''. In this case we have

$$\det(\mathsf{B}'') = \begin{cases} \pm 1 & \text{if } E'' \text{ forms a spanning tree of } G, \\ 0 & \text{otherwise.} \end{cases}$$

4.5 The Matrix-Tree Theorem

This section presents an important interpretation of the adjacency matrix of a simple graph G on n vertices. Assume that G has m edges, unless otherwise stated.

As mentioned in §4.4, when a graph G has no loops, each row of the digraph incidence matrix $\mathsf{B}_{-1}(G)$ has precisely one 1 and one -1 in each column. In addition, if G has no multiple edges, then the adjacency matrix $\mathsf{A}(G)$ is also a binary matrix. Hence, given a simple graph, it is natural to ask about the relationship between the digraph incidence matrix and the adjacency matrix. We begin with some definitions.

Definition 4.23 Let G be a graph and $V(G) = \{u_1, \ldots, u_n\}$ be a labeling of its vertices. The diagonal matrix $\mathsf{D}(G)$, where the ith diagonal entry is the degree of u_i, is called the *degree matrix* of G with respect to the labeling $V(G)$.

Viewing the digraph incidence matrix $\mathsf{B}_{-1}(G)$ from §4.4 as a row matrix

$$\mathsf{B}_{-1}(G) = \begin{bmatrix} \widetilde{b}_1 \\ \widetilde{b}_2 \\ \vdots \\ \widetilde{b}_n \end{bmatrix},$$

we see that the (i, j)-th entry in the $n \times n$ matrix $\mathsf{B}_{-1}(G) \cdot \mathsf{B}_{-1}(G)^t$ is given by the dot product $\widetilde{b}_i \cdot \widetilde{b}_j$. In particular, we have that the ith diagonal entry is $d_G(u_i)$ for each vertex u_i of G. Also, if $i \neq j$, then $\widetilde{b}_i \cdot \widetilde{b}_j = 0$, if the vertices u_i and u_j are not connected. However, $\widetilde{b}_i \cdot \widetilde{b}_j$ equals $-$(the number of edges connecting u_i and u_j). From this we have the following relation between the adjacency matrix $\mathsf{A}(G)$ and the digraph incidence matrix $\mathsf{B}_{-1}(G)$ of G.

Observation 4.24 For a loopless graph G, we have

$$\mathsf{A}(G) + \mathsf{B}_{-1}(G) \cdot \mathsf{B}_{-1}(G)^t = \mathsf{D}(G).$$

The following *Binet-Cauchy Theorem* is needed in order to prove some interesting properties of the matrix $\mathsf{D}(G) - \mathsf{A}(G)$ from Observation 4.24.

Theorem 4.25 **Binet-Cauchy**

Let $m, n \in \mathbb{N}$ where $m \geq n$. Let X be an $n \times m$ matrix and Y an $m \times n$ matrix. For each $S = \{i_1, \ldots, i_n\} \subseteq \{1, \ldots, m\}$, let X_S (respectively, Y_S) be the $n \times n$ square matrix obtained by choosing the columns (respectively, rows) number i_1 through i_n from X (respectively, Y). In this case we have

$$\det(\mathsf{XY}) = \sum_{S \subseteq \{1,\ldots,m\}} \det(\mathsf{X}_S) \det(\mathsf{Y}_S),$$

where the sum is taken over all $\binom{m}{n}$ subsets S of $\{1, \ldots, m\}$ that contain precisely n elements.

Before we prove the Binet-Cauchy Theorem, we give an example to illustrate how it works.

Example 4.26 Let

$$\mathsf{X} = \begin{bmatrix} 1 & 2 & -1 \\ 0 & 1 & 3 \end{bmatrix} \text{ and } \mathsf{Y} = \begin{bmatrix} 0 & 2 \\ 4 & -1 \\ 5 & 0 \end{bmatrix}.$$

In this case $n = 2$, $m = 3$, and $\binom{3}{2} = 3$. So, we have that $\det(\mathsf{XY})$ is given by the following:

$$\begin{aligned} \det(\mathsf{XY}) &= \det\begin{bmatrix} 1 & 2 \\ 0 & 1 \end{bmatrix} \cdot \det\begin{bmatrix} 0 & 2 \\ 4 & -1 \end{bmatrix} \\ &\quad + \det\begin{bmatrix} 1 & -1 \\ 0 & 3 \end{bmatrix} \cdot \det\begin{bmatrix} 0 & 2 \\ 5 & 0 \end{bmatrix} \\ &\quad + \det\begin{bmatrix} 2 & -1 \\ 1 & 3 \end{bmatrix} \cdot \det\begin{bmatrix} 4 & -1 \\ 5 & 0 \end{bmatrix} \\ &= 1 \cdot (-8) + 3 \cdot (-10) + 7 \cdot 5 \\ &= -3. \end{aligned}$$

Now, note that

$$\mathsf{XY} = \begin{bmatrix} 1 & 2 & -1 \\ 0 & 1 & 3 \end{bmatrix} \cdot \begin{bmatrix} 0 & 2 \\ 4 & -1 \\ 5 & 0 \end{bmatrix} = \begin{bmatrix} 3 & 0 \\ 19 & -1 \end{bmatrix},$$

and hence

$$\det(\mathsf{XY}) = \det\begin{bmatrix} 3 & 0 \\ 19 & -1 \end{bmatrix} = -3,$$

which is consistent with what we obtained earlier.

We will use the following technical lemma to prove the Binet-Cauchy Theorem. It can, in fact, be viewed as a generalization of the Binet-Cauchy Theorem.

Lemma 4.27 Let X and Y be $n \times m$ and $m \times n$ matrices, respectively. Let $T \subseteq \{1, \ldots, m\}$ and let P_T be the $m \times m$ diagonal matrix whose i^{th} diagonal entry is 1 if $i \in T$ and 0 otherwise. Let M be the $(m + n) \times (m + n)$ matrix defined by

$$\mathsf{M} = \begin{bmatrix} -\mathsf{P}_T & \mathsf{Y} \\ \mathsf{X} & \mathsf{0} \end{bmatrix}.$$

Let $\overline{T} = \{1, \ldots, m\} \setminus T$ be the complement of T. Using the notation from Theorem 4.25,

$$\det(\mathsf{M}) = (-1)^m \sum_{\overline{T} \subseteq S \subseteq \{1,\ldots,m\}} \det(\mathsf{X}_S)\det(\mathsf{Y}_S),$$

where the sum is taken over all $\binom{|T|}{n-|\overline{T}|}$ subsets S of $\{1, \ldots, m\}$ that contain $\overline{T}$ and have exactly n elements.

Proof First note that if $n > m$, then the rows from $m+1$ through $m+n$ in M are linearly dependent. Also if $|T| < m-n$, then the rows (or columns) in M corresponding to $\overline{T}$ are linearly dependent as well. Hence, in these cases the lemma holds, since the sum is the empty sum equals zero.

Otherwise we have $0 < m - n \leq |T|$, so we can pick an $i \in T$. By expanding the determinant of M by any row (or column) numbered $i \in T$, we get

$$\det(\mathsf{M}) = -\det\begin{bmatrix} -\mathsf{P}_{T;i} & \mathsf{Y}_i \\ \mathsf{X}_i & 0 \end{bmatrix} + \det\begin{bmatrix} -\mathsf{P}_{T'} & \mathsf{Y} \\ \mathsf{X} & 0 \end{bmatrix}.$$

Here the first matrix has dimension $(m+n-1) \times (m+n-1)$, where $\mathsf{P}_{T;i}$ is the $(m-1) \times (m-1)$ matrix obtained from P_T by removing row and column number i, and X_i (respectively, Y_i) is obtained from X (respectively, Y) by removing column (respectively, row) number i from X (respectively, Y). Likewise, the second matrix has dimension $(m+n) \times (m+n)$, where $T' = T \setminus \{i\}$.

We proceed by induction on $m + |T|$ and obtain

$$\begin{aligned} \det(\mathsf{M}) &= -(-1)^{m-1} \sum_{\overline{T} \subseteq S_1 \subseteq \{1,\dots,m\} \setminus \{i\}} \det(\mathsf{X}_{S_1}) \det(\mathsf{Y}_{S_1}) \\ &+ (-1)^m \sum_{\overline{T} \cup \{i\} \subseteq S_2 \subseteq \{1,\dots,m\}} \det(\mathsf{X}_{S_2}) \det(\mathsf{Y}_{S_2}). \end{aligned}$$

Here S_1 ranges over all subsets of $\{1, \dots, m\} \setminus \{i\}$ that contain $\overline{T}$ as a subset and have precisely n elements, and S_2 ranges over all subsets of $\{1, \dots m\}$ that contain $\overline{T} \cup \{i\}$ as a subset and have precisely n elements. Since each $S \subseteq \{1, \dots, m\}$ that contains $\overline{T}$ as a subset and has precisely n elements either contains i or not, we obtain that

$$\det(\mathsf{M}) = (-1)^m \sum_{\overline{T} \subseteq S \subseteq \{1,\dots,m\}} \det(\mathsf{X}_S) \det(\mathsf{Y}_S),$$

which completes our proof. ■

The proof of Theorem 4.25 now follows from Lemma 4.27.

Proof (Theorem 4.25) Note that

$$\begin{bmatrix} \mathsf{I}_m & 0 \\ \mathsf{X} & \mathsf{I}_n \end{bmatrix} \cdot \begin{bmatrix} -\mathsf{I}_m & \mathsf{Y} \\ \mathsf{X} & 0 \end{bmatrix} = \begin{bmatrix} -\mathsf{I}_m & \mathsf{Y} \\ 0 & \mathsf{XY} \end{bmatrix},$$

where I_α is the $\alpha \times \alpha$ identity matrix for $\alpha = m, n$. Therefore,

$$\det\begin{bmatrix} -\mathsf{I}_m & \mathsf{Y} \\ \mathsf{X} & 0 \end{bmatrix} = \det\begin{bmatrix} -\mathsf{I}_m & \mathsf{Y} \\ 0 & \mathsf{XY} \end{bmatrix} = (-1)^m \det(\mathsf{XY}),$$

and hence

$$\det(\mathsf{XY}) = (-1)^m \det\begin{bmatrix} -\mathsf{I}_m & \mathsf{Y} \\ \mathsf{X} & 0 \end{bmatrix}. \tag{4.2}$$

Applying Lemma 4.27 on the right-hand side of (4.2) completes the proof. ■

The Matrix-Tree Theorem describes an important property of the $n \times n$ matrix $\mathsf{D}(G) - \mathsf{A}(G)$ for any simple graph G on n vertices.

Recall that for an $n \times n$ matrix X, the (i, j)-th *cofactor* of M is given by $(-1)^{i+j}\det(\mathsf{X}_{i,j})$, where X_{ij} is the matrix obtained from X by removing row i and column j.

Theorem 4.28 **Matrix-Tree**

For a loopless graph G, all of the cofactors of $\mathsf{D}(G) - \mathsf{A}(G)$ are equal to $\tau(G)$, the number of spanning trees of G.

Proof By Observation 4.24 we have that

$$\mathsf{D}(G) - \mathsf{A}(G) = \mathsf{B}_{-1}(G) \cdot \mathsf{B}_{-1}(G)^t.$$

From this we note that the $(i, i)^{\text{th}}$ cofactor of $\mathsf{D}(G) - \mathsf{A}(G)$ is obtained by the matrix product $\mathsf{B}_{-1;i}(G) \cdot \mathsf{B}_{-1;i}(G)^t$, where $\mathsf{B}_{-1;i}(G)$ is obtained from $\mathsf{B}_{-1}(G)$ by removing the ith row. By the Binet-Cauchy Theorem 4.25 we have that

$$\det(\mathsf{B}_{-1;i}(G) \cdot \mathsf{B}_{-1;i}(G)^t) = \sum \det(\mathsf{B}') \cdot \det(\mathsf{B}'^t),$$

where B' is a nonsingular $(n-1) \times (n-1)$ submatrix of $\mathsf{B}_{-1;i}(G)$. By Theorem 4.21 there are precisely $\tau(G)$ such summands, and each summand is equal to $(\pm 1)^2 = 1$. Therefore, each (i, i)-th cofactor of $\mathsf{D}(G) - \mathsf{A}(G)$ is equal to $\tau(G)$.

To show that every (i, j)-th cofactor is equal to $\tau(G)$, we may for simplicity assume that $j > i$. First note that the sum of all the rows in $\mathsf{T} = \mathsf{D}(G) - \mathsf{A}(G)$ is the zero vector. Consider the submatrices $\mathsf{T}_{i,j}$ and $\mathsf{T}_{j,j}$ obtained from T by removing row i and column j, and row j and column j, respectively. The corresponding cofactors are

$$\begin{aligned} C_{i,j} &= (-1)^{i+j}\det(\mathsf{T}_{i,j}), \\ C_{j,j} &= (-1)^{j+j}\det(\mathsf{T}_{j,j}) = \det(\mathsf{T}_{j,j}). \end{aligned}$$

Recall that a determinant remains unaltered when a multiple of one row is added to another row. Hence, adding rows number $\ell \in \{1, \ldots, n\} \setminus \{i, j\}$ to the jth row in T, and thereby obtaining T', will alter neither the (j, j)-th cofactor nor the (i, j)-th cofactor, that is to say, the (j, j)-th and the (i, j)-th cofactors $C'_{j,j}$ and $C'_{i,j}$ of T' satisfy

$$C'_{j,j} = C_{j,j} \quad \text{and} \quad C'_{i,j} = C_{i,j}.$$

Since all the rows in T add up to zero, the jth row of T' equals the negative of the ith row of T'. This means precisely that $\mathsf{T}'_{i,j}$ can be obtained from $\mathsf{T}'_{j,j}$ by first multiplying the ith row of $\mathsf{T}'_{j,j}$ by -1, and then moving it downward $(j-i)-1$ steps—that is, by $(j-i)-1$ interchanges of rows. Since each interchange reverses the sign of the determinant, we finally have that

$$\begin{aligned} C_{i,j} &= C'_{i,j} \\ &= (-1)^{i+j}\det(\mathsf{T}'_{i,j}) \\ &= (-1)^{i+j}((-1)^{(j-i)-1}(-\det(\mathsf{T}'_{j,j}))) \\ &= (-1)^{i+j}((-1)^{j-i-1}(-C'_{j,j})) \\ &= C_{j,j}, \end{aligned}$$

which completes the proof of the theorem. ■

If we let $\mathsf{T}(G) = \mathsf{D}(G) - \mathsf{A}(G)$, then the formula

$$\tau(G) = (-1)^{i+j} \det(\mathsf{T}(G)_{i,j})$$

implied by Theorem 4.28 is called the *Matrix-Tree Formula*. This gives us a second method to compute the number of spanning trees in any loopless graph. Recall that the first method was given in Theorem 4.4. Moreover, the Matrix-Tree Formula is mechanical and well suited for computer computations when graphs have been implemented as adjacency or incidence matrices.

The following example gives us an alternative way to prove Cayley's Theorem 4.7.

Example 4.29 For the complete graph K_n on $n \geq 2$ vertices, we have that

$$\mathsf{D}(K_n) - \mathsf{A}(K_n) = \begin{bmatrix} n-1 & -1 & -1 & \cdots & -1 \\ -1 & n-1 & -1 & \cdots & -1 \\ -1 & -1 & n-1 & \cdots & -1 \\ \vdots & \vdots & \vdots & \ddots & \vdots \\ -1 & -1 & -1 & \cdots & n-1 \end{bmatrix}.$$

By the Matrix-Tree Theorem 4.28, we have that $\tau(K_n)$ is equal to any cofactor of the preceding matrix. Considering the $(1, 1)$-th cofactor $C_{1,1}$ and then subtracting the first row from all the remaining rows, we get that $\tau(K_n)$ is given by the following $(n-1) \times (n-1)$ determinant:

$$\tau(K_n) = \det \begin{bmatrix} n-1 & -1 & \cdots & -1 \\ -1 & n-1 & \cdots & -1 \\ \vdots & \vdots & \ddots & \vdots \\ -1 & -1 & \cdots & n-1 \end{bmatrix}$$

$$= \det \begin{bmatrix} n-1 & -1 & \cdots & -1 \\ -n & & & \\ \vdots & & n \cdot \mathsf{I}_{n-2} & \\ -n & & & \end{bmatrix},$$

where I_{n-2} denotes the $(n-2) \times (n-2)$ identity matrix. Adding columns $2, \ldots, n-1$ to the first column, we finally get

$$\tau(K_n) = \det \begin{bmatrix} 1 & -1 & \cdots & -1 \\ 0 & & & \\ \vdots & & n \cdot \mathsf{I}_{n-2} & \\ 0 & & & \end{bmatrix} = n^{n-2}.$$

This completes the alternative proof of Cayley's Theorem.

4.6 An Application to Electrical Networks

In this section we discuss an application for electrical network systems, where the total resistance between two points in the network system is found. While prior knowledge in electrical physics is not assumed, such knowledge will help the reader appreciate this section. We provide a brief overview of the necessary background.

We consider the type of networks displayed in Figure 4.6. Our electrical circuits are provided with a power supply having a given potential between two nodes u and v, denoted by V_{uv}, or simply V when there is no danger of ambiguity. We assume the electrical devices are connected to a power supply in an arbitrary fashion with combined resistance R. In this case the total electrical current I flowing from one node to the other is given by the formula

$$I = \frac{V_{uv}}{R}.$$

Hence, if we can compute the total resistance R, we can calculate the total flow I of electrical current. There are two basic rules to calculate the total resistance between two nodes u and v in a given network. First, if electrical devices are connected in a series as shown in Figure 4.7, then the total resistance R_t is given by the sum of all the resistors:

$$R_t = R_1 + R_2 + \cdots + R_n. \tag{4.3}$$

Second, if the resistors are connected in parallel as shown in Figure 4.8, then the total resistance R_t satisfies the formula

$$\frac{1}{R_t} = \frac{1}{R_1} + \frac{1}{R_2} + \cdots + \frac{1}{R_n}. \tag{4.4}$$

Any introductory book on physics explains why (4.3) and (4.4) hold. There are many other physical phenomena that behave in a similar fashion, but for simplicity's sake we shall restrict ourselves to networks involving only a potential, electrical current, and resistance.

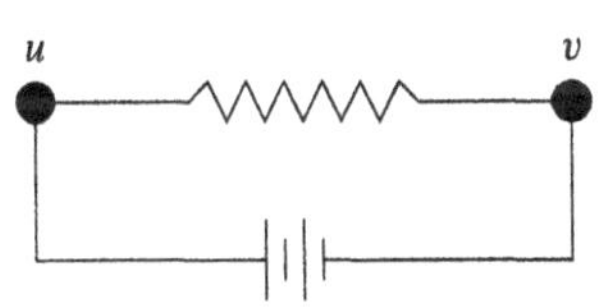

Figure 4.6: An electrical circuit with a given potential and a system of resistors, shown as one combined resistor.

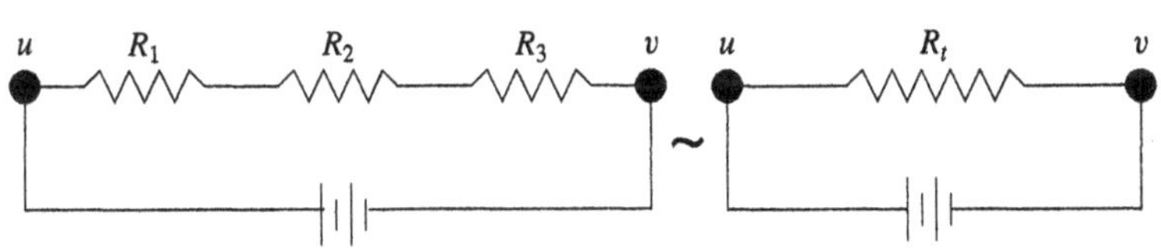

Figure 4.7: Resistors connected in a series.

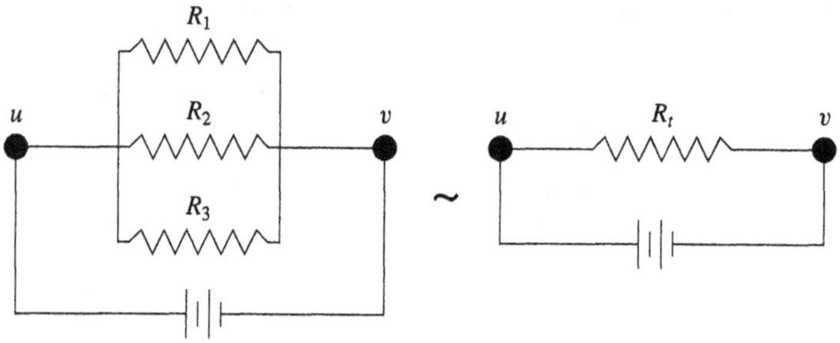

Figure 4.8: Resistors connected in parallel.

The salient point to notice is that the total resistance of many resistor network systems can be calculated by repeated use of (4.3) and (4.4). However, in countless other systems they fail to be of any use without some geometrical insight or physical tricks. The following theorem, proved in Thomassen [52, Proposition 2.3], can be used to compute the total resistance in a completely mechanical way between two nodes u and v in many cases without requiring any prior knowledge in physics.

Theorem 4.30 Let G be a (general) connected loopless graph. Let u and v be adjacent vertices of G. Assume that each edge in G corresponds to the resistance of one ohm (1 Ω). In this case the total resistance of the network between u and v is given by

$$R_t = \frac{\tau_{uv}(G)}{\tau(G)},$$

where $\tau_{uv}(G)$ is the number of spanning trees that include the edge $\{u, v\}$.

Note 4.31 As seen from the proof of Theorem 4.4, the number $\tau(G \cdot e)$ is precisely the number of spanning trees of G that contain the edge e. Hence, from Theorem 4.4, we can write R_t as

$$R_t = \frac{\tau(G \cdot \{u, v\})}{\tau(G)} = \frac{\tau(G \cdot \{u, v\})}{\tau(G \cdot \{u, v\}) + \tau(G \setminus \{u, v\})}.$$

Example 4.32 Consider the graph G shown on the left in Figure 4.9 and the corresponding electrical network, where each edge represents a resistance of one ohm.

Here we can easily compute the total resistance between u and v using the classical formulas (4.3) and (4.4). By these the total resistance R_t satisfies

$$\frac{1}{R_t} = \frac{1}{1+1} + \frac{1}{1},$$

and hence we get $R_t = 2/3\ \Omega$.

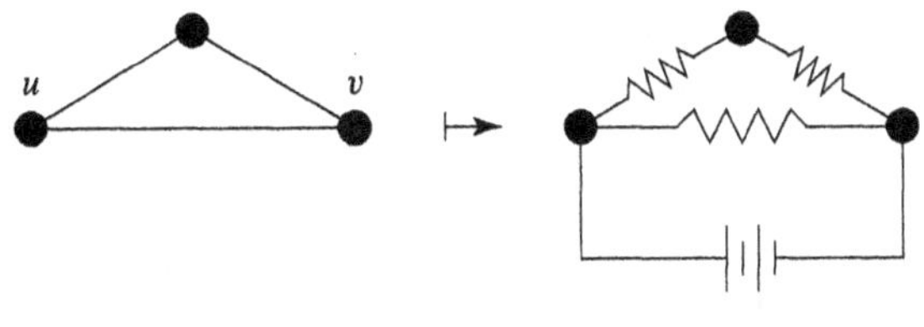

Figure 4.9: A graph and the corresponding electrical network.

Since the vertices u and v are neighbors in G, by Theorem 4.30, we get

$$R_t = \frac{\tau_{uv}(G)}{\tau(G)} = \frac{2}{3},$$

which is in complete agreement with the classical formulas.

Theorem 4.30 gives the resistance between two points, provided that the corresponding vertices in the graph are neighbors. If the vertices u and v are not neighbors, then in addition to Theorem 4.30 we use (4.4), as shown in the following example.

Example 4.33 Consider the graph G on the left in Figure 4.10. The vertices u and v are not adjacent. A direct application of the classical formulas will not be sufficient.

Adding the edge $e = \{u, v\}$, we obtain the graph G' and the corresponding electrical network displayed on the right in Figure 4.10. Let G'' be the graph we obtain from G by collapsing u and v into one vertex. Now, if R_t is the total resistance of the network corresponding to G and R_t' is the total resistance corresponding to G', we have by (4.4) that

$$\frac{1}{R_t'} = \frac{1}{R_t} + \frac{1}{1}.$$

Since $R_t' = \tau_{uv}(G')/\tau(G')$ by Theorem 4.30, we have by Theorem 4.4 that

$$R_t = \frac{\tau_{uv}(G')}{\tau(G') - \tau_{uv}(G')} = \frac{\tau_{uv}(G')}{\tau(G' \setminus \{u, v\})} = \frac{\tau(G'')}{\tau(G)}.$$

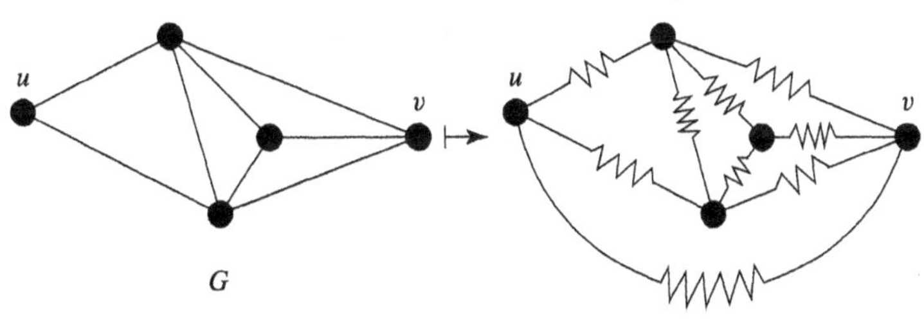

Figure 4.10: A graph G and a corresponding electrical network.

We have observed that Theorem 4.30 reduces the tricky problem of determining the total resistance between two given points of a large class of electrical networks to that of computing the number of spanning trees of corresponding graphs. The Matrix-Tree Theorem 4.28 reduces the problem of computing the number of spanning trees of graphs to the problem of evaluating determinants. Since there are many efficient ways of evaluating determinants of square matrices, we have a mechanical and efficient way of computing the total resistance of general electrical networks, where each edge of the corresponding graphs represents one ohm.

What about general networks where the edges represent different resistance? In this case there is a partial algorithmic solution. Suppose we have an electrical network of resistors where each resistor can represent any a ohms, where $a \in \mathbb{N}$. Looking at the corresponding graph for the network, replace each edge representing a ohms with a simple path of length a (see Figure 4.11). By this subdivision all edges represent a resistance of one ohm, and the total resistance between u and v is a Ω.

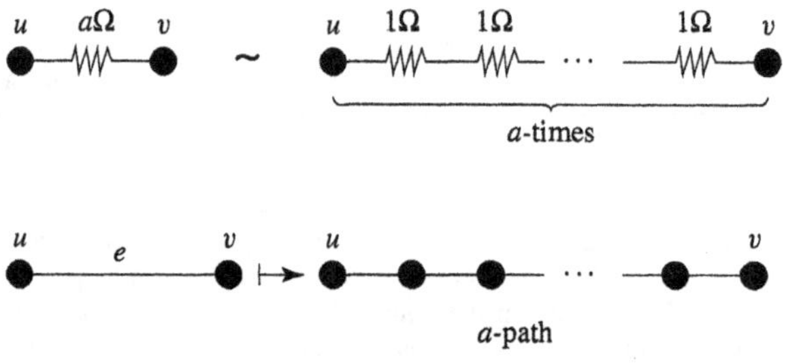

Figure 4.11: An a-Ω resistor corresponds to a subdivided edge.

Suppose now that we have an electrical network of resistors where each resistor can represent any $1/b$ Ω, where $b \in \mathbb{N}$. Looking at the corresponding graph for the network, replace each edge representing $1/b\Omega$ with b copies of parallel edges (see Figure 4.12). The same graph would be obtained by adding $b - 1$ parallel edges to the one edge representing the $1/b$ Ω resistor. By this procedure of generating a multigraph from our original graph, all edges in this multigraph represent a resistance of one ohm, and the total resistance between u and v is $1/b$ Ω.

We now combine these two methods. In a network of resistors, where each resistor can represent any rational resistance $a/b\Omega$, replacing the corresponding edge by b parallel a-paths (see Figure 4.13) yields a graph where each edge corresponds to one ohm, and the resistance between u and v is $a/b\Omega$.

We summarize with the following observation.

Observation 4.34 The total resistance of an electrical network of rational resistors can be computed by applying the Matrix-Tree Theorem to the "altered" corresponding graph.

Note 4.35 The procedure described for computing the total resistance of a given electrical network of resistors does not provide an *exact* solution if some resistors have irrational resistance x ohms, where $x \in \mathbb{R} \setminus \mathbb{Q}$, although we are able to compute the real resistance to arbitrary

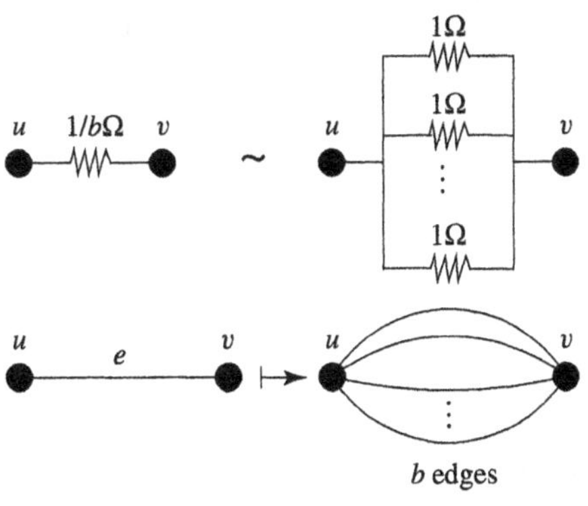

Figure 4.12: An $1/b\Omega$ resistor corresponds to b parallel edges.

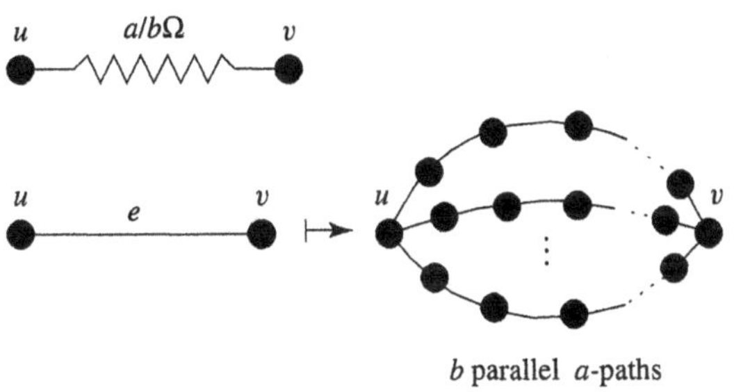

Figure 4.13: An $a/b\Omega$ resistor corresponds to b parallel a-paths.

precision. However, from a practical point of view, our method does provide a computational method for any real-world network, since every resistor is given by a floating number of finitely many decimal (or binary) digits that represents a rational number.

A more detailed treatment of the analysis of electrical networks can be found in Deo [11].

4.7 Minimum Cost Spanning Trees

So far, we have discussed spanning trees only in graphs where, between every pair of vertices, there is either an edge or not. We have not taken any cost or *weight* of the edges into account.

We conclude this chapter by briefly discussing how to find the *minimum cost spanning tree* in a graph G. After some preliminary definitions, we present the problem formally.

Definition 4.36 Let G be a graph and $W : E(G) \to \mathbb{R}$ a function.

1. The ordered tuple (G, W) is called a *weighted graph*. The function W is called a *weight* or *weight function* of (G, W).
2. If G' is a subgraph of G, then

$$W(G') = \sum_{e \in E(G')} W(e)$$

is called the *weight* of G' in G.

When there is no possibility of ambiguity, we talk about the weighted graph G instead of the tuple (G, W). In most applications we also assume that $W(e) \geq 0$ for all $e \in E(G)$.

For a weighted connected graph G, when we ask for a minimum cost connected subgraph $G' \subseteq G$ that covers all the vertices, $V(G') = V(G)$, we can assume that G' has no cycle, since the connectedness of G' is not altered by removing one edge from the cycle of G' when $V(e) \geq 0$. Repeating this process, removing one edge from each cycle of G', we end up with a connected subtree of G.

Definition 4.37 For a connected weighted graph (G, W), a spanning tree T with the minimum weight $W(T)$ is called a *minimum cost spanning tree.*

Example 4.38 Suppose we have a collection of n cities $c_1, \ldots, c_n$ in a certain state, where the cost of building roads between the cities varies. Assume we have the task of building a road system of lowest possible cost, such that between any two distinct cities c_i and c_j, there is always a road (not necessarily a direct road) from c_i to c_j. This road might go through some other cities on the way. Denote the cost, in millions of dollars, of building the road between c_i and c_j, by w_{ij}. Evidently for some $i \neq j$ the cost w_{ij} could be extremely high, due perhaps to high mountains or to wet swamp land between the cities. In some cases, due to governmental regulations regarding national parks or whatnot, it might in fact be impossible to build a direct road between c_i and c_j, in which case we set $w_{ij} = \infty$.

Here, the cheapest road system would be given by a minimum cost spanning tree for the weighted complete graph (K_n, W), where the vertices and the weight function W are given by

$$\begin{aligned} V(K_n) &= \{c_1, \ldots, c_n\}, \\ W(\{c_i, c_j\}) &= w_{ij}. \end{aligned}$$

In the cases where $w_{ij} = \infty$, since ∞ cannot be implemented in a precise manner in any computer program, we see that it suffices to choose $w_{ij} = L$ for all such pairs, where L is a fixed number that satisfies

$$L \geq \sum_{w_{ij} < \infty} |w_{ij}|,$$

where the sum is taken over all proper (that is, "non"-infinity) weights w_{ij}.

The following algorithm, called KRUSKAL'S ALGORITHM, computes a minimum cost spanning tree in any connected weighted graph.

KRUSKAL'S ALGORITHM
`Input:` A connected weighted graph (G, W) on n vertices.
`Output:` A minimum cost spanning tree T of G.
begin
$T_1 = \varnothing$;
for $i = 1$ **to** $n - 1$ **do** {
let $e_i \in E(G) \setminus E(T_i)$ be a minimum weight edge such
that $T_i \cup \{e_i\}$ is a forest;
$T_{i+1} = T_i \cup \{e_i\}$;   // e_i with its other endpoint added
}
output $T = T_n$;
end.

We consider an example of KRUSKAL'S ALGORITHM.

Example 4.39 Consider the weighted graph G with five vertices shown in Figure 4.14. Note that some pairs of edges are not connected, in which case we let the weight of the corresponding "imaginary" edges be infinity ∞. The weights are given in the symmetric 5×5 Table 4.1. In the table the (i, j)-th entry denotes the weight $W(\{u_i, u_j\})$. Note that in the case of multiple edges between a pair of vertices, we list only one edge with the minimum cost among all the edges between these vertices. Also, the weight of a loop contributes nothing and is therefore not listed.

KRUSKAL'S ALGORITHM begins by selecting a minimum cost edge to include in the forest it is constructing. We start with the empty tree $T_1 = \varnothing$.

In what follows we represent a tree by a set of edges. The algorithm first picks a minimum cost edge, either $\{u_1, u_3\}$ or $\{u_4, u_5\}$, say the first one, so $T_2 = \{\{u_1, u_3\}\}$.

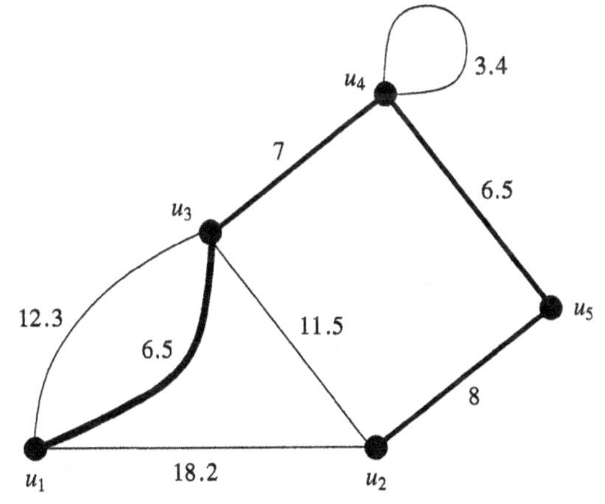

Figure 4.14: Minimum cost spanning tree in a weighted graph G.

	u_1	u_2	u_3	u_4	u_5
u_1	–	18.2	6.5	∞	∞
u_2	18.2	–	11.5	∞	8
u_3	6.5	11.5	–	7	∞
u_4	∞	∞	7	–	6.5
u_5	∞	8	∞	6.5	–

Table 4.1: The weights of the graph used in Examples 4.39 and 4.41.

The algorithm next picks a minimum cost edge from the remaining edges that does not form a cycle when added to T_2. In this case the edge is $\{u_4, u_5\}$, so $T_3 = \{\{u_1, u_3\}, \{u_4, u_5\}\}$.

The algorithm next picks a minimum cost edge from the remaining edges that does not form a cycle when added to T_3. In this case the edge is $\{u_3, u_4\}$, so $T_4 = \{\{u_1, u_3\}, \{u_4, u_5\}, \{u_3, u_4\}\}$.

Since G has only five vertices, the algorithm finally picks a minimum cost edge from the remaining edges that does not form a cycle when added to T_4. In this case the edge is $\{u_2, u_5\}$, so $T_5 = \{\{u_1, u_3\}, \{u_4, u_5\}, \{u_3, u_4\}, \{u_2, u_5\}\}$.

Hence, we have here that

$$T = T_5 = \{\{u_1, u_3\}, \{u_4, u_5\}, \{u_3, u_4\}, \{u_2, u_5\}\},$$

as shown in thick edges in Figure 4.14.

Note that at each step in KRUSKAL'S ALGORITHM we have a forest, but not necessarily a connected tree. For a large weighted graph G, KRUSKAL'S ALGORITHM can have T_i consisting of many little scattered trees from G at each phase, but after the final step we can rest assured that we end up with a *connected* forest, namely the minimum cost spanning tree of G.

The cost here of the minimum cost spanning tree T is given by $W(T) = 6.5 + 6.5 + 7 + 8 = 28$.

The following theorem shows that KRUSKAL'S ALGORITHM always computes a minimum cost spanning tree.

Theorem 4.40 Let (G, W) be a connected weighted graph. KRUSKAL'S ALGORITHM computes a minimum cost spanning tree T of G.

Proof It is clear that KRUSKAL'S ALGORITHM builds a spanning tree T of the connected graph G on n vertices, because it constructs a simple graph having $n - 1$ edges. We need to show that if T_1 is another spanning tree of G, then $W(T) \leq W(T_1)$.

Assume we have labelings $e_1, \ldots, e_{n-1}$ (respectively, $e'_1, \ldots, e'_{n-1}$) of the edges of T (respectively, T_1) such that $W(e_1) \leq \cdots \leq W(e_{n-1})$ (respectively, $W(e'_1) \leq \cdots \leq W(e'_{n-1})$). Since T_1 and T are not equal, there is a least number k such that $e_k \notin \{e'_1, \ldots, e'_{n-1}\}$ but $e_{k-1} \in \{e'_1, \ldots, e'_k\}$.

Notice that $T_1 \cup e_k$ has a unique cycle C that is not completely contained in T. Hence, there is an edge e' in C that is not in T. In this case we have that

$$T_2 = (T_1 \cup e_k) \setminus e'$$

is also a spanning tree of G. In particular, by minimality of k, the edges $e_1, \ldots, e_{k-1}$ and e' are all in T, and hence e' does not form a cycle together with the edges $e_1, \ldots, e_{k-1}$. Hence, by the algorithm's choice of e_k, we have that $W(e_k) \leq W(e')$ and hence, $W(T_2) \leq W(T_1)$.

Note now that T_2 has one more edge in common with T than T_1 does. If T_2 and T are not equal, we continue in the same fashion and obtain a spanning tree T_3 with $W(T_3) \leq W(T_2)$ and where T_3 has one more edge in common with T than T_2 does.

In this way we obtain a sequence of trees $T_1, \ldots, T_r$ with $W(T_r) \leq \cdots \leq W(T_1)$ and where $T_r = T$. In particular we have $W(T) \leq W(T_1)$ and we have the theorem. ∎

The next algorithm, called PRIM'S ALGORITHM, also yields a minimum cost spanning tree in any connected weighted graph, but it constructs a tree in a different way.

```
PRIM'S ALGORITHM
Input: A connected weighted graph (G, W) on n vertices.
Output: A minimum cost spanning tree T of G.
begin
   T_1 = ({u_1}, ∅);          // u_i arbitrary initial vertex
   for i = 1 to n do {
      let e_i ∈ E(G) \ E(T_i) be a minimum weight edge such
         that |V(T_i) ∩ e_i| = 1;
      T_{i+1} = T_i ∪ {e_i};        // e_i with its other endpoint added
   }
   output T = T_n;
end.
```

We now informally argue the correctness of PRIM'S ALGORITHM as follows:

At each phase of the algorithm, the edges obtained thus far form a spanning tree of the subgraph of G induced by these edges. This subtree is a minimum cost spanning tree in the corresponding weighted subgraph of G. Therefore, PRIM'S ALGORITHM computes a minimum cost spanning tree of G. We ask for a formal proof in Exercise 4.40.

Here is an example of PRIM'S ALGORITHM.

Example 4.41 Consider the graph depicted in Figure 4.14 with the weights shown in Table 4.1.

PRIM'S ALGORITHM starts from a given vertex u_1—that is, $T_1 = \{u_1\}$, the single vertex tree.

The algorithm then selects a minimum weight edge that has u_1 as one of its endvertices. Thus it picks the edge corresponding to a smallest entry of the first row in Table 4.1, which in this case is 6.5, corresponding to $W(\{u_1, u_3\})$. So here $T_2 = \{\{u_1, u_3\}\}$. Note that again we represent a tree by its set of edges.

The algorithm then selects a minimum weight edge that has exactly one endvertex in T_2—that is, has u_1 or u_3 as an endvertex, and does not form a cycle when added to T_2. Thus it picks the edge corresponding to the smallest unused entry of the first and the third row in Table 4.1, which in this case is 7, corresponding to $W(\{u_3, u_4\})$. So here $T_3 = \{\{u_1, u_3\}, \{u_3, u_4\}\}$.

The algorithm then selects a minimum weight edge that has exactly one endvertex in T_3—that is, has u_1, u_3, or u_4 as an endvertex, and does not form a cycle when added to T_3. Thus it picks the edge corresponding to the smallest entry of the first, third, and fourth row that is not in the first, third, or fourth column, from Table 4.1. In this case it is 6.5, corresponding to $W(\{u_4, u_5\})$. So here $T_4 = \{\{u_1, u_3\}, \{u_3, u_4\}, \{u_4, u_5\}\}$.

The algorithm finally selects a minimum weight edge that has exactly one endvertex in T_4—that is, has u_1, u_3, u_4, or u_5 as an endvertex, and does not form a cycle when added to T_4. Thus it picks the edge corresponding to the smallest entry of the first, third, fourth, and fifth row that is not in the first, third, fourth, or fifth column, from Table 4.1. In this case it is 8, corresponding to $W(\{u_2, u_5\})$. So here $T_5 = \{\{u_1, u_3\}, \{u_3, u_4\}, \{u_4, u_5\}, \{u_2, u_5\}\}$, and hence $T = T_5$ is a minimum cost spanning tree of G, with the minimum cost of $W(T) = 6.5 + 6.5 + 7 + 8 = 28$.

Note that at each step in PRIM'S ALGORITHM T_i is a connected tree.

It is interesting to compare Examples 4.39 and 4.41 to see how KRUSKAL'S and PRIM'S algorithms relate. KRUSKAL'S ALGORITHM builds a forest that may consist of many components. On the other hand, PRIM'S ALGORITHM maintains a tree at each phase of its execution. In this case both algorithms computed the same spanning tree. Both algorithms are examples of *greedy* algorithms, since at each step in the **for**-loop they select the cheapest edge. Both are in fact reasonably efficient algorithms in terms of n, the number of vertices in G.

In the exercises we ask the reader to design a connected weighted graph where the algorithms construct different minimum cost spanning trees.

4.8 Exercises

4.1. In Example 4.5 we computed $\tau(G) = 3$. List the three spanning trees for G in this example. Are any of them isomorphic?

4.2. Use Theorem 4.4 repeatedly to determine the number of spanning trees of the graph shown in Figure 4.15.

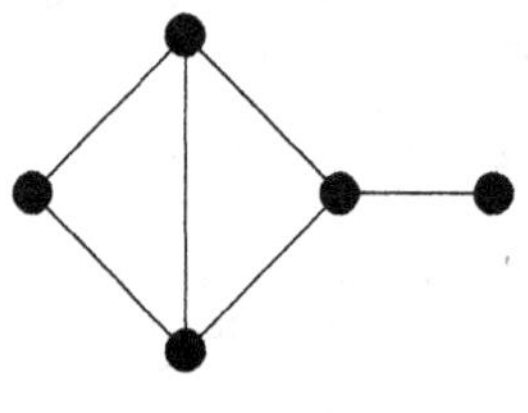

Figure 4.15: A simple graph used in Exercises 4.2 and 4.3.

4.3. Draw all spanning trees of the graph G shown in Figure 4.15.

4.4. Show that every edge incident to a leaf in a connected graph G is an edge in every spanning tree of G.

4.5. Prove Observation 4.3.

4.6. Let T be a spanning tree of K_n. How many edges does K_n have that are not in T? Does this number depend on the spanning tree T?

4.7. Let G be a simple connected graph on n vertices and m edges. Show that $C = m - n + 1$ is the number of edges in $E(G) \setminus E(T)$ for any spanning tree T of G. Show further that the value of c does not change even if a vertex is added in the middle of some edge, or when a degree-two vertex is contracted.

4.8. Let G be a connected graph on n vertices and T a spanning tree of G. Let C be a cycle of length $k \leq n$ in G and denote by $c(n, k)$ the number of edges contained in C and not in T. Show that $c(n, k) \geq 1$ and that $c(n, k)$ can take any of the values between 1 and $k - 1$. Can we obtain a tighter bound on $c(n, k)$ if we know that C is a spanning cycle of G—that is, a cycle that includes all the vertices of G?

4.9. For $n \geq 2$, compute $\tau(C_n)$ recursively.

4.10. For $n \geq 1$, compute $\tau(K_{2,n})$ recursively. [Can you compute $\tau(K_{3,n})$ recursively? Do you see a possible pattern in your recursive computations?]

4.11. For $n \geq 2$, let u be a fixed vertex of the complete graph K_n. Compute the number $\tau_u(n)$ of spanning trees of K_n that contain the vertex u as a leaf. Use this to show that the probability a vertex in a tree on n vertices is a leaf is approximately $1/e$, where $e = 2.71828182\ldots$ is the base number for the natural logarithm.

4.12. Draw the tree whose Prüfer code is $(1, 1, 1, 1, 6, 5)$.

4.13. What is the Prüfer code corresponding to the tree shown in Figure 4.3 if you swap numbers one and six?

4.14. In Example 4.8 we verified for the tree T on eight vertices that

$$T \mapsto \text{six-tuple} \mapsto T$$

gave the same tree back again. Verify that the other procedure

$$\text{six-tuple} \mapsto T \mapsto \text{six-tuple}$$

yields the same six-tuple $(4, 4, 2, 5, 4, 2)$ with which we started. (This verifies that the number of labeled trees is at least that of the codes.)

4.15. Let G be a connected graph and $\mathcal{T}(G)$ be the set of all the spanning trees of G. Show that $\partial : \mathcal{T}(G) \to \mathbb{R}$ defined by

$$\partial(T, T') = |E(T) \triangle E(T')|$$

is a metric on $\mathcal{T}(G)$. Show further that $\partial(T, T')$ is always an even number. [*Hint:* Show first that $|A \triangle C| \leq |A \triangle B| + |B \triangle C|$ for any finite sets A, B, and C.]

4.16. Assume the graph G has two components G_1 and G_2. Show there is a labeling of the vertices of G such that the adjacency matrix of G has the form

$$\mathsf{A}(G) = \begin{bmatrix} \mathsf{A}(G_1) & 0 \\ 0 & \mathsf{A}(G_2) \end{bmatrix},$$

where 0 denotes rectangular matrices with zero in every entry. Generalize this result for G having an arbitrary number of components.

4.17. Given two permutations σ and ρ on $\{1, \ldots, n\}$, show that the corresponding permutation matrices satisfy $\mathsf{P}_\sigma \mathsf{P}_\rho = \mathsf{P}_{\sigma\rho}$. In particular, show that $\mathsf{P}_\sigma \mathsf{P}_{\sigma^{-1}} = \mathsf{I}_n$ is the $n \times n$ identity matrix.

4.18. Use Exercise 4.17 to prove Observation 4.13.

4.19. Let G be a simple graph. Show that the diagonal entry $a_{ii}^{(2)}$ in the square $\mathsf{A}(G)^2$ of the adjacency matrix of G with respect to the labeling $V(G) = \{u_1, \ldots u_n\}$ is given by

$$a_{ii}^{(2)} = d_G(u_i).$$

Does this hold for general graphs? What can you say when every two distinct pairs of vertices have precisely k edges between them?

4.20. Prove Theorem 4.16. [*Hint:* Use induction on k, the length of the walks.]

4.21. Let G be a connected graph with adjacency matrix $\mathsf{A}(G)$ with respect to the vertex labeling $V(G) = \{u_1, \ldots, u_n\}$. Prove that the distance between two distinct vertices u_i and u_j is the smallest k for which $a_{ij}^{(k)} \neq 0$. [*Hint:* Use Theorem 4.16.]

4.22. Let G be a graph with adjacency matrix $\mathsf{A}(G)$ with respect to the vertex labeling $V(G) = \{u_1, \ldots, u_n\}$. Let

$$\mathsf{Y} = \mathsf{A}(G) + \mathsf{A}(G)^2 + \mathsf{A}(G)^3 + \cdots + \mathsf{A}(G)^{n-1}.$$

Show that G is connected if, and only if, all entries of Y are nonzero. [*Hint:* Use Theorem 4.16.]

4.23. For $n \in \mathbb{N}$, what are the adjacency matrix and the incidence matrix for the null graph N_n on n vertices? What about the complete graph K_n on n vertices?

4.24. For $n, m \in \mathbb{N}$, find the adjacency matrix and the incidence matrix for the complete bipartite graph $K_{m,n}$ for a suitable labeling of the vertices and edges.

4.25. Show that for a loopless graph G with n vertices and k components, we have

$$\mathrm{rank}(\mathsf{B}_{-1}(G)) = n - k.$$

[*Hint:* Use Theorem 4.21.]

4.26. Show that G has two or more components G_1 and G_2 if, and only if, there is a labeling of $V(G)$ and $E(G)$ such that the incidence matrix of G has the form

$$\mathsf{B}(G) = \begin{bmatrix} \mathsf{B}(G_1) & 0 \\ 0 & \mathsf{B}(G_2) \end{bmatrix}.$$

Generalize this result for G with k components and discuss the sizes of the blocks in $\mathsf{B}(G)$.

4.27. Suppose P and Q are $n \times n$ and $m \times m$ permutation matrices, respectively. Without using Observation 4.20, prove directly that $\mathsf{X}' = \mathsf{P}^t\mathsf{X}\mathsf{Q}$ defines an equivalence relation among all $n \times m$ matrices.

4.28. Show that if P and Q are permutation matrices, then the rank of X and $\mathsf{P}'\mathsf{X}\mathsf{Q}$ is the same. [*Note!* This means that the rank of an incidence matrix of a graph is always the same, regardless of the labelings of the vertices and edges. In particular, all isomorphic graphs have the same rank in this sense.]

4.29. Assume G has a cut-vertex u, such that $G - u = G_1 \cup G_2$, where G_1 and G_2 are each connected, but not to each other. For the labeling of the vertices of G

$$V(G) = \{u_1, \ldots, u_{n_1}\} \cup \{u\} \cup \{v_1, \ldots, v_{n_2}\},$$

where the u_i's are the vertices of G_1 and the v_j the vertices of G_2, what can be said about the form of $\mathsf{A}(G)$? If, further, the edges of G are labeled as

$$E(G) = \{e_1, \ldots, e_{m_1}\} \cup \{f_1, \ldots, f_{m_2}\},$$

where the e_i's are the edges of the subgraph induced by G_1 and u, and the f_j's are the edges of the subgraph induced by G_2 and u, what can be said about the form of the incidence matrix $\mathsf{B}(G)$?

4.30. Use the Matrix-Tree Formula to compute $\tau(K_{2,n})$ for any $n \geq 1$. Can you find a combinatorial argument for your answer?

4.31. Use the Matrix-Tree Formula to compute $\tau(K_{3,n})$ for any $n \geq 1$.

4.32. Use the Matrix-Tree Formula to compute $\tau(K_{m,n})$ for any $m, n \in \mathbb{N}$. Note that the result should, of course, be symmetric in m and n.

4.33. Use the Matrix-Tree Formula to compute $\tau(K_n - e)$ for any $n \geq 1$, where $K_n - e$ is a graph obtained from K_n by removing one edge e.

4.34. Compute $\tau(W_n)$, where W_n is the *wheel* on $n + 1$ vertices. The wheel W_n is C_n modified by adding one additional vertex that is adjacent to all of the vertices in C_n. Placing this additional vertex in the center of C_n creates a wheel. [*Hint:* If you use the Matrix-Tree Theorem, choose your cofactor carefully.]

4.35. Compute the total resistance of the electrical network of resistors shown in Figure 4.16, where the vertical resistor is two ohms and the remaining resistors are one ohm.

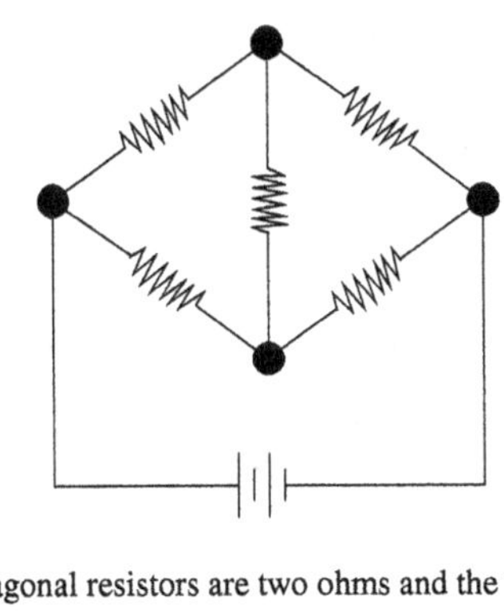

Figure 4.16: The four diagonal resistors are two ohms and the vertical resistor is one ohm.

4.36. Compute the total resistance between u and v in the electrical network of resistors corresponding to the graph of the cube Q_3 shown in Figure 4.17, where each edge represents one ohm. (Often this problem has been used as a brain teaser in college freshman physics classes. Without the machinery we have developed, one must rely on insight in electrical physics.)

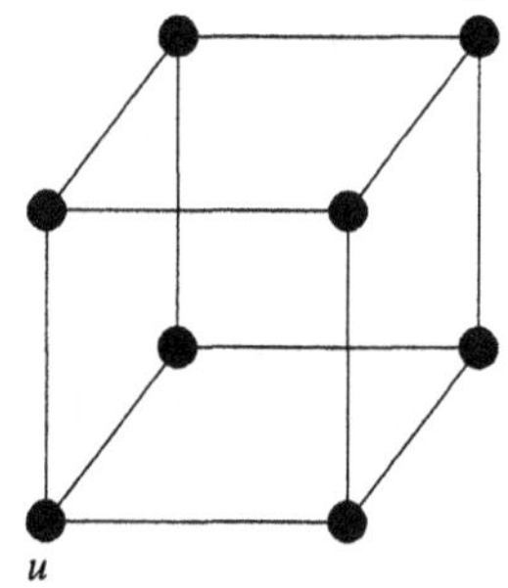

Figure 4.17: Each edge of the cube Q_3 represents a one-ohm resistor.

4.37. Compute the total resistance between u and v in the electrical network of resistors shown in Figure 4.18, where one diagonal resistor is two ohms, the other diagonal resistor is three ohms, and the remaining resistors are one ohm.

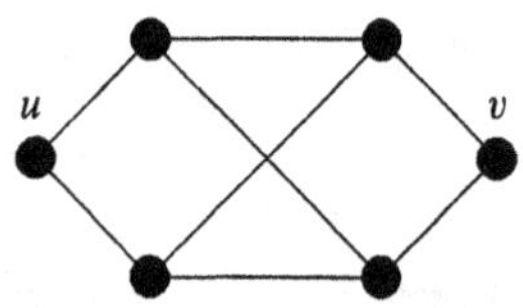

Figure 4.18: Each edge represents a resistor. Horizontal resistors are one ohm; one diagonal resistor is two ohms; the other is three ohms.

4.38. If a weighted graph (G, W) has a unique edge e_m of minimum weight—that is, $W(e_m) < W(e)$ for all $e \neq e_m$—show that any minimum cost spanning tree of G must contain e_m.

4.39. Let (G, W) be a simple connected weighted graph. Let $e_m \in E(G)$ be as in Exercise 4.38, so that $W(e_m) < W(e)$ for $e \neq e_m$. As we have seen, every minimum cost spanning tree of G must contain e_m. Now let $e'_m \in E(G)$ be such that $W(e'_m) < W(e)$ for all $e \notin \{e_m, e'_m\}$. Must every minimum cost spanning tree contain the edge e'_m? If $e''_m \in E(G)$ is such that $W(e''_m) < E(e)$ for all $e \notin \{e_m, e'_m, e''_m\}$, must every minimum cost spanning tree contain the edge e''_m?

4.40. Prove that PRIM'S ALGORITHM yields a minimum cost spanning tree for a given connected weighted graph G. [*Hint:* Consider the very first step where PRIM'S ALGORITHM goes wrong, assuming that it does not produce a minimum cost spanning tree.]

4.41. Draw a connected weighted graph for which KRUSKAL'S ALGORITHM and PRIM'S ALGORITHM find different minimum cost spanning trees. Draw the minimum cost spanning trees the algorithms construct.

CHAPTER

5

Fundamental Properties of Graphs and Digraphs

Chapter Goals

Study the structure of bipartite graphs.

Define and characterize Eulerian graphs.

Define and study Hamiltonian graphs.

Define and briefly discuss the traveling salesman problem.

Prove interesting properties of tournament digraphs.

Discuss the pecking order among hens in a hen house.

Present and discuss the adjacency matrix for a digraph.

Define and study acyclic digraphs.

Show how acyclic digraphs naturally associate with posets.

Describe the form of the adjacency matrix for an acyclic digraph.

5.1 Bipartite Graphs

The main theme of this chapter involves cycles and circuits in graphs, or the complete lack thereof. We start by demonstrating how cycles are used for describing the structure of bipartite graphs.

In Chapters 3 and 4 we discussed in some depth the important class of acyclic graphs (i.e., graphs containing no cycles), namely trees and forests. If T is a tree, we can pick an arbitrary root $r \in V(T)$ and then let $X \subseteq V(T)$ consist of all the vertices on the even levels $0, 2, 4, \ldots$, and $Y \subseteq V(T)$ consist of all the vertices on the odd levels $1, 3, 5, \ldots$ In this way we see that $V(T)$ can be partitioned into two parts, $V(T) = X \cup Y$, in such a way that every edge has one endvertex in one part X and the other endvertex in the other part Y. Equivalently, no two vertices in the same part, X or Y, are adjacent. This is a special property that every tree and forest satisfies. Clearly, there are graphs, which are not trees, having such a partition of their vertices. The 4-cycle C_4, for example, has such a partition, where X consists of two nonadjacent vertices, and Y of the other nonadjacent pair of vertices. Is there something structural we can say in general about graphs that have such a partition of their vertices?

In this case there is, but before describing the structure in detail, we begin with a formal definition.

Definition 5.1 A graph G is a *bipartite graph* if there are $X, Y \subseteq V(G)$ meeting the following conditions:

1. $V(G) = X \cup Y$,
2. $X \cap Y = \varnothing$,
3. $G[X]$ and $G[Y]$ are both null graphs.

The word "bipartite" means "two parts." If you keep this in mind, you will have no trouble remembering Definition 5.1.

Bipartite graphs are often used to model some common real-world situations, where the vertices in one part X represent agents and the vertices in the other part Y represent projects which are assigned to the agents by means of edges between X and Y. These situations will be discussed in more detail in Chapters 10 and 11.

Note 5.2 Several importants points are worth noting.

- A bipartite graph G with partition $V(G) = X \cup Y$ implies that each edge of G has one endvertex in X and the other in Y.
- Every null graph N_n is bipartite; any partition $V(N_n) = X \cup Y$ demonstrates this fact.
- A bipartite graph cannot have any loops.
- A graph is bipartite if, and only if, each of its components is bipartite.
- For all $m, n \in \mathbb{N}$, the complete bipartite graph $K_{m,n}$ is a simple bipartite graph.

It is not hard to see that every cycle in a bipartite graph must have an even length. (See Exercise 5.2.) This property is, in fact, a characterizing property of bipartite graphs, as next stated in the main theorem of this section.

Theorem 5.3 For a graph G the following statements are equivalent:

1. G is bipartite.
2. Every cycle in G has an even length.

Moreover, if G is connected and satisfies either 1 or 2, then the partition $V(G) = X \cup Y$ is unique.

Before proving this Theorem 5.3, we need the following observation.

Observation 5.4 If G contains a closed walk of odd length, then it contains a cycle of odd length.

Proof Assume that $p = (u_0, e_1, u_1, \ldots, e_n, u_n)$ is closed walk of shortest possible odd length n. We show that p is actually a cycle. If p is not a cycle, then there are two indices j and k with $0 \le j < k \le n$, $(j, k) \ne (0, n)$, and $u_j = u_k$. In this case we have two closed walks $p_1 = (u_0, e_1, u_1, \ldots, u_j, e_k, u_{k+1}, \ldots, u_n)$ and $p_2 = (u_j, e_j, u_{j+1}, \ldots, e_{k-1}, u_k)$. The lengths of p_1 and p_2 are both positive and sum up to n. Hence, one of these lengths must be odd. Thus we have a closed walk of odd length less than p, which contradicts our initial assumption. Therefore, p must be a cycle. ■

Note 5.5 There are graphs containing closed walks or circuits of even lengths that do not contain any cycles of even lengths.

We now use Observation 5.4 to prove Theorem 5.3.

Proof (Theorem 5.3) Without loss of generality, we can assume that G is connected.

Assume first that $V(G) = X \cup Y$ is a partition of the vertices such that both $G[X]$ and $G[Y]$ are null graphs. Let $p = (u_0, e_1, u_1, \ldots, e_n, u_n)$ be a cycle. Since each edge in G has one endvertex in X and the other in Y, we see that for each i the vertices u_i and u_{i+1} of the cycle p are always in opposite parts, that is, one is in X and the other is in Y. Since p is a closed walk, $u_0 = u_n$ must be in the same part X or Y, so n must be even.

For the second half of the theorem, assume that each cycle of G has even length. We show that G is bipartite. Choose a fixed vertex $x \in V(G)$. Consider an arbitrary vertex $y \in V(G)$. If p and q are two distinct paths from x to y of respective lengths n and m, then pq^{-1} is a closed walk in G of length $n + m$. Combining our assumption that each cycle of G has even length with the contrapositive of Observation 5.4, we can conclude that $n + m$ is even. Hence, either n and m are both even or they are both odd. Therefore, we can form a partition of $V(G)$ by letting

$$X = \{y \in V(G) : \text{the length of each path from } x \text{ to } y \text{ is even}\},$$
$$Y = \{y \in V(G) : \text{the length of each path from } x \text{ to } y \text{ is odd}\}.$$

When G is connected, we have $V(G) = X \cup Y$ and $X \cap Y = \varnothing$. Lastly, consider an edge e of G with endvertices u and v. We need to show that they are not contained in the same part. Suppose p is a x, u-path of length ℓ and q is an x, v-path of length ℓ'. In this case $(u, e, v)q^{-1}p$ is a closed walk of length $\ell + \ell' + 1$, which by Observation 5.4 must be even. Hence, ℓ and ℓ' must be of opposite parity modulo 2. Hence, one of u and v is contained in X, and the other in Y. Hence, $G[X]$ and $G[Y]$ are both null graphs.

Since G is connected, every two vertices are joined by a simple path. Hence, the partition $V(G) = X \cup Y$ is uniquely determined. ■

We can easily extend the definition of a bipartite graph: A graph G for which $V(G)$ can be partitioned into $k \ge 2$ sets

$$V(G) = X_1 \cup X_2 \cup \cdots \cup X_k,$$

where $X_i \cap X_j = \varnothing$ for each $i \ne j$ and each of the induced subgraphs $G[X_i]$ is a null graph, is called a *k-partite graph.* If G is k-partite for some k, then G is called a *multipartite graph.* However, the nice characterization for bipartite graphs given in Theorem 5.3 cannot be extended for $k \ge 3$. Multipartite graphs will be discussed further in Chapter 8.

5.2 Eulerian Graphs

As we saw in Chapter 4, a subgraph that minimally connects all of the vertices in a connected graph is given by a spanning tree of the graph. We also saw that every connected graph has at least one spanning tree.

In this and the following two sections we consider various other "spanning" properties of graphs. Is there a circuit or a trail in the graph that spans or covers all of the edges of the graph? Is there a simple path or a cycle that covers all of the vertices of the graph? For a digraph, is there a directed cycle that covers all of the vertices of the digraph? Unlike the notion of a spanning tree, where every connected graph has at least one spanning tree, we cannot answer these questions positively for every connected graph or digraph.

We start by considering the first question in detail, a question that prompted Euler to write the first paper on graph theory, as we mentioned in Chapter 1. In that paper from 1736, Euler solved the Kőnigsberg Bridge problem, stated in Problem 1.1.

Euler asked and solved a more general question—namely, what type of a connected graph has a circuit that contains all of the edges of the graph?

Definition 5.6 Let G be a graph. A trail of G that contains each edge of G is called an *Eulerian trail* of G. A circuit of G that contains each edge of G is called an *Eulerian circuit* of G. If G has an Eulerian circuit, then G is called an *Eulerian graph.*

Note that an Eulerian graph, with no isolated vertices, is necessarily a connected graph, since a circuit going through each edge of G must connect all the nonisolated vertices of G.

The following theorem was the main result of the first paper on graph theory by Euler from 1736.

Theorem 5.7 A connected graph G is Eulerian if, and only if, each vertex in G has even degree.

Proof We start with the easier part first and assume that G has an Eulerian circuit c. Starting at a given vertex in c and traversing along c, we leave each vertex $u \in V(G)$ for each time we enter it. Since we enter and leave u each time along different edges, the number of edges in c that have u as an endvertex must be even. Since c covers all the edges of G, then $d_G(u)$ must be even.

Assume now that G is connected where each vertex has an even positive degree. We show by induction on $|E(G)|$ that G has an Eulerian circuit. We start at an arbitrary vertex $u_0 \in V(G)$.

1. Assume that we have constructed a simple path $p_i = (u_0, e_1, u_1 \ldots, e_i, u_i)$ in G of length $i \geq 0$. Since $d_G(u_i) > 0$ is even, there is an edge $e_{i+1} \notin \{e_1, \ldots, e_i\}$ with u_i as one endvertex and u_{i+1} as the other endvertex.
   - If $u_{i+1} \notin \{u_0, \ldots, u_i\}$, then $p_{i+1} = p_i(u_i, e_{i+1}, u_{i+1})$ is a simple path of length $i+1$. Change i to $i+1$, and go to 1.
   - If $u_{i+1} = u_j \in \{u_0, \ldots, u_i\}$, then $c' = (u_j, e_{j+1}, \ldots, e_{i+1}, u_{i+1})$ is a simple cycle of a positive length, so let $E_{c'} = \{e_{j+1}, \ldots, e_{i+1}\}$, and go to 2.
2. Let $G' = G - E_{c'}$.

Note that G' has strictly fewer edges than G and each vertex in G' has an even degree. If the components of G' are $H_1, \ldots, H_k$, then by the induction hypothesis, each H_i has an Eulerian circuit c_i that starts and ends at a vertex v_i on the cycle c'. Since we can write $c' = p_1 p_2 \cdots p_{k-1} p_k$, where p_i is a trail on c' from v_i to v_{i+1} if $1 \leq i \leq k-1$, and p_k is a trail from v_k to v_1, then c given by

$$c = c_1\ p_1\ c_2\ p_2 \cdots p_{k-1}\ c_k\ p_k$$

is an Eulerian circuit of G, which completes the proof. ■

Clearly, the vertices of the graph shown in Figure 5.1 do not all have even degree (in fact, they *all* have odd degree!). Hence, the graph does not have an Eulerian circuit; and therefore, the original question of the Kőnigsberg Bridge problem, whether one can start on one bank, and walk over every bridge exactly once and end up on the same spot, is by Theorem 5.7 answered in the negative.

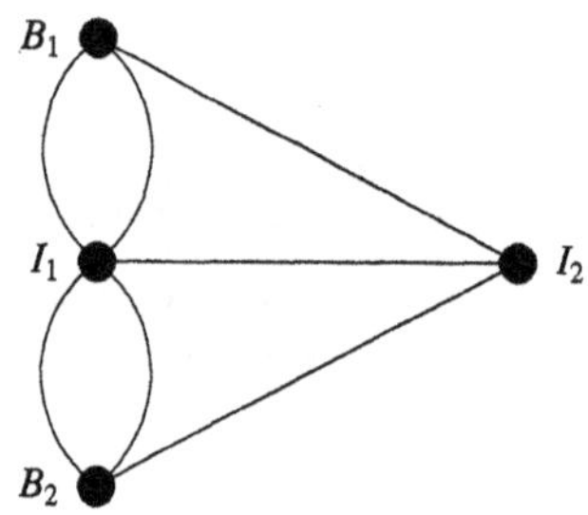

Figure 5.1: Graph of the Kőnigsberg Bridge problem.

There are several corollaries we can deduce from Theorem 5.7.

Corollary 5.8 A connected graph G has an Eulerian trail if, and only if, all except two vertices in G have an even degree.

Proof If G has an Eulerian trail t, then all of the vertices of G except the initial and final vertex of t have an even degree.

If all of the vertices in G, except u and v, have even degree, then by connecting u and v by an extra edge e we obtain a connected graph G' with all its vertices of even degree. By Theorem 5.7 the graph G' has an Eulerian circuit c, and hence G has an Eulerian trail $t = c - e$. ■

Looking at the graph in Figure 5.1, we see that it does not have an Eulerian trail either.

Note that the proof of Theorem 5.7 was based on the fact that if G is connected and has all of its vertices of even degree, then we are *always* able to find a cycle in G of positive length. By induction on $|E(G)|$, we find that in this case G is a union of edge-disjoint cycles in G. Clearly, if G is an edge-disjoint union of cycles in G, then each vertex has an even degree. Hence, we obtain the following corollary.

Corollary 5.9 For a graph G the following statements are equivalent:

1. All of the vertices have a positive even degree.
2. G is a disjoint union of edge-disjoint cycles in G.

In particular, a connected graph G is Eulerian if, and only if, it is a disjoint union of cycles in G.

In many drawing games from puzzle magazines, we find that knowledge of Theorem 5.7 and Corollary 5.8 can save us a sleepless night or two! The following example illustrates this very point.

Example 5.10 Consider the graph depicted in Figure 5.2. Is it possible to draw this graph starting and ending at the same vertex, without lifting one's pencil from the paper? We assume that once a line (that is, an edge) has been drawn, one cannot retrace it. So, the question of whether the graph can be drawn in this manner is the same as finding an Eulerian circuit of the graph.

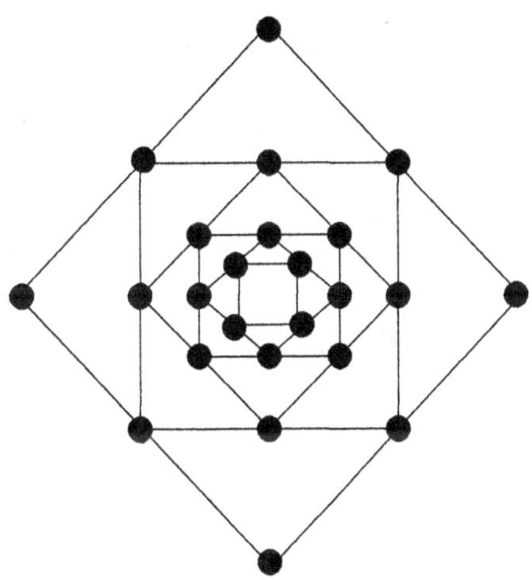

Figure 5.2: An Eulerian graph.

We see that the graph in Figure 5.2 has all even-degree vertices. Hence by Theorem 5.7, it does have an Eulerian circuit. One could inductively construct an Eulerian circuit by removing one cycle at a time from the graph. However, we provide a better method in ALGORITHM EULERIAN CIRCUIT that also produces an Eulerian circuit in a connected graph, where all of the vertices of G are even. Meanwhile, using symmetry, the reader can construct an Eulerian circuit for graph in Figure 5.2.

Example 5.11 Consider the graph shown in Figure 5.3. Here, we note that there are two vertices that do not have even degree (each has a degree of three). Hence, by Theorem 5.7, starting and ending at the same point, it is impossible to draw this graph without lifting one's pencil, no matter how hard one tries.

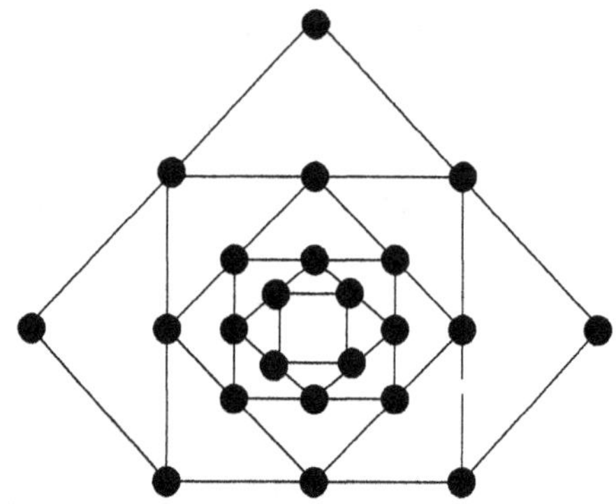

Figure 5.3: A graph that is not Eulerian.

However, since there are only two vertices of odd degree, the graph has by Corollary 5.8 an Eulerian trail. Thus, it is possible to draw this graph without lifting one's pencil, if one can start at one point and end at another.

For an Eulerian graph G, let $u \in V(G)$ be any vertex in G. Since G has an Eulerian circuit, we should be able to construct one that starts at u. The following algorithm does precisely that.

ALGORITHM EULERIAN CIRCUIT, sometimes also called FLEURY'S ALGORITHM, shows how to construct an Eulerian circuit of a graph G. Note that we use // to indicate a comment in an algorithm.

```
ALGORITHM EULERIAN CIRCUIT
Input: An Eulerian graph G = (V, E, φ).
Output: An Eulerian circuit of G represented by a list L of edges.

begin
   currentVertex = u ∈ V;
   L = ∅;       // edges are appended to the list L
   edges = E;
   while edges ≠ ∅ do {
        pick e ∈ edges with currentVertex ∈ φ(e)
             such that e is not a bridge in G[edges],
             unless it is the only choice;
        L = Le;
        if |φ(e)| = 2 then
             currentVertex = v ∈ φ(e) \ {currentVertex};
        edges = edges \ {e};
        }
   output L;
end.
```

The proof that ALGORITHM EULERIAN CIRCUIT actually produces an Eulerian circuit is left to Exercise 5.12.

5.3 Hamiltonian Graphs

In §5.2 we found necessary and sufficient conditions for a graph G to contain an Eulerian circuit. An Eulerian circuit visits each *edge* exactly once, but could potentially visit some vertices more than once (usually that is indeed the case). In this section, we ask if there is a "round trip" through a given graph G such that every *vertex* is visited exactly once. As we will see, this is a much harder question than the one for edges addressed previously.

As for Eulerian graphs, the original question about a round trip in the graph G, visiting each vertex exactly once, is also due to a well-known Irish mathematician, Sir William Rowan Hamilton.

Example 5.12 Consider the dodecahedron and its graph shown in Figure 5.4. There the graph has vertices consisting of the corner points of the dodecahedron, and edges consisting of the lines where two adjacent faces meet, connecting two corner points.

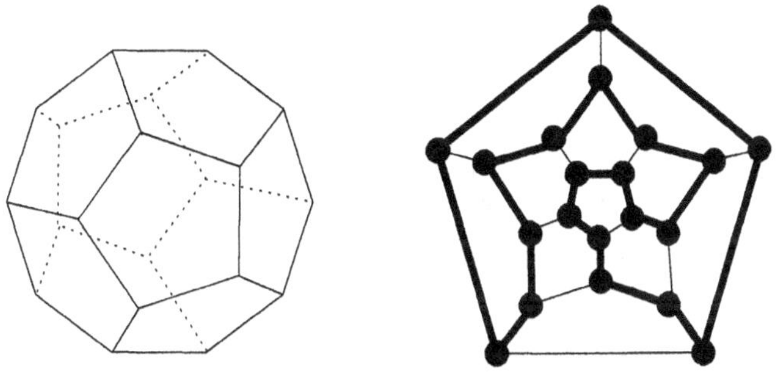

Figure 5.4: A dodecahedron and the corresponding graph.

In 1859, Hamilton wanted to market a game where the object was to start at any corner of the dodecahedron and go along the edges, visiting each corner point exactly once, and return to the original starting corner point. As we see in Figure 5.4, such a route exists for the dodecahedron, and it is shown in thick lines in the graph on the right.

Example 5.12 motivates the following definition.

Definition 5.13 Let G be a graph. A path in G that includes every vertex in G is called a *Hamiltonian path* of G. A cycle in G that includes every vertex in G is called a *Hamiltonian cycle* of G. If G contains a Hamiltonian cycle, then G is called a *Hamiltonian graph.*

The graph presented in Figure 5.5 is a Hamiltonian graph, since it contains a Hamiltonian cycle shown by thick edges.

Remark 5.14 Every Hamiltonian cycle of a Hamiltonian graph G on n vertices has exactly n vertices and n edges. Hence, most likely, not all of the edges of G are included in a Hamiltonian cycle of G. This is fundamentally different from an Eulerian graph, where an Eulerian circuit covers all of the edges and therefore all of the vertices as well! From this point alone, one

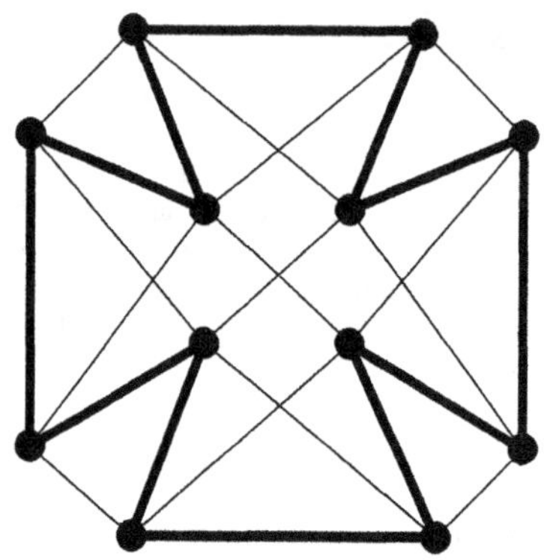

Figure 5.5: A Hamiltonian graph and a Hamiltonian cycle shown by thick edges.

can intuitively see that much less can be said about the structure of a graph when we know that it is Hamiltonian than when we know that it is Eulerian.

Note 5.15

- Every Hamiltonian graph must be connected.
- No tree is Hamiltonian.
- For each $n \geq 3$, the cycle graph C_n is Hamiltonian.
- For each $n \geq 3$, the complete graph K_n is Hamiltonian.
- For each $n \geq 2$, the complete bipartite graph $K_{n,n}$ is Hamiltonian. (Exercise 5.15.)

As for Eulerian graphs, it is natural to ask: What are the necessary and sufficient conditions a graph G must satisfy in order to be Hamiltonian? As we just saw, G must be connected, but that is not sufficient, since trees are connected but clearly not Hamiltonian. In addition, a graph G is Hamiltonian if, and only if, the graph G', where all the loops and multiple edges of G have been removed, is Hamiltonian. Hence, when determining whether or not a graph G is Hamiltonian, we can assume G is simple and connected.

The following theorem shows us a condition that is necessary for a graph to be Hamiltonian. We will discuss some sufficient conditions later.

Theorem 5.16 If G is a simple Hamiltonian graph, then for each $S \subseteq V(G)$, the number of components of $G - S$ is at most $|S|$.

Proof Let C be a Hamiltonian cycle of G, viewed as a subgraph of G. Because C is a subgraph of G that includes all of the vertices of G, the number of components of $G - S$ is at most the number of components of $C - S$. Since C is a cycle, the number of components of $C - S$ is at most $|S|$, which completes the proof. ■

Theorem 5.16 can be used to prove that neither of the two graphs shown in Figure 5.6 is Hamiltonian. The proofs are left as Exercise 5.14. Theorem 5.16 provided an easy necessary condition for a graph to be Hamiltonian—easy in the sense that if we can find a subset $S \subseteq V(G)$ with more than $|S|$ components of $G - S$, then G is not Hamiltonian. It can be hard, however, for large graphs, to find such a subset S, even if it exists.

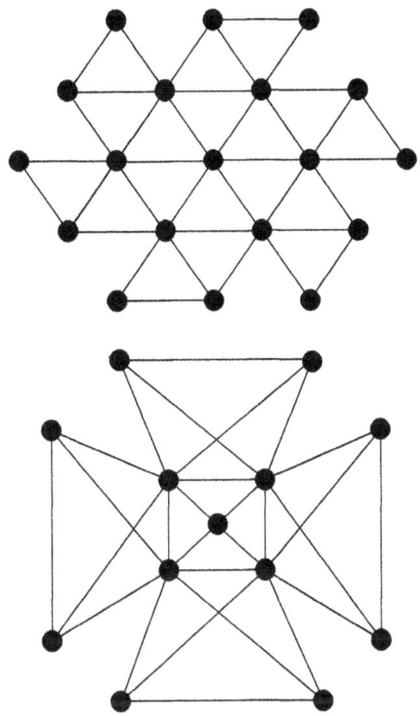

Figure 5.6: Graphs that are not Hamiltonian.

Example 5.17 Let the complete bipartite graph $K_{2,3}$ be presented on the vertices

$$V(K_{2,3}) = \{u_1, u_2, u_3\} \cup \{v_1, v_2\},$$

where each u_i is connected to each v_j. If we let $S = \{v_1, v_2\}$, then $|S| = 2$ but $G - S$ is the graph consisting of three isolated vertices u_1, u_2, and u_3, and hence $G - S$ has three components, one more than the elements of S. By Theorem 5.16 the graph G is not Hamiltonian.

Example 5.18 For $n \in \mathbb{N}$, let K_{n+1} be the complete graph with vertices $\{u, u_1, \ldots, u_n\}$. For each $i \in \{1, \ldots, n\}$ connect an additional vertex v_i to u and u_i. Let G be the resulting graph on

$2n + 1$ vertices $\{u, u_1, \ldots, u_n, v_1, \ldots, v_n\}$. Note that each vertex v_i has degree 2 in G. Since a potential Hamiltonian cycle c of G must contain each vertex v_i, we see that c must contain the edges $\{u_i, v_i\}$ and $\{v_i, u\}$ for each i. So c cannot be a cycle and G is not Hamiltonian.

On the other hand, one can see (see Exercise 5.17) that for each $S \subseteq V(G)$ the number of components of $G - S$ is at most $|S|$.

Therefore, there are infinitely many graphs satisfying the necessary condition of Theorem 5.16 that are not Hamiltonian.

We now want to discuss some sufficient conditions that ensure us a graph G is Hamiltonian. Unfortunately, no complete description is known to determine exactly when a graph is Hamiltonian and when it is not.

Suppose G is a simple graph, and u and v are two nonadjacent vertices in G. We denote by $G \cup \{u, v\}$ the simple graph obtained by adding the edge between u and v to G—that is,

$$G \cup \{u, v\} = (V(G), E(G) \cup \{\{u, v\}\}).$$

With this convention in mind, we have the following lemma.

Lemma 5.19 **Bondy, Chvátal 1976**

Let G be a simple graph on n vertices. Suppose $u, v \in V(G)$ are such that

1. $\{u, v\} \notin E(G)$,
2. $d_G(u) + d_G(v) \geq n$.

In this case, G is Hamiltonian if, and only if, $G \cup \{u, v\}$ is Hamiltonian.

Proof Clearly, if G is Hamiltonian, then so is $G \cup \{u, v\}$, so one implication is a tautology.

For the other implication, assume that $G \cup \{u, v\}$ is Hamiltonian with a Hamiltonian cycle C. If C does not include the edge $\{u, v\}$, then C is a Hamiltonian cycle in G, so we can assume further that C includes the edge $\{u, v\}$. Let us write C as $C = (u_0, u_1, \ldots, u_n)$, where $u_0 = u_n = u$ and $u_1 = v$. Define the sets U and V by

$$\begin{aligned} U &= \{i : \{u, u_i\} \in E(G),\ 2 \leq i \leq n-2\}, \\ V &= \{i : \{v, u_{i+1}\} \in E(G),\ 2 \leq i \leq n-2\}. \end{aligned}$$

Since $\{u, v\} \notin E(G)$, we see that $|U| = d_G(u) - 1$ and $|V| = d_G(v) - 1$. By our assumption on the degrees of u and v, we therefore have that

$$n - 2 \leq |U| + |V| = |U \cup V| + |U \cap V| \leq (n-3) + |U \cap V|,$$

from which we deduce that $|U \cap V| \geq 1$, and hence $U \cap V \neq \varnothing$. If $j \in U \cap V$, then we can form a Hamiltonian cycle

$$C' = (u_0, u_j, u_{j-1}, \ldots, u_1, u_{j+1}, u_{j+2}, \ldots, u_{n-1}, u_n)$$

that does not include the edge $\{u, v\}$, as shown in Figure 5.7. Therefore, C' is a Hamiltonian cycle of G, and hence we have the lemma. ∎

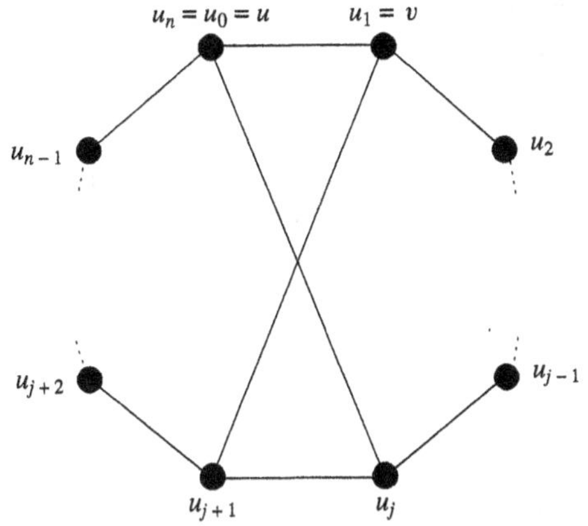

Figure 5.7: C' is a Hamiltonian cycle in G.

From Lemma 5.19, we can derive a well-known theorem by Ore [43] as an easy corollary.

Corollary 5.20 Let G be a simple graph on $n \geq 3$ vertices. If for every pair of nonadjacent vertices u and v in G we have

$$d_G(u) + d_G(v) \geq n,$$

then G is Hamiltonian.

Proof Let $G_0 = G$. If u and v are not adjacent in G_0, then let $G_1 = G_0 \cup \{u, v\}$. By Lemma 5.19, we have that G_0 is Hamiltonian if, and only if, G_1 is Hamiltonian. If G_1 is a complete graph, then it is clearly Hamiltonian. Therefore, $G_0 = G$ is also Hamiltonian and we are done. Otherwise, we continue to add edges in this fashion and form a sequence

$$G = G_0 \subseteq G_1 \subseteq \cdots \subseteq G_k = K_n.$$

We eventually reach the complete graph K_n. Here each graph G_i is a subgraph of G_{i+1} that has one more edge than G_i. By Lemma 5.19, G_i is Hamiltonian if, and only if, G_{i+1} is Hamiltonian. Since K_n is Hamiltonian, then so is $G_0 = G$, which completes the proof. ■

The following simpler version is a direct consequence of Corollary 5.20 and is due to Dirac [13].

Corollary 5.21 Let G be a simple graph on $n \geq 3$ vertices. If for every $u \in V(G)$ we have

$$d_G(u) \geq \frac{n}{2}$$

then G is Hamiltonian.

Note 5.22 Both Corollaries 5.20 and 5.21 provide a mechanical way to determine a sufficient condition for a graph to be Hamiltonian. More important, though, the way that we proved Lemma 5.19 gives us the following algorithm to find a Hamiltonian cycle in a graph G, if G satisfies the conditions of either Corollary 5.20 or 5.21.

ALGORITHM RESTRICTED HAMILTONIAN CYCLE constructs a Hamiltonian cycle in a simple graph G, provided that G satisfies the conditions of either Corollary 5.20 or 5.21.

```
ALGORITHM RESTRICTED HAMILTONIAN CYCLE
Input: A simple graph G = (V, E) on n vertices satisfying the conditions of
either Corollary 5.20 or 5.21.
Output: A Hamiltonian cycle of G.
begin
   G_0 = G;
   k = (n - 1)n/2 - |E|;
   for i = 1 to k do        // e_1, ..., e_k are new edges not in E
      G_i = G_{i-1} ∪ e_i;      // in particular, G_k = K_n
   let C_k denote an arbitrary Hamiltonian cycle of G_k;
   for i = k downto 0 do {
      if E(C_i) ∩ {e_1, ..., e_i} = ∅
        then output C_i;
        else let j ∈ {1, ..., i} be the largest index such
              that e_j is contained in the Hamiltonian cycle C_i;
      assume C_i = (u_0, u_1, ..., u_n), where e_j = {u_0, u_1};
      find l ∈ {2, ..., n - 2} such that
        {u_0, u_ℓ}, {u_1, u_ℓ+1} ∈ E(G_{i-1});
      form the Hamiltonian cycle C_{i-1} of G_{i-1} by letting
        C_{i-1} = (u_0, u_ℓ, u_ℓ-1, ..., u_1, u_ℓ+1, u_ℓ+2, ..., u_n);
      }
end.
```

5.4 Hamiltonian Cycles in Weighted Graphs

In this section we briefly discuss Hamiltonian cycles in weighted graphs (G, W) on n vertices. Recall that if C is a cycle in G, then

$$W(C) = \sum_{e \in E(C)} W(e)$$

is the weight of the cycle C in G. When seeking the minimum cost Hamiltonian cycle in G, we can assume G is a complete graph on n vertices—that is, $G = K_n$—simply by putting $W(\{u, v\}) = \infty$ whenever u and v are not adjacent in G. In this way, the weight of the minimum cost Hamiltonian cycle of G is infinite if, and only if, G has no Hamiltonian cycle.

The algorithmic problem of finding a minimum cost Hamiltonian cycle in a weighted graph G is a well-known and very hard problem, which usually goes by the name of the *traveling salesman problem.*

Example 5.23 Traveling Salesman Problem

A traveling salesman wants to make a round trip through n cities, $c_1, \ldots, c_n$, to sell his goods. He starts in his home city c_1, visits each remaining city c_i exactly once, and ends in his home city c_1, where he started the trip. If he knows the distances between every pair of cities c_i and c_j, how should he plan his round trip to make the total round-trip distance as short as possible?

Here, we can represent the cities as vertices in the complete graph K_n and let the weight between two distinct cities be the distance in miles:

$$V(K_n) = \{c_1, \ldots, c_n\},$$
$$W(\{c_i, c_j\}) = \text{ the distance between cities } c_i \text{ and } c_j \text{ in miles.}$$

In this way, the problem of finding the shortest route is precisely that of finding a minimum weight Hamiltonian cycle of the weighted completed graph K_n.

In theory, we could just check all of the possible round trips and pick the one that has the minimum weight among them. This is, however, not computationally feasible, since the total number of round trips for n cities is given by $(n-1)!/2$ (see Exercise 5.22), which quickly becomes too large for exhaustively checking as n grows.

To this day, a large amount of research focuses on developing efficient algorithms to find the minimum cost round trip to solve the traveling salesman problem. Approximations of the minimum cost round trip have been studied extensively—that is, how to find a round trip that we know for sure is at most, say, 30% longer than the optimal minimum round trip.

5.5 Eulerian and Hamiltonian Digraphs

In the sections to follow, we discuss routing problems for digraphs, which are similar to those that we have discussed so far for Eulerian and Hamiltonian graphs. For a given digraph $\vec{G}$, one can also ask the same questions, whether or not one can find either a directed trail or circuit that covers all the edges, or a directed path or cycle that covers all of the vertices of $\vec{G}$. Although similar, we make the following definition to erase all possible ambiguities.

Definition 5.24 Let $\vec{G}$ be a digraph.

1. A directed trail of $\vec{G}$ that contains exactly one copy of each directed edge of $\vec{G}$ is called a *directed Eulerian trail.*
2. A closed directed Eulerian trail of $\vec{G}$ is called a *directed Eulerian circuit.* A digraph that has a directed Eulerian circuit is called an *Eulerian digraph.*
3. A directed path of $\vec{G}$ that contains all of the vertices of $\vec{G}$ is called a *directed Hamiltonian path.*
4. A directed cycle of $\vec{G}$ that contains all of the vertices of $\vec{G}$ is called a *directed Hamiltonian cycle.* A digraph containing a directed Hamiltonian cycle is called a *Hamiltonian digraph.*

The following theorem states exactly when a weakly connected digraph is Eulerian and when it is not. It is similar to Theorem 5.7, and its proof is, in fact, easier than that of Theorem 5.7, and is therefore left as Exercise 5.31.

Theorem 5.25 Let $\vec{G}$ be a weakly connected digraph.

1. $\vec{G}$ is an Eulerian digraph if, and only if, for each $u \in V(\vec{G})$, we have $d^-(u) = d^+(u)$.
2. $\vec{G}$ has a directed Eulerian trail if, and only if, at most two vertices have their indegree and outdegree unequal, in which case for one of them, $u \in V(\vec{G})$, we have $d^+(u) - d^-(u) = 1$.

Remark 5.26 By Theorem 1.44, the directed version of the Hand-Shaking Theorem, we have that if the number of vertices having their indegree and outdegree unequal is at most two, then their number is either 0 or 2. In the case they are two, $u, v \in V(\vec{G})$ and $d^+(u) - d^-(u) = 1$, then we must have $d^-(v) - d^+(v) = 1$.

Example 5.27 Consider the digraphs in Figure 5.8. Their underlying graphs are isomorphic and in fact all their vertices have even degree. Viewing them as digraphs, we see by Theorem 5.25 that the one on the right is Eulerian, since each vertex has equal indegree and outdegree. The digraph on the left, though, is not Eulerian and therefore (by Exercise 5.32) does not even have an Eulerian directed trail.

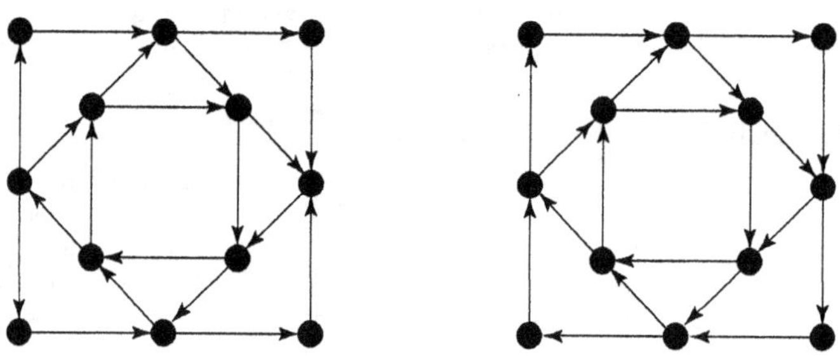

Figure 5.8: One digraph is Eulerian and the other is not.

As for the undirected case, finding necessary and sufficient conditions for a connected digraph to possess a directed Hamiltonian cycle is much harder than for the directed Eulerian circuit. The following directed version of Corollary 5.20 is due to Ghouilá-Houri, and the proof is from [19].

Theorem 5.28 A strongly connected digraph $\vec{G}$ on n vertices with $d^-(u) + d^+(u) \geq n$ for all $u \in V(\vec{G})$ is a Hamiltonian digraph.

Instead of further discussing the necessary and sufficient conditions for a general digraph to be Hamiltonian, we restrict ourselves to digraphs whose underlying graphs are complete. We should note that a balanced digraph is strongly connected if, and only if, it is weakly connected (Exercise 5.34).

5.6 Tournament Digraphs

In this section we focus on very special kinds of digraphs called *tournament digraphs.*

Definition 5.29 A digraph $\vec{G}$ is a *tournament digraph* if its underlying graph G is isomorphic to the complete graph K_n for some $n \in \mathbb{N}$.

Note 5.30 For a tournament digraph $\vec{G}$ we have the following:

- For every pair of distinct vertices $u, v \in V(\vec{G})$, precisely one of (u, v) and (v, u) is in $E(\vec{G})$.
- If $\vec{G}$ has n vertices, then $\vec{G}$ has precisely $n(n-1)/2$ directed edges.
- For a given set $\{u_1, \ldots, u_n\}$, there are precisely $2^{n(n-1)/2}$ tournament digraphs with these vertices.

The name "tournament" is fitting, since every pairwise competition among n competitors, where there is no draw, can be described completely by a tournament digraph. For example, the outcome of a boxing competition among n boxers $b_1, \ldots, b_n$, where everyone fights everyone else, is fully described by a tournament digraph $\vec{G}$ with vertex set consisting of

$$V(\vec{G}) = \{b_1, \ldots, b_n\},$$

and where the directed edge $(b_i, b_j) \in E(\vec{G})$ if, and only if, boxer b_i defeated boxer b_j in their match. As all boxing fans know, the overall outcome can often be quite surprising, and a clear-cut winner is not always present! One reason that might in part explain this oft-occurring disagreement on the winner of a boxing competition is stated in the following theorem.

Theorem 5.31 Let $\vec{G}$ be a tournament digraph on n vertices. We have the following:

1. $\vec{G}$ always has a directed Hamiltonian path.
2. If $\vec{G}$ is strongly connected, then $\vec{G}$ has a directed cycle of length k, for every $k \in \{3, \ldots, n\}$. In particular, $\vec{G}$ is a directed Hamiltonian digraph in this case.

Proof We prove the first statement by induction on n, the number of vertices of $\vec{G}$.

The theorem is clear for $n \leq 3$. Assume we have a tournament digraph $\vec{G}$ on $n+1$ vertices where $n \geq 3$. Pick a vertex $u \in V(\vec{G})$. The digraph $\vec{G}' = \vec{G} - u$ is a tournament digraph on n vertices and by the induction hypothesis has a directed Hamiltonian path $\vec{p}$, which we can assume is given by $\vec{p} = (u_1, \ldots, u_n)$. We now consider three cases:

Case one: $(u, u_1) \in E(\vec{G})$: Here $(u)\vec{p}$ is a directed Hamiltonian path in $\vec{G}$.

Case two: $(u_n, u) \in E(\vec{G})$: Here $\vec{p}(u)$ is a directed Hamiltonian path in $\vec{G}$.

Case three: $(u_1, u), (u, u_n) \in E(\vec{G})$: Here there is a smallest $i \in \{2, \ldots, n-1\}$ such that $(u_{i-1}, u), (u, u_i) \in E(\vec{G})$, and here $(u_1, \ldots, u_{i-1}, u, u_i, \ldots, u_n)$ represents a directed Hamiltonian path in $\vec{G}$ (see Figure 5.9).

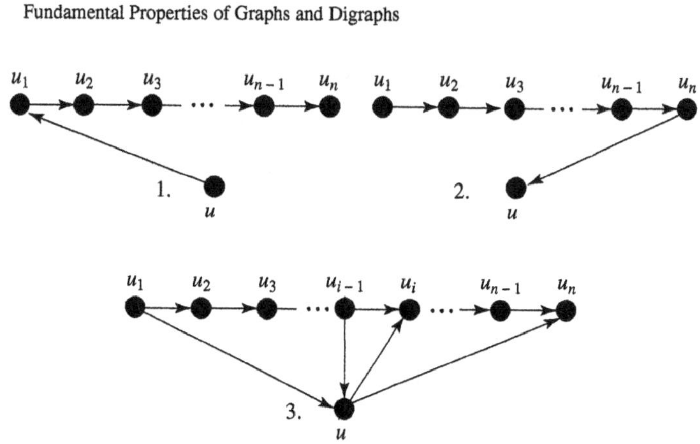

Figure 5.9: Cases one, two, and three in Theorem 5.31.

This completes the proof of the first statement.

We prove the second statement by induction on k. First let us show that there is indeed a directed 3-cycle: For a given vertex $u \in V(\vec{G})$, we clearly have

$$
\begin{aligned}
N^-(u) \cap N^+(u) &= \varnothing, \\
N^-(u) \cup N^+(u) &= V(\vec{G}).
\end{aligned}
$$

Since $\vec{G}$ is strongly connected, there is a directed edge $(v, w) \in E(\vec{G})$ with $v \in N^+(u)$ and $w \in N^-(u)$. In this case (u, v, w, u) is a directed cycle of length three in $\vec{G}$ (see Figure 5.10).

We now show that the existence of a directed cycle $(u_1, \ldots, u_k, u_1)$ of length $k \in \{3, \ldots, n-1\}$ implies the existence of a directed cycle of length $k+1$. First assume there is a vertex $u \in V(\vec{G})$ such that there are i, j with $(u, u_i), (u_j, u) \in E(\vec{G})$. In this case there exists a smallest index ℓ such that $(u_{\ell-1}, u), (u, u_\ell) \in E(\vec{G})$, and hence $(u_1, \ldots, u_{\ell-1}, u, u_\ell, \ldots, u_n)$ is a directed cycle of length $k+1$ in $\vec{G}$. If, however, there is no such vertex, then we can partition the set $V(\vec{G}) \setminus \{u_1, \ldots, u_k\} = A \cup B$, where A is the set of vertices that are heads of the directed edges pointing from $\{u_1, \ldots, u_k\}$, and B is the

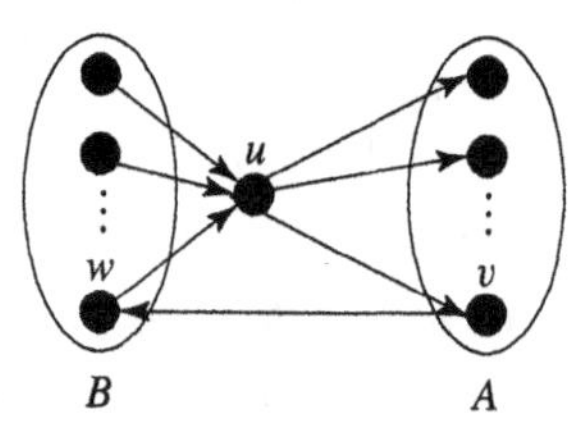

Figure 5.10: A cycle of length three in a tournament digraph.

set of those vertices that are tails of the directed edges pointing into $\{u_1, \ldots, u_k\}$. Since $\vec{G}$ is strongly connected, there must be vertices $v \in A$ and $w \in B$ such that $(v, w) \in E(\vec{G})$. Hence, $(u_1, v, w, u_3, \ldots, u_k, u_1)$ is a directed cycle of length $k + 1$ in $\vec{G}$, which completes our proof. ■

Going back to the dilemma of actually ranking the n boxers in the boxing competition, we now know that the corresponding digraph $\vec{G}$ has at least one Hamiltonian path, which yields one possible ranking

$$b_1 \to b_2 \to b_3 \to \cdots \to b_n,$$

where for each $i \in \{1, \ldots, n-1\}$ boxer b_i defeated boxer b_{i+1}. However, this method of ranking also suffers from some drawbacks. For one, there may be discrepancies between such a ranking and the overall scores of the boxers. Second, a tournament digraph may have more than one directed Hamiltonian path; therefore, several different rankings are possible for the boxers.

Similar context in which tournament digraphs naturally arise is the social study of chickens in a hen house, where the chickens have a delicate hierarchy system of a so-called *pecking order* among them. If one chicken c_1 pecks chicken c_2, then c_2 never pecks c_1. Moreover, for every pair of chickens, one always pecks the other. Therefore, the pecking order among chickens is completely described by a tournament digraph, where the vertices are represented by the chickens and each directed edge (c_i, c_j) indicates that c_i pecks c_j.

Definition 5.32 Let $\vec{G}$ be a tournament digraph. A vertex $u^* \in V(\vec{G})$ such that for every vertex $v \in V(\vec{G})$ we have either

1. $(u^*, v) \in E(\vec{G})$, or
2. there is a directed path (u^*, w, v) of length two in $\vec{G}$,

is called a *king* of $\vec{G}$.

Theorem 5.33 Every tournament digraph has a king. In fact, if $\vec{G}$ is a tournament digraph and $u^* \in V(\vec{G})$ is a vertex with maximum outdegree, then u^* is indeed a king of $\vec{G}$.

Proof Let $u^* \in V(\vec{G})$ be a vertex of $\vec{G}$ with maximum outdegree. Pick a vertex $v \in N^-(u^*)$. If $(v, w) \in E(\vec{G})$ for all $w \in N^+(u^*)$, then $N^+(u^*) \cup \{u^*\} \subseteq N^+(v)$, contradicting the fact that u^* is of maximum outdegree. Hence, there is a vertex $w \in N^+(u^*)$ with $(w, v) \in E(\vec{G})$, and hence, (u^*, w, v) is a directed path of length two. ■

By Theorem 5.33 we obtain the following corollary that holds for any enclosed collection of chickens.

Corollary 5.34 Every nonempty hen house has a chicken-king.

5.7 On the Adjacency Matrix of a Digraph

In the last two sections of this chapter, we study digraphs that have no cycles; this is in some sense the opposite of what we have done so far. But even in this case, a lot can be said about structure, showing again that cycles and circuits play a key role when studying structures of graphs and digraphs. First, we briefly discuss matrix representations of digraphs.

As for undirected graphs, one of the most convenient ways to represent a digraph in computer programs is by means of a square *adjacency matrix*. The definition for digraphs is similar to the undirected case.

Definition 5.35 Let $\vec{G}$ be a digraph on n vertices labeled by $V(\vec{G}) = \{u_1, \ldots, u_n\}$. For each $i, j \in \{1, \ldots, n\}$ define the entry a_{ij} by

$$a_{ij} = |\{e \in E(\vec{G}) : \eta(e) = (u_i, u_j)\}|.$$

That is, a_{ij} is the number of directed edges from u_i to u_j. The *adjacency matrix* of $\vec{G}$ with respect to the labeling of $V(\vec{G})$ is then defined to be the $n \times n$ matrix

$$\mathsf{A}(\vec{G}) = [a_{ij}]_{i,j \in \{1,\ldots,n\}}.$$

Example 5.36 Consider the digraph $\vec{G}$ shown in Figure 5.11. The matrix $\mathsf{A}(\vec{G})$ corresponding to the labeling of the vertices given by $V(\vec{G}) = \{u_1, u_2, u_3, u_4, u_5\}$ is as follows:

$$\mathsf{A}(\vec{G}) = \begin{bmatrix} 0&1&1&0&0 \\ 0&0&1&0&0 \\ 1&0&0&1&0 \\ 0&0&0&1&0 \\ 0&0&0&0&0 \end{bmatrix}.$$

Note here that each entry is either 0 or 1, since the graph has no parallel directed edges. Also, observe that in contrast to the undirected case, $\mathsf{A}(\vec{G})$ is not necessarily a symmetric matrix.

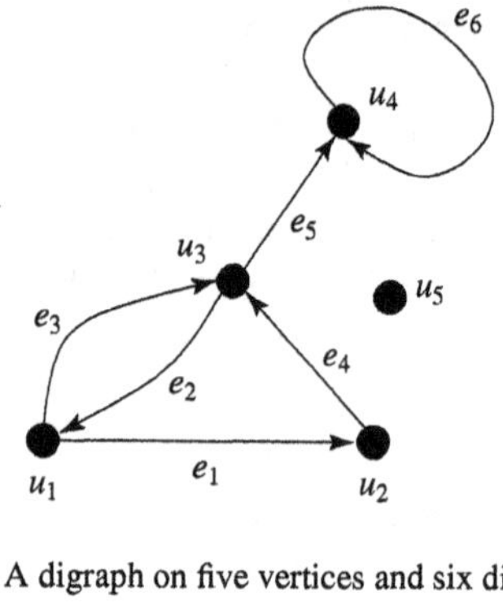

Figure 5.11: A digraph on five vertices and six directed edges.

Remark 5.37

- For any digraph $\vec{G}$, and any listing of its vertices, the adjacency matrix $\mathsf{A}(\vec{G})$ is a square matrix on nonnegative integers. Moreover, any square $n \times n$ matrix A with a_{ij} a nonnegative integer is an adjacency matrix for some digraph $\vec{G}$ on n vertices.
- If $\vec{G}$ is a digraph as in Example 5.36, with no parallel directed edges, then $\mathsf{A}(\vec{G})$ is a binary matrix. If, in addition, $\vec{G}$ has no directed loops, then $a_{ii} = 0$ for each i.
- The number of directed edges in $\vec{G}$ is the sum of all of the entries in $\mathsf{A}(\vec{G})$—that is,

$$|E(\vec{G})| = \sum_{i,j} a_{ij}.$$

- For each i the indegree $d^-(u_i)$ is the sum of all of the entries in column number i in $\mathsf{A}(\vec{G})$—that is,

$$d^-(u_i) = \sum_{\ell=1}^{n} a_{\ell i}.$$

- For each i the outdegree $d^+(u_i)$ is the sum of all of the entries in row number i in $\mathsf{A}(\vec{G})$—that is

$$d^+(u_i) = \sum_{\ell=1}^{n} a_{i\ell}.$$

As for the adjacency matrix for graphs, we have the following observation, whose proof we leave as Exercise 5.36.

Observation 5.38 Let $\vec{G}$ and $\vec{G}'$ be digraphs. The following statements are equivalent:

1. The digraphs $\vec{G}$ and $\vec{G}'$ are isomorphic.
2. Their adjacency matrices $\mathsf{A}(\vec{G})$ and $\mathsf{A}(\vec{G}')$ are orthogonally equivalent with respect to any fixed labeling of their vertices.

Also, similarly to Theorem 4.16, we have the analog for digraphs. The proofs of both the following theorem and its corollary are left as Exercises 5.37 and 5.38.

Theorem 5.39 Let $\vec{G}$ be a digraph with a given labeling $V(\vec{G}) = \{u_1, \ldots, u_n\}$ of its vertices and $k \geq 1$ a fixed integer. Then the entry $a_{ij}^{(k)}$, in row i and column j in the kth power $\mathsf{A}(\vec{G})^k$ of the adjacency matrix of $\vec{G}$, is the number of distinct directed walks from u_i to u_j of length k in $\vec{G}$.

Corollary 5.40 Let $\vec{G}$ be a digraph on $n \geq 2$ vertices. Let

$$\mathsf{Y} = \mathsf{A}(\vec{G}) + \mathsf{A}(\vec{G})^2 + \mathsf{A}(\vec{G})^3 + \cdots + \mathsf{A}(\vec{G})^{n-1}.$$

Then $\vec{G}$ is strongly connected if, and only if, all entries of Y are nonzero.

Needless to say, Corollary 5.40 can be quite a cumbersome calculation when n is large, since it simply lists all possible walks from any given vertex. This exhaustive case analysis, of determining whether there exists a directed walk from one vertex to another by listing all possibilities, can be very time consuming for n large. However, on the bright side, the information we obtain from each power is not only whether a directed walk exists or not, but precisely how many.

Note again that, by Theorem 5.39, the entry $a_{ij}^{(k)}$ in the kth power of $\mathsf{A}(\vec{G})$ counts the number of directed *walks* from u_i to u_j. We have, however, no simple way of determining which of these directed walks are actually directed trails and which are directed paths, and which are neither.

Example 5.41 Consider the graph shown in Figure 5.11. The third power of this adjacency matrix is given by

$$\mathsf{A}(\vec{G})^3 = \begin{bmatrix} 1 & 1 & 1 & 2 & 0 \\ 0 & 1 & 1 & 1 & 0 \\ 1 & 0 & 1 & 2 & 0 \\ 0 & 0 & 0 & 1 & 0 \\ 0 & 0 & 0 & 0 & 0 \end{bmatrix}.$$

The entry $a_{34}^{(3)} = 2$ represents two directed walks of length three from u_3 to u_4:

$$\begin{aligned} \vec{w}_1 &= (e_2, e_3, e_5), \text{ is a directed trail,} \\ \vec{w}_2 &= (e_5, e_6, e_6), \text{ is not a directed trail.} \end{aligned}$$

The reader is encouraged to examine the remaining entries in $\mathsf{A}(\vec{G})^3$.

5.8 Acyclic Digraphs and Posets

Any binary relation $\sim$ on a given set can be represented as a simple digraph with loops with the given set as the vertex set, where a directed edge from u to v is presented in the digraph if, and only if, $u \sim v$. Hence, a simple digraph can be viewed as a set of vertices together with a binary relation. An important class of simple digraphs consists of those digraphs that have no directed cycles. Such digraphs come up in various situations, such as family trees (which many times are not trees, if related people have children!) and various other schemes, describing some sort of interaction between agents over discrete moments in time.

Definition 5.42 A digraph $\vec{G}$ with no directed cycles is called an *acyclic* digraph.

One of the most important aspects of finite acyclic digraphs is given by the following theorem, which we hinted at in Exercise 1.33 in Chapter 1. In it we view a single vertex as a simple directed path of length zero. Recall Definition 1.34 of a partial order.

Theorem 5.43 Any acyclic digraph $\vec{G}$ yields a partial order $\preceq$ on $V(\vec{G})$ given by

$$u \preceq v \Leftrightarrow \text{there is a directed path from } v \text{ to } u \text{ in } \vec{G}. \tag{5.1}$$

Conversely, given a partial order $\preceq$ on a finite set V, the digraph $\vec{G}^*$ given by

$$V(\vec{G}^*) = V,$$
$$E(\vec{G}^*) = \{(v, u) : u \neq v, \text{ and } u \preceq v\}$$

is an acyclic digraph.

Proof Clearly the relation $\preceq$ given by (5.1) is reflexive, since the zero-length directed path from u to u is in $\vec{G}$, for all $u \in V(\vec{G})$.

Also, $\preceq$ is transitive, since if $\vec{p}$ is a directed path from w to v, and $\vec{q}$ a directed path from v to u, then the composite $\vec{p}\vec{q}$ is a directed path from w to u.

If now $u \neq v$ and $\vec{p}$ is a directed path from v to u, then $\vec{p}$ has a positive length. If now $\vec{r}$ is a directed path from u to v then $\vec{p}\vec{r}$ is a directed circuit that starts and ends at $v \in V(\vec{G})$. Since every directed circuit of positive length contains a directed cycle (Exercise 5.43), then $\vec{G}$ must contain a directed cycle. This contradicts the acyclicity of $\vec{G}$. Hence, if $u \preceq v$ and $v \preceq u$, then $u = v$ must hold.

The rest of the theorem is clear. ∎

In Figure 5.12 we now see that an acyclic digraph naturally yields a partial order. There the maximal vertices are on the 0th level and the minimal elements are on the shallowest or deepest level, in this case the second level.

Remark 5.44 Note that if we start with an acyclic digraph $\vec{G}$, then form the partial order $\preceq$ from this digraph by (5.1), and then finally form a digraph $\vec{G}^*$ from the partial order $\preceq$, then $\vec{G}$ and $\vec{G}^*$ are in general not equal. In fact, $\vec{G} = \vec{G}^*$ if, and only if, the original digraph is *transitive* in the sense that whenever $(u, v), (v, w) \in E(\vec{G})$, then $(u, w) \in E(\vec{G})$. (See Exercise 5.42.)

For example, if $\vec{G}$ is the directed path on three vertices and two directed edges as follows:

$$V(\vec{G}) = \{u_1, u_2, u_3\},$$
$$E(\vec{G}) = \{(u_1, u_2), (u_2, u_3)\},$$

then $\vec{G}^*$ turns out to be the transitive acyclic digraph, whose underlying graph is K_3,

$$V(\vec{G}^*) = \{u_1, u_2, u_3\},$$
$$E(\vec{G}^*) = \{(u_1, u_2), (u_2, u_3), (u_1, u_3)\}.$$

The digraph $\vec{G}^*$ is sometimes called the *transitive closure* of $\vec{G}$.

From Theorem 5.43 we obtain some interesting corollaries.

Corollary 5.45 If $\vec{G}$ is an acyclic digraph, then $\vec{G}$ contains vertices u and v with

$$d^-(u) = d^+(v) = 0.$$

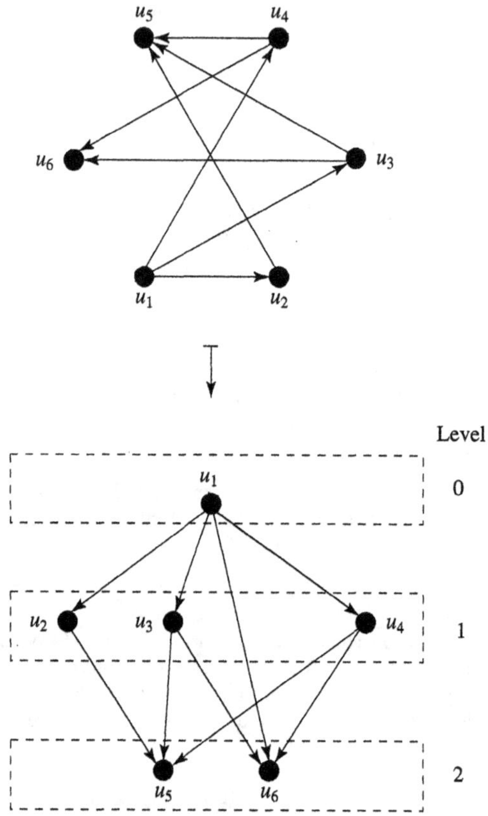

Figure 5.12: An acyclic digraph separated into levels.

Proof Let u be a maximal element with respect to the partial order $\preceq$ and v a minimal element; the theorem follows. ∎

Corollary 5.46 A digraph $\vec{G}$ is acyclic if, and only if, there is a labeling

$$V(\vec{G}) = \{u_1, \ldots, u_n\}$$

such that the corresponding adjacency matrix $\mathsf{A}(\vec{G})$ is an upper (or lower) triangular matrix.

Proof Assume first that $\mathsf{A}(\vec{G})$ is upper triangular—that is, $a_{ij} = 0$ for $i \geq j$. Since all powers $\mathsf{A}(\vec{G})^k$ are also upper triangular with zeroes on their diagonals (this can be shown directly; see Exercise 5.44), then we have that $\vec{G}$ has no cycles and is therefore acyclic.

Assume now that $\vec{G}$ is an acyclic digraph. Pick a labeling $V(\vec{G}) = \{u_1, \ldots, u_n\}$ such that for each $i \geq 1$, the vertex u_i is a maximal element of $V(\vec{G}) \setminus \{u_1, \ldots, u_{i-1}\}$ with respect to $\preceq$. By this labeling $\mathsf{A}(\vec{G})$ is an upper triangular matrix with zero on its diagonal.

That there is a labeling such that $\mathsf{A}(\vec{G})$ is a lower triangular matrix, we leave as Exercise 5.45. ■

Remark 5.47 The ordering of vertices in the proof of Corollary 5.46 is what computer scientists call a *topological sorting* of the vertices.

5.9 Exercises

5.1. For what values of n are the graphs N_n, P_n, and K_n bipartite?

5.2. Using Definition 5.1 of a bipartite graph and without using Theorem 5.3, show directly that for $n \geq 3$, the n-cycle C_n is bipartite if, and only if, n is even.

5.3. For what values of n are the graphs N_n, P_n, C_n, and K_n 3-partite?

5.4. Show directly, from the definition, that every subgraph of a bipartite graph is also bipartite.

5.5. Let G be a simple bipartite graph. Show that G is a subgraph of $K_{m,n}$ for suitable $m, n \in \mathbb{N}$.

5.6. Show that if $k \leq k' \leq n$, then any k-partite graph on n vertices is also k'-partite.

5.7. Using your favorite programming language, implement an algorithm that determines whether or not a graph is bipartite. You should be able to enter a graph, suitably coded, as input to your algorithm. Test your program on P_6, C_5, K_4, and $K_{3,3}$.

5.8. Determine whether the graphs shown in Figure 5.13 are Eulerian. If they are, find an Eulerian circuit through them. If they are not, then prove it. Do they possess an Eulerian trail? For those that do, find such an Eulerian trail, or else prove that no Eulerian trail exists.

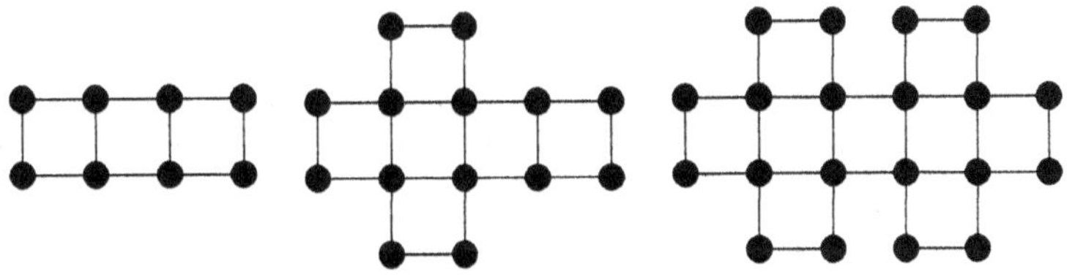

Figure 5.13: Eulerian graphs?

5.9. Can you draw the graphs shown in Figure 5.14 without lifting your pen or pencil from the paper and without redrawing a line already drawn? Can you do it starting and ending at the same point? Can you do it starting and ending at distinct points? When possible, construct such a continuous line, or prove that it cannot be done.

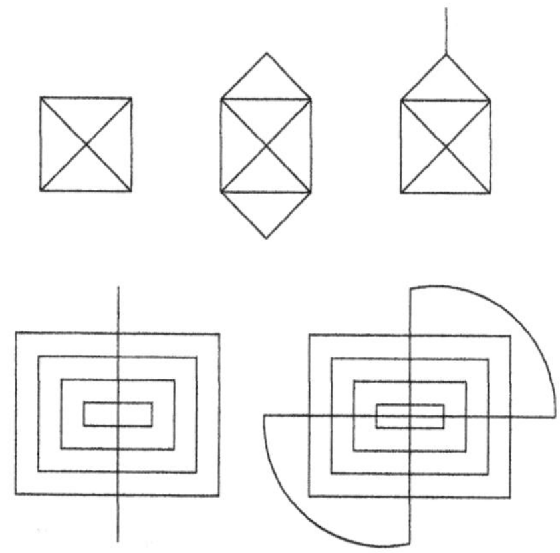

Figure 5.14: Continuously drawn figures?

5.10. Implement the algorithm implied by the proof of Theorem 5.7 using your favorite programming language. You should be able to enter a graph, suitably coded, as input to your algorithm. Test your program on the graphs shown in Figure 5.13. (Vertices occur at end points, "bends," or intersections.)

5.11. For $k \in \mathbb{N}$, let G be a connected graph that contains $2k$ vertices of odd degree. Show that there exist k edge-disjoint subgraphs $G_1, \ldots, G_k$, such that

(a) $E(G) = E(G_1) \cup \cdots \cup E(G_k)$,
(b) each G_i has an Eulerian trail.

[*Hint:* Consider the graph when you add k edges to form an Eulerian graph.]

5.12. Prove that ALGORITHM EULERIAN CIRCUIT (also known as FLEURY'S ALGORITHM) on page 138 outputs an Eulerian circuit when provided with an Eulerian graph as input. [*Hint:* Show that the algorithm stops/crashes with a circuit c_m covering m edges. Assuming that c_m does not cover all the edges, let $U \subseteq V(G)$ be the vertices of positive degree in $G - E(c_m)$. Show that the least edge in c_m going from U to $\overline{U}$ is a bridge. Derive a contradiction.]

5.13. Implement ALGORITHM EULERIAN CIRCUIT using your favorite programming language. You should be able to enter a graph, suitably coded, as input to your algorithm. Test your program on the graphs C_5, K_5, and $K_{2,4}$.

5.14. Show that neither of the graphs illustrated in Figure 5.6 is Hamiltonian. [*Hint:* Use Theorem 5.16.]

5.15. For $n \geq 2$, show that the complete bipartite graph $K_{n,n}$ is a Hamiltonian graph.

5.16. Determine all $m, n \in \mathbb{N}$ such that the complete bipartite graph $K_{m,n}$ is Hamiltonian.

5.17. Show that the graph G from Example 5.18 on the vertices $V(G) = \{u, u_1, \ldots, u_n, v_1, \ldots, v_n\}$ is such that for any $S \subseteq V(G)$ the number of components of $G - S$ is at most $|S|$.

5.18. Let m and n be positive integers. Let $[m] = \{1, \ldots, m\}$ and $[n] = \{1, \ldots, n\}$. Consider the *grid graph* $[m] \times [n]$ with vertex set $\{(i, j) : i \in [m] \text{ and } j \in [n]\}$ and where we connect every two ordered tuples (i, j) and (i', j') if, and only if, $|i - i'| + |j - j'| = 1$. Find necessary and sufficient conditions that m and n must satisfy in order for the graph $[m] \times [n]$ to be Hamiltonian.

5.19. Adopting the terminology from Exercise 5.18, we can make the following generalization: If k is a positive integer and $n_1, \ldots, n_k$ are also positive integers, the *k-box graph* $[n_1] \times \cdots \times [n_k]$ is the graph with vertex set $\{(i_1, \ldots, i_k) : i_l \in [n_\ell] \text{ for each } \ell \in [k]\}$ and where we connect every two ordered k-tuples $(i_1, \ldots, i_k)$ and $(i'_1, \ldots, i'_k)$ if, and only if, $|i_1 - i'_1| + \cdots + |i_k - i'_k| = 1$. For each fixed k, find necessary and sufficient conditions that the numbers $n_1, \ldots, n_k$ must satisfy in order for the corresponding k-box graph to be Hamiltonian.

5.20. Let G be a simple graph on n vertices and k components. Show that

$$d_G(u) + d_G(v) \leq n - k,$$

for all $u, v \in V(G)$. Show further that the upper bound of $n - k$ can be reached for all n and k.

5.21. Let G be a simple graph on n vertices such that

$$d_G(u) + d_G(v) \geq n - 1,$$

for all nonadjacent vertices $u, v \in V(G)$. Show that G has a Hamiltonian path. [*Hint:* Form a new graph G' on $n + 1$ vertices by connecting a new vertex u' to every vertex of G. Use Corollary 5.20 on G' and then obtain a Hamiltonian path in G.]

5.22. Let $c_1, \ldots, c_n$ be a collection of n cities. Show that a round trip though the cities that starts and ends in the same city and visits each city exactly once can be given uniquely by an ordered sequence $(c_{i_1}, c_{i_2}, \ldots, c_{i_{n-1}})$, where all the indices $i_1, \ldots, i_{n-1}$ from $\{1, \ldots, n\}$ are distinct. Show further that for each round trip given by such sequence, there is a unique and equivalent (equally as long!) opposite round trip. Use this to show that the total number of distinct Hamiltonian cycles in K_n is given by $(n - 1)!/2$.

5.23. Evaluate the formula $(n-1)!/2$ for the following values: 5, 10, 50, 100, 200, 300, 500, and 1,000. Assuming it takes a computer .001 second to evaluate a single Hamiltonian cycle in an instance of the traveling salesman problem, how long would it take to run instances of the problem of these sizes? For which values of n would you say this is feasible?

5.24. Consider an instance of the traveling salesman problem that uses driving distances between cities as the transportation costs between them. What would be a solution for the traveling salesman problem involving the following cities: Boston, Chicago, Los Angeles, Miami, New Orleans, Savannah, and Seattle? Use AAA mileage estimates and report the values you used.

5.25. Consider an instance of the traveling salesman problem that uses Delta Airlines' flying costs between cities as the transportation costs between them. What would be a solution for the traveling salesman problem involving the following cities: Boston, Chicago, Los Angeles, Miami, New Orleans, Savannah, and Seattle? Report the dollar values you used for the routes between cities. (If a direct flight between the cities is not available, use the cost of the cheapest alternative route.)

5.26. Let K_n be a weighted graph with n cities $c_1, \ldots, c_n$ as vertices. Show that for a minimum cost Hamiltonian cycle $C^* \subseteq K_n$ and a minimum cost spanning tree $T \subseteq K_n$, we have $W(C^*) > W(T)$.

5.27. Continuing from Exercise 5.26, let T' be a graph where every edge in a minimum cost spanning tree T of K_n has been doubled. Show that T' is Eulerian and $W(T') = 2W(T)$.

5.28. Continuing from Exercise 5.27, let the weight of each edge of K_n be given by $W(\{c_i, c_j\}) = w_{ij}$, and assume further that K_n satisfies the triangle inequality

$$w_{ik} \leq w_{ij} + w_{jk}. \tag{5.2}$$

(This is quite natural to do when measuring distances between cities!) Under this assumption, show that if C^* is a minimum cost Hamiltonian cycle of K_n then $W(C^*) \leq W(T')$, where T' is given in Exercise 5.27. [*Hint:* Obtain a Hamiltonian cycle (not necessarily minimum cost) from the Eulerian circuit of T' by "jumping" over each city you have already visited.]

5.29. Use Exercises 5.26, 5.27, and 5.28 to prove the following theorem:

Theorem 5.48 For a weighted complete graph K_n satisfying the triangle inequality (5.2), there is an efficient algorithm to find a weighted Hamiltonian cycle C such that $W(C) \leq 2W(C^*)$, where C^* is a minimum cost Hamiltonian cycle of K_n.

5.30. An editor from the *New York Times* has contacted you to write a one-page article for the *Times* that makes the ideas behind the traveling salesman problem understandable to the masses. Develop such an article, being sure to include lots of intuition about the problem.

5.31. Prove Theorem 5.25.

5.32. Let $\vec{G}$ be a digraph that is not Eulerian, but where its underlying graph G is Eulerian. Show that $\vec{G}$ does not have a directed Eulerian trail. [*Note!* This explains why the graph on the left in Example 5.27 does not have a directed Eulerian trail.]

5.33. Give an example of a strongly connected digraph, whose underlying graph is not a Hamiltonian digraph.

5.34. Show that a balanced digraph $\vec{G}$ is strongly connected if, and only if, it is weakly connected.

5.35. Consider the following definition:

Definition 5.49 Let $\vec{G}$ be a tournament digraph on n vertices. The *outdegree sequence* of $\vec{G}$ is the sequence $(d_1^+, \ldots, d_n^+)$ of outdegrees of vertices of $\vec{G}$ such that $d_1^+ \leq \cdots \leq d_n^+$.

Prove the following:

(a)

$$d_1^+ + \cdots + d_n^+ = \frac{n(n-1)}{2}.$$

(b) For each $k < n$, we have

$$d_1^+ + \cdots + d_k^+ \geq \frac{k(k-1)}{2}.$$

Moreover, the inequality is strict for all k if, and only if, $\vec{G}$ is strongly connected.

(c) $\vec{G}$ represents a transitive relation if, and only if, $d_i^+ = i - 1$ for all $i \in \{1, \ldots, n\}$.

5.36. Prove Observation 5.38. [*Hint:* Mimic the given proof for the undirected case.]

5.37. Prove Theorem 5.39. [*Hint:* Mimic the given proof for the undirected case.]

5.38. Prove Corollary 5.40.

5.39. Show that a digraph whose underlying graph is a tree is always acyclic.

5.40. For $n \geq 3$, how many acyclic digraphs have C_n as underlying graph?

5.41. Show that a digraph $\vec{G}$ is acyclic if, and only if, every directed walk in $\vec{G}$ is a directed path in $\vec{G}$. [*Note!* This shows that for a given labeling $V(\vec{G}) = \{u_1, \ldots, u_n\}$, the (i, j)-th entry of $\mathsf{A}(\vec{G})^k$ gives the number of distinct directed *paths* of length k from u_i to u_j in $\vec{G}$.]

5.42. Show that for $\vec{G}^*$, obtained from the acyclic digraph $\vec{G}$ as described in Remark 5.44 after Theorem 5.43, we have $\vec{G}^* = \vec{G}$ if, and only if, $\vec{G}$ is a transitive acyclic digraph.

5.43. Show that if a digraph $\vec{G}$ contains a directed circuit of positive length, then $\vec{G}$ must contain a cycle.

5.44. Show that if A is an upper triangular $n \times n$ matrix with zero on its diagonal, then

(a) X^k is upper triangular for all k.
(b) $\mathsf{X}^k = \mathsf{0}$ for all $k \geq n$.

5.45. Show that $\mathsf{A}(\vec{G})$ is a lower triangular matrix if, and only if, $\vec{G}$ is an acyclic digraph.. [*Hint:* Pick the labeling $V(\vec{G}) = \{u_1, \ldots, u_n\}$ where each u_i is *minimal* among the ones left.]

5.46. Let $\vec{G}$ be a digraph on n vertices, and let I_n be the $n \times n$ identity matrix. Show that $\vec{G}$ is acyclic if, and only if, $\mathsf{I}_n - \mathsf{A}(\vec{G})$ is an invertible matrix. [*Hint:* Use Theorem 5.43. Also, note that if $\tilde{x}$ is an $n \times 1$ column vector corresponding to a cycle C in $\vec{G}$ (that is, $x_i = 1$ if u_i is in C and $x_i = 0$ otherwise), then $\mathsf{A}(\vec{G})\tilde{x} = \tilde{x}$].

CHAPTER

6

Connectivity and Flow

Chapter Goals

Define edge cuts and vertex cuts for a graph, and discuss their properties.

Present a theorem of Robbins describing when a graph can be oriented into a strongly connected digraph.

Define edge connectivity and connectivity of a graph, and discuss their properties.

Define blocks in a separable graph G and the corresponding block-cutpoint graph of G.

Discuss general properties of weighted digraphs.

Present the Max-Flow-Min-Cut Theorem of Ford and Fulkerson.

Derive Menger's Theorems from properties of flows in networks.

Use one of Menger's Theorems to describe a property of k-connected graphs.

6.1 Edge Cuts

As we saw in Chapter 4, every connected graph has a spanning tree. A tree is connected, but by Theorem 3.7, the removal of one edge from it results in a disconnected forest. However, if a connected graph G is Hamiltonian, then G has a spanning cycle. Hence, the removal of one edge from G does not disconnect G, because the removal of one edge from a cycle does not disconnect the cycle and hence not the graph G either. In this sense cycles are better connected than trees.

In many real-life situations we need to know how "robust" a graph that represents a certain network is, how many edges or vertices can be removed without completely destroying the overall connectivity. For example, a network of connected computers (like the Internet) where all communication must go through a specific connection or central computer is more vulnerable than a network where every two computers are connected via several distinct computers or connections. This is precisely the main theme of this chapter—to study various levels of connectivity of graphs, and explore structural properties of graphs where we know a good deal about how well they are connected.

Definition 6.1 Let G be a connected graph. An *edge cut* or simply a *cut* S of G is a set of edges $S \subseteq E(G)$ such that the graph $G - S$ satisfies the following:

1. $G - S$ is disconnected.
2. $G - S'$ is connected for any proper subset $S' \subset S$.

Remark 6.2

- Note that an edge cut is a *minimal* set of edges that disconnects a connected graph.
- An edge cut S is a set of edges. The induced subgraph $G[S]$ of G consisting of the edges of S together with their endvertices is uniquely determined by S and vice versa. Hence, we sometimes talk about and view an edge cut as a subgraph of G.
- If e is a bridge, then the set $\{e\}$ consisting of e alone is an edge cut.
- Clearly, it is easy to extend the concept of an edge cut to a disconnected graph G, since in that case $\varnothing \subseteq E(G)$ is *the unique* minimal edge set that makes G disconnected.

Example 6.3 Consider the graph given in Figure 6.1. There the set $S = \{e_4, e_9\}$ is an edge cut of G, since the removal of both e_4 and e_9 from G makes it disconnected, whereas the removal of only one of these edges does not disconnect G.

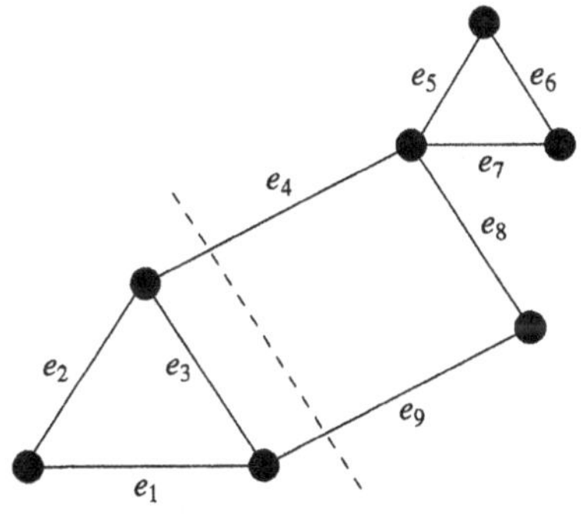

Figure 6.1: The removal of edges e_4 and e_9 disconnects the graph, but the removal of either one of these edges separately does not.

Example 6.4 Assume that the vertices $V(G)$ of the graph G in Figure 6.1 represent computers and the edges $E(G)$ represent direct connections between the computers, in the sense that if two computers are connected by an edge, then they can communicate directly.

If now the connection indicated by e_4 is lost, the network of computers can still work simply by routing the data using the edge e_9 "detour." However, if the connection indicated by e_9 is also faulty, then the network is broken and some computers will not be able to share data with some of the other computers.

This *toy scenario* often happens in the Internet, which is a very large network of computers: Sometimes when one sends an email, one later obtains a warning message stating that the mail has not been delivered yet, although the email address is a legitimate one. Such messages can be direct results of connections between computers being down. In this case the mail server tries to "reroute" the mail, which can take a long time, especially if many connections are down at that particular time.

We next discuss some basic properties of edge cuts in general. The first one is an easy lemma, whose proof we leave as Exercise 6.4. Note that here we talk about an edge cut of graphs that are not necessarily connected according to the previous remarks.

Lemma 6.5 If S is an edge cut of a connected graph G, then $G - S$ has precisely two components.

Theorem 6.6 Let G be a connected graph. For a set $S \subseteq E(G)$, the following are equivalent:

1. $G - S$ is disconnected.
2. Every spanning tree of G has an edge in S.

Proof Let T be a spanning tree of G. If $S \cap E(T) = \varnothing$, then the edges of $G - S$ contain $E(T)$ and hence T as a subgraph of $G - S$, thereby making $G - S$ connected.

If, however, S is such that $G - S$ is connected, then $G - S$ has a spanning tree T, which is also a spanning tree of G, since $V(G - S) = V(G)$. Hence, there is a spanning tree T with no edges in S. ■

From Theorem 6.6 we obtain the following corollary, which gives a different description of an edge cut $S \subseteq E(G)$ for a connected graph G, simply by considering *minimal* subsets $S \subseteq E(G)$ satisfying the two equivalent conditions stated in the theorem.

Corollary 6.7 Let G be a connected graph. A subset $S \subseteq E(G)$ is an edge cut if, and only if, S is a minimal set containing an edge of every spanning tree of G.

As mentioned in the beginning of this section, the set containing a single edge of a tree is an edge cut of the tree. Likewise, no set containing a single edge is an edge cut of a cycle. Hence, a set consisting of a single edge of a connected graph G that lies on a cycle in G is never an edge cut of G. The following theorem provides yet another property of cycles and edge cuts in a connected graph.

Theorem 6.8 Let G be a connected graph. If C is a cycle in G and S is an edge cut of G, then $|E(C) \cap S|$ is always even.

Proof Since S is an edge cut, by Lemma 6.5, $G - S$ divides into exactly two components with vertex sets V_1 and V_2, respectively. Note that S consists of those edges in G with one endvertex in V_1 and the other in V_2. Traversing along C, we must start and end at the same vertex, which we may assume is in V_1. Hence, for each time we go along an edge in S out of V_1 to V_2, there must be one edge in S on C out of V_2 back to V_1. Therefore, $|E(C) \cap S|$ must be even. ■

Note that if e is a bridge in G, then by Theorem 6.8 $|\{e\} \cap C|$ must be even for every cycle C in G, which can happen only when e is not an edge in C for any cycle C in G. Hence, we have the following corollary, which characterizes the bridges of G.

Corollary 6.9 For a graph G and $e \in E(G)$, the following are equivalent:

1. The edge e is a bridge of G.
2. The edge e is not an edge of any cycle in G.

At this point we can use Corollary 6.9 to answer the following natural question: For a given connected graph G can one direct its edges to obtain a strongly connected digraph $\vec{G}$? This question is answered by the following theorem due to Robbins [46].

Theorem 6.10 For a connected graph G the following are equivalent:

1. G has no bridges.
2. The edges of G can be directed in such a way that the resulting digraph $\vec{G}$ is strongly connected.

Proof Clearly if G has a bridge $e = \{u, v\}$, then if the direction (u, v) is assigned to e in $\vec{G}$, there is no directed walk from v to u in $\vec{G}$. Similarly, if the direction (v, u) is assigned to e, there is no directed walk from u to v in $\vec{G}$. Therefore, the first condition is necessary.

To show the sufficiency of the first condition is more involved. We assume G has more than one vertex, no bridges, and at least one edge. By Corollary 6.9, every edge of G must be on a cycle $C \subseteq G$. We define a sequence $G_1 \subseteq G_2 \subseteq \cdots \subseteq G_k$ of subgraphs of G in the following way: We let $G_1 = C$. While $G_i \neq G$, there is an edge e that is not in G_i but has one endvertex in G_i. Since e is not a bridge in G, e must be on a cycle in G and hence there is a simple path p_i' from the other endvertex to some vertex in G_i, whose edge set is disjoint from that of G_i. If $E_i = \{e\} \cup E(p_i')$, then $p_i = G[E_i]$ is a simple path with both its initial and final vertices in G_i, and p_i is edge-disjoint from G_i. Let $G_{i+1} = G_i \cup p_i$. Since $G_n = G$, we now can direct the edges of G:

1. Let $G_1 = C$ be a directed cycle. In this way $\vec{G}_1$ is strongly connected.
2. Assume $\vec{G}_i$ is strongly connected. Direct the edges of p_i in such a way that p_i becomes a directed simple path $\vec{p}_i$. In this way $\vec{G}_{i+1}$ is strongly connected. (See Figure 6.2.)

In this way $\vec{G} = \vec{G}_k$ becomes a strongly connected digraph. ∎

6.2 Edge Connectivity and Connectivity

As we mentioned earlier, the removal of one edge from a tree disconnects the tree. Likewise, the removal of one edge from a cycle does not disconnect the cycle, but the removal of any two edges from a cycle disconnects the cycle. In this case we say that the *edge connectivity* of a tree is one, and the edge connectivity of a cycle is two.

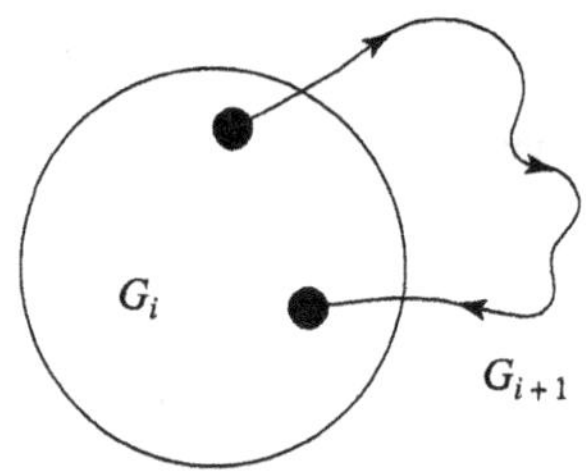

Figure 6.2: The formation of G_{i+1} from G_i in Theorem 6.10.

In general, we have the following definition.

Definition 6.11 Let G be a graph on two or more vertices. The smallest cardinality of an edge cut S of G is called the *edge connectivity* of G, and is denoted by $\kappa'(G)$. If $k \leq \kappa'(G)$, we say that G is *k-edge-connected.*

Remark 6.12 Note the following about edge connectivity:

- For any disconnected graph G, we have $\kappa'(G) = 0$.
- A connected graph G has a bridge if, and only if, we have $\kappa'(G) = 1$.
- If G is a graph and G' is the graph we get from G by removing all of its loops, then

$$\kappa'(G) = \kappa'(G').$$

In Definition 6.11 we assumed G to have at least two vertices. This is to guarantee that *some* $S \subseteq E(G)$ is an edge cut for G. If G is an isolated vertex, then G has no edge cut and hence the minimum cardinality could only be defined as ∞ in this case.

Example 6.13 Consider the graph G given in Figure 6.1. As we saw in Example 6.3, the set $S = \{e_4, e_9\}$ is an edge cut of G, and hence $\kappa'(G) \leq |S| = 2$. Since G has no bridges—that is, G has no edge cuts containing a single edge—we conclude that $\kappa'(G) = 2$ in this case.

If a connected graph has a cut-vertex, then the minimum number of vertices whose removal disconnects the graph is one. We can generalize connectivity for vertices as well.

Definition 6.14 Let G be a graph. The minimum number of vertices of G, whose removal disconnects G, or creates a graph with a single vertex, is called the *connectivity* of G and is denoted by $\kappa(G)$. If $k \leq \kappa(G)$, then we say that G is *k-connected.*

Example 6.15 Consider the graphs G_1 and G_2 in Figure 6.3. Since G_1 has a bridge, it has a cut-vertex, since the removal of either of the endvertices of the cut edge disconnects the graph G_1. So, we have here that $\kappa(G_1) = 1$.

However, the graph G_2 is isomorphic to the graph in Figure 6.1 and therefore has no bridges, but G_2 has a cut-vertex, so $\kappa(G_2) = 1$. Both G_1 and G_2 are therefore 1-connected.

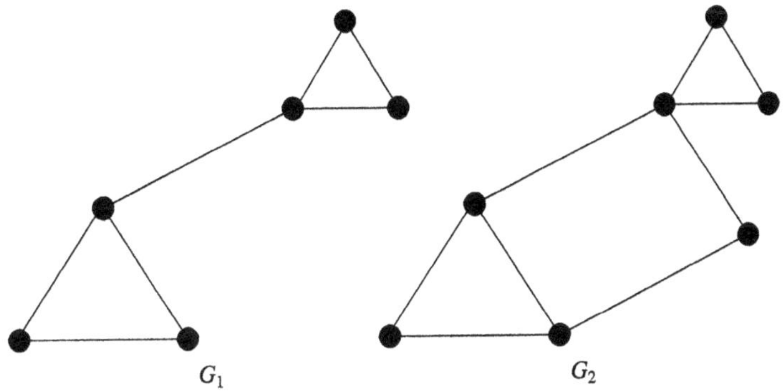

Figure 6.3: G_1 has a bridge and a cut-vertex; G_2 has no bridge but has a cut-vertex.

Remark 6.16 Several observations about vertex connectivity in general are worth noting.

- For any disconnected graph G, we have $\kappa(G) = 0$.
- For a graph G, let G' be the graph obtained from G by removing all the loops from G and changing all multiple edges to single ones. In this case we have

$$\kappa(G) = \kappa(G').$$

- The set of vertices removed in order to disconnect the graph or reduce it to a single vertex is called a *vertex cut*. It is instructive to compare and contrast the notion of vertex cut with that of edge cut. (See Definition 6.1.)

Example 6.17 For natural numbers $m, n \geq 2$, we observe the following:

- For the null graph N_n, we have $\kappa(N_n) = 0$.
- For any tree T on two or more vertices, we have $\kappa(T) = 1$.
- For any cycle C_n on $n \geq 3$ vertices, we have $\kappa(C_n) = 2$.
- For the complete graph K_n, we have $\kappa(K_n) = n - 1$.
- For the complete bipartite graph $K_{m,n}$, we have $\kappa(K_{m,n}) = \min(\{m, n\})$.

Example 6.18 Consider a similar situation as in Example 6.4 where the vertices $V(G)$ of a graph G represent computers in a computer network, and the edges $E(G)$ represent direct connections between the computers.

Consider the graphs G_1 and G_2 shown in Figure 6.4. Both G_1 and G_2 have seven vertices and both are connected by means of ten edges. As we mentioned in Example 6.4, the situation when a connection between a pair of computers fails is described by the removal

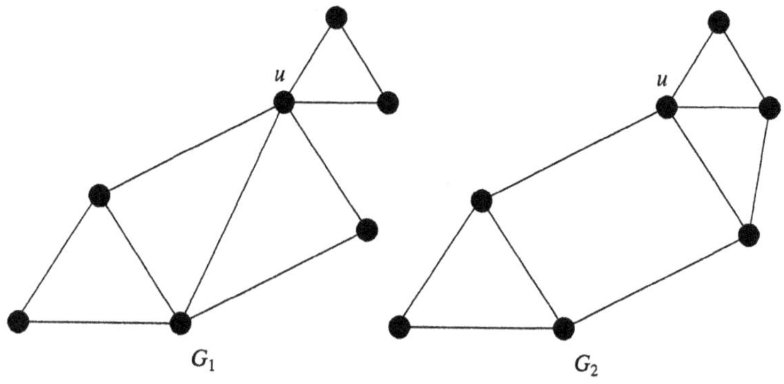

Figure 6.4: G_2 is "better" connected than G_1.

of the corresponding edge. Both G_1 and G_2 are "connection-robust" in the sense that they will remain connected if we remove any one edge (that is, neither has a bridge!). In fact, they both have edge connectivity of two, $\kappa'(G_1) = \kappa'(G_2) = 2$.

However, what happens if some computers break down? That situation is described by the removal of the corresponding *vertices* of either G_1 or G_2. We see that the vertex u in G_1 is a cut-vertex and hence the removal of u from G_1 will disconnect G_1. Hence, $\kappa(G_1) = 1$. Equivalently, the breakdown of the corresponding computer in the computer network will destroy the computer network. The vertex u in G_2 is not a cut-vertex, since the removal of u from G_2 does not disconnect G_2. In fact, the removal of any single vertex from G_2 does not disconnect G_2. Since G_2 has vertices of degree two (two of them), the removal of both neighbors of any degree-two vertex disconnects G_2. Hence, $\kappa(G_2) = 2$. Thus, although G_1 and G_2 are both equally "connection-robust," the connectivity tells us that the computer network corresponding to G_2 is more "computer-station-robust" than the one for G_1.

As we noted in Example 6.18, if a connected graph G has a vertex u of degree d, then the removal of all of the d neighbors from G either disconnects u from the rest of the graph (thereby leaving the graph disconnected), or makes the graph G into a single vertex graph (consisting of u alone). With this in mind, we make the next definition, which we use in the proof of Theorem 6.20.

Definition 6.19 For any graph G let

$$\Delta(G) = \max(\{d_G(u) \mid u \in V(G)\}),$$
$$\delta(G) = \min(\{d_G(u) \mid u \in V(G)\})$$

be the maximum degree and minimum degree, respectively, of any vertex in G. When it is unambiguous, we write Δ (respectively δ) instead of $\Delta(G)$ [respectively, $\delta(G)$].

Theorem 6.20 For a graph G, we have

$$\kappa(G) \leq \kappa'(G) \leq \delta(G).$$

Proof By removing all of the edges with a vertex of degree δ from G, we disconnect the graph, and hence $\kappa'(G) \leq \delta(G)$.

For the other inequality, let S be an edge cut consisting of $\kappa'(G)$ edges. Let $V(G) = V_1 \cup V_2$ be the corresponding partition and let $n_i = |V_i|$. We can assume $n_i \geq 2$ since otherwise the theorem follows easily. By the first part of this theorem, we have

$$\kappa'(G) \leq \delta(G) \leq \Delta(G) \leq |V(G)| - 1 \leq n_1 + n_2 - 1.$$

Since $(n_1 - 1)(n_2 - 1) > 0$, we have that $\kappa'(G) < n_1 n_2$. Hence, there are two vertices $u_1 \in V_1$ and $u_2 \in V_2$ that are not adjacent. For every edge $e \in S$, pick one endvertex of e that is not equal to u_1 nor u_2. Let W denote this set of vertices. Clearly, we have $|W| \leq |S| = \kappa'(G)$, and the removal of W eliminates all edges in S from G. Furthermore, the remaining graph consists of at least two components, one containing u_1 and the other u_2. Therefore, $\kappa(G) \leq |W| = \kappa'(G)$, which completes our proof. ∎

Note 6.21 The bounds in Theorem 6.20 can be reached. They can take any positive integer value, since

$$\kappa(K_n) = \kappa'(K_n) = \delta(K_n) = n - 1$$

for all $n \geq 1$.

Example 6.22 Let n and m be positive fixed integers with $m \geq n - 1$. Let $k = \lfloor 2m/n \rfloor$ and assume k is even. Construct a k-regular graph H on n vertices by connecting each vertex u on the cycle C_n to all vertices of distance $k/2$ or less from u in C_n. This graph is sometimes called the k-regular *Harary graph* on n vertices, due to Frank Harary. The graph shown in Figure 6.5 is the 4-regular Harary graph on eight vertices.

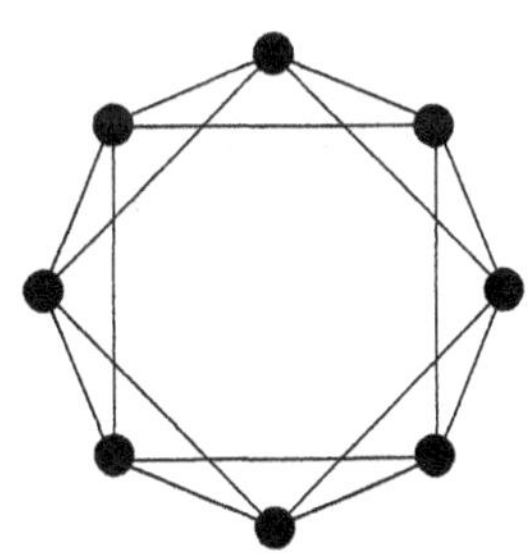

Figure 6.5: The 4-regular Harary graph on eight vertices.

One can argue that the graph resulting from removing any $k - 1$ vertices from H is connected, and hence $\kappa(H) \geq k$. (See Exercise 6.10.) By Theorem 6.20 we have that $\kappa(H) = k$. A similar construction can be developed in the case when k is odd. (See Exercise 6.11.)

It is easy to show that $\kappa(G) \leq \lfloor 2m/n \rfloor$ for all graphs on n vertices and $m \geq n - 1$ edges. (See Exercise 6.8.) Thus, the Harary graph H is the "best" connected graph on n vertices and $m \geq n - 1$ edges.

6.3 Blocks in Separable Graphs

This section focuses on graphs having connectivity one or two. They are of special interest, since their structures are relatively simple to describe. Recall the Definition 2.33 of a cut-vertex. One way to describe a cut-vertex is the following: A vertex u in a graph G is a cut-vertex if, and only if, there are vertices $v, w \in V(G) \setminus \{u\}$ such that every path between v and w contains the vertex u (see Exercise 6.12).

We begin with an important definition.

Definition 6.23 Let G be a connected graph on $n \geq 3$ vertices.

1. G is *separable* if it has at least one cut-vertex. Otherwise, G is *nonseparable*.
2. A maximal nonseparable connected subgraph in G is called a *block* of G.

Remark 6.24

- Note that both K_1 and K_2 are nonseparable graphs, although neither one of them is 2-connected.
- A graph G on $n \geq 3$ vertices is separable if, and only if, $\kappa(G) = 1$.
- A graph G on $n \geq 3$ vertices is nonseparable if, and only if, $\kappa(G) > 1$.
- It makes sense to talk about the blocks of any graph G, even if G is not connected. We just consider each component at a time; a block of a component of G is then a block of G.

Note 6.25 The only nonseparable graphs with connectivity one or less are K_1 and K_2, where $\kappa(K_n) = n - 1$ for *all* $n \in \mathbb{N}$, in particular for $n \in \{1, 2\}$. It is a "Catch-22" with the graph K_1: It is connected, so a priori we would like it to have connectivity of one or greater, yet it is a complete graph and should satisfy the formula $\kappa(K_n) = n - 1$, which is precisely the common convention in the graph-theory literature. So, $G = K_1$ is the *only* connected graph G with $\kappa(G) = 0$.

The next example illustrates Definition 6.23.

Example 6.26 In Figure 6.6 the vertex u is a cut-vertex, and the graph is separable with four blocks B_1, B_2, B_3, and B_4.

The next lemma is needed in the proof of Theorem 6.30, the main theorem of this section.

Lemma 6.27 If B_1 and B_2 are two distinct blocks in a graph G, then $B_1 \cap B_2$ either is empty or consists of a single vertex.

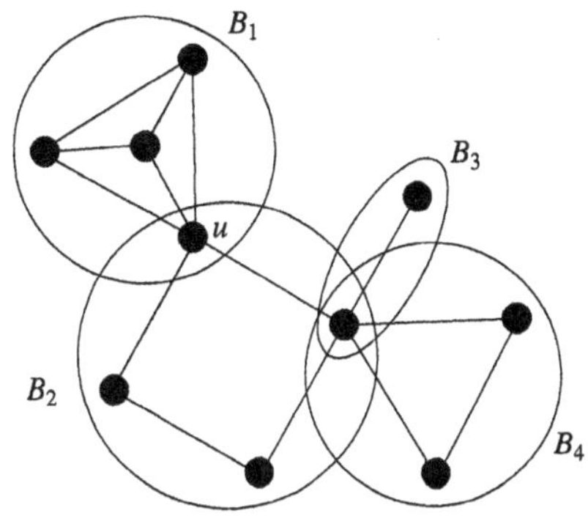

Figure 6.6: A connected graph with four blocks.

Proof Assume that B_1 and B_2 have two distinct vertices u and v in common. We argue that $B_1 \cup B_2$ is nonseparable as follows.

Removing a vertex $w \notin \{u, v\}$ from one of the blocks, say B_1, will leave a subgraph $(B_1 - w) \cup B_2$ of $B_1 \cup B_2$. Since B_1 is a block, then $B_1 - w$ is connected, and so is B_2. If $x \in V(B_1 - w)$ and $y \in V(B_2)$, then there are paths from x to u in $B_1 - w$, and from u to y in B_2, respectively. Hence, $(B_1 - w) \cup B_2$ is connected. Likewise, removing either u or v, say u, from $B_1 \cup B_2$ will yield a subgraph $(B_1 - u) \cup (B_2 - u)$ of $B_1 \cup B_2$. Every vertex in $V(B_1 - u)$ can be connected to every vertex in $V(B_2 - u)$ via the vertex v. So, in any case, the removal of a single vertex from $B_1 \cup B_2$ yields a connected graph. Therefore, $B_1 \cup B_2$, which properly contains both B_1 and B_2, is nonseparable. This contradicts the fact that B_1 and B_2 are blocks of G. ■

Corollary 6.28 If $\mathcal{B}(G)$ is the set of blocks of a loopless graph G, then

$$E(G) = \bigcup_{B \in \mathcal{B}(G)} E(B)$$

is a partition of the edge set of G.

Definition 6.29 Let G be a connected graph. Let $C(G) \subseteq V(G)$ be the set of cut-vertices of G. Let $\mathcal{B}(G)$ be the set of blocks of G. The *block-cutpoint graph* is the graph with vertex set $C(G) \cup \mathcal{B}(G)$ and edge set

$$\{\{u, B\} : u \in C(G), B \in \mathcal{B}(G), \text{ and } u \in V(B)\}.$$

It is denoted by BC(G).

In Figure 6.7 we see the block-cutpoint graph BC(G) of the graph G in Figure 6.6. There the vertices corresponding to blocks are white or hollow, and the vertices corresponding to the cut-vertices are black.

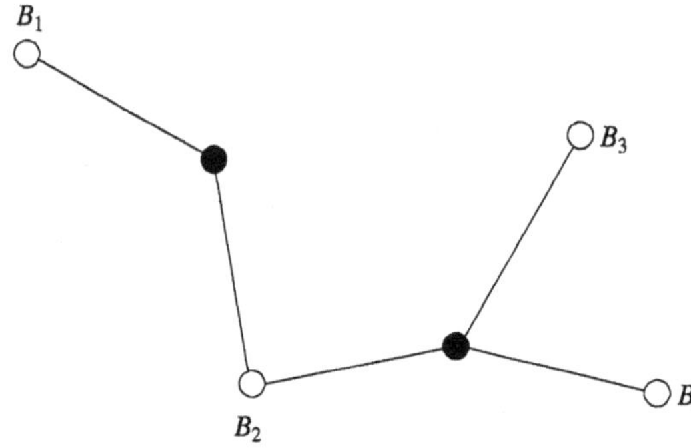

Figure 6.7: The block-cutpoint graph of the graph shown in Figure 6.6.

We note that BC(G) is a bipartite graph with vertex partition $C(G) \cup \mathcal{B}(G)$. Moreover, we have the following.

Theorem 6.30 For a connected graph G, the block-cutpoint graph BC(G) is a tree. In addition, all of the leaves of BC(G) are contained in $\mathcal{B}(G)$.

Proof If G is a connected graph, then clearly BC(G) is also connected. Assume that we have a simple cycle $(u_1, B_1, u_2, B_2, \ldots, u_m, B_m, u_1)$ in BC(G). By Lemma 6.27, we have that $m \geq 3$. Also, u_3 is a cut-vertex of G, which is not contained in B_1. Since $G' = B_2 \cup \cdots \cup B_m$ is a connected graph, there is a path p from u_2 to u_1 contained in G' and containing $u_2, u_3, \ldots, u_m, u_1$. Now the subgraph $B_1 \cup p$ in G is clearly nonseparable and strictly contains both B_1 and the vertex u_3. This contradicts the fact that B_1 is a block of G.

Since each cut-vertex is contained in at least two blocks, every cut-vertex must be an internal vertex of BC(G). Therefore, all of the leaves in BC(G) are contained in $\mathcal{B}(G)$. ■

Remark 6.31 If G is disconnected, then BC(G) is a forest with the same number of components as G. If G has components $G_1, \ldots, G_k$, then

$$\mathrm{BC}(G) = \mathrm{BC}(G_1) \cup \cdots \cup \mathrm{BC}(G_k).$$

By Theorem 3.4, every tree on two or more vertices has at least two leaves, and hence we obtain the following.

Corollary 6.32 Every connected separable graph has at least two blocks containing exactly one cut-vertex of G.

The previous results give us a way to reduce our consideration to the blocks of a given separable graph. In particular, if we can demonstrate a property for each block of given graph G, then we can use induction on the number of blocks of G using Corollary 6.32. But in order to follow this course of action, we need some description of the blocks.

By definition, a block of a connected graph G on three or more vertices has no cut-vertices. Therefore, it is 2-connected. Likewise, each maximal 2-connected subgraph of G is a block of G containing three or more vertices. Hence, a block of G is either one of K_1 or K_2, or a 2-connected subgraph on three or more vertices. So, describing the blocks on three or more vertices of G is the same as describing 2-connected graphs in general.

Theorem 6.33 For a connected graph G on three or more vertices the following statements are equivalent:

1. G is 2-connected.
2. For every pair of distinct vertices, there is a cycle in G passing through both of them.
3. For every pair of distinct vertices, there are two vertex-disjoint paths in G connecting them.

Proof Clearly, we have that statement 3 is true if, and only if, statement 2 is true. For the other equivalences, we proceed as follows:

$2 \Rightarrow 1$: Let x, y, and u be distinct vertices of G, and γ a simple cycle through the two vertices x and y. Then one of the paths from x to y will not contain the vertex u. This holds for arbitrary x and y, so u is not a cut-vertex. Hence, G is 2-connected.

$1 \Rightarrow 2$: Let x and y be vertices of G. Let

$$U_x = \{u \in V(G) : \text{ there is a cycle through } x \text{ and } u \text{ in } G\}.$$

Since G is 2-connected on three or more vertices, it cannot have any bridges. Otherwise, one of the endvertices would be a cut-vertex. But any edge e, which is not a bridge, must by Corollary 6.9 be contained in a cycle in G. Hence, all of the neighbors of x are contained in U_x.

Assume now that $y \notin U_x$. We will derive a contradiction. Let $\alpha = (u_0, e_1, u_1, \dots, e_a, u_a)$ be a path from $u_0 = x$ to $u_a = y$. Let i be the largest index such that $u_i \in U_x$. From above, we have $1 \le i < a$ and $u_i \neq x$. Since G is 2-connected, $G - u_i$ is connected. So there is a path $\beta = (v_0, f_1, v_1, \dots, f_b, v_b)$ from $v_0 = x$ to $v_b = u_{i+1}$ that does not contain u_i. Let γ be a cycle through x and u_i. Let k be the largest index such that v_k is on γ. At this point we construct a cycle though x and u_{i+1} as follows. (See Figure 6.8.)

Let p_1 be the path from x to v_k along the path of γ that does not contain u_i. If $k = 0$, then p_1 consists of just one vertex. Let p_2 be the path from v_k to u_{i+1} along β. Let p_3 be the path (u_{i+1}, e_{i+1}, u_i). Finally, we let p_4 be the path from u_i to x along the path of γ that does not contain v_k. We now have a cycle

$$\gamma' = p_1 p_2 p_3 p_4$$

through both x and u_{i+1}, contradicting the fact that $u_{i+1} \notin U_x$. ■

Note that one can often reduce the study of a particular graph to that of its blocks. That is to say, many properties hold for a given connected graph if, and only if, they hold for each of its blocks. Two graphs that have the same blocks are sometimes said to be *1-isomorphic*. More precisely, two graphs G and G' are 1-isomorphic if, and only if, G has n blocks $B_1, \dots, B_n$ (some of these blocks could be isomorphic!), G' has n blocks $B'_1, \dots, B'_n$, and

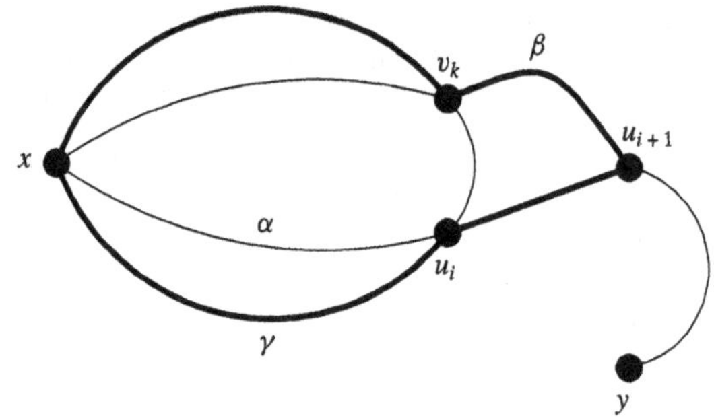

Figure 6.8: A cycle through x and u_{i+1}.

each B_i is isomorphic to B_i'. For example, the graph in Figure 6.9 is 1-isomorphic to the graph in Figure 6.6, since they both have four blocks B_1, B_2, B_3, and B_4. In fact, they both have ten vertices and twelve edges. However, it is not necessary for two 1-isomorphic graphs to have the same number of vertices. In Figure 6.10 we see a graph with eleven vertices that is 1-isomorphic to the graph shown in Figure 6.6.

Consider the following operation called a *1-split*. For a graph G, let u be a cut-vertex of G. Let $B_1, \ldots, B_m$ be the set of blocks containing u.

1. Introduce new vertices $b_1, \ldots, b_m$.
2. In $G - u$ connect each vertex b_i to the neighbors of u in B_i while preserving edge multiplicity. This results in a new graph G'.

The graph G' has one less cut-vertex than G. However, it is 1-isomorphic to G. The 1-split pulls the blocks that are connected by u apart and "splits" u into m copies, such that each block B_i has a copy of u. (See Figure 6.11.)

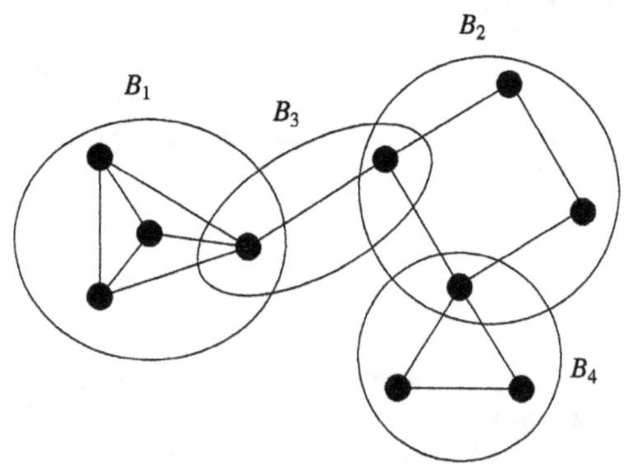

Figure 6.9: A graph with the same blocks as the one in Figure 6.6.

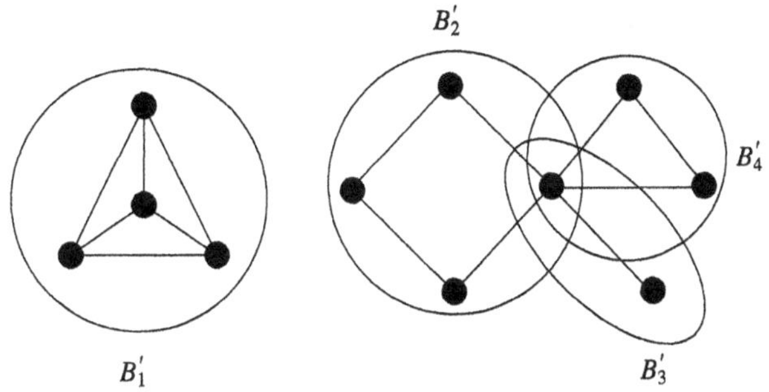

Figure 6.10: G' has the same set of blocks as G from Figure 6.6.

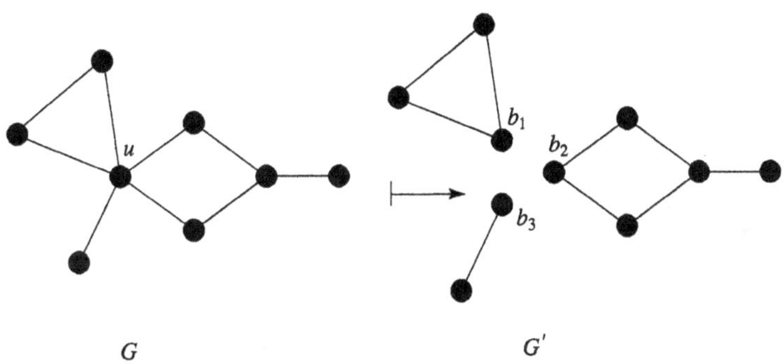

Figure 6.11: G' derived from G by a 1-split of the cut-vertex u.

By repeating the 1-split operation until each block is a component of the graph, we have the following observation.

Observation 6.34 Two graphs are 1-isomorphic if, and only if, they become isomorphic under repeated 1-split operations until each block is a component.

6.4 Flows in Networks

Edge connectivity and connectivity in general are closely related to many practical computational problems that can be presented in terms of certain weighted digraphs called *networks*. Our ultimate goal is to find the maximum *flow* of a network, a concept we will define and discuss in this section.

We should mention that our coverage here is by no means a comprehensive one. In fact, the topic of this section encompasses enough material for several books. Our purpose is to give an introduction, so that we are able to present and prove Menger's Theorems in §6.5.

Recall that a weighted graph is a tuple (G, W), where G is a graph and $W : E(G) \to \mathbb{R}$ is a function. In the same way, we can define a *weighted digraph* as a tuple $(\vec{G}, W)$, where $W : E(\vec{G}) \to \mathbb{R}$ is a function. When there is no danger of ambiguity and the weight W is understood, we simply talk about the weighted digraph $\vec{G}$. For a vertex $u \in V(\vec{G})$, we let $\mathrm{Tl}(u)$ and $\mathrm{Hd}(u)$ be the set of edges with u as a tail and head, respectively. Let

$$W^+(u) = \sum_{e \in \mathrm{Tl}(u)} W(e),$$
$$W^-(u) = \sum_{e \in \mathrm{Hd}(u)} W(e).$$

That is, $W^-(u)$ is the sum of the weights of all directed edges coming into u, and $W^+(u)$ is the sum of the weights of all edges going out of u.

The following two definitions are fundamental for this section.

Definition 6.35 A *network* is a 4-tuple $(\vec{G}, c; s, t)$, where $(\vec{G}, c)$ is a weighted digraph with a nonnegative weight $c \geq 0$, and $s, t \in V(\vec{G})$ are two distinguished vertices, the *source vertex* s and the *sink* or *target vertex* t. When there is no danger of ambiguity, we simply talk about the network $\vec{G}$.

For each $e \in E(\vec{G})$, the weight $c(e)$ is called the *capacity* of the edge e.

Definition 6.36 A *flow* of a network $(\vec{G}, c; s, t)$ is a function $f : E(\vec{G}) \to \mathbb{R}$ such that

1. For each $e \in E(\vec{G})$, we have $0 \leq f(e) \leq c(e)$.
2. For each $u \in V(\vec{G}) \setminus \{s, t\}$, we have

$$f^-(u) = f^+(u). \tag{6.1}$$

The *value* of f, denoted by $\mathrm{val}(f)$, is defined by $\mathrm{val}(f) = f^+(s) - f^-(s)$. By a *maximum flow* f we mean a flow f with $\mathrm{val}(f)$ at maximum—that is, $\mathrm{val}(f) \geq \mathrm{val}(f')$ for all flows f'.

Remark 6.37 The condition (6.1) is often called *Kirchhoff's Law* for the German physicist Gustav Robert Kirchhoff, who was investigating networks of electrical circuits in the 1850s.

Example 6.38 Two countries, Iceland (I) and Norway (N), export cod to two countries, Germany (G) and the United States (U). The supply and demand of fish each week, in metric tons, are given by the following:

$$I: 4{,}000, \quad G: 2{,}000,$$
$$N: 8{,}000, \quad U: 2{,}500.$$

Each week, one refrigerated cargo vessel sails between every pair (I, G), (I, U), (N, G), and (N, U) of the countries as indicated in Table 6.1, where the number stands for their capacity in metric tons.[1] We can now form a network $\vec{G}$ by adding two additional vertices,

[1] All these estimates are according to Agnar Erlingsson, a recently retired naval architect and principal marine surveyor in Iceland. He is moreover the father of one of the authors (can you tell which author?).

	G	U
I	500	1,100
N	1,000	800

Table 6.1: The capacity of the ships in metric tons.

s (the source) and t (the sink), in addition to the four vertices I, N, G, and U representing the countries. This completely sets up the problem of determining the maximum cod transportation load, as shown in Figure 6.12, where the capacity is shown in bold numbers. Finding how to transport the most amount of cod is the same as finding the maximum flow f of the network $\vec{G}$.

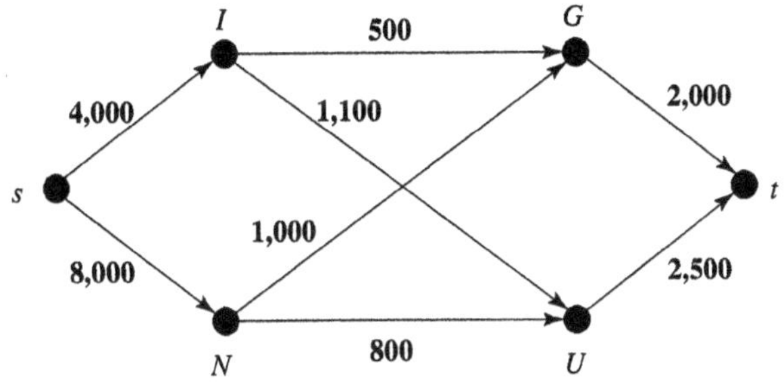

Figure 6.12: The network $\vec{G}$ completely describes the problem of determining the maximum cod transportation load.

The problem of actually finding the maximum flow for a network (if it exists!) can in general be a daunting task. Before we discuss how to compute the maximum flow, we need to introduce some notation and easy results.

For a weighted digraph $(\vec{G}, W)$ and subsets $S, T \subseteq V(\vec{G})$, we let $W(S, T)$ denote the sum of all of the weights of directed edges with their tail in S and head in T—that is,

$$W(S, T) = \sum_{\eta(e) \in S \times T} W(e).$$

If $\overline{X} = V(\vec{G}) \setminus X$ denotes the complement of X in $V(\vec{G})$, we set

$$\begin{aligned} W^-(T) &= W(\overline{T}, T), \\ W^+(S) &= W(S, \overline{S}). \end{aligned}$$

That is, $W^-(T)$ is the sum of the weights of all of the directed edges pointing into T—that is, with heads in T. Similarly, $W^+(S)$ is the sum of the weights of all of the directed edges pointing out of S—that is, the edges with tails in S. (See Figure 6.13.) Note that we have, in particular, that $W^-(\{u\}) = W^-(u)$ and $W^+(\{u\}) = W^+(u)$.

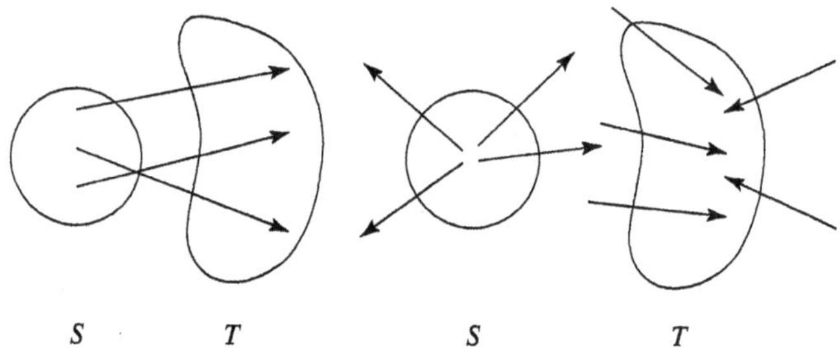

Figure 6.13: Illustration of $W(S, T)$, $W^-(T)$, and $W^+(S)$.

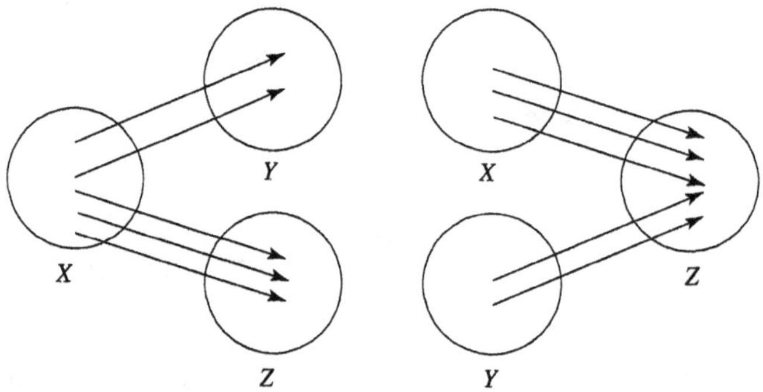

Figure 6.14: Illustration of the situation in Lemma 6.39.

The proof of the following lemma is easy and left as Exercise 6.23. Figure 6.14 illustrates the situation.

Lemma 6.39 Let $(\vec{G}, W)$ be a weighted digraph and $X, Y, Z \subseteq V(\vec{G})$.

1. If $Y \cap Z = \varnothing$, then $W(X, Y \cup Z) = W(X, Y) + W(X, Z)$.
2. If $X \cap Y = \varnothing$, then $W(X \cup Y, Z) = W(X, Z) + W(Y, Z)$.
3. We have $W^-(X) = W^+(\overline{X})$.
4. We have $W^+(X) = W^-(\overline{X})$.

Note that Lemma 6.39 holds for an arbitrary weighted digraph, without assuming W to be a flow satisfiying the Kirchhoff's condition (6.1).

Corollary 6.40 For a weighted digraph $(\vec{G}, W)$ and any subset $X \subseteq V(\vec{G})$, we have

$$W^+(X) - W^-(X) = \sum_{u \in X} (W^+(u) - W^-(u)).$$

Proof For $X \subseteq V(\vec{G})$ and $u \in V(\vec{G}) \setminus X$, let $X' = X \cup \{u\}$. By Lemma 6.39 we get

$$\begin{aligned} W^+(X') &= W(X \cup \{u\}, \overline{X'}) \\ &= W(X, \overline{X'}) + W(\{u\}, \overline{X'}) \\ &= W(X, \overline{X}) - W(X, \{u\}) + W(\{u\}, \overline{X}) - W(\{u\}, \{u\}). \end{aligned}$$

In exactly the same fashion, we obtain

$$W^-(X') = W(\overline{X}, X) - W(\{u\}, X) + W(\overline{X}, \{u\}) - W(\{u\}, \{u\}).$$

Combining these two expressions for $W^+(X')$ and $W^-(X')$, respectively, we obtain that

$$\begin{aligned} W^+(X') - W^-(X') &= W^+(X) - W^-(X) \\ &\quad + [W(\{u\}, \overline{X}) + W(\{u\}, X)] \\ &\quad - [W(X, \{u\}) + W(\overline{X}, \{u\})] \\ &= W^+(X) - W^-(X) + W^+(u) - W^-(u), \end{aligned}$$

from which the corollary now follows by induction on $|X|$. ∎

Remark 6.41 Corollary 6.40 could (with some stretch of the imagination!) be viewed as a special case of a well-known theorem in analysis due to Gauss. Namely, if $X \subseteq V(\vec{G})$, then $W^+(X) - W^-(X)$ really measures the net flow out of X—that is, the total flow going through the *boundary* of X—which we can denote by ∂X. Also, each $W^+(u) - W^-(u)$ measures the total flow through the vertex or *point* $u \in X$—that is, the total "infinitesimal" flow through the point $u \in X$. (See Figure 6.15.) Such an infinitesimal flow is usually denoted by ∂W.

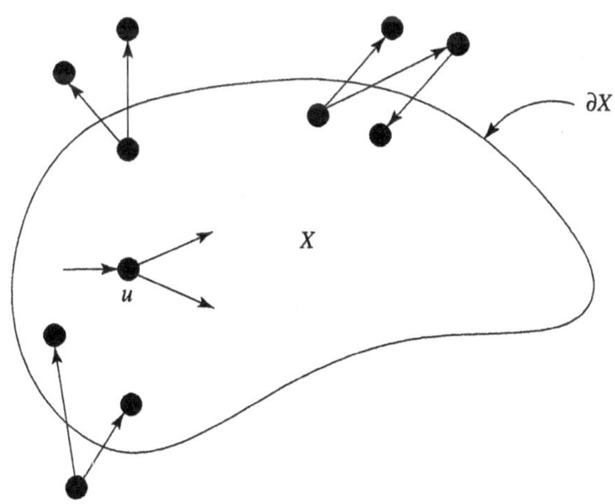

Figure 6.15: Total flow through ∂X is the sum of all "infinitesimal" flows in X.

Now, since the generalized symbol for a sum $\sum$ is the integral sign $\int$, with this notation in mind we have that $W^+(X) - W^-(X) = \int_{\partial X} W$ and $\sum_{u \in X}(W^+(u) - W^-(u)) = \int_X \partial W$. Therefore, Corollary 6.40 can be written as

$$\int_{\partial X} W = \int_X \partial W,$$

which in analysis is the well-known theorem of Gauss, called *Gauss's Law*, for integrating or "summing" a function W over the smooth two-dimensional manifold ∂X in $\mathbb{R}^3$, with interior X. (See Figure 6.16.)

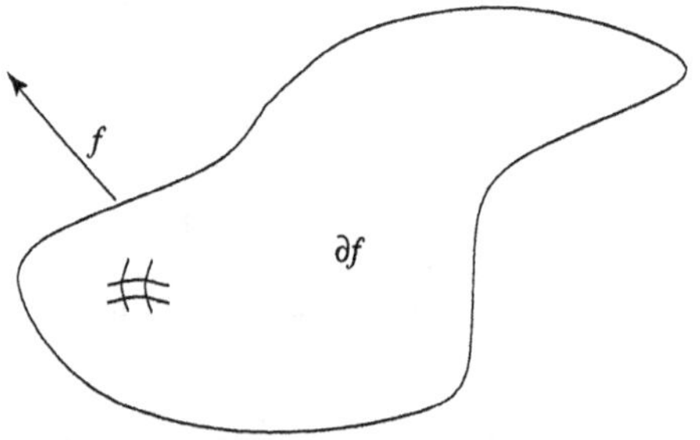

Figure 6.16: Summing ∂f inside X is the same as summing f on ∂X.

Note that it seems as if we have just moved the "∂"-sign from X over to the function W and obtained a valid formula. Such a phenomenon, where the moving of a "∂"-sign does not seem to change the outcome, occurs again and again in mathematics.

We now turn our attention to networks rather than just weighted digraphs.

Definition 6.42 Let $(\vec{G}, c; s, t)$ be a network.

1. A set of vertices $S \subseteq V(\vec{G})$ is called a *source vertex cut* if $s \in S$ and $t \notin S$. The source vertex cut S is *minimum* if its *capacity* $c^+(S)$ is at minimum—that is, $c^+(S) \leq c^+(S')$ for all source vertex cuts S'.
2. A set of vertices $T \subseteq V(\vec{G})$ is called a *sink vertex cut* if $t \in T$ and $s \notin T$. The sink vertex cut T is *minimum* if its *capacity* $c^-(T)$ is at minimum—that is, $c^-(T) \leq c^-(T')$ for all sink vertex cuts T'.

The following observation is a direct consequence of Kirchhoff's Law, Corollary 6.40, and the definition of a flow; its proof is therefore left as Exercise 6.26.

Observation 6.43 Let $(\vec{G}, c; s, t)$ be a network and f a flow.

1. If S is a source vertex cut, then $\mathrm{val}(f) = f^+(S) - f^-(S) \leq c^+(S)$.
2. If T is a sink vertex cut, then $\mathrm{val}(f) = f^-(T) - f^+(T) \leq c^-(T)$.

In particular, we have $\mathrm{val}(f) = f^+(s) - f^-(s) = f^-(t) - f^+(t)$.

From Observation 6.43, we have the following easy-to-prove, yet important theorem.

Theorem 6.44 Let f be a flow for a network $\vec{G}$.

1. If S is a source vertex cut satisfying $\text{val}(f) = c^+(S)$, then f is a maximum flow and S a minimum source vertex cut.
2. If T is a sink vertex cut satisfying $\text{val}(f) = c^-(T)$, then f is a maximum flow and T a minimum sink vertex cut.

Proof Assume that $f_{\max}$ is a maximum flow, and $S_{\min}$ a minimum source vertex cut. Then by definition and Observation 6.43, we have

$$\text{val}(f) \le \text{val}(f_{\max}) \le c^+(S_{\min}) \le c^+(S) = \text{val}(f), \tag{6.2}$$

and hence, equality holds throughout in (6.2). The case for the sink vertex cut T is similar and left as Exercise 6.27. This means that f must be a maximum flow and S a minimum source vertex cut. ■

Example 6.45 Consider the network $\vec{G}$ shown in Figure 6.12 in Example 6.38, where the capacity c is given in bold numbers. Since every source vertex cut S is given by $S = \{s\} \cup A$, where $A \subseteq \{I, N, G, U\}$, there are altogether $2^4 = 16$ possible source vertex cuts for the network $\vec{G}$. Their capacities are given by the sum of all of the weights of the edges going out of S—that is, the edges with tails in S and tail in $V(\vec{G}) \setminus S$.

$$\begin{aligned}
c^+(\{s\}) &= 4{,}000 + 8{,}000 = 12{,}000,\\
c^+(\{s, I\}) &= 8{,}000 + 500 + 1{,}100 = 9{,}600,\\
c^+(\{s, N\}) &= 4{,}000 + 1{,}000 + 800 = 5{,}800,\\
c^+(\{s, G\}) &= 4{,}000 + 8{,}000 + 2{,}000 = 14{,}000,\\
c^+(\{s, U\}) &= 4{,}000 + 8{,}000 + 2{,}500 = 14{,}500,\\
c^+(\{s, I, N\}) &= 500 + 1{,}000 + 1{,}100 + 800 = 3{,}400,\\
c^+(\{s, N, G\}) &= 4{,}000 + 800 + 2{,}000 = 6{,}800,\\
c^+(\{s, G, U\}) &= 4{,}000 + 8{,}000 + 2{,}000 + 2{,}500 = 16{,}500,\\
c^+(\{s, I, U\}) &= 8{,}000 + 500 + 2{,}500 = 11{,}000,\\
c^+(\{s, I, G\}) &= 8{,}000 + 1{,}100 + 2{,}000 = 11{,}100,\\
c^+(\{s, N, U\}) &= 4{,}000 + 1{,}000 + 2{,}500 = 7{,}500,\\
c^+(\{s, I, N, G\}) &= 1{,}100 + 800 + 2{,}000 = 3{,}900,\\
c^+(\{s, I, N, U\}) &= 500 + 1{,}000 + 2{,}500 = 4{,}000,\\
c^+(\{s, I, G, U\}) &= 8{,}000 + 2{,}000 + 2{,}500 = 12{,}500,\\
c^+(\{s, N, G, U\}) &= 4{,}000 + 2{,}000 + 2{,}500 = 8{,}500,\\
c^+(\{s, I, N, G, U\}) &= 2{,}000 + 2{,}500 = 4{,}500.
\end{aligned}$$

From this we see that $S = \{s, I, N\}$ is a minimum source vertex cut with the minimum capacity of $c^+(S) = 3{,}400$.

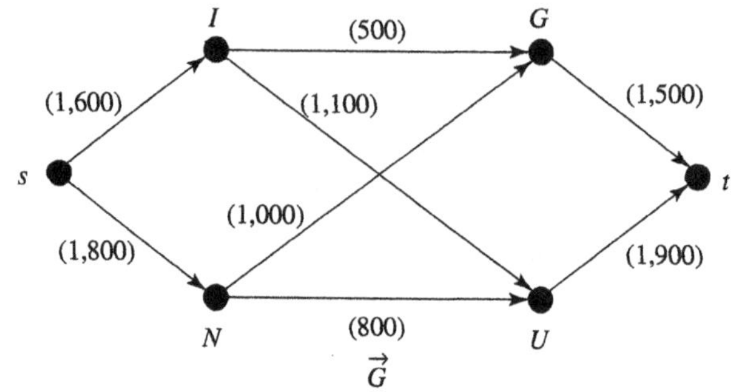

Figure 6.17: A flow f for the network $\vec{G}$ in Figure 6.12 is given by the numbers in parentheses.

Example 6.46 Consider again the network $\vec{G}$ shown in Figure 6.12 in Example 6.38, where the capacity c is given in bold numbers. If a flow f is given by Figure 6.17, then f is indeed a flow, since $f(e) \leq c(e)$ for each directed edge, and f clearly satisfies Kirchhoff's condition (6.1).

In fact, if $S = \{s, I, N\}$ from Example 6.45, then

$$\mathrm{val}(f) = f^+(s) - f^-(s) = 3{,}400 = c^+(S),$$

so by Theorem 6.44, f must be a maximum flow of $\vec{G}$.

Keeping Theorem 6.44 in mind, it is natural to ask if every maximum flow f for a network $\vec{G}$ is always equal to the capacity of a minimum source vertex cut. Note that Theorem 6.44 states only that *if* a flow f is such that its value equals the capacity of a source vertex cut, *then* f must be a maximum flow; it says nothing about other possible maximum flows.

It turns out that there are no maximum flows other than those that are equal to the capacity of any minimum source vertex cut. In order to show that, we first need some new terminology.

Let $(\vec{G}, c; s, t)$ be a network and f a flow. A path (not necessarily directed!) $p = (u_0, e_1, u_1, \ldots, e_k, u_k)$ in $\vec{G}$ such that for each $i \in \{1, \ldots, k\}$ we have

1. $f(e_i) < c(e_i)$, if $\eta(e_i) = (u_{i-1}, u_i)$ and
2. $f(e_i) > 0$, if $\eta(e_i) = (u_i, u_{i-1})$

is called an *augmenting path* of f from u_0 to u_k. For each $i \in \{1, \ldots, k\}$ we let

$$\varepsilon_f(e_i) = \begin{cases} c(e_i) - f(e_i) & \text{if } \eta(e_i) = (u_{i-1}, u_i), \\ f(e_i) & \text{if } \eta(e_i) = (u_i, u_{i-1}), \end{cases}$$

and call the quantity $\delta = \min(\{\varepsilon_f(e_i) \mid 1 \leq i \leq k\})$ the *tolerance* of the path p.

We leave the proof of the following lemma as Exercise 6.28.

Lemma 6.47 Let $(\vec{G}, c; s, t)$ be a network, f a flow, and p an augmenting path of $\vec{G}$ from s to t with a tolerance of $\delta > 0$. Define a new flow f' by

$$f'(e) = \begin{cases} f(e) + \delta & \text{if } \eta(e) = \eta(e_i) = (u_{i-1}, u_i), \\ f(e) - \delta & \text{if } \eta(e) = \eta(e_i) = (u_i, u_{i-1}), \\ f(e) & \text{if } e \text{ is not in } p. \end{cases}$$

In this case f' is a flow of $\vec{G}$ and $\text{val}(f') = \text{val}(f) + \delta$.

Example 6.48 Consider yet again the network $\vec{G}$ shown in Figure 6.12 in Example 6.38, where the capacity c is given in bold numbers. Let f_1 and p be a flow and a path, respectively, given by Figure 6.18, where the flow is specified by the numbers in parentheses and the path is given in bold lines. Clearly, f_1 is not a maximum flow, since $\text{val}(f_1) = 2{,}600$ whereas $c^+(S) = 3{,}400$, if $S = \{s, I, N\}$ as we saw in Example 6.45. The path p given by $p = (s, N, G, I, U, t)$ is an augmenting path for f_1 with a tolerance of $\delta = 100$.

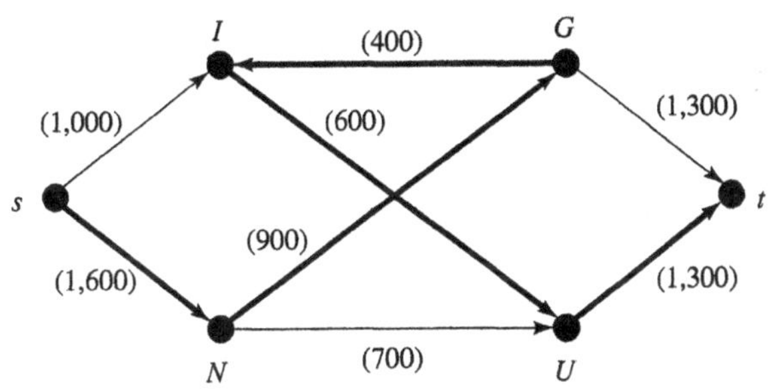

Figure 6.18: A flow f_1 for the network $\vec{G}$ in Figure 6.12 is given by the numbers in parenthesis.

Now, if $f_2 = f_1'$ as in Lemma 6.47, then $\text{val}(f_2) = \text{val}(f_1) + 100 = 2{,}700$, as shown in Figure 6.19, and f_2 is one step (or several steps!) closer to the maximum flow of $\vec{G}$, since it has a value strictly larger than that of f_1.

Theorem 6.49 A flow f in a network $(\vec{G}, c; s, t)$ is a maximum flow if, and only if, f has no augmenting path from s to t.

Proof By Lemma 6.47, a flow f cannot be maximum if it has an augmenting path.

Assume that f has no augmenting path from s to t. We want to show that there is a source vertex cut S with capacity of $\text{val}(f)$, which then, by Theorem 6.44, means that f is a maximum flow.

Let $A \subseteq V(\vec{G})$ be the set of vertices u such that there is an augmenting path from s to u in $\vec{G}$. Then by assumption $t \notin S$, and hence $S = \{s\} \cup A$ is a source vertex cut. Note that for each edge e pointing out of S into $V(\vec{G}) \setminus S$, we have $f(e) = c(e)$, and for each edge e' from $V(\vec{G}) \setminus S$ into S, we have $f(e') = 0$, since otherwise we could make

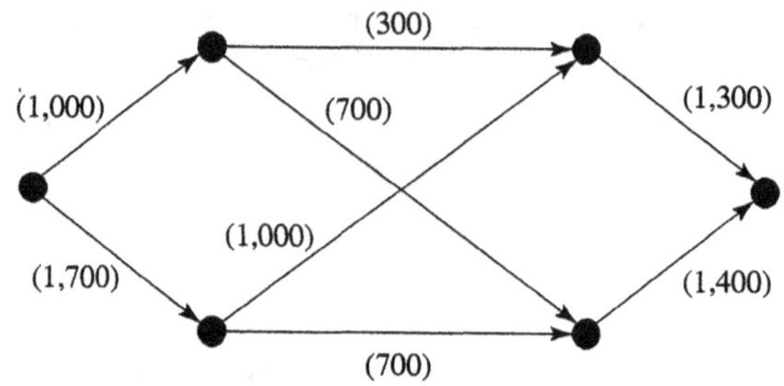

Figure 6.19: The flow f_2 has a value strictly larger than that of f_1.

the set S larger. This, however, implies that $f^+(S) = c^+(S)$ and $f^-(S) = 0$, and hence $\text{val}(f) = f^+(S) - f^-(S) = c^+(S)$. Therefore, f is a maximum flow. ■

That every network has a maximal flow can be seen by using linear programming (see Exercise 6.24), or directly by real mathematical analysis (see Exercise 6.25). Knowing that a maximal flow exists, we obtain as a direct corollary the following main theorem of this section, Theorem 6.50, which is often called the *Max-Flow-Min-Cut Theorem* or the *Ford-Fulkerson Theorem* for its authors L. R. Ford and D. R. Fulkerson.

Theorem 6.50 **Max-Flow-Min-Cut**

The value of a maximum flow in a network $\vec{G}$ is equal to the capacity of a minimum source vertex cut of $\vec{G}$.

If the capacity c for the network $\vec{G}$ in question is integral—that is, $c : E(\vec{G}) \to \mathbb{N} \cup \{0\}$ takes only nonnegative values—then it is possible to compute or "extend" each integral flow f to a maximum flow of $\vec{G}$ by adding an augmenting path p to f, if f is not maximum. The fact that both c and f take integral values on all edges ensures that at each step, the value $\text{val}(f)$ will increase at least by 1 when we "add" an augmenting path p to f. Therefore, this process must eventually stop at a flow, which by Theorem 6.49 must be maximum.

The idea of this method of computing a maximum flow from a given flow f of $\vec{G}$ is given by the following process:

- Let f_0 be a flow of $\vec{G}$ with nonnegative integer values.
- Repeat the following steps:

  1. If f_i has no augmenting path from s to t, then $f_{\max} = f_i$ and we stop.
  2. If f_i has an augmenting path p from s to t, then let $f_{i+1} = f_i'$ as in Lemma 6.47, and go to 1.

- Return $f_{\max} = f_i$ as the maximum flow of $\vec{G}$.

Note that this process will stop, since the sequence of flows $(f_i)_{i \geq 0}$ will eventually stop. This is because it is bounded above by the capacity of a minimum cut S,

$$\text{val}(f_0) < \text{val}(f_1) < \cdots < \text{val}(f_i) \cdots \leq c^+(S).$$

At the point when $\text{val}(f_k) = c^+(S)$, we have $f_k = f_{\max}$ is a maximum flow.

This process is not a precise algorithm, since finding an augmenting path p from the source to the sink of $\vec{G}$ is not given in detail. To give a precise algorithm, one needs to specify in more detail how to find an augmenting path from the source to the sink when it exists.

A general approach is to maintain a set $U \subseteq V(\vec{G})$ of vertices, where initially $U = \{s\}$ contains only the source. Then, for each exiting edge $e = (u, v)$, with $u \in U$ and $v \in \overline{U}$ with $f(e) < c(e)$, we add v to U, and for each incoming edge $e' = (u', v')$, with $u' \in \overline{U}$ and $v' \in U$ with $f(e') > 0$, we add u' to U. We continue in this fashion until the sink t has been reached, and record the corresponding path p from s to t as an augmenting path. Again, this procedure works only because we have *integer*-valued capacities of the edges!

With this idea made precise, one can obtain an algorithm called the FORD-FULKERSON ALGORITHM or the HUNGARIAN ALGORITHM that works for networks with nonnegative integral edge capacities and flows. We leave the proof of the following corollary as Exercise 6.31.

Corollary 6.51 If a network $\vec{G}$ has capacity $c : E(\vec{G}) \to \mathbb{Q}^+ \cup \{0\}$ taking only nonnegative rational values, then the FORD-FULKERSON ALGORITHM will yield a maximum flow $f_{\max}$ of $\vec{G}$, if we start with any nonnegative rational value flow f_0.

Remark 6.52 We have at this point briefly discussed the mere idea of the FORD-FULKERSON ALGORITHM of computing a flow of maximum value, and we have stressed the necessary condition that the edge capacities c of the network $\vec{G}$ in question and the flows under consideration must be nonnegative integer valued $c : E(\vec{G}) \to \mathbb{N} \cup \{0\}$ (or equivalently nonnegative rational valued). However, we have also mentioned (in Exercise 6.24) that *any* network, where the edge capacities c can take any real value, can be computed by solving a corresponding linear program (LP). Therefore, it is quite natural to ask oneself: Why do most books on graph theory (including this one!) emphasize the seemingly restricted FORD-FULKERSON ALGORITHM instead of existing polynomial-time (or poly-time for short) algorithms to solve LP's? A priori they seem to be able to handle a more general network $\vec{G}$ with real edge capacities $c : E(\vec{G}) \to \mathbb{R}$ as opposed to integer valued ones. In what follows we will try to clarify this.

Although this book deals with neither linear programs nor complexity theory, we should mention a few things to help clarify our presentation. A typical *linear program (LP)* asks for

$$\max\{\tilde{c} \cdot \tilde{x} \;:\; \tilde{x} \geq \tilde{0},\ \mathsf{A}\tilde{x} \leq \tilde{b}\}, \tag{6.3}$$

where $\tilde{b} = (b_1, \ldots, b_n)$ and $\tilde{c} = (c_1, \ldots, c_n)$ are given fixed vectors in $\mathbb{R}^n$, $\tilde{x} = (x_1, \ldots, x_n)$ is a real vector of n indeterminates, and $\mathsf{A} = [a_{ij}]$ is an $n \times n$ matrix with real entries. Also, $\tilde{c} \cdot \tilde{x}$ denotes the usual dot product, $\tilde{x} \geq \tilde{0}$ means that each $x_i \geq 0$, and $\mathsf{A}\tilde{x} \leq \tilde{b}$ means that each ith entry of $\mathsf{A}\tilde{x}$ is at most b_i—that is, $a_{i1}x_1 + \cdots + a_{in}x_n \leq b_i$. The word *real* here is very deceiving, since all computers can deal only with rational numbers that are given either as a fraction p/q of two integers or as a terminating decimal (or binary) fraction [that is, $q = 10^d$ (or $q = 2^d$) for some $d \in \mathbb{N}$]. Any poly-time algorithm relies heavily on the fact that the actual arithmetic manipulation of the numbers takes a constant (or a bounded) time in comparison with the repetition needed in a **for**- or **while**-loop of an algorithm. When this arithmetic manipulation of the numbers actually takes much more time than the loops, then it becomes less clear what is meant by a "poly-time" algorithm. Strictly speaking, the *size* of a number, vector, or matrix, denoted by sz($*$), is given by the following:

$$\begin{aligned} \mathrm{sz}(p/q) &= \lg(|p|) + \lg(|q|), \\ \mathrm{sz}(\tilde{b}) &= n + \sum_{i=1}^{n} \mathrm{sz}(c_i), \\ \mathrm{sz}(\mathsf{A}) &= n^2 + \sum_{i,j} \mathrm{sz}(a_{ij}), \end{aligned}$$

where lg is the base-2 logarithm and the numbers b_i, c_i, and a_{ij} are assumed rational. Hence, a *poly-time* algorithm is then an algorithm whose running time is given by a fixed polynomial in the size of the input—that is, polynomial in $\mathrm{sz}(\tilde{b}) + \mathrm{sz}(\tilde{c}) + \mathrm{sz}(\mathsf{A})$. (An algorithm that works in polynomial time in terms of n only is sometimes called *strongly polynomial*. For an introductory discussion see Chapter 13.) Therefore, the fact that the FORD-FULKERSON ALGORITHM assumes integer (or by Exercise 6.31, rational) valued edge capacities and flows of the network $\vec{G}$ in question is not a restriction at all! Still, there currently are several competing approaches of computing maximum flow problems:

Ford-Fulkerson: This algorithm/procedure is also known as the HUNGARIAN METHOD/ALGORITHM or the AUGMENTING PATH ALGORITHM, especially when the network in question yields matchings of maximum size (see Chapter 11 in that case). Under the assumption that the edge capacities and flows are integer valued, this method is efficient and has a fast polynomial running time. However, when we assume rational valued edge capacities and flows, the running time does get progressively worse (that is, the running time increases in terms of the reciprocal of the minimum difference between a pair of rational numbers given to us), although the running time is still poly-time. Note, however, that for integer valued edge capacities and flows this algorithm is purely combinatorial and is therefore the algorithm of choice in combinatorial settings and in graph theory.

The Simplex Method: The SIMPLEX METHOD to solve an LP in (6.3) has a more geometrical rather than combinatorial or graph-theoretic interpretation. This is, moreover, the classical approach to solving LP's. Here the set of all $\tilde{x} \in \mathbb{R}^n$ such that $\tilde{x} \geq \tilde{0}$ and $\mathsf{A}\tilde{x} \leq \tilde{b}$ defines a closed subset of $\mathbb{R}^n$ that is called a *polyhedron*. Such a polyhedron has vertices, edges and faces, just as a cube, a tetrahedron, or an octahedron has in $\mathbb{R}^3$. The idea of the SIMPLEX METHOD to solve an LP is basically to make a trip on the polyhedron, from one vertex to another along the edges, until an optimal solution is reached. This method is quite excellent (fast poly-time) if we manage to start at a *vertex* of the polyhedron. However, it can take a while to actually find a vertex. This method works very well on the average (that is, in most cases) and is therefore the method of choice for many folks in the operations research area. However, the SIMPLEX METHOD is not a poly-time algorithm, since there are a few "bad" cases (specific types of polyhedra) where it can take a long time (an exponentially long time) to find a vertex to start at.

The Ellipsoid Method: The ELLIPSOID METHOD, also known as the KHACHIYAN'S METHOD or the SOVIET METHOD [for obvious reasons, since Khachiyan was one of the Soviets (in fact, he was Armenian) to develop this method in the former Soviet Union during the Cold War Era!], is the first truly poly-time algorithm to solve an LP in (6.3), since it runs in poly-time in every single case. The bad thing is that the average running time is much worse than that of the classical SIMPLEX METHOD. The idea behind the ELLIPSOID METHOD is also geometrical. It is based on finding a sequence of ellipsoids in $\mathbb{R}^n$ (that is, a closed surface in $\mathbb{R}^n$ all of whose plane sections are either ellipses or circles) of decreasing volume, all of which include the polyhedron (which we can assume to be bounded) defined by the conditions in (6.3). For each ellipsoid in the sequence we find an optimal point $\tilde{x}_k$ maximizing $\tilde{c} \cdot \tilde{x}$ from (6.3). The sequence $(\tilde{x}_k)_{k\geq 1}$ of these points will then get "close enough" to the optimal solution $\tilde{x}^*$ on the polyhedron, at which step one can actually determine $\tilde{x}^*$ exactly.

The Interior Point Method: The INTERIOR POINT METHOD is also a true poly-time algorithm to solve an LP as in (6.3). It works much faster than the original ELLIPSOID METHOD. The idea here is also geometrical, where one, at each step, makes a trip through the middle of the solid of the polyhedron, instead of going around along the surface of the solid as in the SIMPLEX METHOD. Like the ELLIPSOID METHOD, the INTERIOR POINT METHOD obtains a point $\tilde{x}_k$ that is within certain bounds from the optimal solution $\tilde{x}^*$ on the polyhedron, at which step there is a poly-time procedure to determine the optimal solution $\tilde{x}^*$ from $\tilde{x}_k$.

All the methods mentioned here above have their pros and cons when computing an optimal solution to an LP in (6.3). But we now see that our restriction of considering only integer or rational edge capacities and flows is not a restriction from a computational point of view. Any one of the four methods above can be used to compute a maximum flow in a network, which therefore can be done in poly-time, provided that the capacities and flows are rational. For a comprehensive discussion on LP's from a thorough theoretical point of view, we refer to [48]. Also, for an excellent treatment of the algorithmic aspect of maximum flow problems, as well as LP's in general, we refer to [44].

6.5 The Theorems of Menger

In this final section we discuss four theorems attributed to Menger. They can be viewed as different versions of the same theorem—this important one that has a wide range of applications. We start with the following result, which deals with a very special network from §6.4.

Theorem 6.53 Let $(\vec{G}, c; s, t)$ be a network where $c(e) = 1$ for all $e \in E(\vec{G})$. Let $f_{\max}$ be a maximum flow and $S_{\min}$ be a minimum source vertex cut of $\vec{G}$.

1. We have that $\text{val}(f_{\max})$ is the maximum number of edge-disjoint directed s, t-paths in $\vec{G}$.
2. We have that $c^+(S_{\min})$ is the minimum number of edges needed to be removed from $\vec{G}$ to destroy all directed s, t-paths.

Proof Since the capacity is integral, we can use the FORD-FULKERSON ALGORITHM to obtain a maximum flow $f_{\max}$ with integral values, and hence $f_{\max}(e) \in \{0, 1\}$ for each

$e \in E(\vec{G})$. Let $\vec{G}'$ be the subnetwork of $\vec{G}$ on the same set of vertices as $\vec{G}$, where the edges e for which $f_{\max}(e) = 0$ have all been removed from $\vec{G}$. By Observation 6.43 and the definition of $\vec{G}$, we have $\text{val}(f_{\max}) = d^+_{\vec{G}}(s) - d^-_{\vec{G}}(s) = d^-_{\vec{G}}(t) - d^+_{\vec{G}}(t)$. If $\text{val}(f_{\max}) = k \geq 0$, then by Kirchhoff's condition (6.1), we have that $d^+_{\vec{G}}(u) = d^-_{\vec{G}}(u)$ for all $u \in V(\vec{G}) \setminus \{s, t\}$. In this case, form a directed graph $\vec{G}''$ by adding k directed edges $e_1, \ldots, e_k$ from t to s in $\vec{G}'$. In this way, the indegree and outdegree of $u \neq s, t$ remain unaltered, but the indegree and outdegree of both s and t are now equal. Hence, we have

$$d^+_{\vec{G}''}(u) = d^-_{\vec{G}''}(u)$$

for all $u \in V(\vec{G}'') = V(\vec{G})$. By Theorem 5.25, $\vec{G}''$ is an Eulerian digraph with a directed Eulerian circuit $\vec{c}$. Assuming that the starting vertex is s; then we can further assume the labeling of the edges e_i to be such that $\vec{c}$ has the form

$$\vec{c} = \vec{p}_1 e_1 \vec{p}_2 e_2 \cdots \vec{p}_k e_k,$$

where each $\vec{p}_i$ is a directed path from s to t that does not include any of the edges e_j. Note that $\vec{p}_i$ is a directed path in $\vec{G}'$, and hence we see that $\vec{G}'$ is an edge-disjoint union of the directed paths p_i—that is,

$$E(\vec{G}') = E(\vec{p}_1) \cup \cdots \cup E(\vec{p}_k).$$

Therefore, there are $k = \text{val}(f_{\max})$ edge-disjoint directed s, t-paths in $\vec{G}$, and hence the maximum number m of edge-disjoint s, t-paths in $\vec{G}$ is at least k—that is, $m \geq k$.

To show that $m \leq k = \text{val}(f_{\max})$, we consider m edge-disjoint s, t-paths in $\vec{G}$ and form a flow f by letting $f(e) = 1$, if e is an edge in one of the m edge-disjoint s, t-paths, and $f(e) = 0$ otherwise. In this way f is a flow of $\vec{G}$ with value $\text{val}(f) = m$. From this we must have

$$k = \text{val}(f_{\max}) \geq \text{val}(f) = m,$$

which completes the proof of the first part of the theorem.

To prove the second part, let ℓ be the least number of directed edges needed to destroy all directed s, t-paths. If $S_{\min}$ is a minimum source vertex cut and D is the set of all directed edges from $S_{\min}$ to $V(\vec{G}) \setminus S_{\min}$, then $c^+(S_{\min}) = |D|$, and the removal of D from $\vec{G}$ will disconnect s from t and hence $|D| \geq \ell$.

For the other inequality, let L be a set of ℓ directed edges needed to be removed from $\vec{G}$ to destroy all directed s, t-paths. Let $S \subseteq V(\vec{G})$ be the set of vertices u such that there is a directed path from s to u in $\vec{G} - L$. In this way we have $s \in S$ and $t \notin S$; therefore, S is a source vertex cut of $\vec{G}$. We now argue that $c^+(S) \leq \ell$ by showing that if a directed edge e has its tail in S and head in $V(\vec{G}) \setminus S$, then we must have $e \in L$: Suppose there is an edge e pointing out of S into $V(\vec{G}) \setminus S$ that is not contained in L. Since there is a directed path p from s to the head of e in $\vec{G} - L$, then there is also a directed path from s to the tail of e, which we obtain from p by attaching e at the end. Thus by definition of S, the tail of e must also be in S. This is, however, a contradiction, since the tail is contained in $V(\vec{G}) \setminus S$. We therefore conclude that no such edge exists. Thus all edges pointing out of S and into $V(\vec{G}) \setminus S$ are contained in L, so $c^+(S) \leq |L| = \ell$. From this we conclude that

$$c^+(S_{\min}) \leq c^+(S) \leq \ell,$$

thereby completing the proof of the theorem. ∎

By the Max-Flow-Min-Cut Theorem 6.50, we immediately obtain the following corollary, which is our first version of Menger's Theorem, the version for directed graphs.

Corollary 6.54 **Menger's Theorem, directed edge version**

Let $\vec{G}$ be a directed graph and $u, v \in V(\vec{G})$. The maximum number of edge-disjoint directed u, v-paths is equal to the minimum number of edges needed to be removed from $\vec{G}$ to destroy all directed u, v-paths.

From Corollary 6.54 we are able to obtain the following version of Menger's Theorem for undirected graphs.

Theorem 6.55 **Menger's Theorem, edge version**

Let G be a graph and $u, v \in V(G)$. The maximum number of edge-disjoint u, v-paths in G is equal to the minimum number of edges needed to be removed from G to disconnect u from v.

Proof We form a directed graph $\vec{G}$ from G by replacing each edge $\{u, v\} \in E(G)$ by two directed edges (u, v) and (v, u), as shown in Figure 6.20.

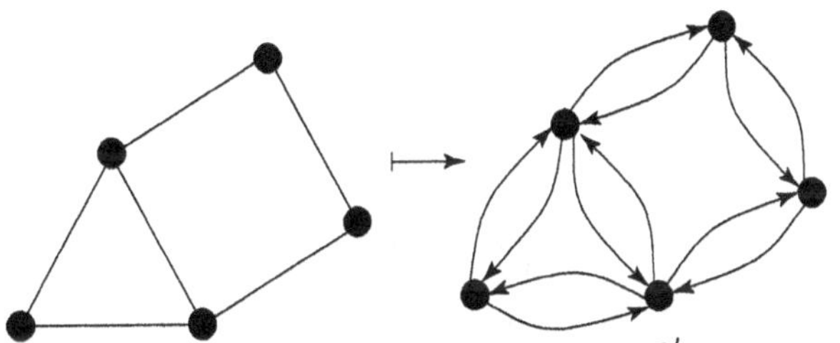

Figure 6.20: We replace each edge by two directed edges to obtain $\vec{G}$.

If there are k edge-disjoint u, v-paths in G, then we easily obtain k edge-disjoint directed u, v-paths in $\vec{G}$.

On the other hand, if we have k edge-disjoint directed u, v-paths in $\vec{G}$, then there might be a pair $\vec{p}_1$ and $\vec{p}_2$ of directed u, v-paths where $\vec{p}_1$ includes a directed edge e_1 and $\vec{p}_2$ a directed edge e_2, where e_1 and e_2 both were formed by the same undirected edge $e = \{x, y\}$ in G. In this case $e_1 = (x, y)$ and $e_2 = (y, x)$ have opposite directions, and hence we may assume that $\vec{p}_1$ and $\vec{p}_2$ have the following forms:

$$\begin{aligned} \vec{p}_1 &= \vec{p}_1{}' e_1 \vec{p}_1{}'', \\ \vec{p}_2 &= \vec{p}_2{}' e_2 \vec{p}_2{}''. \end{aligned}$$

In this case we can obtain two new directed u, v-paths $\vec{q}_1$ and $\vec{q}_2$ by the following:

$$\begin{aligned} \vec{q}_1 &= \vec{p}_1{}' \vec{p}_2{}'', \\ \vec{q}_2 &= \vec{p}_2{}' \vec{p}_1{}''. \end{aligned}$$

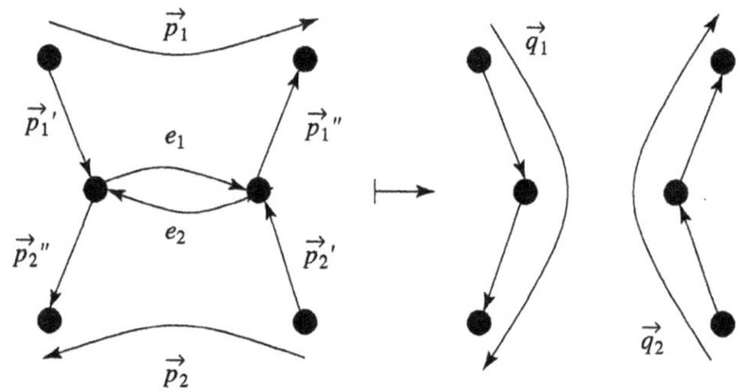

Figure 6.21: We obtain a new pair of directed u, v-paths, which have fewer directed edges in common, stemming from the same undirected edge in G.

Figure 6.21 depicts the situation. These u, v-paths have fewer directed edges in common that are formed by the same undirected edge in G. This process $\vec{p} \mapsto \vec{q}$ can be repeated until we have a collection of k directed edge-disjoint u, v-paths, no two of which contain a directed edge obtained from the same undirected edge in G. This collection yields k edge-disjoint u, v-paths in G. Therefore, the maximum number of edge-disjoint directed u, v-paths in $\vec{G}$ is the same as the maximum number of edge-disjoint u, v-paths in G.

Using this same idea of obtaining a directed u, v-path $\vec{q}$ from $\vec{p}$ as shown in Figure 6.21, one can show (see Exercises 6.40 and 6.41) that the minimum number of edges needed to be removed from G to disconnect u from v in G is the same as the minimum number of directed edges needed to be removed from $\vec{G}$ to destroy all directed paths from u to v.

The theorem now follows directly from Corollary 6.54. ■

We conclude this chapter with the "vertex" versions of Menger's Theorem.

Theorem 6.56 **Menger's Theorem, directed vertex version**

Let $\vec{G}$ be a digraph and $u, v \in V(\vec{G})$. The maximum number of vertex-disjoint directed u, v-paths is equal to the minimum number of vertices, other than u and v, needed to be removed from $\vec{G}$ to destroy all directed u, v-paths.

Proof (Sketch) From the digraph $\vec{G}$, we form a new digraph $\vec{H}$ in the following way:

1. For each vertex $x \in V(\vec{G})$ form two vertices, the tail t_x and the head h_x, and the directed edge (t_x, h_x) in $\vec{H}$.
2. For each directed edge $(x, y) \in E(\vec{G})$, we form the directed edge (h_x, t_y) in $\vec{H}$.

Note that the vertices of $\vec{H}$ are of two types: the h-vertices and the t-vertices. (See Figure 6.22.) There is a one-to-one correspondence between $V(\vec{G})$ and the edges in $E(\vec{H})$ going from an h-vertex to a t-vertex; likewise there is a one-to-one correspondence between $E(\vec{G})$ and the edges in $E(\vec{H})$ going from a t-vertex to an h-vertex. It is also clear that each u, v-path in $\vec{G}$ corresponds to a unique t_u, h_v-path in $\vec{H}$ whose vertices alternate between

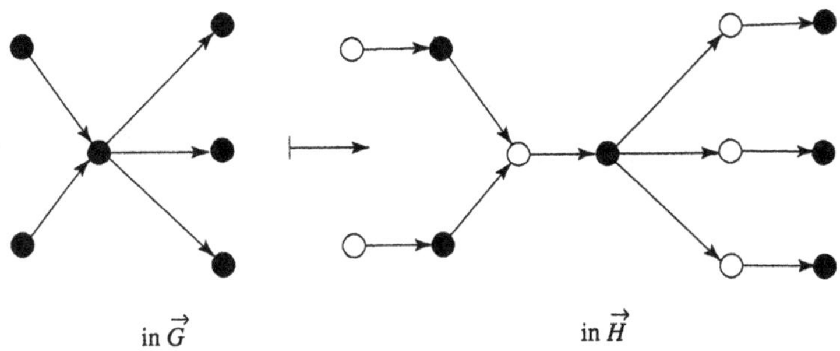

Figure 6.22: The h-vertices are white and the t-vertices are black.

t- and h-vertices. Two u, v-paths in $\vec{G}$ are vertex-disjoint (except for u and v, of course) if, and only if, their corresponding t_u, h_v-paths in $\vec{H}$ have no edge in common. Therefore, the maximum number of vertex-disjoint u, v-paths in $\vec{G}$ is equal to the maximum number of edge-disjoint t_u, h_v-paths in $\vec{H}$.

If $V \subseteq V(\vec{G})$ is a set of vertices the removal of which from $\vec{G}$ will destroy all directed u, v-paths in $\vec{G}$, then the removal of the corresponding edges in $\vec{H}$ will destroy all directed t_u, h_v-paths in $\vec{H}$. Hence, the minimum number of vertices of $\vec{G}$ needed to be removed to destroy all directed u, v-paths in $\vec{G}$ is at least the minimum number of edges of $\vec{H}$ needed to be removed to destroy all directed t_u, h_v-paths in $\vec{H}$.

Note that $d^+(t_x) = 1$ for all tail vertices $t_x \in V(\vec{H})$. Therefore, if $F \subseteq E(\vec{H})$ is a set of edges the removal of which will destroy all directed t_u, h_v-paths in $\vec{H}$, then by replacing each ht-edge by the succeeding th-edge [that is, (h_x, t_y) replaced by (t_y, h_y)], we obtain another set F' possibly smaller than F (since distinct ht-edges might yield the same th-edge!), the removal of which will also destroy all directed t_u, h_v-paths in $\vec{H}$. Since F' corresponds to a set of vertices in $\vec{G}$, the removal of these corresponding vertices in $\vec{G}$ will destroy all the directed u, v-paths in $\vec{G}$. This shows that the minimum number of vertices needed to be removed from $\vec{G}$ to destroy all directed u, v-paths in $\vec{G}$ is the same as the minimum number of edges needed to be removed from $\vec{H}$ to destroy all directed t_u, h_v-paths in $\vec{H}$. By Corollary 6.54, we have the theorem. ■

For a given graph G we can form a corresponding digraph $\vec{G}$ as in the proof of Theorem 6.55. Applying Theorem 6.56 to $\vec{G}$, we obtain the following vertex versions of Menger's Theorem.

Theorem 6.57 **Menger's Theorem, vertex version**

Let G be a graph and $u, v \in V(G)$. The maximum number of vertex-disjoint u, v-paths in G is equal to the minimum number of vertices needed to be removed from G to disconnect u from v.

Theorem 6.57, the vertex version of Menger's Theorem, can be viewed as a generalization of Theorem 6.33 for k-connected graphs, where $k \geq 2$ as follows.

Theorem 6.58 A connected graph G is k-connected if, and only if, there are at least k vertex-disjoint (excluding endvertices) paths between each pair of G's vertices.

A similar description holds for k-edge-connected graphs as a direct consequence of Theorem 6.55.

Theorem 6.59 A connected graph G is k-edge-connected if, and only if, there are at least k edge-disjoint paths between each pair of G's vertices.

6.6 Exercises

6.1. Let G be a simple graph and $\overline{G}$ its complement. Show that a cut-vertex in G is not a cut-vertex in $\overline{G}$.

6.2. For $n \geq 1$, show that any graph on n vertices has at most $n - 2$ cut-vertices. Describe the graphs on n vertices with precisely $n - 2$ cut-vertices.

6.3. Let e be a bridge of a graph G. Show that if G has k components, then $G - e$ has $k + 1$ components. [*Hint:* The endvertices of e have one endvertex in one component and another endvertex in a different component.]

6.4. Prove Lemma 6.5: Show that for any edge cut S of a connected graph G, the number of components of $G - S$ is precisely two.

6.5. Let G be a graph and $e \in E(G)$. Show that the following are equivalent:

(a) The edge e is a bridge.
(b) There exist two vertices $u, v \in V(G)$, such that e is on every path between u and v in G.

6.6. Let G be a connected graph. Show that if $C \subseteq E(G)$ has an even number of edges in common with every edge cut of G, then C is an edge-disjoint union of cycles.

6.7. Let G be a connected graph for which there are at most μ edges between any pair of distinct vertices. If G' is the simple graph obtained from G by removing all loops and replacing all multiple edges by a single edge, then show that

$$\kappa'(G') \leq \kappa'(G) \leq \mu\kappa'(G').$$

6.8. Show that if G is a graph on n vertices and m edges, we have

$$\kappa(G) \leq \kappa'(G) \leq \left\lfloor \frac{2m}{n} \right\rfloor.$$

[*Hint:* Note that $\delta(G) \leq d_G(u)$ for all $u \in V(G)$ and use the Hand-Shaking Theorem (Theorem 1.26). Complete the proof by using Theorem 6.20.]

6.9. Let G be a graph on n vertices and m edges, where between each pair of adjacent vertices in G there are at most μ edges and at least ν edges. Express the optimal upper bound for $\kappa'(G)$ in terms of n, m, μ, and ν. [*Hint:* Consider first the case where there are exactly μ edges between every pair of adjacent vertices.]

6.10. Show that when k is even, the k-regular Harary graph in Example 6.22 is k-connected.

6.11. Assume n and m are such that $m \geq n - 1$, and $k = \lfloor 2m/n \rfloor$ is an odd integer, where $k \leq n - 1$. Start by constructing the $(k - 1)$-regular Harary graph on n vertices from C_n as in Example 6.22. If n is even, then add the $n/2$ edges connecting pairs of vertices at distance $n/2$. This graph is also called a k-regular Harary graph. If n is odd, then label the vertices cyclically by $0, 1, \ldots, n - 1$ and add the $(n + 1)/2$ edges between i and $i + (n - 1)/2$ for each $i \in \{0, 1, \ldots, (n - 1)/2\}$. This is not a regular graph but is also called a Harary graph. Show that both of these types of Harary graphs are k-connected. This illustrates that the upper bound in Exercise 6.8 can be reached for any given n and e having the stated properties.

6.12. Show that a vertex u in a graph G is a cut-vertex if, and only if, there are vertices $v, w \in V(G) \setminus \{u\}$ such that every path between v and w contains the vertex u.

6.13. For $n \geq 1$, show that any connected graph on n vertices has at most $n - 1$ blocks. Moreover, show that such a graph has exactly $n - 1$ blocks if, and only if, it is a tree on n vertices.

6.14. For a graph G on $n \geq 1$ vertices, what is the maximum number of vertices of its block-cutpoint graph $\mathrm{BC}(G)$?

6.15. For a tree T on n vertices and with maximum degree Δ, what is the minimum number of vertices in its block-cutpoint graph $\mathrm{BC}(T)$? [*Hint:* A vertex of T is a cut-vertex if, and only if, it is not a leaf of T.]

6.16. Let G be a connected graph. For each $u \in V(G)$, let $b(u)$ denote the number of blocks in G that contain the vertex u. Show that the number of blocks in G is given by

$$|\mathcal{B}(G)| = 1 + \sum_{u \in V(G)} (b(u) - 1).$$

6.17. For a graph G, define its *block graph* $B(G)$ as the graph with vertex set $\mathcal{B}(G)$, the blocks of G, and where two blocks B_1 and B_2 are connected if, and only if, $B_1 \cap B_2 \neq \varnothing$. Show that each block in $B(G)$ is isomorphic to the complete graph K_n for some $n \geq 2$. Moreover, show that if G' is a graph, where each of its blocks is a complete graph, then there is a graph G such that $G' \cong B(G)$. [*Hint:* Form G from $B(H)$ by adding leaves to some of the vertices of $B(H)$.]

6.18. Draw two graphs on six vertices that are 1-isomorphic but are not isomorphic.

6.19. Show that two 1-isomorphic graphs have the same number of edges.

6.20. Show that two 1-isomorphic connected graphs have the same number of vertices.

6.21. Show that two isomorphic graphs are also 1-isomorphic.

6.22. Let G and G' be 1-isomorphic graphs, on n and n' vertices and with c and c' compo nents, respectively. Show that $n - c = n' - c'$.

6.23. Prove Lemma 6.39.

6.24. Show that any network has a maximum flow. [*Hint:* If $E(\vec{G}) = \{e_1, \ldots, e_m\}$, then a flow f is uniquely given by the m-tuple $(x_1, \ldots, x_m) = (f(e_1), \ldots, f(e_m)) \in \mathbb{R}^m$. Express Kirchhoff's Law as linear equations, and then use linear programming—i.e., the fact that every linear functional on a bounded compact region in $\mathbb{R}^m$, which is determined by finitely many linear equations, always takes a maximum value.]

6.25. (*For those who know some real analysis.*) For a given network $\vec{G}$, let $\text{Val}(\vec{G})$ be the set of possible values of flows of $\vec{G}$—that is,

$$\text{Val}(\vec{G}) = \{x \in \mathbb{R} : \text{val}(f) = x \text{ for some flow } f \text{ of } \vec{G}\}.$$

Let $x^* = \sup(\text{Val}(\vec{G}))$ be the supremum of $\text{Val}(\vec{G})$. Show that $x^* = \text{val}(f_{\max})$ is the value of some maximal flow $f_{\max}$, thereby showing that every network has a maximum flow directly by real analysis. [*Hint:* First show that $\text{Val}(\vec{G})$ is a bounded set of $\mathbb{R}$. To show that $\text{Val}(\vec{G})$ is a closed set of the $\mathbb{R}$ (and hence compact), assume $(x_n)_{n\geq 1}$ is a sequence of real numbers in $\text{Val}(\vec{G})$ with $\lim_{n\to\infty} x_n = x^*$. For each n let f_n be a corresponding flow of $\vec{G}$ with $\text{val}(f_n) = x_n$. Since $f_n(e)$ is bounded for each $e \in E(\vec{G})$, there is a subsequence $(n_i)_{i\geq 1}$ of the sequence $(n)_{n\geq 1}$ of natural numbers such that for each edge e the sequence $(f_{n_i}(e))_{i\geq 1}$ is convergent. Show that a flow f^* on $\vec{G}$ with $\text{val}(f^*) = x^*$ can be obtained by letting $f^*(e) = \lim_{i\to\infty} f_{n_i}(e)$, which therefore must be a maximal flow of $\vec{G}$.]

6.26. Prove Observation 6.43. [*Hint:* Note that a flow is always less than the capacity $f \leq c$.]

6.27. Complete the proof of Theorem 6.44.

6.28. Prove Lemma 6.47, thereby showing that if a flow f has an augmenting path p, then f cannot be a maximum flow.

6.29. Give an example of a network $\vec{G}$ with a unique maximum flow f.

6.30. Give an example of a network $\vec{G}$ with more than one maximum flow. Is it possible to have precisely two maximum flows? In general, is it possible to have finitely many maximum flows if the maximum flow is not unique?

6.31. Consider the FORD-FULKERSON ALGORITHM when the capacities of a network $\vec{G}$ are *rational* numbers. Show that Corollary 6.51 still holds in this case. [*Hint:* Consider the common denominator for all the fractions $c(e)$, where $e \in V(\vec{G})$, and call it M. Show that at each step the value of a flow increases by at least $1/M$.]

6.32. Let $\alpha \in \mathbb{R}$ be a real number, and $\vec{G}$ a network such that $c(e) = n\alpha$, an integer multiple of α, for each $e \in E(\vec{G})$. Show that the value of a maximum flow $f_{\max}$ is also an integer multiple of α—that is, $\text{val}(f_{\max}) \in \alpha\mathbb{N}$.

6.33. Consider the network $\vec{G}$ with vertices $V(\vec{G}) = \{s, t, a, b, c, d, e, f\}$ given in Figure 6.23. Use the FORD-FULKERSON ALGORITHM to find a maximum flow for $\vec{G}$. Prove that your final flow f is indeed maximum by finding a source vertex cut S such that $\text{val}(f) = c^+(S)$. [*Note!* By Exercise 6.32 one could consider $\alpha = 10$, divide each capacity by 10, and work with that corresponding network. The maximum flow would then have to be multiplied by 10 at the end.]

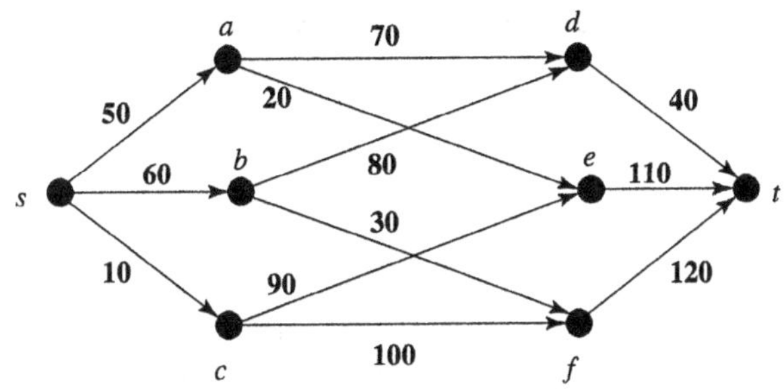

Figure 6.23: A network for Exercise 6.33.

6.34. Consider the following possible generalization of Exercise 6.32: Let $\alpha, \beta \in \mathbb{R}$ be real numbers, and $\vec{G}$ a network such that the capacity of each edge has the form $c(e) = n\alpha + m\beta$, for each $e \in E(\vec{G})$, where $m, n \in \mathbb{Z}$. Assume that the real numbers α and β are such that

$$n\alpha + m\beta \geq \epsilon > 0, \tag{6.4}$$

where ϵ is a fixed real number. Show that the value of a maximum flow $f_{\max}$ is also of the form $n\alpha + m\beta$ for some integers m and n—that is, $\text{val}(f_{\max}) \in \alpha\mathbb{N} + \beta\mathbb{N}$. Why can we not conclude this if (6.4) does not hold? [*Note!* One can show that (6.4) holds if, and only if, α/β is a rational number, thereby indicating that this does not generalize Exercise 6.32.]

6.35. Let $\vec{G}$ be a network with capacity c. Let f_1 and f_2 be two flows of $\vec{G}$ such that $f_1(e)+f_2(e) \leq c(e)$ for every $e \in E(\vec{G})$. Prove that $f' = f_1+f_2$ is also a flow of $\vec{G}$ and that

$$\text{val}(f') = \text{val}(f_1) + \text{val}(f_2).$$

6.36. Let $\vec{G}$ be a network with capacity c. Let f be a flow of $\vec{G}$ and $\alpha > 1$ be a real number such that $\alpha f(e) \leq c(e)$ for every $e \in E(\vec{G})$. Prove that $f' = \alpha f$ is also a flow of $\vec{G}$ and that

$$\text{val}(f') = \alpha \ \text{val}(f).$$

6.37. Let $f_1, \ldots, f_k$ be flows for a network $\vec{G}$. Define the *convex hull* of these flows by

$$\mathcal{F}_{\vec{G}} = \{\alpha_1 f_1 + \cdots + \alpha_k f_k : \alpha_i \geq 0, \ \ \alpha_1 + \cdots + \alpha_k = 1\}.$$

Use Exercises 6.35 and 6.36 to show that every $f \in \mathcal{F}_{\vec{G}}$ is also a flow.

6.38. Let $\mathcal{F}_{\vec{G}}$ be as in Exercise 6.37. Show that if each f_i is a maximum flow of $\vec{G}$, then so is every $f \in \mathcal{F}_{\vec{G}}$.

6.39. Let $\vec{G}$ be an Eulerian digraph and let $u, v \in \vec{G}$ be connected by exactly k parallel directed edges with tail u and head v. Let $\vec{G}'$ be the digraph obtained from $\vec{G}$ by removing the k parallel edges from u to v. Show that $\vec{G}'$ is a disjoint union of k directed u, v-paths.

6.40. Let $\vec{G}$ be the digraph obtained from the graph G, as in the proof of Theorem 6.55. Show that the minimum number of directed edges needed to be removed from $\vec{G}$ to destroy all directed u, v-paths is at least the minimum number of edges needed to be removed from G to disconnect u from v in G. [*Hint:* For a minimum collection of directed edges in $\vec{G}$ needed to be removed to destroy all directed u, v-paths, show that the removal of the corresponding collection of undirected edges will disconnect u from v in G.]

6.41. Let $\vec{G}$ be the digraph obtained from the graph G, as in the proof of Theorem 6.55. Show that the minimum number of directed edges needed to be removed from $\vec{G}$ to destroy all directed u, v-paths is at most the minimum number of edges needed to be removed from G to disconnect u from v in G. [*Hint:* If M is a minimum set of edges needed to be removed from G to disconnect u from v, then each $e \in M$ is on a u, v-path in $G - (M \setminus \{e\})$, which yields a unique directed u, v-path in $\vec{G}$ and contains precisely one of the directed edges corresponding to e. Let M_d be the set of all of these directed edges. Show that the removal of M_d from $\vec{G}$ will destroy all the directed u, v-paths by assuming the contrary, that there is a directed u, v-path p in $\vec{G} - M_d$. If p contains no directed edge that corresponds to the same undirected edge as a directed edge in M_d, then p yields a u, v-path in G, a contradiction. Otherwise, assume that p has a minimum number of such matching directed edges of M_d. Let e_1 be a directed edge of p yielding the same undirected edge as $e_2 \in M_d$. Using the same formation as indicated in Figure 6.21, obtain a new directed u, v-path p' with strictly less matching directed edges of M_d than p, thereby obtaining a contradiction of the minimality of p.]

6.42. Let G be a k-connected graph. Let G^* be the graph obtained from G by connecting a new vertex w to exactly k vertices of G. Show that G^* is k-connected as well. [*Hint:* For a set $S \subseteq V(G^*)$, the removal of which disconnects G^*, show directly that $|S| \geq k$ in all cases.]

6.43. For a graph G, use previous Exercise 6.42 and Menger's Theorem 6.57, the undirected vertex version, to show that the following are equivalent:

(a) G is k-connected.

(b) G has at least $k+1$ vertices and, for every choice of vertices $v \in V(G)$ and $W \subseteq V(G)$ where $|W| \geq k$, there are paths $p_1, \ldots, p_k$ from v into W that are vertex disjoint except for the vertex v.

[*Note!* This is sometimes called the *Fan Lemma* and is due to Dirac from 1960.]

6.44. For $k \geq 2$ let G be a k-connected graph. Show that for any k vertices of G there is a cycle in G containing them. [*Hint:* Let $u_1, \ldots, u_k \in V(G)$ and assume this to be true for $k-1$, so there is a cycle C in G containing $u_1, \ldots, u_{k-1}$. Either $|V(C)| = k-1$ or $|V(C)| \geq k$. In the latter case, by the k-connectivity of G and previous Exercise 6.43, there are k vertex-disjoint paths from u_k into $V(C)$. Use the Pigeon-Hole Principle to deduce that there is an i such that the arc on C from u_i to u_{i+1} (computed $\pmod{k-1}$, where the index k is identified with the index 1) contains two of the end points of the mentioned k paths. Use this to complete the induction in this case.]

CHAPTER

7

Planar Graphs

Chapter Goals

Define and discuss various surfaces.

Discuss the meaning of planar embeddings of planar graphs.

Present Euler's Formula for planar graphs and examine its consequences.

Present a combinatorial characterization of planar graphs in the Kuratowski-Wagner Theorem.

Discuss duality for plane graphs and define an abstract dual for a graph.

Present a combinatorial characterization of planar graphs in terms of abstract duality.

Define the genus γ of a graph and discuss embeddings in the γ-torus T_γ.

Present the generalized Euler's formula for graphs of genus γ.

Present the classical theorem by Ringel, Young, and Mayer.

Discuss the crossing number of a graph.

7.1 Embeddings in Surfaces

Throughout this book we have represented vertices by dots and edges by lines or curves in the "plane," which here has been the flat surface of the page. The dots help us distinguish vertices from the intersection of edges. Some graphs may be drawn in the plane in such a way that any intersection of two distinct edges occurs only at a vertex of the graphs. Such graphs are called *planar*. Other times, we may try to draw a given graph in this manner without any success. An example of such a graph is $K_{3,3}$. As we shall see, it doesn't matter how we connect the vertices with edges as lines, there will always be at least one crossing of the edges, which does not occur at the vertices themselves.

Determining whether or not a graph is planar is not only an interesting theoretical problem, but it also has many practical applications. For example, a well-known practical problem, where planar graphs occur naturally, is the problem of *Very Large-Scale Integrated Design*, or VLSI design for short. The objective in VLSI design is to construct an electronic printed circuit board in such a way that each pair of soldering points that are to be connected by a conductor will be connected with no two conductors overlapping. This is very important in the production of computer chips, where the thickness of each chip should be as small as possible—that is, the chips should be as thin as possible. Such an optimal thinness of a chip can be achieved by a planar layout of the conductors on the chip.

In order to investigate planarity of graphs in detail, we first need to discuss the notion of various surfaces, such as the plane, and what exactly is meant by drawing, or *embedding*, a graph into a particular surface. In the following, we assume familiarity with the set of real numbers $\mathbb{R}$ and the two- and three-dimensional Cartesian products, $\mathbb{R}^2$ and $\mathbb{R}^3$. The next convention addresses the four types of surfaces we discuss in this chapter.

Convention 7.1 The *plane* surface is given by $P = \mathbb{R} \times \mathbb{R} = \mathbb{R}^2$, as depicted in Figure 7.1.

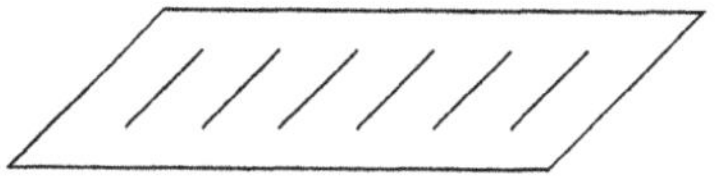

Figure 7.1: The plane surface P.

The *sphere* surface is given by $S^2 = \{(x, y, z) \in \mathbb{R}^3 : x^2 + y^2 + z^2 = 1\}$, as depicted in Figure 7.2.

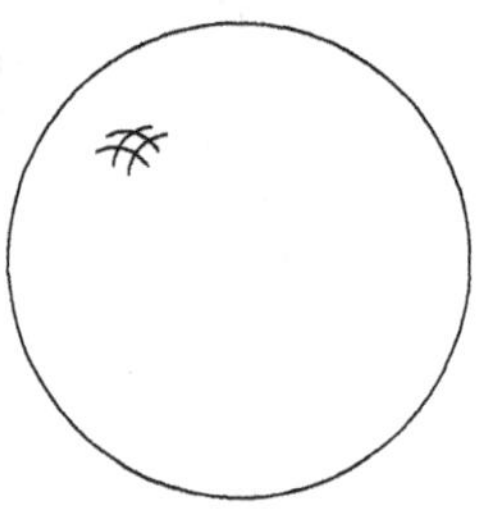

Figure 7.2: The sphere surface S^2.

The *torus* is the surface $T = \{(x, y, z) \in \mathbb{R}^3 : (x - 2\cos\theta)^2 + (y - 2\sin\theta)^2 + z^2 = 1\}$, where $\cos\theta = x/\sqrt{x^2 + y^2}$ and $\sin\theta = y/\sqrt{x^2 + y^2}$, as depicted in Figure 7.3.

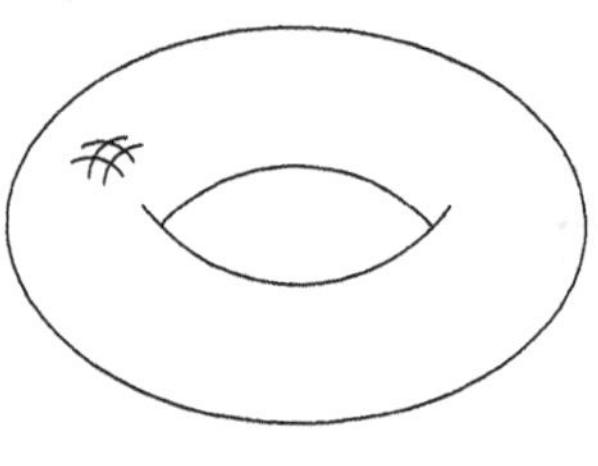

Figure 7.3: The torus T.

More generally, for any $g \in \mathbb{N}$, the *g-torus* surface T_g denotes the g-fold torus embedded in $\mathbb{R}^3$, as depicted in Figure 7.4.

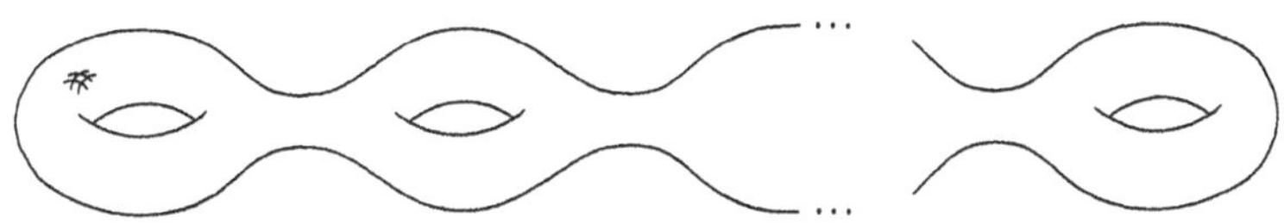

Figure 7.4: The g-torus T_g.

Remark 7.2 The usual torus T is really T_1 in the general setting. The surface T_g is a g-fold torus with g smooth holes in it. In topology, the natural number g is called the *genus* of the surface T_g. Later we will define the genus of a *graph* in terms of the surface T_g.

The next definition makes precise the notion of drawing a graph in a particular type of surface.

Definition 7.3 Let G be a graph and S a surface. We say that G is *embeddable* in S if G has a representation in S, by maps $V(G) \to S$ and $E(G) \to \mathbf{P}(S)$ respectively, in the following way:

1. The vertices in $V(G)$ are represented by distinct points in S.
2. Each edge in $E(G)$ is represented by a continuous curve $\xi : [0; 1] \to S$, where the end points $\xi(0)$ and $\xi(1)$ represent the endvertices of the edge. The restricted curve $\xi : [0; 1[\to S$ is always injective.
3. Two curves, each representing an edge in $E(G)$, intersect if, and only if, their corresponding edges in G are adjacent. In that case, the intersection points of the curves are precisely the end points representing the common endvertices of the corresponding edges.

We say that G is *planar* if it can be embedded in the plane P. The actual embedding is called a *planar embedding*. A *plane graph* of G is a fixed embedding of G in P.

Remark 7.4

- Since our surfaces always sit in the real three-dimensional space $\mathbb{R}^3$, a curve $\xi : [0; 1] \to S$ is continuous if, and only if, it is continuous when we view it as a function $[0; 1] \to \mathbb{R}^3$.
- Note that one can view the 2-torus as a sphere, where we have added two "handles" on it, as shown in Figure 7.5. Likewise, one can topologically deform the g-torus into a sphere with g handles in a continuous fashion (that is, without cutting or ripping the surface or closing any holes) and vice versa. Hence, a graph is embeddable in T_g if, and only if, it is embeddable in a sphere with g attached handles.

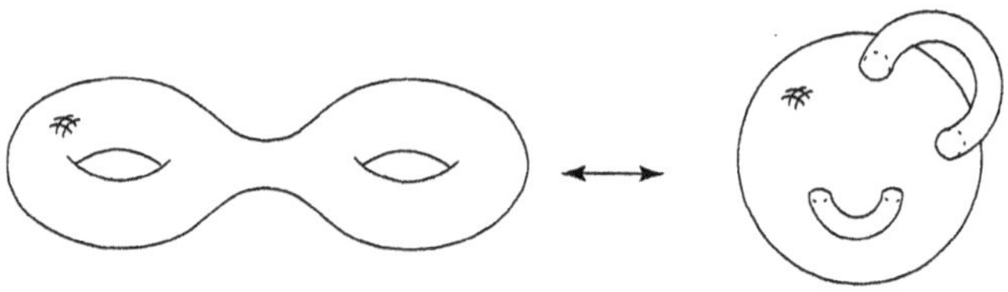

Figure 7.5: The 2-torus can be viewed as a sphere with two handles attached to it.

For the most part, we will be concerned with planar graphs and their embeddings in the plane P. For planar graphs, we have the following theorem.

Theorem 7.5 A graph is planar if, and only if, it is embeddable in the sphere S^2.

Proof Imagine that we have the sphere S^2 lying on the plane P, as shown in Figure 7.6. The sphere and the plane have exactly one point in common. This point is called the "south pole" and denoted by $\tilde{s}$. The opposite point of the sphere is called the "north pole" and denoted by $\tilde{n}$. For each point x on the sphere, not including the north pole $\tilde{n}$, the line segment from $\tilde{n}$ through x intersects the plane P in exactly one point x', so we have a well-defined mapping called the *stereographic projection* π, defined by

$$\begin{aligned} \pi : S^2 \setminus \{\tilde{n}\} &\to P, \\ \pi(x) &= x'. \end{aligned}$$

This map is bijective, and moreover, both π and its inverse π^{-1} are continuous. This implies that if ξ is a continuous curve in S^2, then $\pi(\xi)$ is a continuous curve in P and vice versa.

If a graph G has an embedding G' in S^2, then clearly we can move the embedding G' in such a way that G' is contained in $S^2 \setminus \{\tilde{n}\}$. In this case $\pi(G')$ is an embedding of G in

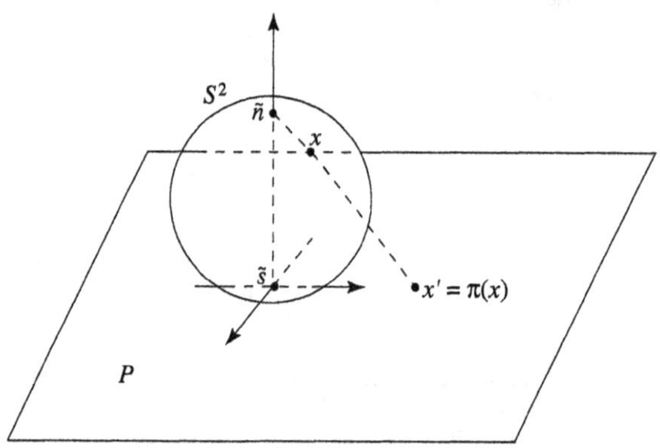

Figure 7.6: The stereographic projection.

the plane P. Thus G is planar. Likewise, if G has a planar embedding G'' in the plane P, then $\pi^{-1}(G'')$ is an embedding of G in $S^2 \setminus \{\tilde{n}\}$. In particular, this is an embedding of G in S^2. Hence, we have the theorem. ■

Theorem 7.5 shows that for our purposes the plane P and the sphere S^2 are equivalent.

Assume that a graph G is embeddable in a surface S. In this case, we have an embedding G' of G in S. All of the graph-theoretic information of G can be interpreted from the geometric embedding G', despite the fact that G and G' are not the same thing, since G is an abstract graph and G' is a collection of subsets in $\mathbb{R}^3$.

Convention 7.6 For a graph G that is embeddable in a surface S, we call the actual embedding G' of G in S the *surface-S graph* of G, or just a *surface graph* when the type is not important. In particular, if S is the plane P, then an embedding of a planar graph in P is a plane graph. If S is the torus, then an embedding is called a *torus graph*.

7.2 More on Planar Embeddings

Let S be a surface. As one can imagine, a continuous curve $\xi : [0; 1] \to S$ can be arbitrarily long. It can have many twists and turns, so the image of a continuous curve in the plane P can be quite complicated. Hence, an embedding of a planar graph in the plane can, for that matter, contain line segments that represent the edges as extremely complicated continuous curves. However, it turns out that every simple planar graph G can be embedded in the plane P in such a way that each edge is represented by a straight line segment. The proof of this result, due to Fary [15], uses quite involved topology and will therefore be omitted here. Figure 7.7 depicts a simple planar graph and a straight line embedding of the graph.

Note that from a given plane graph G we can obtain all of the graph-theoretic information we need about the graph G. However, we actually get much more, since a plane graph is in the plane P, which has many properties of its own that cannot be described in graph-theoretic terms. Keeping in mind that each planar graph has a straight line representation, the following definition, describing the additional properties of plane graphs, should be easy to comprehend. In it we view the plane graph G as a subset of the plane, consisting of points and straight line segments.

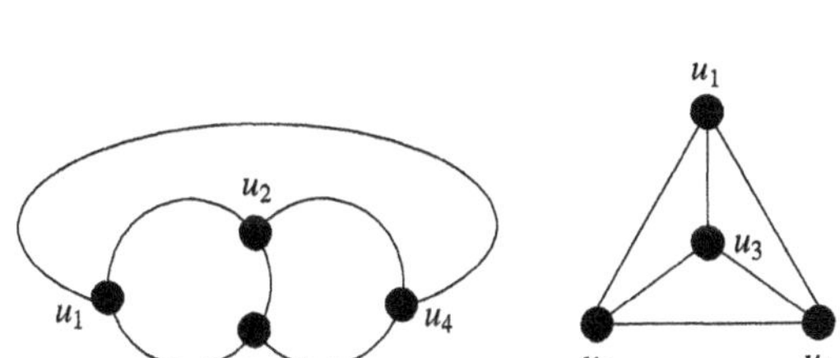

Figure 7.7: Two embeddings of the same graph; the one on the right has every edge represented as a straight line segment.

Definition 7.7 For a plane graph G in P, there are an $f \in \mathbb{N}$ and regions $r_1, \ldots, r_f$ in P such that $P \setminus G = r_1 \cup \cdots \cup r_f$ and the following conditions are satisfied:

1. The sets $r_1, \ldots, r_f$ are pairwise disjoint. That is, $r_i \cap r_j = \emptyset$ for all $i \neq j$.
2. Each r_i is *path connected.* That is, for every two points $x, y \in r_i$, there is a continuous curve from x to y that does not intersect G.
3. Each continuous curve from $x \in r_i$ to $y \in r_j$, where $i \neq j$, intersects G at some point.

Each region r_i is called a *face* of the plane graph G.

Note 7.8 Definition 7.7 can made for any surface S in $\mathbb{R}^3$. That is, we can talk about faces of sphere graphs, torus graphs, and general g-torus graphs in the same way as we do about plane graphs.

Example 7.9 Consider the connected plane graph shown in Figure 7.8. It has $n = 4$ vertices, $m = 6$ edges, and $f = 4$ faces. Note that $n - m + f = 2$ for this plane graph. We will see in Theorem 7.14 that this relation between the number of vertices, edges, and faces of a plane graphs always holds. Any embedding of the underlying graph of a plane graph always has the same number of faces. In this case there are four.

In the plane graph shown in Figure 7.8 there is exactly one face, namely r_4, that is *unbounded.* By this we mean that it can contain the interior region of a square of an arbitrary size. This face is called the *infinite face* of the plane graph. The other faces are all *bounded*—that is, each can be contained within a square of a certain fixed size.

Consider a plane graph G in the plane P with f faces $r_1, \ldots, r_f$, where r_1 is the infinite face. Let $i \in \{1, \ldots, f\}$ and let us put the sphere S^2 on P such that the south pole $\tilde{s}$ of S^2 touches a point inside the face r_i. Now apply the inverse stereographic projection π^{-1} and obtain an embedding of G on the sphere S^2. Next roll the sphere S^2, with the embedding of G on it, in such a way that the former south pole $\tilde{s}$ becomes the north pole $\tilde{n}$. At this

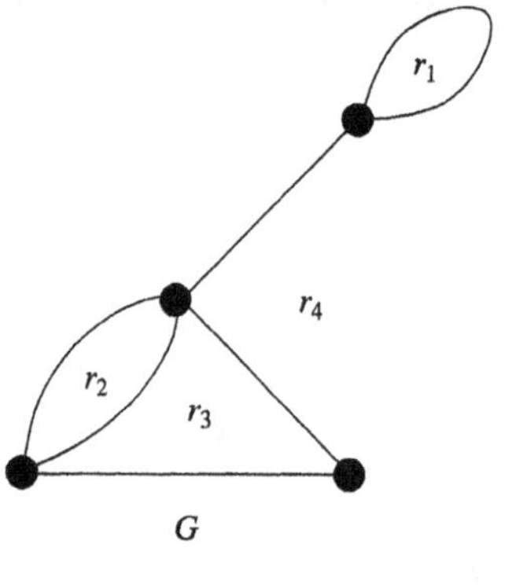

Figure 7.8: The plane graph G has four faces: r_1, r_2, r_3, and r_4.

point apply the stereographic projection π and get a plane graph G', which is isomorphic to our original plane graph G. In this new plane graph G', the face r_i has become the unique infinite face and the prior infinite face r_1 has become a bounded face. We summarize in the following.

Observation 7.10 For every plane graph G and a face r of G, there is a plane embedding of G such that r is the unique infinite face.

Note 7.11 For all surfaces S that we have considered, other than the plane, the faces of a surface-S graph are all bounded in the sense that they can each be contained in a box of a certain size. In particular, every face of a graph embedded in a torus is bounded.

We conclude this chapter with two useful theorems, which will be used in the next section. Their proofs use involved mathematical analysis and topology and will therefore be omitted.

Theorem 7.12 For a plane graph G we have the following:

1. A tree has only one face.
2. *Jordan Curve Theorem* (restricted version): If G is a plane embedding of a cycle graph, then G has precisely two faces. One face is formed by the region "inside" the cycle and one face is formed by the region "outside" the cycle.

Our next theorem is the general version of the Jordan Curve Theorem, which is hard to prove.

Theorem 7.13 **Jordan Curve**

If $\xi : [0; 1] \rightarrow P$ is a continuous and *closed* curve—that is, $\xi(0) = \xi(1)$ such that ξ restricted to $[0; 1[= [0; 1] \setminus \{1\}$ is injective—then $P \setminus \xi([0; 1])$ is a disjoint union of two connected regions S_1 and S_2; that is, every pair of points in each of S_i can be connected by a continuous path, which is completely within S_i. (See Munkres [42] for further details.)

7.3 Euler's Formula and Consequences

One of the most important properties of planar graphs is given in Theorem 7.14. This result is due to Euler.

Theorem 7.14 **Euler's Formula**

Every connected plane graph with n vertices, m edges, and f faces satisfies the following equation:

$$n - m + f = 2.$$

Proof We use induction on f, the number of faces of G, to prove the result.

If $f = 1$, then by Theorem 7.12 G cannot contain any cycles. Since G is connected, by definition it must be a tree. By Theorem 3.7, $m = n - 1$, and we have

$$n - m + f = n - (n - 1) + 1 = 2.$$

If $f > 1$, then by Theorem 7.12(1) G cannot be a tree, and hence it has a cycle. Let e be an edge on the cycle. By Theorem 7.12(2), e is on the boundary of two distinct faces S and S'. By removing the edge e, the two faces S and S' merge and form a new face S''. Since $G - e$ now has n vertices, $m' = m - 1$ edges, and $f' = f - 1$ faces, the induction hypothesis applies to $G - e$, and we have

$$2 = n - m' + f' = n - m + f,$$

which completes the proof of the theorem. ■

Euler's Formula generates some interesting corollaries.

Corollary 7.15 Let G be a simple planar graph on n vertices having m edges. Then the following properties hold:

1. $m \leq 3n - 6$.
2. $m \leq 2n - 4$, if G has no 3-cycles.

Proof Let $r_1, \ldots, r_f$ be the faces of a plane embedding of G. If $f = 1$, then G is a forest and hence $m \leq n - 1$. This implies the corollary. Otherwise, let m_i be the number of edges that bound the face r_i for each $i \in \{1, \ldots, f\}$. Since each edge of G bounds either one or two faces, when we sum up for all the f faces, the number of edges each face bounds, we count each edge in G either once or twice. Hence, we have

$$\sum_{i=1}^{f} m_i \leq 2m. \tag{7.1}$$

Since G is simple, then each face r_i is bounded by at least three edges—that is, $m_i \geq 3$—and hence

$$\sum_{i=1}^{f} m_i \geq 3f. \tag{7.2}$$

From (7.1) and (7.2), we have

$$3f \leq 2m. \tag{7.3}$$

By (7.3) and Theorem 7.14, we obtain

$$m = 3m - 2m \leq 3(m - f) = 3(n - 2) = 3n - 6,$$

proving the first inequality of the theorem.

For the second part, if G has no 3-cycles, then each face r_i is bounded by at least four edges—that is, $m_i \geq 4$—and hence

$$\sum_{i=1}^{f} m_i \geq 4f.$$

From this and (7.1), we obtain $4f \leq 2m$ and hence $2f \leq m$, which together with Theorem 7.14 yields

$$m = 2m - m \leq 2(m - f) = 2(n - 2) = 2n - 4,$$

proving the second inequality and thereby completing the proof. ■

The next example illustrates how Corollary 7.15 can be used to prove that certain graphs are not planar.

Example 7.16 Consider the complete graph K_5. Since it has $n = 5$ vertices and $m = 5(5 - 1)/2 = 10$ edges, we have that

$$m = 10 > 9 = 3n - 6.$$

By Corollary 7.15, K_5 is not planar.

In the same manner, consider the complete bipartite graph $K_{3,3}$. Since it has $n = 6$ and $m = 3 \cdot 3 = 9$ edges, we have

$$m = 9 > 8 = 2n - 4.$$

Since $K_{3,3}$ has no 3-cycles, it is not planar either.

As we shall soon see, the graphs K_5 and $K_{3,3}$ are in some sense the fundamental nonplanar graphs. For emphasis, we make the following observation, as proved in Example 7.16.

Observation 7.17 The complete graph K_5 and the complete bipartite graph $K_{3,3}$ are both nonplanar.

Clearly, every subgraph of a planar graph is also planar. By Observation 7.17, we have that for $n \geq 5$, the complete graph K_n is also nonplanar. This is because it contains the nonplanar graph K_5 as a subgraph. Also, for $m, n \geq 3$, the complete bipartite graph $K_{m,n}$ is also nonplanar, since it contains $K_{3,3}$ as a subgraph. However, the graph K_4 and the graphs $K_{2,n}$ *are* planar for all $n \in \mathbb{N}$, as shown in Figure 7.9.

If a planar graph G on n vertices has exactly $3n - 6$ edges, then equality must hold in (7.1) and (7.2). This means that each face of G is bounded by precisely three edges, or a *triangle*. Clearly, the other direction holds as well. In this case we say that G is a *maximal planar graph* or a *triangulation*.

The following corollary illustrates another interesting property of simple, connected planar graphs.

Corollary 7.18 Every simple, connected planar graph contains a vertex of degree five or less.

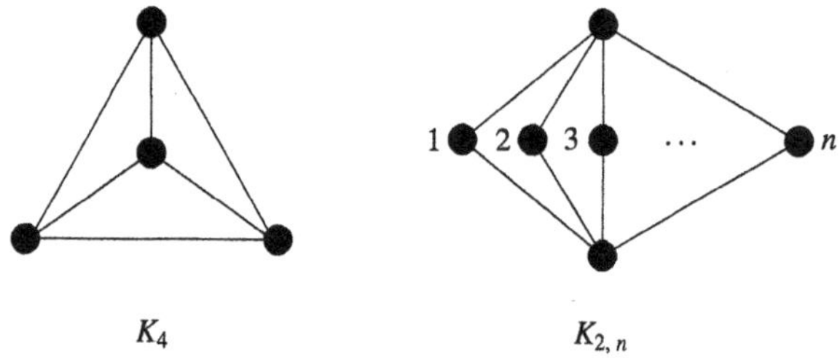

Figure 7.9: K_4 is planar. $K_{2,n}$ is planar for all n greater than zero.

Proof Let G be a simple, connected planar graph on n vertices and with m edges. Let δ be the minimum degree in G. Then, by the Hand-Shaking Theorem 1.26 and Corollary 7.15, we have

$$n\delta \leq \sum_{u \in V(G)} d_G(u) = 2m \leq 2(3n-6) < 6n.$$

Since $\delta \in \mathbb{N}$ we must have $\delta \leq 5$. ■

Later we will use Corollary 7.18 in inductive proofs regarding vertex colorings of planar graphs.

We conclude this section with a theorem showing that when determining the planarity of a graph, it suffices to consider graphs with no cut-edges and no cut-vertices. That is, we can assume the graph is 2-connected.

Theorem 7.19 A connected graph G is planar if, and only if, each block of G is planar.

Proof We prove this by induction on the number of blocks of G. If G is 2-connected, then G has only one block, consisting of G itself, so the theorem is a tautology.

Assume that G has more than one block. Let B be a block that corresponds to a leaf in the block-cutpoint graph BC(G) of G. Let H be the union of the other blocks. We now have

$$\begin{aligned} B \cup H &= G, \\ B \cap H &= \{u\}, \end{aligned}$$

where u is a cutpoint in B. Now, clearly G is planar if, and only if, both B and H are planar. (See Figure 7.10.) Since H has one fewer blocks than G, the induction hypothesis applies. We conclude that H is planar if, and only if, all the blocks of H are planar. Because the set of blocks of G consists of B, together with the set of blocks of H, we have the theorem. ■

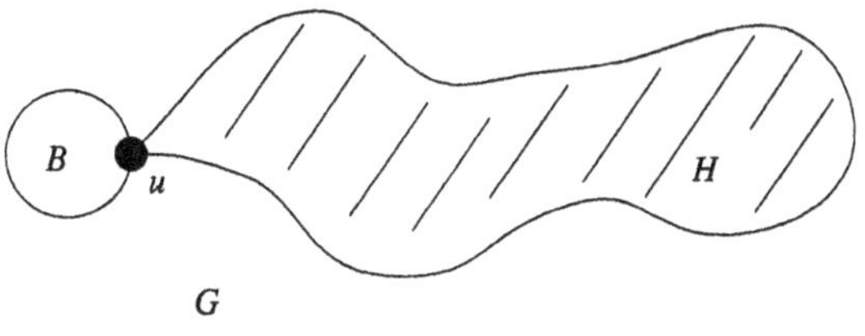

Figure 7.10: The graph G is planar if, and only if, both the block B and the subgraph H are planar.

7.4 Characterization of Planar Graphs

Consider the following question: Is there an automatic method or procedure that one can apply to determine whether a given graph is planar or not? In this and the next section we describe a way to detect exactly when a given graph is planar. These results, due to Kuratowski and Wagner, are considered one of the cornerstones of graph theory. Before we dive into this theory, let us note some easy facts about planarity in general.

Clearly, if a graph on n vertices has more than $3n - 6$ edges, then by Corollary 7.15 it cannot be planar. However, there are many nonplanar graphs on n vertices with $3n - 6$ or fewer edges.

For a given graph G, there are some easy assumptions one can make when determining its planarity. First of all, we may assume that our graph G is simple, since planarity of G is determined neither by whether G has multiple edges nor by whether it has loops. More specifically, if G is not simple, then G is planar if, and only if, G' is planar, where G' is the simple graph formed by removing all loops from G and replacing all multiple edges in G by a single edge. Second, we may assume G is connected, since a graph is planar if, and only if, each of its connected components is. Third, by Theorem 7.19, we may assume G is 2-connected, since a connected graph is planar if, and only if, each of its blocks is planar. For a given simple 2-connected graph G, we further have that G is planar if, and only if, G'' is planar, where G'' is the graph obtained from G by replacing all pairs of adjacent edges to a degree-2 vertex with a single edge. That is, replace all paths (u, w, v) of length two, where $d_G(w) = 2$, by a single edge $\{u, v\}$. Note that by this procedure, the graph G'' might not be simple, in which case we repeat the whole process just discussed, which must eventually stop.

Even with these reductions, we can end up with a *very* complicated 2-connected graph that has a minimum degree of at least three, for which it is a priori hard to determine whether such a graph is planar or not. So, something more is needed to be able to determine whether G is planar or not in the general case.

Consider the last easy observation, that we can replace every pair of adjacent edges to a degree-2 vertex by a single edge. The "reverse" of this procedure is given in the following.

Definition 7.20 Let G be a graph and $e = \{u, v\}$ an edge of G. A *subdivision of e* is the replacement of the edge e by a simple path $(u_0, u_1, \ldots, u_k)$, where $u_0 = u$ and $u_k = v$ are the only vertices of the path in $V(G)$. We say that G' is a *subdivision* of G, if G' is obtained from G by a sequence of subdivisions of edges in G.

Two graphs G' and G'' are *homeomorphic* or *topologically equivalent* if both G' and G'' are subdivisions of the same graph G.

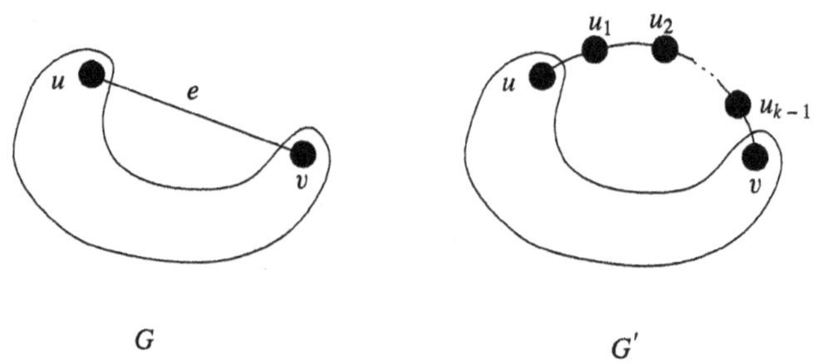

Figure 7.11: A subdivision of e in G.

Figure 7.11 illustrates the concept of subdivision.

Note 7.21 If G is a planar graph, then so is any subdivision G'. We can obtain the graph G back from G' by a sequence of simple contractions (recall Definition 2.34), where we contract every subdivided edge back to a single edge. Since any simple contraction of a planar graph is planar, we see that if G' is planar, so is G. We summarize in the following observation.

Observation 7.22 A graph G is planar if, and only if, every graph homeomorphic to G is planar.

Example 7.23 The two graphs shown in Figure 7.12 are homeomorphic, since they are both subdivisions of K_5.

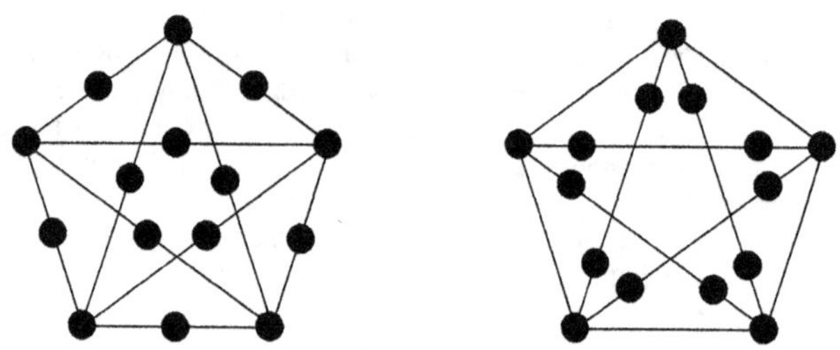

Figure 7.12: Homeomorphic graphs.

Note 7.24 If G and G' are two homeomorphic graphs, then there is a graph G'' that is a subdivision of both G and G'. This property is equivalent to the definition of homeomorphism. (See Exercise 7.20.)

The next example illustrates nonhomeomorphic graphs.

Example 7.25 In contrast to the graphs shown in Figure 7.12, no two of the three graphs G_1, G_2, and G_3 shown in Figure 7.13 are homeomorphic.

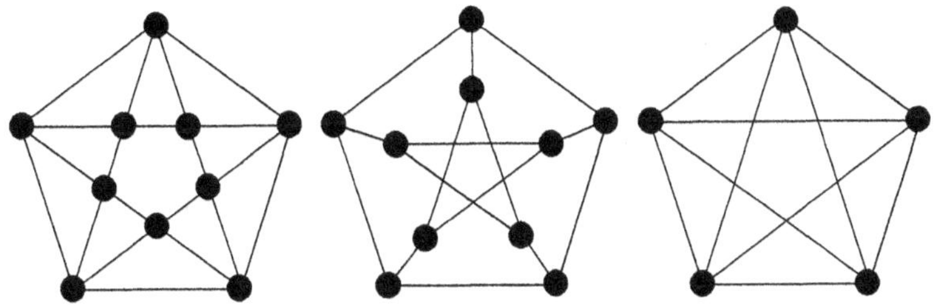

G_1 G_2 G_3

Figure 7.13: Three nonhomeomorphic graphs.

Subdivision does not change the number of vertices having degrees other than two. Hence, if two graphs are homeomorphic, they must have the same number of vertices of degree i for each $i \neq 2$. Thus, none of the graphs G_1, G_2, and G_3 are homeomorphic: The graph G_1 has all of its ten vertices of degree four, the Petersen graph G_2 has no vertices of degree four, and the complete graph G_3 has five vertices of degree four.

As we saw earlier in Observation 7.17, neither K_5 nor $K_{3,3}$ is a planar graph. Also, any graph containing K_5 or $K_{3,3}$ as a subgraph cannot be planar, since every subgraph of a planar graph is also planar. By Observation 7.22, any graph homeomorphic to either K_5 or $K_{3,3}$ is not planar. Therefore, any graph containing a subgraph that is homeomorphic to either K_5 or $K_{3,3}$ is not planar either. Remarkably, this is precisely the *characterization* of planar graphs stated in the following theorem by Kuratowski from 1930. Kuratowski's proof [33] is based on an exhaustive case analysis and is fairly elementary but long. Many different proofs have appeared since, and a readable proof can, for instance, be found in West [57].

Theorem 7.26 **Kuratowski**

A graph G is planar if, and only if, it has no subgraphs homeomorphic to either K_5 or $K_{3,3}$.

The operation of subdivision preserves planarity of graphs. Also, a simple contraction of an edge in a planar graph yields a planar graph. Recalling the notion of a minor of a graph, we summarize in the following.

Observation 7.27 Let G be a simple connected planar graph.

1. Each graph G' homeomorphic to a subgraph of G is planar.
2. Each graph G'' that is a minor of G is planar.

We also have the following observation, whose proof is required in Exercise 7.21.

Observation 7.28 Consider a collection of finite simple connected graphs $\mathcal{G}$. We have the following:

1. Homeomorphism defines an equivalence relation $\approx$ on $\mathcal{G}$. That is, $G \approx G'$ if, and only if, G and G' are homeomorphic.
2. Contraction defines a partial order $\preceq$ on $\mathcal{G}$. That is, $G' \preceq G$ if, and only if, G' is a contraction of G.

Note that condition 2 in Observation 7.28 differs from condition 1 by the relation $\preceq$ lacking symmetry.

We consider the subdivision and contraction operations further: Let G be a simple connected graph that is homeomorphic to G', where G' has no vertex of degree two. In this case G' is not a subdivision of any graph, so G must itself be a subdivision of G'. In particular, G' must be a contraction of G. As we saw from Observation 7.22, G is planar if, and only if, G' is planar. However, this does not hold for contraction, since a planar graph can be a contraction of a nonplanar graph. For example, K_4 is planar and a contraction of the nonplanar K_5.

The reversal of a subdivision is a sequence of contraction of edges, each with one endvertex of degree exactly two. Hence, contraction can be viewed as a *more general* operation.

Note 7.29 If a graph G has a subgraph that is homeomorphic to a given graph H, then H is a contraction of that very subgraph. If, on the other hand, H is a minor of G, then it is not necessarily the case that G has a subgraph homeomorphic to H.

The next example illustrates this.

Example 7.30 Consider the Petersen graph. As we see in Figure 7.14, the Petersen graph has K_5 as a minor. However, the Petersen graph has no subgraph that is homeomorphic to K_5, since all of the vertices of the Petersen graph have degree three. A graph that is homeomorphic to K_5 must have at least five vertices of degree four.

The fact that the Petersen graph has a subgraph homeomorphic to $K_{3,3}$ is not so clear from a first glance at the standard geometric representation of it. This can, however, be seen in the following way. The upper left part of Figure 7.15 shows the special geometric

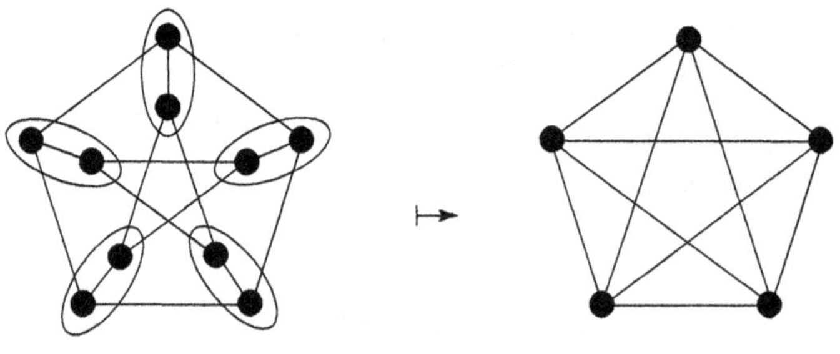

Figure 7.14: K_5 is a contraction of the Petersen graph.

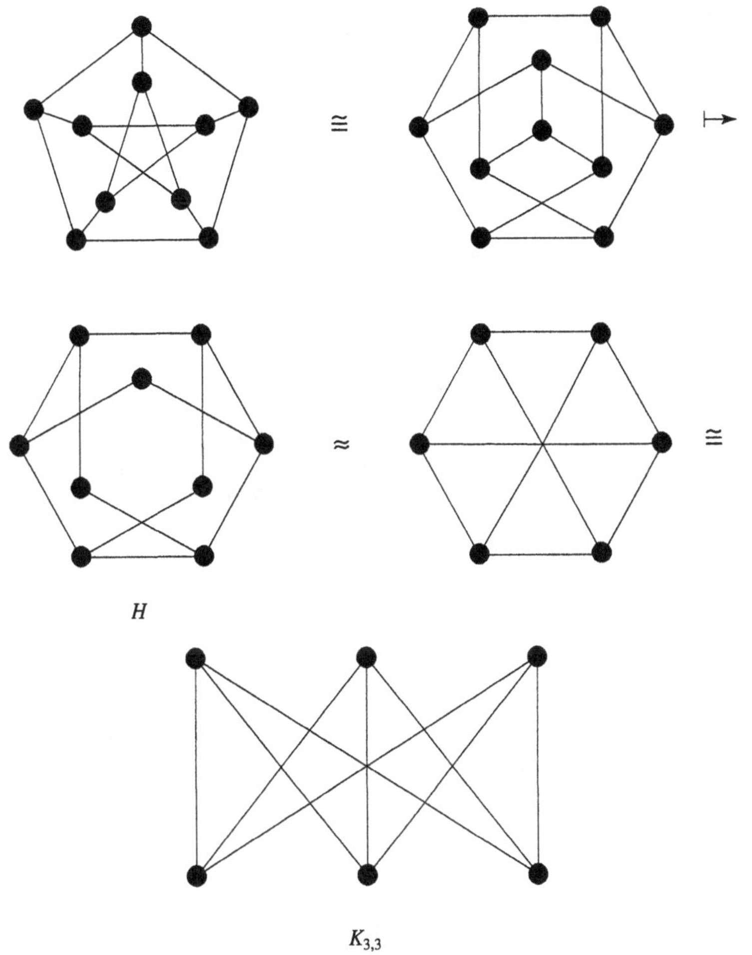

Figure 7.15: The Petersen graph has a subgraph H homeomorphic to $K_{3,3}$.

representation of the Petersen graph from Figure 2.7 in Chapter 2. By removing one vertex from the Petersen graph, we get a subgraph H that is homeomorphic to $K_{3,3}$. The fact that $K_{3,3}$ is homeomorphic to a subgraph of the Petersen graph implies by Kuratowski's Theorem 7.26 that the Petersen graph is not planar.

As we see from Example 7.30, contraction is "easier" to deal with than seeking a subgraph that is homeomorphic to some graph. It is precisely this generalization of Kuratowski's Theorem that was proved by Wagner in 1937. We present the theorem in the next section.

7.5 Kuratowski and Wagner's Theorem

The next theorem provides another classification of planar graphs.

Theorem 7.31 **Kuratowski and Wagner's**

The following statements are equivalent for a graph G:

1. The graph G is planar.
2. Neither K_5 nor $K_{3,3}$ is homeomorphic to any subgraph of G.
3. Neither K_5 nor $K_{3,3}$ is a minor of G.

Proof We sketch the proof of Wagner's part, that the third condition is equivalent to the second condition.

If a subgraph H of G is homeomorphic to a given graph M with no vertices of degree two, then H must be a subdivision of M. Hence, M is a contraction of H. Since neither K_5 nor $K_{3,3}$ has any vertices of degree two, condition three implies condition two.

To prove the other direction, we need to show that if either K_5 or $K_{3,3}$ is a minor of G, then G must contain a subgraph that is homeomorphic to either K_5 or $K_{3,3}$. We consider each case separately.

CASE ONE: $K_{3,3}$ is a contraction of a subgraph H of G. In this case there are six disjoint subsets H_1, H_2, H_3, M_1, M_2, and M_3 of $V(G)$ such that

- Each induces a connected simple subgraph in G, and
- For each $i, j \in \{1, 2, 3\}$ there is at least one edge in G with one endvertex $u_{i,j} \in H_i$ and the other $v_{i,j} \in M_j$.

Consider one of these sets, say H_i. This set contains the vertices $u_{i,1}$, $u_{i,2}$, and $u_{i,3}$ that are not necessarily distinct. Since the induced subgraph $G[H_i]$ connects all of these three vertices together, there is a subtree T_i of G with exactly three leaves $v_{1,i}$, $v_{2,i}$, and $v_{3,i}$ connecting all the vertices $u_{i,1}, u_{i,2}, u_{i,3}, v_{1,i}, v_{2,i}$, and $v_{3,i}$ together. Here the only vertices in T_i that are not in H_i are the $v_{i,\ell}$, where $\ell \in \{1, 2, 3\}$. It is easy to see that a tree with exactly three leaves is homeomorphic to $K_{1,3}$ as displayed in Figure 7.16, where a "zigzag" line indicates a simple path. This also holds for all of the other sets M_j and the corresponding trees T'_j. Hence, G has a subgraph G' that is homeomorphic to $K_{3,3}$ (Figure 7.17) and consists of the union of these six trees:

$$G' = T_1 \cup T_2 \cup T_3 \cup T'_1 \cup T'_2 \cup T'_3.$$

CASE TWO: K_5 is a contraction of a subgraph H of G. In this case there are five disjoint subsets H_1, H_2, H_3, H_4, and H_5 of $V(G)$ such that each H_i induces a connected simple subgraph in G. Also, for each $i \neq j$, there is an edge e_{ij} with endvertices $u_{i;\{i,j\}} \in H_i$ and $u_{j;\{i,j\}} \in H_j$. As in the first case, each H_i contains four vertices $u_{i;\{i,\alpha\}}, u_{i;\{i,\beta\}}, u_{i;\{i,\gamma\}}$, and $u_{i;\{i,\delta\}}$, where $\{\alpha, \beta, \gamma, \delta\} = \{1, 2, 3, 4, 5\} \setminus \{i\}$. Also, G has a subtree T_i with exactly four leaves $u_{\alpha;\{i,\alpha\}}, u_{\beta;\{i,\beta\}}, u_{\gamma;\{i,\gamma\}}$, and $u_{\delta;\{i,\delta\}}$ that connects all of the vertices $u_{i;\{i,\ell\}}$ and $u_{\ell,\{i,\ell\}}$ together, where $\ell \in \{\alpha, \beta, \gamma, \delta\}$. The only vertices not in H_i are the $u_{\ell;\{i,\ell\}}$, as illustrated in Figure 7.18.

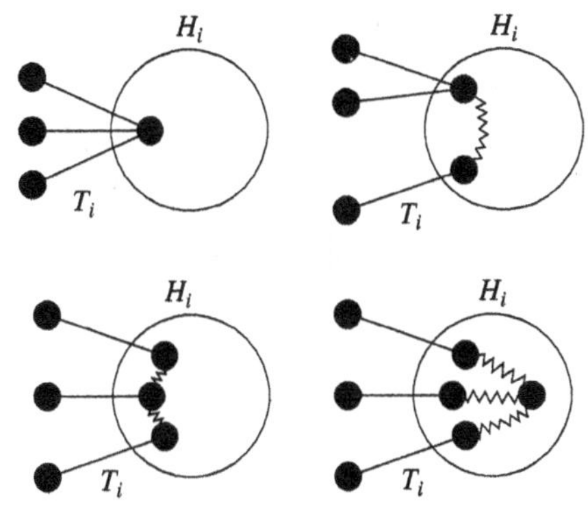

Figure 7.16: Each of these cases for T_i is homeomorphic to $K_{1,3}$.

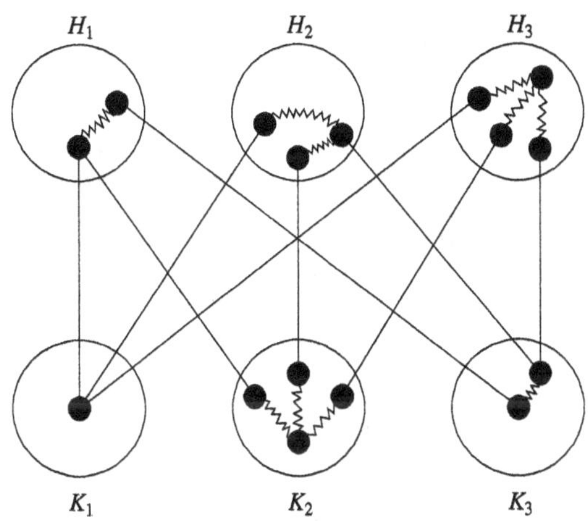

Figure 7.17: The subgraph G' of G is homeomorphic to $K_{3,3}$.

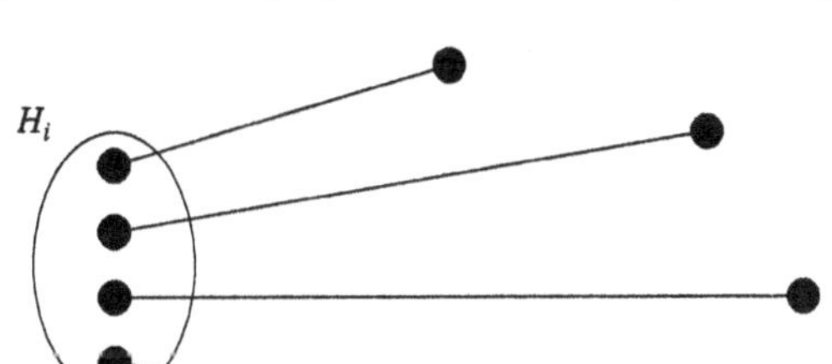

Figure 7.18: The subtree T_i of G connects the eight vertices shown.

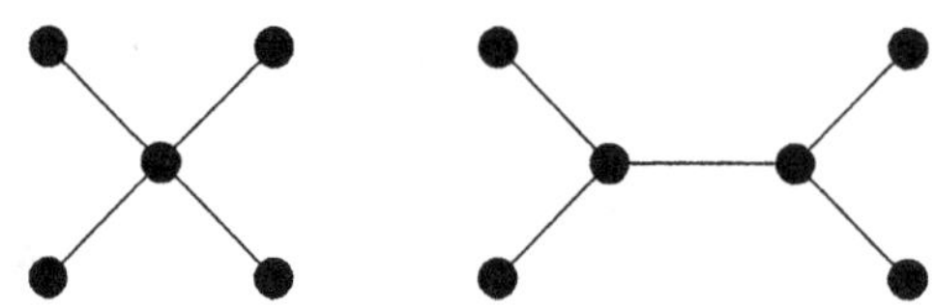

Figure 7.19: Each tree with exactly four leaves is homeomorphic to one of these trees.

The following lemma, which we need only in this proof, is left to the reader in Exercise 7.25, and is illustrated in Figure 7.19.

Lemma 7.32 A tree with exactly four leaves is homeomorphic to either $K_{1,4}$ or the graph composed of an edge with two leaves connected to each endvertex of that edge.

Hence, T_i is homeomorphic to one of the two trees in Lemma 7.32. We need to consider two subcases.

SUBCASE ONE: If all of the trees T_i, where $i \in \{1, 2, 3, 4, 5\}$, are homeomorphic to $K_{1,4}$, then G has a subgraph

$$G' = T_1 \cup T_2 \cup T_3 \cup T_4 \cup T_5$$

that is homeomorphic to K_5, as illustrated in Figure 7.20.

SUBCASE TWO: There is at least one T_i that is homeomorphic to the second tree mentioned in Lemma 7.32. This tree consists of an edge with two leaves connected to each endvertex of the edge. Assume $i = 1$ for the simplicity of indexing. Restricting our attention to T_1 and the edges e_{ij} together with their endvertices, where $i \in \{2, 3\}$ and $j \in \{4, 5\}$,

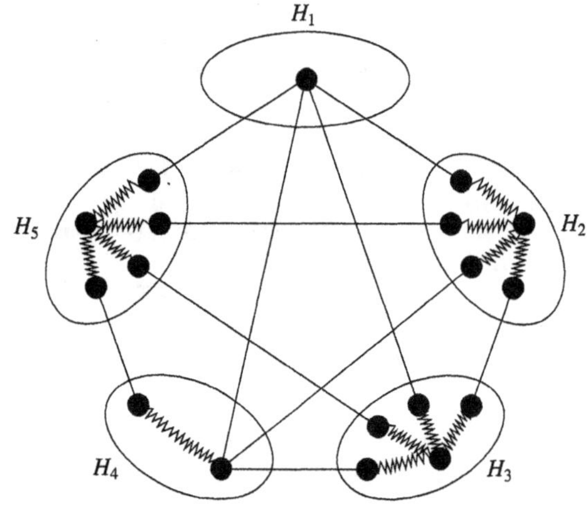

Figure 7.20: The graph G' is homeomorphic to K_5, since each T_i is homeomorphic to $K_{1,3}$.

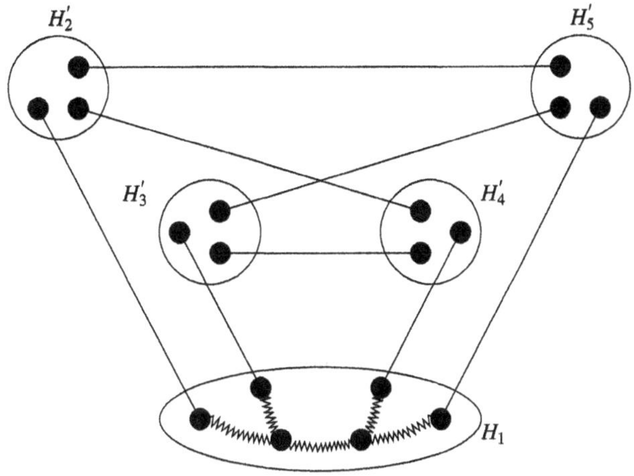

Figure 7.21: The tree T_1 together with the four sets of vertices H_2', H_3', H_4', and H_5'.

we have the situation as displayed in Figure 7.21. Here we are considering $H_i' \subseteq H_i$, where $i \in \{2, 3, 4, 5\}$, each containing three vertices as listed next.

$$\begin{aligned} H_2' &= \{u_{2;\{1,2\}}, u_{2;\{2,4\}}, u_{2;\{2,5\}}\}, \\ H_3' &= \{u_{3;\{1,3\}}, u_{3;\{3,4\}}, u_{3;\{3,5\}}\}, \\ H_4' &= \{u_{4;\{1,4\}}, u_{4;\{2,4\}}, u_{4;\{3,4\}}\}, \\ H_5' &= \{u_{5;\{1,5\}}, u_{5;\{2,5\}}, u_{5;\{3,5\}}\}. \end{aligned}$$

As in the first case, every tree with exactly three leaves is homeomorphic to $K_{1,3}$. Hence, G has a subgraph G' homeomorphic to the graph shown in Figure 7.22, which is isomorphic to $K_{3,3}$, since it has bipartition

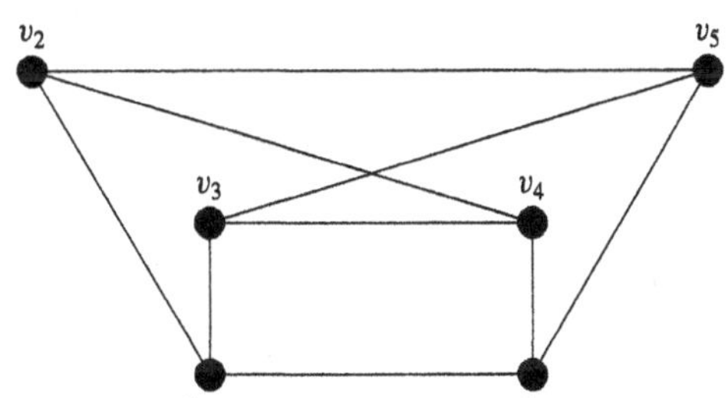

Figure 7.22: The subgraph G' of G is homeomorphic to this graph, which is in turn isomorphic to $K_{3,3}$.

$$\{v_2, v_3, v_{1,2}\} \cup \{v_4, v_5, v_{1,1}\}.$$

So, in this case, G has a subgraph that is homeomorphic to $K_{3,3}$.

This completes the argument that the condition of Wagner implies the condition of Kuratowski, and we have the theorem. ∎

From this proof, we extract the following interesting corollary.

Corollary 7.33 Let G be a graph.

1. If $K_{3,3}$ is a minor of G, then G contains a subgraph homeomorphic to $K_{3,3}$.
2. If K_5 is a minor of G, then G contains a subgraph homeomorphic to either K_5 or $K_{3,3}$.

Note 7.34 For $n \geq 3$ the graph $K_{1,n-1}$, used in the proof of Theorem 7.31, is a tree and is also called a *star* on n vertices, or an *n-star*.

Recall that if a graph G is planar, then so is every induced subgraph of G. A type of a graph-theoretic property that holds for all induced subgraphs is sometimes called a *hereditary* property. Thus planarity is a hereditary property of graphs. By Theorem 7.31 planarity can be characterized as those graphs not having any of the graphs in the set $\{K_5, K_{3,3}\}$ as a minor. These are called the *forbidden* minors for planarity. The minimal forbidden minors are sometimes called *obstructions*.

A deep theorem called a *Minor Theorem* by Robertson and Seymour, published as a series of many papers during the years from 1986 to 1997, implies that for *any* hereditary property $\mathbf{P}$ of graphs, there exists a finite collection $C(\mathbf{P})$ of obstructions such that a graph G has property $\mathbf{P}$ if, and only if, G does not have any graphs in $C(\mathbf{P})$ as a minor. In particular, if $\mathbf{P}$ is the hereditary property of a graph being embeddable in a certain surface S, we have the following corollary.

Corollary 7.35 For every surface S, there exists a finite collection $\{H_1, \ldots, H_k\}$ of graphs such that a graph G is embeddable in S if, and only if, no H_i is a minor of G.

If the surface S is not the plane, then even to this day, the collection of forbidden minors corresponding to the surface S remains unknown. The only set known is the one corresponding to the plane given by Theorem 7.31. We discuss embeddings on other surfaces S in §7.7.

7.6 Plane Duality

In this section we discuss another combinatorial characterization of planar graphs in terms of abstract duality. First, though, we discuss the dual of plane graphs.

The idea of a dual of a plane graph is very simple and most likely stems from viewing a geographical map of countries and their boundaries as a graph, where we view each country as a vertex and then connect two vertices if their corresponding countries have a common border line. More precisely, if a connected plane graph G has vertices, edges, and faces given by

$$\begin{aligned} V(G) &= \{u_1, \ldots, u_n\}, \\ E(G) &= \{e_1, \ldots, e_m\}, \\ F(G) &= \{r_1, \ldots, r_f\}, \end{aligned}$$

then the *plane dual* G^* of G is the plane graph with vertices, edges, and faces given by

$$\begin{aligned} V(G^*) &= \{u_1^*, \ldots, u_f^*\}, \\ E(G^*) &= \{e_1^*, \ldots, e_m^*\}, \\ F(G^*) &= \{r_1^*, \ldots, r_n^*\}, \end{aligned}$$

which is constructed in the following way:

- Embed each vertex u_i^* in the face r_i^*.
- For every edge e_k that borders both r_i and r_j in G, draw an edge connecting the vertices u_i^* and u_j^* that intersects e_k exactly once.
- For every edge e_ℓ that borders just one face r_i in G, draw a loop e_ℓ^* at u_i^* that intersects e_ℓ exactly once.

Note that each face of G^* has precisely one vertex of G in it, since all the faces in G that border a vertex form a cycle in G^* surrounding that vertex. Hence, the number of faces of G^* is indeed n.

Example 7.36 Consider the graph G shown in Figure 7.23. Here G has four faces r_1, r_2, r_3, and the infinite face r_4. Hence, its plane dual G^* has four vertices corresponding to the four faces of G. Note that there is a one-to-one correspondence between $E(G)$ and $E(G^*)$, since we can pair up the edges, $e \leftrightarrow e^*$, of G and G^*, by pairing e^* with the unique edge e that it crosses exactly once. Also note that we have a one-to-one correspondence between the faces $F(G^*)$ of G^* and the vertices $V(G)$ of G, since there is exactly one vertex of G in each face of G^*.

Looking at G^* in Example 7.36 and viewing the vertices of G as representatives for the faces of G^*, we see that the vertices of G are connected by exactly the same rules as for the plane dual of G: For each edge $e^* \in E(G^*)$ on the common border of the two faces of G^* corresponding to u and v in $V(G)$, draw one edge e between u and v that crosses e^* exactly once. Hence, we have the following observation.

Observation 7.37 If G is a connected plane graph and G^* is its plane dual, then G is the plane dual of G^*—that is to say,

$$G^{**} = G.$$

The above Observation 7.37 will be generalized later in Theorem 7.43, the proof of which will not use any planarity whatsoever.

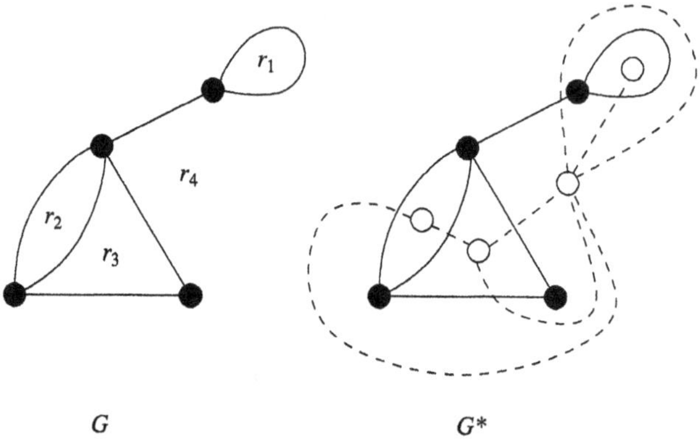

Figure 7.23: A plane graph G and its plane dual G^*.

Example 7.38 As defined, the plane dual of a plane graph depends heavily on the actual plane embedding. Unfortunately, two embeddings of the same planar graph could result in two different plane duals, which are not even isomorphic! Consider the plane graph H in Figure 7.24, which is just a different plane embedding of the plane graph G from Example 7.36. Here we see that the plane dual of H^* is not isomorphic to the plane dual G^* of G from Figure 7.23, since G^* has a vertex of degree six, but H^* does not.

For a fixed plane graph G, the plane dual G^* is, of course, a uniquely determined plane graph also. It is natural to examine how G^* changes, if we, in a planar way, alter G. That is to say, if we make some planar transformation $G \mapsto G'$ of G, how will the dual graph $G^* \mapsto G'^*$ change by that transformation?

First, consider the transformation of G, where we take one edge $e \in E(G)$ and subdivide it into a path of length $k \geq 2$. Assuming that e^* has endvertices u_i^* and u_j^* (so e is on the boundary of two faces r_i and r_j of G), then each of the k edges in the subdivision of e is also on the boundary of the faces r_i and r_j of G. Therefore, after the subdivision, u_i^* and u_j^* must

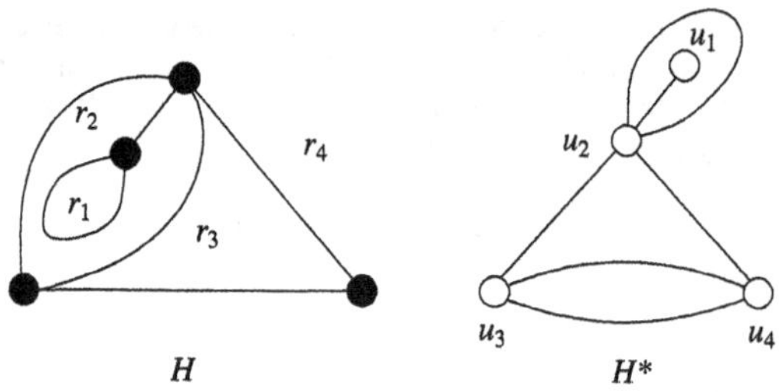

Figure 7.24: The plane graph H is isomorphic to G in Figure 7.23, but their plane duals H^* and G^* are not isomorphic.

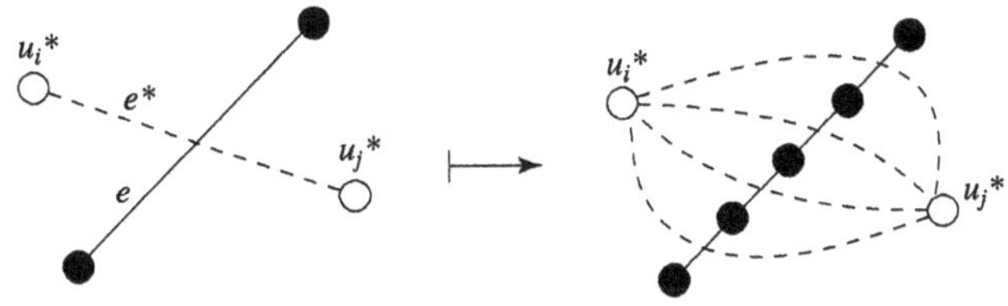

Figure 7.25: Subdividing an edge e of G into a path of length k yields that e^* will become k-fold.

be connected with exactly k multiple edges, corresponding to each edge in the subdivision, as depicted in Figure 7.25.

Second, consider the transformation of G, where we remove one edge $e \in E(G)$ from G. If e is on the border of two faces r_i and r_j, then by removing e, these two faces will become one face r'_{ij}. This yields that the corresponding vertices u_i^* and u_j^* in the dual graph G^* will merge into one new vertex $u'_{ij}{}^*$. Since u_i^* and u_j^* were connected by e^*, this is the same as saying that G^* has been contracted by e^* to $G^* \cdot e^*$, as depicted in Figure 7.26. If e borders just one face, then the removal of e will result in the contraction of the loop e^*, which is simply the same as the removal of e^*.

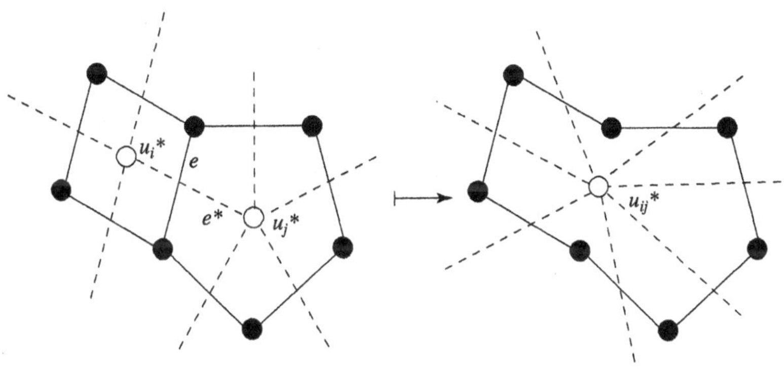

Figure 7.26: Removing an edge e of G yields that G^* will be contracted by e^* to $G^* \cdot e^*$.

We summarize in the following observation.

Observation 7.39 Let G be a connected planar graph and $e \in E(G)$ an edge of G. Further, let G^* be the plane dual of G and $e^* \in E(G^*)$ be the edge corresponding to e.

1. The transformation of removing e from G will result in the contraction of G^* by e^*. In short, we have

$$(G - e)^* = G^* \cdot e^*.$$

2. The transformation of subdividing e into a simple path of length k, thereby obtaining the graph we denote by $G_{e/k}$, results in edges e^* becoming k-fold in G^*, thereby obtaining the graph we denote by $(G^*)_{ke^*}$. In short, we have

$$(G_{e/k})^* = (G^*)_{ke^*}.$$

Consider the plane graph G and its plane dual G^* shown in Figure 7.23. As we have seen, there is a one-to-one correspondence $e \leftrightarrow e^*$ between the edges of G and G^*. But there is, in fact, more to this correspondence: Every cycle in G "cuts off" some vertices of G^*, namely those representing the faces inside the cycle, from the other vertices of G^*, namely those representing the faces outside the cycle. This means that each subset of $E(G)$ that forms a cycle in G yields an edge cut of $E(G^*)$, namely those edges crossing the edges in G forming the cycle. Hence, if $\mathcal{C}(G)$ is the set of cycles in G and $\mathcal{S}(G^*)$ is the set of edge cuts in G^*, then we have an injective map

$$\mathcal{C}(G) \to \mathcal{S}(G^*),$$

which is induced by the map $E(G) \to E(G^*)$, where $e \mapsto e^*$.

On the other hand, if we have an edge cut S of G, then there are faces $r_1, \ldots, r_s$, such that r_1 is the infinite face and each consecutive pair r_i and r_{i+1} (modulo s—that is, index $s+1$ is identified with index 1) of faces have exactly one edge e_i on their common border. Hence, $(u_1^*, e_1^*, u_2^*, e_2^*, \ldots, u_s^*, e_s^*, u_1^*)$ forms a cycle in G^*. Applying this argument to G^*, we obtain by Observation 7.37 that we have an injective map

$$\mathcal{S}(G^*) \to \mathcal{C}(G^{**}) = \mathcal{C}(G).$$

Since both sets $\mathcal{C}(G)$ and $\mathcal{S}(G^*)$ are finite, we have that the mapping $e \mapsto e^*$ induces a one-to-one correspondence between them. This correspondence motivates the following definition of an *abstract dual* of a graph G, which does not assume any planarity at all!

Definition 7.40 Let G be a graph. We say that a graph G' is an *abstract dual graph* of G, if there is bijection $\phi : E(G) \to E(G')$ such that ϕ induces a one-to-one correspondence between the set of cycles $\mathcal{C}(G)$ of G and the set of edge cuts $\mathcal{S}(G')$ of G'. That is to say, ϕ yields a bijection

$$\tilde{\phi} : \mathcal{C}(G) \to \mathcal{S}(G').$$

Keeping in mind Observation 7.39, which holds for plane graphs, we can obtain a similar result for an abstract dual of a graph G.

Lemma 7.41 If G is a graph, which has an abstract dual G', then we have

1. Every subgraph of G has an abstract dual.
2. Every graph homeomorphic to G has an abstract dual.

Proof For the first part, it suffices to show that for any edge $e \in E(G)$ the graph $G - e$ has an abstract dual. This is sufficient because by repeatedly removing edges, we eventually obtain our subgraph in question, possibly together with some isolated vertices. But whether or not a graph has an abstract dual is not affected by its isolated vertices.

Since G' is the abstract dual of G, there is a bijection $\phi : E(G) \to E(G')$ that yields a one-to-one correspondence between cycles in G and edge cuts in G'. Note that $S \subseteq E(G') \setminus \{\phi(e)\}$ is an edge cut in G' if, and only if, it is an edge cut in $G' \cdot \phi(e)$. Hence, ϕ induces a one-to-one correspondence between cycles in $G - e$ and edge cuts in $G' \cdot \phi(e)$, which therefore is the abstract dual of $G - e$.

For the second part, it suffices to show that the graph $G_{e/k}$ obtained from G by subdividing the edge $e \in E(G)$ into a simple path of length $k \geq 2$, has an abstract dual, if G has one.

Again, since G has an abstract dual G', then we obtain an abstract dual of $G_{e/k}$ simply by making the edge e^* a k-fold one, thereby obtaining the graph $(G^*)_{ke^*}$. From ϕ, we obtain a bijection

$$\phi' : E(G_{e/k}) \to E((G^*)_{ke^*}),$$

which clearly induces a bijection between the cycles of $G_{e/k}$ and the edge cuts of $(G^*)_{ke^*}$. This is because every cycle of $G_{e/k}$ must contain all the edges of the k-path, which e was subdivided into, or none at all; and hence, each edge cut contains either all the k parallel edges, corresponding to e^*, or none at all. This completes the proof. ■

From Lemma 7.41 we obtain an interesting corollary.

Corollary 7.42 If G is a graph that has an abstract dual, then so does each graph homeomorphic to a subgraph of G.

As for Observation 7.37, we have the following generalization, which can be proven without the use of planarity at all.

Theorem 7.43 If G' is an abstract dual of G, then G is an abstract dual of G'.

Proof Let $\phi : E(G) \to E(G')$ be the bijection that induces the one-to-one correspondence between the cycles of G and the edge cuts of G'. If S is an edge cut of G, then by Theorem 6.8, S has an even number of edges in common with any cycle of G. Therefore, $\phi(S) \subseteq E(G')$ has an even number in common with any edge cut of G'. By Exercise 6.6 in Chapter 6, $\phi(S)$ must be an edge-disjoint union of cycles. But, in a similar fashion, it is easy to see that an edge-disjoint union of at least two cycles in G' correspond to an edge-disjoint union of at least two edge cuts in G, which contradicts the fact that we started with a single edge cut S. Hence, $\phi(S)$ must be a simple cycle, which completes the proof. ■

Theorem 7.43 shows that the one-to-one correspondence between cycles of G and edge cuts of G' *implies* a one-to-one correspondence between edge cuts of G and cycles of G'. We use this in the proof of the following lemma.

Lemma 7.44 Neither of the graphs K_5 nor $K_{3,3}$ has an abstract dual.

Proof Assume K_5 has an abstract dual G'. In this case, G' must have ten edges, as K_5 does. Since K_5 is simple, the degree of each vertex of G' must be at least three. Clearly an edge cut of K_5 has either four or six edges, and hence G' contains a 6-cycle. Note that at most three edges can be added to a 6-cycle, without creating parallel edges, or a 3-cycle. But since K_5 has no vertex of degree two, nor an edge cut consisting of three edges, G' must have at least seven vertices. Since each vertex in G' has a degree of three or more, then by the Hand-Shaking Theorem, G' has at least $\lceil 7 \cdot 3/2 \rceil = 11$ edges, which is a contradiction. Hence, K_5 has no abstract dual.

Now, assume $K_{3,3}$ has an abstract dual G''. Again, note that a cycle of $K_{3,3}$ has four or six edges, and hence each vertex of G'' has a degree of at least four. Since all edge cuts of $K_{3,3}$ have at least three edges, then each cycle of G'' has a length of at least three. In particular, G'' has no loops or parallel edges and must therefore be simple. But now, by the Hand-Shaking Theorem, the number n'' of vertices of G'' satisfies

$$4n'' \leq 2|E(G'')| = 2 \cdot 9 = 18;$$

and hence $n'' \leq \lfloor 18/4 \rfloor = 4$, which contradicts the fact that all vertices have degree at least four. Hence, $K_{3,3}$ cannot have an abstract dual. ∎

By Kuratowski's Theorem 7.26, Corollary 7.42, and Lemma 7.44, we obtain the following theorem, one of the main results of this section.

Theorem 7.45 A graph G is planar if, and only if, it has an abstract dual.

Theorem 7.45 can viewed as yet another characterization of planar graphs. Note that the abstract dual is a completely combinatorial phenomenon, which did not rely on any plane topology.

7.7 Higher Genus

In this section we discuss embeddings of graphs in surfaces S other than the plane. Recall that a graph is planar if, and only if, it can be embedded in the sphere S^2. Also note that if we can embed a graph on the sphere, then it can be embedded in the torus T. More generally, we can view the plane as the 0-torus and the usual torus as the 1-torus. So, if a graph is embeddable in the g-torus T_g, then it is also embeddable in the g'-torus $T_{g'}$ for any $g' \geq g$.

Recall that g is called the genus of the surface T_g, the g-fold torus. The plane and the torus have genus (or genera) zero and one, respectively. Recall that the surface T_g can be viewed as a g-fold torus or as a sphere with g attached handles, as indicated in Figure 7.5.

Theorem 7.46 Every graph on n vertices is embeddable in a surface of genus g, where $g \geq n(n-1)/2$. In particular, every graph is embeddable in some surface T_g.

Proof Draw the n vertices on a sphere in an arbitrary fashion. For every two neighboring vertices, we add a handle to the sphere and draw all the corresponding connecting edges along that handle. There are at most $\binom{n}{2} = n(n-1)/2$ such pairs of vertices. Hence, we need to add at most $n(n-1)/2$ handles to the sphere in order to draw all the edges in a nonintersecting way. ∎

We will see later in this section that the number $n(n-1)/2$ of handles in Theorem 7.46 is not a bad bound! At this point, however, we merely want to establish the fact that all graphs are embeddable in some surface. Theorem 7.46 justifies the following definition.

Definition 7.47 The *genus* of a graph G is the least number γ such that G is embeddable in T_γ. The genus of G is denoted by $\gamma(G)$.

The genus can be viewed as a parameter used to measure how far a graph is from being planar.

We have that a graph G is planar if, and only if, $\gamma(G) = 0$. If $\gamma(G) = 1$, then G is a *toroidal* graph. Since K_5 and $K_{3,3}$ are nonplanar, we have that both have genus of one or larger. To show that K_5 and $K_{3,3}$ are toroidal graphs, one must show that there is an embedding of each on the torus. When considering specific embeddings on the torus T, it is often a good idea to cut T up and flatten T out to a square, as shown in Figure 7.27. In this figure the flat version of T has the top side AB and the bottom side CD identified as the arrows indicate. Likewise, the left side AC and the right side BD are also identified as the arrows indicate.

We can draw an edge from a vertex in the square representing the torus, as a continuous curve down through the side CD, which then continues down from the side AB. The intersection point of the edge and the side CD has the same distances from the corner points C and D as the intersection point of the edge and the side AB has from the corner points A and B, respectively. This we depict in Figure 7.28, where the vertex a is adjacent to vertex d in both K_5 and $K_{3,3}$. In a similar fashion, we can go right and appear left again. As an example, vertex b is adjacent to vertex c in Figure 7.28. These drawings of K_5 and $K_{3,3}$ illustrate that the genuses of these graphs are at most one. Hence, K_5 and $K_{3,3}$ are both toroidal graphs with genus of exactly one.

The next example shows that the Petersen graph is also toroidal.

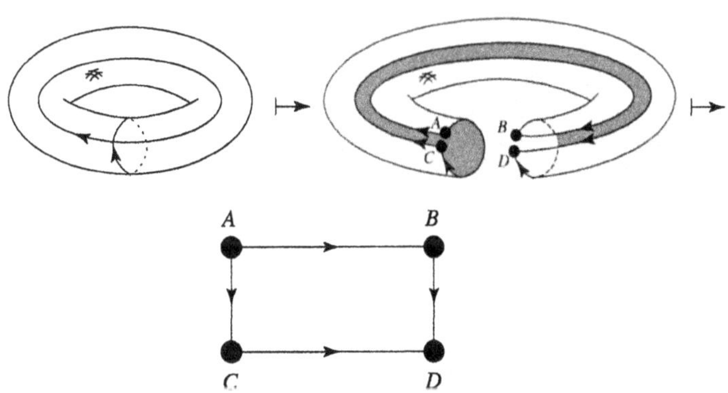

Figure 7.27: A representation of the torus T as a square with the top, bottom, left, and right sides identified.

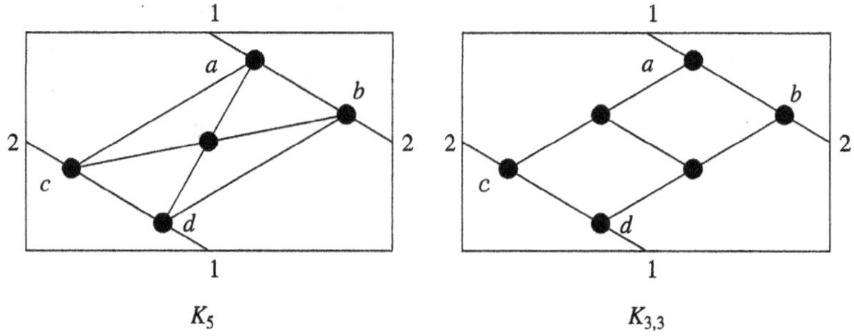

Figure 7.28: Embeddings of K_5 and $K_{3,3}$ on the torus.

Example 7.48 Recall from Example 7.30 that the Petersen graph has K_5 as a minor, so by Theorem 7.31 it is a nonplanar graph. Hence, the Petersen graph has a genus of at least one. In Figure 7.29 we depict an embedding of the Petersen graph on the torus. Thus, it is a toroidal graph.

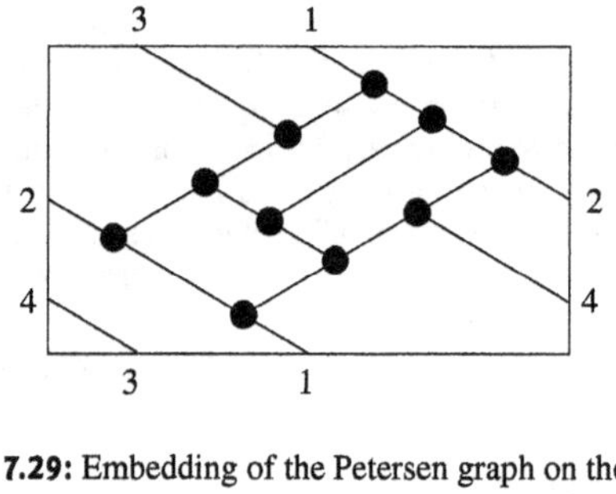

Figure 7.29: Embedding of the Petersen graph on the torus.

Note that in Figures 7.28 and 7.29 the corresponding points where the edges disappear and reappear are labeled with numbers to avoid confusion.

Suppose $g \geq 2$. For a surface graph embedded in T_g, we can talk about faces just as for plane graphs in the plane. We call a face r a *2-cell* if it is *simply connected* in the sense that we can pull every closed curve $\xi : [0; 1] \to r$ into one point in r in a continuous manner. (See Figure 7.30.) Loosely speaking, this means that a 2-cell is a face with no "holes" in it. Let

$$E = \{(x, y) : x^2 + y^2 < 1\}$$

denote the open disk in $\mathbb{R}^2$. If r denotes a face of a surface graph, without its boundary in $\mathbb{R}^3$, then a precise definition of a 2-cell can be given as follows.

Definition 7.49 Assuming the usual topology on $\mathbb{R}^k$ for $k \in \{2, 3\}$, a face r is a *2-cell* if, and only if, there is a continuous bijective map $f : E \to r$ with a continuous inverse $f^{-1} : r \to E$.

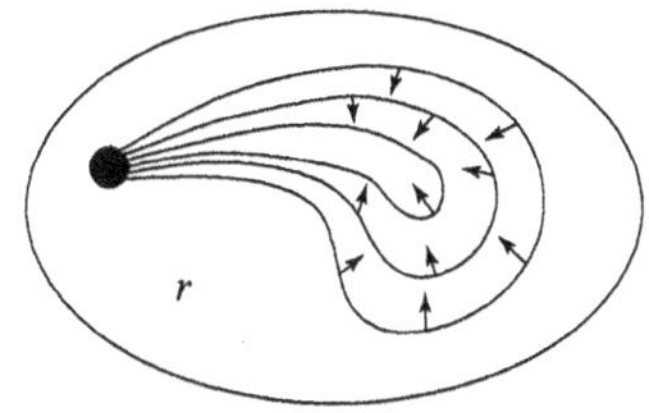

Figure 7.30: Every closed curve in a 2-cell r can be pulled into one point in r.

With this in mind, we have the following theorem characterizing the faces of a graph embedded in a surface.

Theorem 7.50 In an embedding of a connected graph G on $T_{\gamma(G)}$, every face is a 2-cell.

Proof (Sketch) Assume we have an embedding of G in $T_{\gamma(G)}$ such that there is a face r which is not a 2-cell. Then r must contain a closed curve $\zeta : [0; 1] \to r$ that cannot be pulled together into one point in r. We can cut the face r along the curve ζ and glue two circular disks along the two boundaries that have just been formed by this cutting. We have two possibilities to consider.

CASE ONE: The face r falls apart into two disconnected parts, as shown in Figure 7.31. In this case we have two components, A and B. Since the graph G is connected, then so is its embedding. Thus, its embedding is either all in A or all in B. Suppose it is in A. Now B cannot be a surface of genus zero, since otherwise we could have pulled ζ together into one point. Therefore, B must have at least one handle. That is, it has genus of one or higher. This means that we have an embedding of G in A that has at most $\gamma(G) - 1$ handles. This contradicts the definition of $\gamma(G)$.

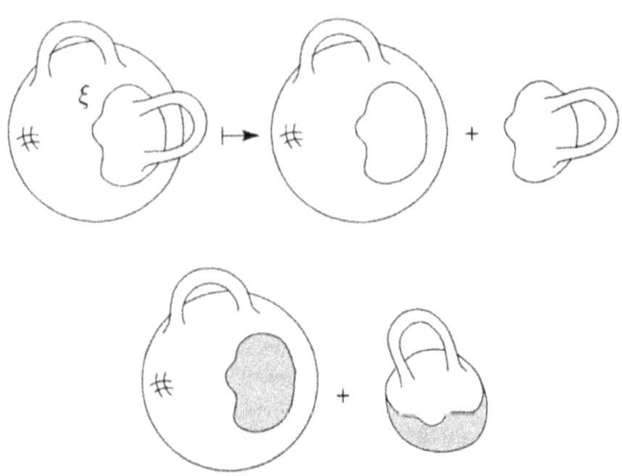

Figure 7.31: The face r falls apart into two disconnected parts.

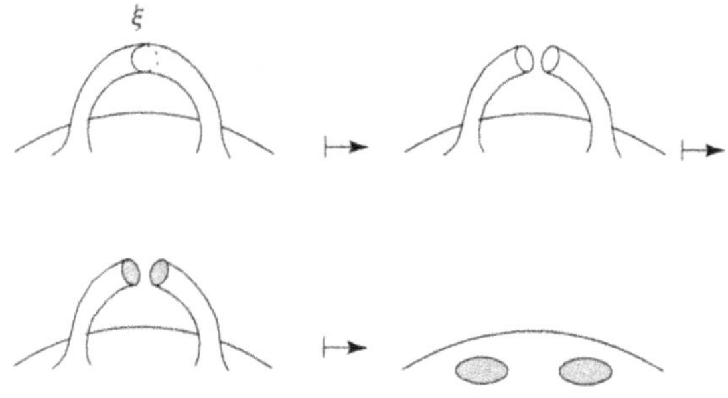

Figure 7.32: The face r does not fall apart when one of its handles is cut.

CASE TWO: The face r does not fall apart. In this case cutting the surface along ζ cuts up one handle of the surface, as shown in Figure 7.32. By this procedure one handle has been eliminated. Hence, we have reduced the genus $\gamma(G)$ by one, again contradicting the definition of $\gamma(G)$.

Therefore, in every embedding of a graph G in $T_{\gamma(G)}$, all of the faces are 2-cells. ∎

Remark 7.51 From Theorem 7.50 we deduce, in particular, that every graph G can be embedded in some surface in such a way that every face is a simply connected 2-cell.

7.8 Generalization of Euler's Formula

Let G be a connected graph on n vertices and m edges. Assume that we have an embedding of G in T_γ, where $\gamma = \gamma(G)$ is the genus of the graph. In this embedding let f be the number of the faces, which according to Theorem 7.50 are all 2-cells. With this framework, we have the following generalization of Euler's Formula in Theorem 7.14.

Theorem 7.52 Let γ be the genus of a connected graph G. We then have

$$n - m + f = 2 - 2\gamma. \tag{7.4}$$

In particular, if G is a nonplanar toroidal graph, then $n - m + f = 0$.

Proof (Sketch) We prove this theorem by induction on the genus γ.

If $\gamma = 0$, then this is the classical Euler's Formula from Theorem 7.14, which we have already proved.

Let $\gamma \geq 1$ and assume that G is a fixed surface graph of genus γ embedded in T_γ, which we view as a sphere with γ attached handles. By topological maneuvering, we can assume that the curves where the handles meet the sphere are in fact cycles of the surface graph G. At this point we form a new graph G' in the following "cut and glue" way. Consider one such cycle C of G with a handle of the surface as shown in Figure 7.33.

First, we cut along the cycle C. We make two copies of C, one along the boundary of the sphere and the other on the end of the handle. Next, we glue two circular disks both into

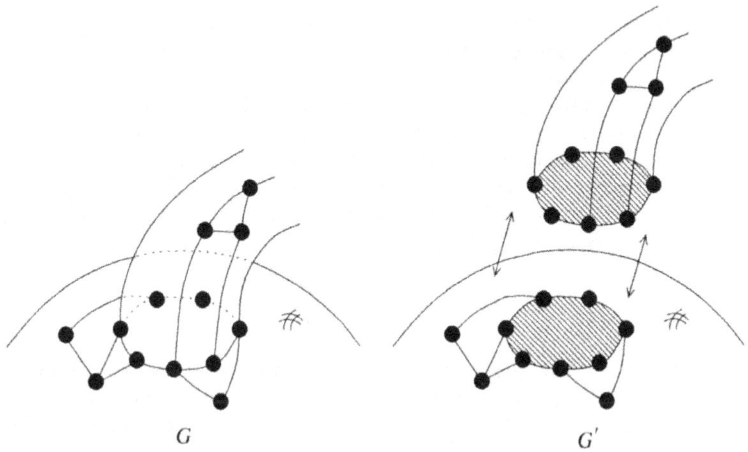

Figure 7.33: The graph G' obtained by cutting one of the handles along a cycle of G.

the hole of the sphere and into the end of the handle. Hence, we have formed a new surface graph G' on a surface having one fewer handles. Clearly the genus of G' is $\gamma - 1$ or less. However, it cannot be strictly less than $\gamma - 1$, since by reversing this process, we could obtain an embedding of G in a surface of genus $\gamma - 1$. Therefore, $\gamma' = \gamma(G') = \gamma - 1$, and the induction hypothesis applies.

If C has i vertices, then C also has i edges. Hence, the number n' of vertices, the number m' of edges, and the number f' of 2-cells of G' are given by the following:

$$\begin{aligned} n' &= n + i, \\ m' &= m + i, \\ f' &= f + 2. \end{aligned}$$

By the induction hypothesis, we have $n' - m' + f' = 2 - 2\gamma'$. So,

$$\begin{aligned} n - m + f &= (n' - i) - (m' - i) + (f' - 2) \\ &= (n' - m' + f') - 2 \\ &= (2 - 2\gamma') - 2 \\ &= 2 - 2(\gamma' + 1) \\ &= 2 - 2\gamma, \end{aligned}$$

which completes the proof of our theorem. ∎

Remark 7.53 This generalized version of Euler's Formula is a special case of a much more general fact in algebraic topology due to Poincaré from the 1890s. The parameter $2 - 2\gamma$ on the right of the generalized Euler's Formula 7.4 is called the *Euler characteristic* of the γ-torus T_γ. This parameter depends only on the surface S itself and is denoted by $\chi(S)$. There are many surfaces that are not mentioned in this chapter. For example, there are surfaces that only have one side. These surfaces are unlike the surfaces T_g that we have discussed, which have an "inside" and an "outside" on which we have embedded our graphs. This is why we call surfaces with two sides *orientable*, and surfaces with only one side (like the Möbius strip) *nonorientable*. It can be shown that the Euler characteristic for a nonorientable surface S with γ handles is given by $\chi(S) = 2-\gamma$. Hence, a graph that is embedded in a nonorientable surface S on n vertices, m edges, and f 2-cells would satisfy $n - m + f = 2 - \gamma$.

More generally, if we have a surface-S graph G on f_0 vertices or 0-cells, f_1 edges or 1-cells, and f_2 faces, which are all simply connected 2-cells, then we get

$$f_0 - f_1 + f_2 = \sum_{i=0}^{2}(-1)^i f_i = \chi(S),$$

for that surface-S graph. This alternating sum formula can again be generalized to higher dimensions than two-dimensional surfaces. For a good overview of the Euler characteristic of certain topological spaces, we refer the reader to Greenberg and Harper [23].

From Theorem 7.52 we get a couple of nice corollaries.

Corollary 7.54 If G is a simple connected graph on n vertices and m edges, then the genus of G satisfies

$$\gamma(G) \geq \left\lceil \frac{m-3n}{6} \right\rceil + 1.$$

Proof Assume G is embedded in a surface of genus γ. By Theorem 7.50 every face is a 2-cell and bounded by at least three edges. Let f be the number of 2-cells and $r_1, \ldots, r_f$ be the complete list of the 2-cells. Let m_i be the number of edges that bounds r_i. Since each edge bounds at most two 2-cells, we have

$$2m \geq \sum_{i=1}^{f} m_i \geq 3f.$$

Hence, by Theorem 7.52, we have

$$\begin{aligned}\gamma(G) &= 1 - n/2 + m/2 - f/2 \\ &\geq 1 - n/2 + m/2 - m/3 \\ &= (m - 3n)/6 + 1.\end{aligned}$$

Since $\gamma(G)$ is a natural number, the corollary follows. ■

From this we can prove one direction of a classical result by Ringel, Youngs, and Mayer from 1968.

Theorem 7.55 **Ringel, Youngs, and Mayer**

For the complete graph on $n \geq 2$, we have

$$\gamma(K_n) = \left\lceil \frac{(n-3)(n-4)}{12} \right\rceil.$$

Proof We will prove only the $\geq$-part. Since K_n has n vertices and $n(n-1)/2$ edges, we get by Corollary 7.54 that

$$\begin{aligned}\gamma(K_n) &\geq \left\lceil \frac{n(n-1)/2 - 3n}{6} \right\rceil + 1 \\ &= \left\lceil \frac{n(n-1)/2 - 3n}{6} + 1 \right\rceil \\ &= \left\lceil \frac{(n-3)(n-4)}{12} \right\rceil.\end{aligned}$$

This proves the lower bound for $\gamma(K_n)$.

To prove the upper bound, one must obtain an explicit embedding of K_n in T_γ, where $\gamma = \lceil (n-3)(n-4)/12 \rceil$. Such an embedding is given in the proof of Ringel, Youngs, and Mayer from 1968. This complicated and involved proof is omitted here. ∎

Using the generalized Euler Formula for various surface graphs, one can obtain formulas for the maximum number of edges and 2-cells just as we did for planar graphs. (See Exercises 7.34 and 7.35.)

7.9 Crossing Number

As we saw in §7.7, we could use the genus $\gamma(G)$ of a graph G as a parameter to measure how far it is from being planar. There are other such parameters. One of them is the *crossing number* of the graph. In a planar embedding of a nonplanar graph, there are one or more pairs of edges that cross, not at the vertices of the graph. But there are some embeddings that have fewer crossings of edges than others. This motivates the following definition:

Definition 7.56 The least number of edge-crossings of a graph G, among all planar embeddings of G, is called the *crossing number* of G and is denoted by $\nu(G)$.

Clearly, every planar graph has a crossing number of zero. It is not hard to see that the crossing number of each of the graphs K_5 and $K_{3,3}$ is one, since we need only one pair of edges to cross in the plane in order to embed them in the plane. It is, however, very hard to determine the exact crossing number of a given graph in general.

The next example examines the crossing number for complete bipartite graphs, a well-known problem in graph theory.

Example 7.57 Suppose $m, n \geq 2$. Consider the complete bipartite graph $K_{m,n}$. We place the n vertices of $K_{m,n}$ in the real plane $\mathbb{R}^2$ on the x-axis with x-coordinates

$$-\lfloor n/2 \rfloor, -\lfloor n/2 \rfloor + 1, \ldots, -2, -1, 1, 2, \ldots, \lceil n/2 \rceil - 1, \lceil n/2 \rceil.$$

Similarly, we place the m vertices of $K_{m,n}$ on the y-axis with y-coordinates

$$-\lfloor m/2 \rfloor, -\lfloor m/2 \rfloor + 1, \ldots, -2, -1, 1, 2, \ldots, \lceil m/2 \rceil - 1, \lceil m/2 \rceil.$$

Consider two straight lines l_1 and l_2 in the plane that intersect at a point p. Assume we have two points a_1 and a_2 on l_1 on the same side with respect to p, and two points b_1 and b_2 on l_2 on the same side with respect to p as the a_i's. If we connect each a_i to each b_j with a

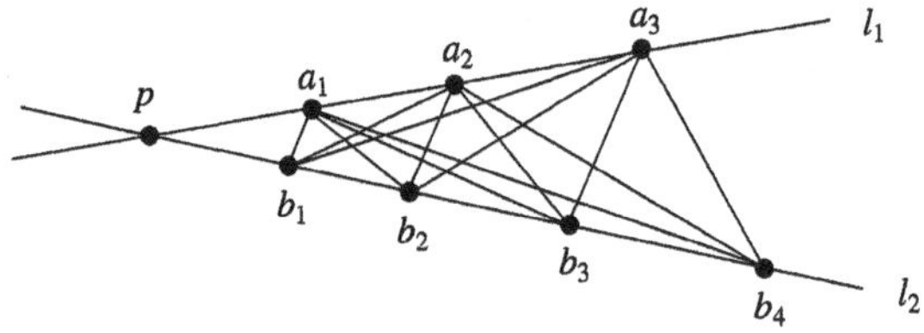

Figure 7.34: The total number of crossings between three points on one line and four points on the other is $\binom{3}{2}\binom{4}{2} = 3 \cdot 6 = 18$.

straight line segment, then we get precisely one crossing among these four line segments. In a similar fashion, if we connect α points on the same side of l_1 to β points on the same side of l_2 with straight line segments, then we have one crossing for every four points, two on l_1 and two on l_2. Hence, the number of crossings is $\binom{\alpha}{2}\binom{\beta}{2}$. In Figure 7.34 we give an example, where $\alpha = 3$ and $\beta = 4$.

We can connect all of the n points on the x-axis to all of the m points on the y-axis in this naive way with straight line segments. When we add up all of the crossings in each of the quadrants of the real plane, we get an upper bound on the crossing number for $K_{m,n}$. That is,

$$\begin{aligned} \nu(K_{m,n}) &\leq \binom{\lfloor m/2 \rfloor}{2}\binom{\lfloor n/2 \rfloor}{2} + \binom{\lfloor m/2 \rfloor}{2}\binom{\lceil n/2 \rceil}{2} \\ &\quad + \binom{\lceil m/2 \rceil}{2}\binom{\lfloor n/2 \rfloor}{2} + \binom{\lceil m/2 \rceil}{2}\binom{\lceil n/2 \rceil}{2} \\ &= \lfloor m/2 \rfloor \lfloor (m-1)/2 \rfloor \lfloor n/2 \rfloor \lfloor (n-1)/2 \rfloor . \end{aligned}$$

We ask the reader to verify this last equality in Exercise 7.38. This result was first obtained by Zarankiewicz in 1954. It has since been conjectured that this is the actual crossing number for $K_{m,n}$ in general. While some partial results have appeared, there is no complete proof of this conjecture.

7.10 Exercises

7.1. Redraw the graph shown in Figure 7.8 three times such that each of the faces r_2, r_3, and r_4 becomes the infinite face.

7.2. Use the following observation to show that K_5 is not planar.

Observation 7.58 Let C_4 be the 4-cycle embedded in the plane. If we connect each pair of opposite vertices with a continuous curve, drawing both in the same face of C_4, then these two curves must intersect.

7.3. Let $i \in \{1, 2, 3\}$. Consider a simple graph G on six vertices u_{11}, u_{12}, u_{21}, u_{22}, u_{31}, and u_{32}, where each pair of vertices is connected except for the three pairs $\{u_{i1}, u_{i2}\}$. Is G a planar graph?

7.4. Prove Euler's Formula in Theorem 7.14 inductively by using contraction of an edge in the graph.

7.5. Let G be a graph on n vertices, m edges, f faces, and with k components. Show that

$$n - m + f = k + 1.$$

7.6. Show that a planar graph on n vertices can have at most $2n - 5$ bounded faces.

7.7. If G is a simple planar graph, does G contain a vertex of degree at most four? Prove or provide a counter example.

7.8. Let G be a simple planar graph with no 3-cycles. Show that G has a vertex of degree at most three.

7.9. Let G be a connected simple plane graph with n vertices, m edges, and f faces. Assume that each face is bounded by at least k edges. Modify (7.2) and use this to show that

$$(k-2)m \leq k(n-2).$$

7.10. Let G be a planar graph on $n \geq 4$ vertices. Show that G has at least four vertices of degree at most five.

7.11. Show that if a simple planar graph is regular, then it must be r-regular where $r \in \{1, 2, 3, 4, 5\}$.

7.12. For each $r \in \{3, 4, 5\}$, construct an infinite collection of simple planar nonisomorphic r-regular graphs. [*Hint:* Use the cycle graph C_n for the construction. For $r = 4$ make n even, connect on the inside of the cycle all vertices with even indices and on the outside all the vertices with odd indices. Use the construction of $r = 4$ for $r = 5$.]

7.13. Show that a simple and connected 3-regular planar graph, where each face is bounded by exactly three edges, must be the complete graph K_4 on four vertices.

7.14. Show that a simple and connected 4-regular planar graph, where each face is bounded by exactly three edges, must have six vertices, twelve edges, and eight faces. Show further that this graph must be the graph of the octahedron, a polyhedron which is one of the Platonic solids.

7.15. Show that a simple and connected 5-regular planar graph, where each face is bounded by exactly three edges, must have twelve vertices, thirty edges, and twenty faces.

7.16. Show that a planar graph from Exercise 7.15, a 5-regular planar graph where each face is bounded by exactly three edges, must be the icosahedron, a polyhedron which is one of the Platonic solids. [*Hint:* In a triangle with endvertices u_1, u_2, and u_3 in a clockwise order, the neighbors of u_1 must be $u_2, u_{12}, v_1, u_{13}, u_3$, in clockwise order, where u_{12}, v_1, and u_{13} are inside the triangle formed by the u_i's. Do similarly for the other vertices u_i, so that we have nine vertices u_{ij} and v_i inside the triangle, where all of them must be distinct. Clearly u_i, u_j, and u_{ij} form a triangle and also each u_i, u_{ij}, and v_i. So far, we have 18 edges. Since each u_{ij} currently has degree of four, for each of them there must be an additional neighbor w_{ij} inside the hexagon formed by the v_i and u_{ij}. They must all be distinct. Use the fact that we have all twelve vertices to complete the argument.]

7.17. Show that a simple and connected 3-regular planar graph, where each face is bounded by exactly four edges, must have eight vertices, twelve edges, and six faces. Show further that this must be the graph of the three-dimensional cube, a polyhedron which is one of the Platonic solids. [*Hint:* (i) By adding a diagonal in each face the graph is fully triangulated, and hence $m + f = 3n - 6$. (ii) Draw a square u_1, u_2, u_3, and u_4 in clockwise order. Argue that each u_i has an additional neighbor v_i, and that all of these eight vertices must be distinct. Use planarity and Observation 7.58 to complete the argument.]

7.18. Show that a simple and connected 3-regular planar graph, where each face is bounded by exactly five edges, must have twenty vertices, thirty edges, and twelve faces. [*Hint:* By adding two edges inside each 5-face we have a fully triangulated graph, and hence $m + 2f = 3n - 6$.]

7.19. Show that a planar graph from Exercise 7.18, a simple and connected 3-regular planar graph, where each face is bounded by exactly five edges, must be the dodecahedron, a polyhedron which is one of the Platonic solids. [*Hint:* In a pentagon with endvertices u_1, u_2, u_3, u_4, and u_5 in a clockwise order, each of the u_i must have a unique extra neighbor v_i inside the pentagon, which all must be distinct. Each pair of consecutive v_i and v_{i+1} (modulo five) must have an additional common neighbor $w_{i+1/2}$, since u_i, u_{i+1}, v_i, and v_{i+1} are on the boundary of a 5-face. By 3-regularity, each of the $w_{i+1/2}$ have an additional neighbor $x_{i+1/2}$ distinct from the other vertices, and distinct from each other. Use the fact that each face is bounded by five edges to complete the proof.]

7.20. Show that two graphs G and G' are homeomorphic if, and only if, there is a graph G'' that is a common subdivision of both of G and G'.

7.21. Prove Observation 7.28.

7.22. Show that for two homeomorphic graphs G and G', we have

$$|V(G)| - |E(G)| = |V(G')| - |E(G')|.$$

7.23. Show that if a graph is homeomorphic to an Euler graph, it is also an Euler graph. Is a contraction of an Euler graph also an Euler graph?

7.24. Find an example of a nonplanar graph G such that neither K_5 nor $K_{3,3}$ is a contraction of G. Find an example of a nonplanar graph G that is not homeomorphic to K_5 or $K_{3,3}$. Why does this not contradict Theorem 7.31?

7.25. Prove Lemma 7.32.

7.26. Let G be a graph such that for every two vertices u and v, there are at most two vertex-disjoint paths of positive lengths from u to v. Show that G is planar.

7.27. Show that if K_4 is a contraction of a graph G, then G has a subgraph that is homeomorphic to K_4. Does the same hold for K_5?

7.28. A graph G is called *outerplanar* if G has an embedding in the plane in such a way that each vertex bounds the infinite face. Prove the following Kuratowski-Wagner-like theorem.

Theorem 7.59 The following are equivalent for a graph G.

1. The graph G is outerplanar.
2. Neither K_4 nor $K_{2,3}$ is homeomorphic to any subgraph of G.
3. Neither K_4 nor $K_{2,3}$ is a minor of G.

7.29. Let G be a plane graph. If $e \in G$, what is the plane dual of the contraction $(G \cdot e)^*$ in terms of the plane dual G^* of G?

7.30. Let G be a planar graph with an abstract dual G'. If $e \in E(G)$, what is the abstract dual of the contraction $(G \cdot e)^*$ in terms of G'? [*Hint:* Your answer should be the same as for the plane graph in Exercise 7.29.]

7.31. Show that every simple toroidal graph having seven or more vertices has a vertex of degree at most six.

7.32. Show that K_7 is a toroidal graph by drawing in a square representing the torus, where top and bottom sides are identified and likewise the right and left sides. [*Hint:* Start by drawing the cycle C_6 in the square together with a center vertex inside the 6-cycle, which is connected to all of the six vertices on C_6.]

7.33. Show that K_n is toroidal if, and only if, $n \in \{5, 6, 7\}$.

7.34. Find a closed formula for the maximum number of edges in a graph with n vertices and genus γ.

7.35. Find a closed formula for the maximum number of 2-cells that a graph can have if it has n vertices and genus γ.

7.36. Show that for the complete bipartite graph $K_{m,n}$, we have

$$\gamma(K_{m,n}) \geq \left\lceil \frac{(m-2)(n-2)}{4} \right\rceil.$$

In fact, Ringel proved that equality holds (see Wilson [59]).

7.37. Show that the crossing number of the Petersen graph is two. [*Hint:* To show that the crossing number is not one, show that the graph G, obtained by removing one edge from the Petersen graph, is nonplanar. In fact, as we saw late in §7.4, the removal of even one vertex leaves a nonplanar graph.]

7.38. Verify the form of the lower bound of the crossing number of $K_{m,n}$ given in Example 7.57. [*Hint:* One way might be to consider cases when both m and n are even or odd.]

CHAPTER

8

Graph Coloring

Chapter Goals

Define and discuss the chromatic number of a graph.

Define the complete k-partite graph and Turán graphs.

Study Brooks' Theorem and related coloring algorithms.

Present and prove the Five Color Theorem for planar graphs.

Present and discuss the Four Color Theorem.

Give Heawood's Formula for graphs of an arbitrary genus.

Define edge coloring of a graph and present Vizing's Theorem.

Present Tait's Equivalence Theorem for 3-regular plane graphs.

8.1 The Chromatic Number of a Graph

Numerous problems involving discrete objects can be phrased in terms of *vertex coloring* of graphs. Situations where one needs to avoid certain conflicts or collisions of items can usually be stated as simple graph coloring problems. In Problem 1.4, storing volatile chemicals in a number of warehouses is a graph coloring problem. Simply stated problems are often by no means easy to solve. For example, we mention the most famous, or infamous, problem of finding a proof for the *Four Color Theorem*, a theorem which simply states that the regions of any geographical map can be colored with at most four colors, in such a way that no two regions sharing a common border have the same color. We will discuss the Four Color Theorem later in this chapter.

First, we need to state some definitions to make precise what we are up against. We illustrate the ideas with simple examples throughout. The most widely used form of graph coloring is the *vertex coloring* of a graph, since in most problems, vertices represent objects to be partitioned or manipulated.

Definition 8.1 Let $G = (V, E, \phi)$ be a graph. A *vertex coloring* or a *proper vertex coloring* of G is a function $c : V(G) \to \mathbb{N}$ satisfying

$$c(u) = c(v) \Rightarrow \{u, v\} \notin \phi(E(G)) \tag{8.1}$$

for all $u, v \in V(G)$; that is, u and v are not adjacent in G. When no ambiguity is possible, a *coloring* means a *vertex coloring*.

For $k \in \mathbb{N}$, a *k-vertex coloring*, or a *proper k-vertex coloring* of G is a proper vertex coloring $c : V(G) \to \{1, \ldots, k\}$.

We say that G is *k-vertex colorable*, or simply *k-colorable*, if G has a proper k-vertex coloring.

The next example should clarify Definition 8.1.

Example 8.2 Consider the simple graph G on $n = 5$ vertices shown in Figure 8.1. The five vertices of G are colored in two different ways using five and three colors, respectively. Both of these colorings are proper, since in both cases no two adjacent vertices receive the same color.

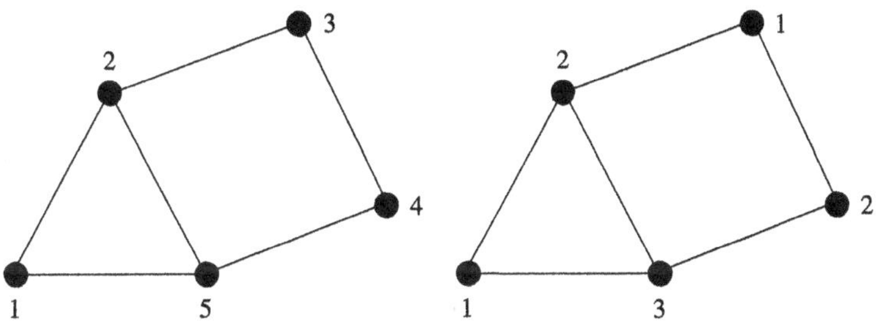

Figure 8.1: A simple graph G and two proper vertex colorings of it.

Assigning one unique color to each vertex in a simple graph always results in a proper vertex coloring of the graph. However, the objective is to use as *few* colors as possible for any given graph.

Remark 8.3 A few words are in order regarding Definition 8.1.

- A map $c : V(G) \to \mathbb{N}$, where we do not necessarily assume (8.1) to hold, is called a *nonproper coloring* of G.
- By definition, any graph with loops has no proper vertex coloring.
- Clearly, if a loopless graph G has n vertices, then G is n-colorable. In particular, every loopless graph is k-colorable for some k.
- If G is a k-colorable graph, then any subgraph G' of G is also k-colorable.

- Any loopless graph with multiple edges is k-colorable if, and only if, the corresponding simple graph, where all the multiple edges have been replaced by a single edge, is k-colorable. For a simple graph, the condition (8.1) becomes

$$c(u) = c(v) \Rightarrow \{u, v\} \notin E(G),$$

where there is no need for any edge map.

Because of the last itemized statement of Remark 8.3, we mostly assume that the graph in question is a simple graph, since all we care about is whether or not a pair of vertices are connected by an edge, but not how many edges. Therefore, in the rest of this chapter, we assume that all of our graphs are simple, unless otherwise stated.

As we saw in Problem 1.4, we could easily have assigned different colors to all of the vertices in question, which corresponds to building a warehouse for each volatile chemical, a very expensive project indeed! Hence, as mentioned in Example 8.2, the objective with any proper graph coloring is to be able to vertex color the graph in question with as few colors as possible. This motivates the following definition.

Definition 8.4 Let G be a simple graph. The least $k \in \mathbb{N}$ such that G is k-vertex colorable is called the *chromatic number* of G and is denoted by $\chi(G)$.

The following example illustrates the concept of chromatic number.

Example 8.5 Consider the simple graph G shown in Figure 8.1. Since G has a 3-coloring, we have $\chi(G) \leq 3$. Note that G has K_3 as a subgraph, so by Remark 8.3, if G is 2-colorable, then so is K_3. But in K_3, any pair of vertices are adjacent, so the vertices in K_3 must all receive different colors. Hence, we have

$$\chi(G) \geq \chi(K_3) = 3,$$

and so $\chi(G) = 3$ in this case.

Clearly, the same argument would hold for K_n for any $n \in \mathbb{N}$; that is, in a proper vertex coloring of K_n no two vertices can have the same color, so $\chi(K_n) = n$.

An interesting way to view vertex coloring and chromatic numbers is via graph homomorphisms. The proof is left as Exercise 8.3.

Observation 8.6 For a simple graph G, the chromatic number $\chi(G)$ is the smallest $k \in \mathbb{N}$ such that there exists a graph homomorphism $\phi : G \to K_k$, the complete graph on k vertices.

Let G be a simple graph. From the above Observation 8.6 we see that the question "*For $k \in \mathbb{N}$, is $\chi(G) = k$?*" is a very special case of a much more general question: "*For a fixed simple graph H, does there exist a graph homomorphism $\phi : G \to H$?*" since the second question reduces to the first one for $H = K_k$. For a comprehensive yet manageable study of graph homomorphisms we refer the reader to a recent and excellent book by Pavol Hell and Jaroslav Nešetřil [28].

However, we will in this chapter concentrate on the special case of vertex colorings, which are hard and general enough for our purposes. We leave the proof of the following observation as Exercise 8.14.

Observation 8.7 If the simple graph G has k components $C_1, \ldots, C_k$, then

$$\chi(G) = \max(\{\chi(C_i) \mid 1 \le i \le k\}).$$

Note that if $\chi(G) \le k$, then by definition we can find a proper k-vertex coloring $c : V(G) \to \{1, \ldots, k\}$ of G. If $\sigma : \{1, \ldots, k\} \to \{1, \ldots, k\}$ is some permutation, then clearly $c \circ \sigma : V(G) \to \{1, \ldots, k\}$ is also a proper k-vertex coloring of G. This implies the following observation.

Observation 8.8 Let G be a simple graph and $u \in V(G)$. If $\chi(G) \le k$ and $i \in \{1, \ldots, k\}$, then there is a proper k-vertex coloring of G where u receives the color i.

A result similar to Observation 8.7 is obtained for a connected graph by looking at its blocks.

Theorem 8.9 If the simple graph G has k blocks $B_1, \ldots, B_k$, then

$$\chi(G) = \max(\{\chi(B_i) \mid 1 \le i \le k\}).$$

Proof Since each block B_i of G is a subgraph of G, then by Remark 8.3, we have

$$\max(\{\chi(B_i) \mid 1 \le i \le k\}) \le \chi(G).$$

For the other inequality, we use induction on k. The theorem is a tautology for $k = 1$.

Assume the other inequality holds for all graphs having $k - 1$ or fewer blocks, and let G be a graph on k blocks $B_1, \ldots, B_k$. We may assume that B_k is a block corresponding to a leaf in the block-cutpoint graph BC(G), as shown in Figure 8.2. If G' is the union of the first $k - 1$ blocks, then by the induction hypothesis

$$\chi(G') \le \max(\{\chi(B_1), \ldots, \chi(B_{k-1})\}).$$

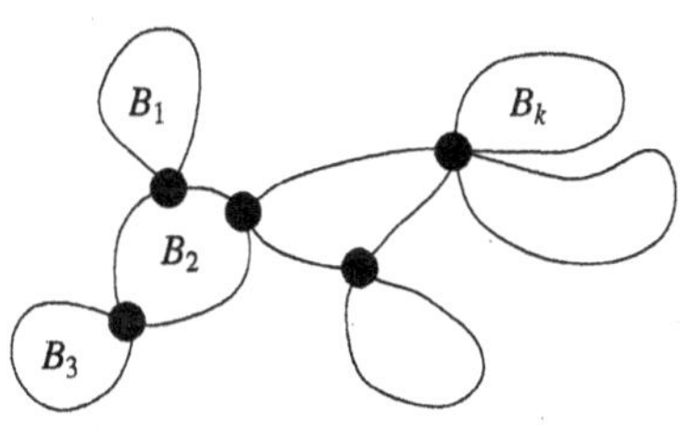

Figure 8.2: A simple graph with k blocks, where B_k corresponds to a leaf in the block-cutpoint graph BC(G).

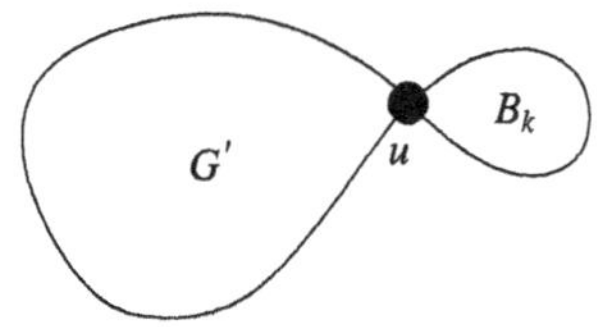

Figure 8.3: Since G' and the block B_k have only one cut-vertex u in common, the coloring of G' can be extended to the entire graph.

Moreover, G' and B_k have only one vertex u in common, as shown in Figure 8.3. For any coloring of G' using $\chi(G')$ colors, by Observation 8.8, we can extend this coloring of G' to a proper coloring of $G = G' \cup B_k$ using $\max(\{\chi(G'), \chi(B_k)\})$ colors. Thus, we have

$$\chi(G) \leq \max(\{\chi(G'), \chi(B_k)\}) \leq \max(\{\chi(B_i) \mid 1 \leq i \leq k\}),$$

which completes the proof. ■

8.2 Multipartite Graphs

In this section we characterize bipartite graphs and consider some special kinds of multipartite graphs with the constraint of having a bounded chromatic number. We start with an example.

Example 8.10 The following results on chromatic numbers are immediate from Definition 8.4.

- For every $n \in \mathbb{N}$, we have for the null graph $\chi(N_n) = 1$. In fact, $\chi(G) = 1$ if, and only if, G is a null graph.
- For every $n \in \mathbb{N}$, we have for the complete graph $\chi(K_n) = n$. In fact, if G has n vertices, then $\chi(G) = n$ if, and only if, G is complete. (See Exercise 8.5.)
- For every $n \geq 2$, we have for the path graph $\chi(P_n) = 2$.
- More generally, for every tree T on $n \geq 2$ vertices, we have $\chi(T) = 2$. We elaborate on this right after this example.
- For $n \geq 3$, we have

$$\chi(C_n) = \begin{cases} 2 & \text{if } n \text{ is even,} \\ 3 & \text{if } n \text{ is odd.} \end{cases}$$

The fact that every tree on two or more vertices has chromatic number two can be seen in various ways. We could use induction together with Theorem 3.4, or we could pick a root and then color the vertices on the odd levels with one color and the vertices on the even levels with the other color. (See Figure 8.4.) From this we see that when we are using two colors for a coloring of a tree T, and one vertex (the root) is assigned a color, then the coloring of the rest of the vertices is uniquely determined. Hence, T has precisely two colorings using two colors.

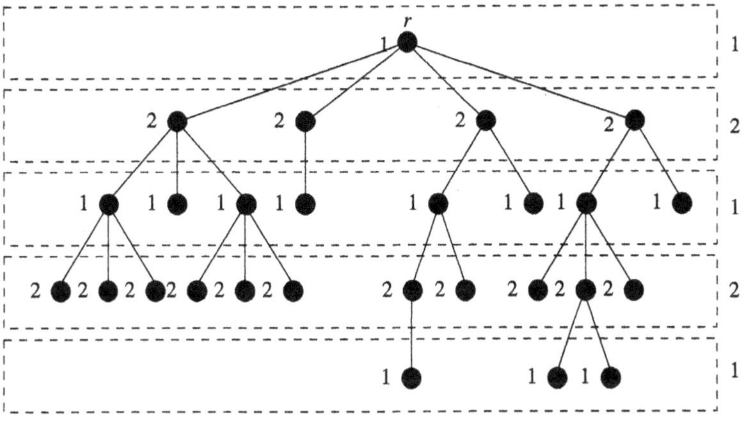

Figure 8.4: A 2-coloring of any tree is uniquely determined by the color of its root r.

Since every tree is bipartite, it is also 2-colorable, since we can assign one color to the vertices in one part and a different color to the remaining vertices. The opposite direction is also valid; that is, if a graph G has a 2-coloring, then we have a bipartition of $V(G)$ by letting $V_1 \subseteq V(G)$ be the set of vertices of the first color, and $V_2 \subseteq V(G)$ the set of vertices of the second color. In this case $V(G) = V_1 \cup V_2$ and each edge has one endvertex in V_1 and the other in V_2. By Theorem 5.3, we have completely characterized bipartite graphs, and so this gives us a complete description of graphs with chromatic number of two. We summarize in the following.

Corollary 8.11 For a connected simple graph G on two or more vertices that has at least one edge, the following are equivalent:

1. $\chi(G) = 2$.
2. G is a bipartite graph.
3. G has no cycles of odd length.

Note that part three in Corollary 8.11 is a characterization of bipartite graphs or graphs with chromatic number of two. Such a characterization is particular for the case of bipartite graphs. The complete characterization of graphs with chromatic number greater than two is much harder to find. In fact, there is *no* simple way of characterizing a graph having chromatic number greater than two. From Corollary 8.11, the condition $\chi(G) = 2$ can be used to *define* a bipartite graph. In fact, if $k \geq 1$ and G is a graph with $\chi(G) = k$, then we have a partition

$$V(G) = V_1 \cup \cdots \cup V_k$$

of the vertices of G such that for each $i \in \{1, \ldots, k\}$ the set V_i consists of those vertices of color i for some fixed optimal k-coloring of G. Therefore, no two vertices in V_i are adjacent. Hence, every edge has its endvertices in distinct parts of $V_1, \ldots, V_k$. Graphs of this kind are called *k-partite*. The other way works similarly. A graph G is k-partite if, and only if, $\chi(G) \le k$. The *multipartite graphs* are precisely the graphs that are k-partite graphs for some k. Of course, this includes *all* loopless graphs! However, for a given k, not all graphs are k-partite. We call a graph multipartite when we are emphasizing the existence of a k-partition for some given fixed k. We now define a special type of k-partite graph.

Definition 8.12 Let $k \in \mathbb{N}$. A *complete k-partite graph* is a simple graph G satisfying:

1. $V(G)$ can be partitioned into a disjoint union of k nonempty subsets

$$V(G) = V_1 \cup \cdots \cup V_k.$$

2. Every pair of vertices $u_i \in V_i$ and $v_j \in V_j$ where $i \neq j$ are distinct indices from $\{1, \ldots, k\}$ are adjacent.
3. No pair of vertices from the same V_i are adjacent.

If $|V_i| = n_i$, then we denote the graph G by $K_{n_1,\ldots,n_k}$. If additionally $n_1 = \cdots = n_k = n$, then we denote the graph by K_n^k.

Note 8.13 For $n_1, \ldots, n_k \in \mathbb{N}$, there is up to isomorphism only one complete k-partite graph having k parts with sizes $n_1, \ldots, n_k$. Furthermore, we have

$$|E(K_{n_1,\ldots,n_k})| = \sum_{1 \le i < j \le k} n_i n_j.$$

Each complete k-partite graph G on $n > k$ vertices has $\chi(G) = k$. But if we add one more nonparallel edge to G, then we obtain a complete subgraph on $k+1$ vertices. Hence, the chromatic number of the new graph increases by at least one. (See Exercise 8.25.)

Let $n \in \mathbb{N}$ and consider the complete k-partite graph $K_{n_1,\ldots,n_k}$, where

$$\sum_{i=1}^{k} n_i = n.$$

For each i, denote the part of size n_i by V_i—that is, $|V_i| = n_i$. Assume that $n_1 < n_2$. Form another complete k-partite graph by moving a vertex $u \in V_2$ to V_1. By this procedure we have increased n_1 by one and reduced n_2 by one. Let $K_{n'_1,\ldots,n'_k}$ be the resulting complete k-partite graph, where $n'_1 = n_1 + 1$, $n'_2 = n_2 - 1$, and $n'_i = n_i$ for each remaining $i \ge 3$. By assumption we have $n_1 + 1 \le n_2$, and hence

$$n_1 n_2 \le (n_1 + 1)(n_2 - 1) = n'_1 n'_2.$$

In addition, we know $n_1 + n_2 = n'_1 + n'_2$, and so by Note 8.13 we obtain

$$\begin{aligned} |E(K_{n_1,\ldots,n_k})| &= n_1 n_2 + (n_1 + n_2) \sum_{3 \le i \le k} n_i + \sum_{3 \le i < j \le k} n_i n_j \\ &\le n'_1 n'_2 + (n'_1 + n'_2) \sum_{3 \le i \le k} n'_i + \sum_{3 \le i < j \le k} n'_i n'_j \\ &= |E(K_{n'_1,\ldots,n'_k})|. \end{aligned}$$

Note that equality holds if, and only if, $n_1n_2 = n'_1n'_2$, which again holds if, and only if, $n_1+1 = n_2$. We conclude that among the complete k-partite graphs on n vertices the graphs with the maximum number of edges are those satisfying

$$|n_i - n_j| \leq 1$$

for all $i, j \in \{1, \ldots, k\}$. For a given n there is, up to isomorphism, exactly one such complete k-partite graph. For various k and n, these graphs form an important class among the complete k-partite graphs, called the *Turán graphs*. They were introduced in 1941 by Pál Turán, a Hungarian graph theorist from the "Erdős school" who additionally was a friend and a collaborator of Paul Erdős.

Definition 8.14 Let $n, k \in \mathbb{N}$ and let $n = qk + r$, where $r \in \{0, 1, \ldots, k-1\}$. The *Turán-k graph* $T_k(n)$ on n vertices is the complete k-partite graph with $k - r$ parts of size q and r parts of size $q + 1$.

An example of the Turán graph $T_3(8)$ is given in Figure 8.5. Each edge in a Turán graph $T_k(n)$ either has both its endvertices in parts of size q, has one endvertex in a part of size q and the other in a part of size $q + 1$, or has both endvertices in parts of size $q + 1$, where $n = qk + r$. Hence, we have the following.

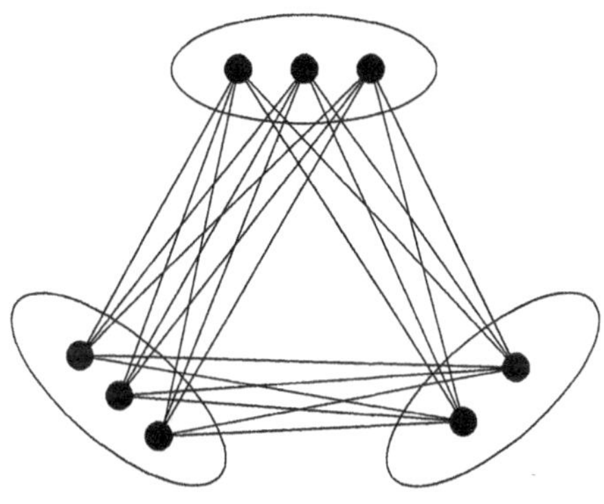

Figure 8.5: The Turán graph $T_3(8)$ with 8 vertices and 21 edges.

Observation 8.15 For $n, k \in \mathbb{N}$ let $n = qk + r$, where $r \in \{0, 1, \ldots, k-1\}$. The number $t_k(n)$ of edges in the Turán graph $T_k(n)$ is given by

$$t_k(n) = |E(T_k(n))| = \binom{k-r}{2}q^2 + \binom{r}{2}(q+1)^2 + r(k-r)q(q+1).$$

The following observation is typical for *extremal graph theory*, where certain constraints are given and we would like to maximize some other parameters of the graph.

Observation 8.16 Let $k, n \in \mathbb{N}$. The maximum number of edges that a simple graph G on at most n vertices can have, which satisfies $\chi(G) \leq k$, is given by $t_k(n)$. This bound is obtained if, and only if, $G = T_k(n)$, the Turán-k graph on n vertices.

In Exercise 8.26 we ask for another proof than the one given right before Definition 8.14, of the fact that graphs with the maximum number of edges on km vertices and chromatic number at most k are among the Turán-k graphs.

In extremal graph theory, the actual formula for a certain parameter of a given graph may not be very informative. Rather, one likes to know the behavior of the parameter when the number of vertices tends to infinity. Sometimes this can be more valuable for certain existence theorems in graph theory. Here is an example.

Example 8.17 Let $k \in \mathbb{N}$ be given. When n tends to infinity, we can obtain the asymptotic behavior of $t_k(n)$ by computing the limit

$$\lim_{n\to\infty} \frac{t_k(n)}{n^2}.$$

Since $q/n = (1/k) - (r/(nk))$, we obtain the following limit as n tends to infinity.

$$\begin{aligned}
\frac{t_k(n)}{n^2} &= \binom{k-r}{2}\left(\frac{q}{n}\right)^2 + \binom{r}{2}\left(\frac{q+1}{n}\right)^2 + r(k-r)\left(\frac{q}{n}\right)\left(\frac{q+1}{n}\right) \\
&\to \frac{1}{k^2}\left(\binom{k-r}{2} + \binom{r}{2} + r(k-r)\right) \\
&= \frac{1}{2} - \frac{1}{2k}, \qquad \text{when } n \to \infty.
\end{aligned}$$

Hence, for large n, the approximate number $t_k(n)$ of edges in the Turán-k graph $T_k(n)$ is given by

$$t_k(n) \approx \left(\frac{1}{2} - \frac{1}{2k}\right) n^2.$$

The Turán graphs are extremely useful in another setting. They are also the graphs with the maximum number of edges among all graphs on n vertices that do not contain K_{k+1} as a subgraph for a given $k \in \mathbb{N}$. This theorem is formally stated in what follows, but we omit its proof. We refer the interested reader to Bollobás [5, chapter IV] or West [57, chapter 5] for readable proofs.

Theorem 8.18 Let $n, k \in \mathbb{N}$. The Turán graph $T_k(n)$ is the unique graph having the maximum number of edges among all graphs G that satisfies either of the following:

1. $|V(G)| \leq n$ and $\chi(G) \leq k$.
2. $|V(G)| \leq n$ and G does not have K_{k+1} as a subgraph.

8.3 Results for General Graphs

To find the chromatic number of a graph in general, the brute-force method of using all possible combinations can always be applied. However, brute force is not satisfactory because of the explosion in the number of combinations as soon as the size of the graph increases beyond a few dozen vertices. This section continues our discussion on the chromatic number for a general graph. We show how the maximum degree of a graph relates naturally to the chromatic number. Brooks' Theorem is presented along with a coloring algorithm due to Welsh and Powell [56]. We begin with a theorem that bounds the chromatic number in terms of the maximum degree of a graph.

Theorem 8.19 For any simple graph G with a maximum degree $\Delta(G)$, we have

$$\chi(G) \leq \Delta(G) + 1.$$

Proof We use induction on n, the number of vertices of G. The theorem is clearly true for $n = 1$.

Assume the theorem to be true for any graph on $n-1$ or fewer vertices. Let G be a graph on n vertices. For any vertex $u \in V(G)$, the graph $G - u$ has maximum degree of at most $\Delta = \Delta(G)$, since $d_{G-u}(v) \leq d_G(v)$ holds for any vertex $v \neq u$. Hence, by the induction hypothesis, $G - u$ has a proper coloring using colors from the set $C = \{1, \ldots, \Delta + 1\}$. Since u has at most Δ neighbors in G, there is at least one color in C distinct from the colors of all the neighbors of u in the coloring of $G - u$. This color can be used to color u and thus complete the coloring of G using colors from C. Therefore, we have a proper Δ-vertex coloring for G. ■

The proof of Theorem 8.19 yields the following algorithm to obtain a proper coloring of G using at most $\Delta(G) + 1$ colors. In the algorithm below, $N_i \subseteq V(G)$ denotes the set of neighbors of u_i among the previously labeled ones $\{u_1, \ldots, u_{i-1}\}$. Note that $N_1 = \varnothing$.

ALGORITHM DELTA PLUS ONE COLORING

`Input:` A labeling $V(G) = \{u_1, \ldots, u_n\}$ of the vertices of G and a list $(N_1, \ldots, N_n)$ of lower-numbered neighbors.

`Output:` A list $S = (c_1, \ldots, c_n)$ of colors, where each c_i is the color assigned to the vertex u_i of G.

begin
  $S_1 = (1)$;   // initial list containing the one color 1
  for $i = 2$ **to** n **do** {
    $c_i = \min(\{1, \ldots, \Delta + 1\} \setminus \{c_j : u_j \in N_i\})$;
    $S_i =$ append c_i to the $(i-1)$-tuple S_{i-1};
    }
  output S_n;
end.

Note 8.20 This is an example of an algorithm type called "greedy algorithm" where we choose the least of some "small" entity at each step and disregard the global structure of the graph. Later in this section we delve into an improved greedy coloring algorithm. Note that this coloring algorithm depends heavily on the ordering of the vertices $u_1, \ldots, u_n$. For a given graph G, there might actually be a specific ordering, which can yield an optimal coloring by the above algorithm. This we will discuss in more detail in Chapter 9 in connection with *chordal graphs*.

The next example shows how to reach the upper bound for Theorem 8.19.

Example 8.21

- If $n \in \mathbb{N}$, then $\chi(K_n) = n = \Delta(K_n) + 1$.
- If $n \geq 2$ is odd, then $\chi(C_n) = 3 = \Delta(C_n) + 1$.

It is interesting that the graphs in Example 8.21 are the *only* examples where the chromatic number reaches the upper bound of Theorem 8.19. Although the following theorem improves the chromatic number by only one, its proof requirements are by no means "proportional to the improvement"!

Theorem 8.22 **Brooks' Theorem**

For a connected simple graph G, which is neither complete nor a cycle of odd length, we have

$$\chi(G) \leq \Delta(G).$$

Proof Let G be a graph that is neither complete nor an odd cycle. Assume first that G is not 2-connected and has blocks $B_1, \ldots, B_k$, where $k \geq 2$. Assume the theorem is correct for each block B_i. For each i we either have $\chi(B_i) \leq \Delta(B_i) \leq \Delta(G)$, or B_i is complete or an odd cycle. Since $k \geq 2$, in the latter case, B_i has a cut-vertex u in G. Hence, u's degree in B_i is strictly less than its degree in G. Therefore,

$$\chi(B_i) = \Delta(B_i) + 1 = d_{B_i}(u) + 1 \leq \Delta(G)$$

and so

$$\chi(G) = \max(\{\chi(B_i) \mid 1 \leq i \leq k\}) \leq \Delta(G).$$

Thus, if G is a graph for which the theorem fails, with the least number of vertices, then G is 2-connected. If $\Delta(G) \leq 2$, then G is a cycle or a path. In fact, it must be a cycle, since a path is not 2-connected. For a cycle the theorem is clearly true, so $\Delta(G) \geq 3$. Furthermore, if there is a vertex $u \in V(G)$ with $d_G(u) < \Delta(G)$, then, by assumption, $G - u$ has a proper $\Delta(G)$-coloring, which can be extended to a $\Delta(G)$-coloring for G. Hence, we can assume G is a $\Delta(G)$-regular graph.

We first show G has vertices u, v, and w such that

1. The vertex u is connected to both v and w, but v and w are not connected.
2. The graph $G - \{v, w\}$ is connected.

CASE ONE: The graph $G - x$ is 2-connected for every vertex $x \in V(G)$. In this case it suffices to find u, v, and w such that condition 1 holds. Since G is not complete, there are two nonadjacent vertices. Let $(u_0, u_1, \ldots, u_k)$ be a simple path between them, and let i be the largest index such that u_0 and u_i are adjacent. The vertices $v = u_0$, $u = v_i$, and $w = v_{i+1}$ will do in this case.

CASE TWO: There is a vertex u such that $G - u$ is connected but not 2-connected. In this case there are two distinct blocks B_1 and B_2 in $G - u$ that both correspond to leaves in the block-cutpoint graph $\mathrm{BC}(G - u)$. This is shown in Figure 8.6.

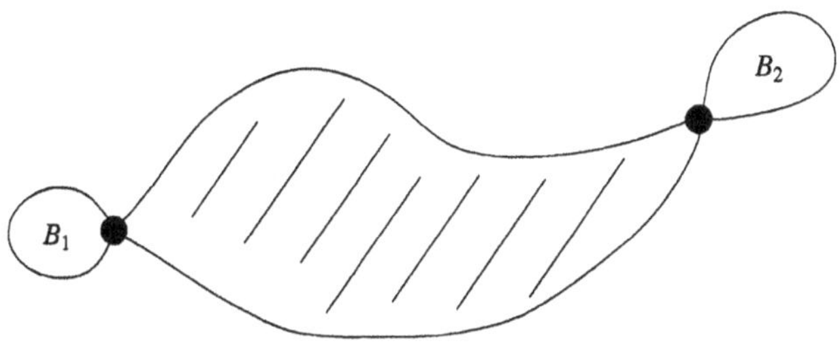

Figure 8.6: The blocks B_1 and B_2 in $G - u$ correspond to leaves.

The vertex u must be adjacent to vertices $v \in V(B_1)$ and $w \in V(B_2)$, neither of which is a cut-vertex of $G - u$ (since otherwise, either B_1 or B_2 would be a block in G). Therefore, the graph $G - \{u, v, w\}$ is connected. Since $d_G(u) \geq 3$, then u must be adjacent to a vertex of $G - u$ that is distinct from both v and w, as depicted in Figure 8.7. Hence, the graph $G - \{v, w\}$ is connected and we have our vertices u, v, and w in this case as well.

Now u, v, and w satisfy both conditions from above. Label the vertices of G by putting $v_1 = v$ and $v_2 = w$, and the remaining vertices of $H = G - \{v, w\}$ by $V(H) = \{v_3, \ldots, v_n\}$ such that $v_n = u$, and such that the distances in H satisfy

$$\partial_H(v_3, u) \geq \partial_H(v_4, u) \geq \cdots \geq \partial_H(v_{n-1}, u). \tag{8.2}$$

We assign the color 1 to both v_1 and v_2. By (8.2), each v_i with $3 \leq i < n$ must be adjacent to some v_j with $j > i$. So v_i has at most $\Delta(G) - 1$ neighbors among the previously labeled $v_1, \ldots, v_{i-1}$. Therefore, we can assign colors to $v_3, \ldots, v_{n-1}$ in this order from the set $C = \{1, \ldots, \Delta(G)\}$ that have not been assigned to lower-labeled neighbors. This can be done in a greedy fashion. Finally, v_n has two neighbors v_1 and v_2 with color 1, and thus at most $\Delta(G) - 1$ colors have been used on the neighbors of v_n. So, there is at least one color left from C for v_n. This completes the proof. ■

Note 8.23 Although we have reduced the general upper bound for $\chi(G)$ by one, unlike the proof of Theorem 8.19 the proof of Brooks' Theorem 8.22 does not yield a direct greedy algorithm to assign the colors to the vertices of G from the set $\{1, \ldots, \Delta(G)\}$. Nevertheless, it is a mathematically interesting result to know that such a coloring exists.

We would like to conclude this section with an algorithm to color a general graph. We begin with a definition.

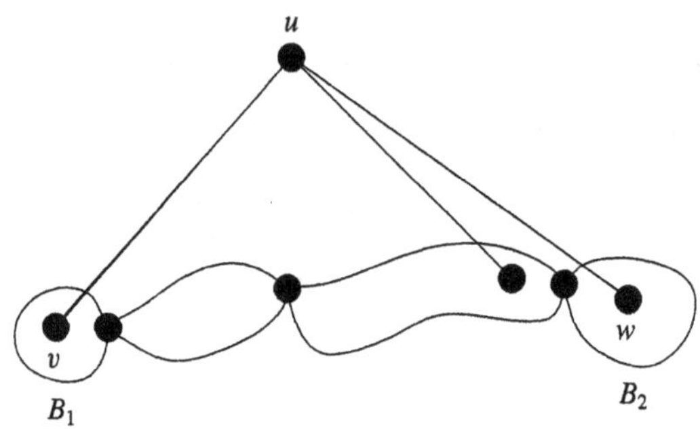

Figure 8.7: The vertex u must be connected to a third vertex in $G - u$.

Definition 8.24 Let G be a graph. A sequence $(d_1, \ldots, d_n)$ is called a *degree sequence* of G, if $d_1 \geq \cdots \geq d_n$ and there is labeling $V(G) = \{u_1, \ldots, u_n\}$ of the vertices of G with $d_i = d_G(u_i)$ for each $i \in \{1, \ldots, n\}$.

The following theorem summarizes the result of the coloring algorithm.

Theorem 8.25 If a graph G has a degree sequence $(d_1, \ldots, d_n)$, then there is a greedy vertex coloring of G that uses at most $\max(\{\min(\{i, d_i + 1\}) \mid 1 \leq i \leq n\})$ colors. In particular,

$$\chi(G) \leq \max(\{\min(\{i, d_i + 1\}) \mid 1 \leq i \leq n\}).$$

The next algorithm provides a sketch of the proof of Theorem 8.25. The algorithm is due to Powell and Welsh. ALGORITHM GREEDY COLORING assigns colors to the n vertices of a graph G using at most $\max(\{\min(\{i, d_i + 1\}) : 1 \leq i \leq n\})$ colors. In the algorithm $N_i \subseteq V(G)$ denotes the set of neighbors of u_i among the previous ones $\{u_1, \ldots, u_{i-1}\}$. Note that $N_1 = \varnothing$.

```
ALGORITHM GREEDY COLORING
Input: A simple graph G with a vertex labeling {u_1, ..., u_n} together with
a degree sequence (d_1, ..., d_n), and a list (N_1, ..., N_n) of lower numbered
neighbors.
Output: A list S = (c_1, ..., c_n) of colors, where each c_i is the color assigned
to the vertex u_i of degree d_i.
begin
   S_1 = (1);        // initial list containing the one color 1
   for i = 2 to n do {
      C_i = {c_j : u_j ∈ N_i};
      c_i = min({1, ..., min({i, d_i + 1})} \ C_i);
      S_i = append c_i to the (i − 1)-tuple S_{i−1};
      }
   output S_n;
end.
```

Needless to say, ALGORITHM GREEDY COLORING is yet another example of an algorithm of the greedy type, since at each step the algorithm chooses the best color based only on the colors of the neighboring vertices, rather than on all the vertices.

The proof that ALGORITHM GREEDY COLORING actually yields a proper vertex coloring for G is left as Exercise 8.33. As with ALGORITHM DELTA PLUS ONE COLORING, the above ALGORITHM GREEDY COLORING also depends heavily on the ordering $u_1, \ldots, u_n$ of the vertices. However, although the ordering satisfies the condition that $d_G(u_i) \geq d_G(u_{i+1})$ for all i, there is still room for a variety of different orderings, some yielding better colorings than others by the above coloring algorithm.

Example 8.26 Consider the graph G shown in Figure 8.8. Note that $\Delta(G) = 3$, so by Theorem 8.19, there is a proper vertex coloring using at most $\Delta(G) + 1 = 4$ colors. In fact, the proof of Theorem 8.19 yields ALGORITHM DELTA PLUS ONE to obtain a 4-coloring of G. Since G is neither complete nor an odd cycle, Brooks' Theorem 8.22 states that there is actually a 3-coloring of G. The theorem does not provide us with a method for obtaining that coloring. However, since G has the degree sequence (3, 3, 2, 2, 2), which corresponds to the labeling $V(G) = \{u_1, u_2, u_3, u_4, u_5\}$ given in Figure 8.8, then Theorem 8.25 states that there is a proper 3-coloring for G, and moreover, ALGORITHM GREEDY COLORING provides such a 3-coloring:

$$(c_1, c_2, c_3, c_4, c_5) = (1, 2, 1, 2, 3).$$

This means that the color map $c : V(G) \to \{1, 2, 3\}$ is given by $c(u_i) = c_i$; that is,

$$c(u_1) = 1,\ c(u_2) = 2,\ c(u_3) = 1,\ c(u_4) = 2,\ c(u_5) = 3.$$

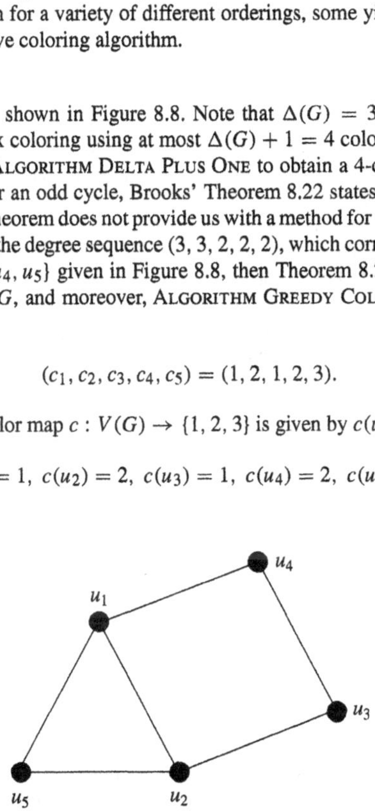

Figure 8.8: A graph G with a vertex labeling corresponding to its degree sequence.

8.4 Planar Graphs and Other Surface Graphs

In this section we consider the chromatic numbers of planar graphs and graphs of a given general genus γ. Graph theorists have focused on proper vertex colorings of planar graphs for a long time. This problem has had a profound impact on the actual systematic development of graph theory itself. Although it is not our purpose to discuss the complete history of planar graphs and their vertex colorings, it is necessary to include some of their rich historical background, since many of the ideas in the following proofs have historical significance.

Consider a geographical map M with countries and clearly marked borders. Suppose M has been drawn either on a globe (that is, the sphere S^2) or on a piece of paper (that is, the plane surface P). We can view M as a plane graph without any bridges, where the edges denote the borders between the countries. The countries of the map correspond to the faces of M, and two countries have a common border if, and only if, the corresponding faces share a bounding edge. We want to color the countries of M with the minimum number of colors, such that any two countries with a common border receive different colors. The problem sounds simple enough.

In 1852 Francis Guthrie, a mathematics student from University College in Victorian London, thought as much when he noticed only four colors were needed to color every map he drew. It is easy to see there are maps where four colors are needed. The smallest such example is shown in Figure 8.9. Francis was not able to prove that four colors always were sufficient. Hence, he later told his younger brother, Frederick Guthrie, who also was a mathematics student at the University College in London, about the problem. It was Frederick who posed this problem to his mathematics professor Augustus De Morgan (1806–1871) at the University College on October 23, 1852. De Morgan became quite fascinated by this challenging problem but was not able to prove it either. He passed the problem on to Sir William Hamilton, also in London, in a letter that same year. The 1852 letter is the earliest documentation of this coloring problem.

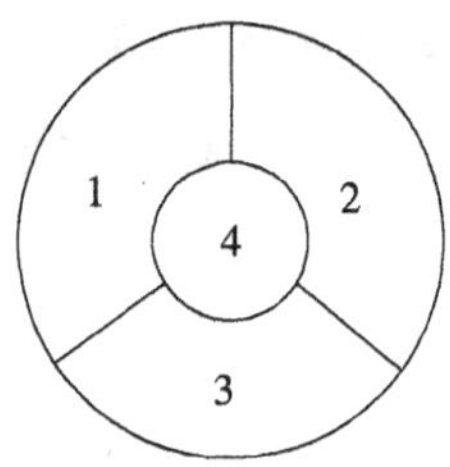

Figure 8.9: The smallest geographical map where four colors are necessary.

Some, however, believe that August Ferdinand Möbius (1790–1868) was the first to state this coloring problem formally, in a talk he gave around 1840. But the first known *written* posing of this four color problem is in the De Morgan letter from 1852, in which De Morgan discussed the problem with Sir Hamilton and attributed it to his student Frederick Guthrie. After this, the word spread among mathematicians and other scholars, who were drawn to its simplicity, and at the same time baffled by the difficulty in its solution. In 1879 the problem became well known, when Arthur Cayley published it in the very first volume of the *Proceedings of the Royal Geographic Society*. He posed the four color problem in a slightly different way, namely in terms of simple graphs, as we now describe.

By considering the plane dual $G = M^*$ of M, where we view M as a simple plane graph, each face in M corresponds to a vertex in G, and two vertices in G are adjacent if, and only if, the corresponding faces have a common bounding edge. Since M is a plane

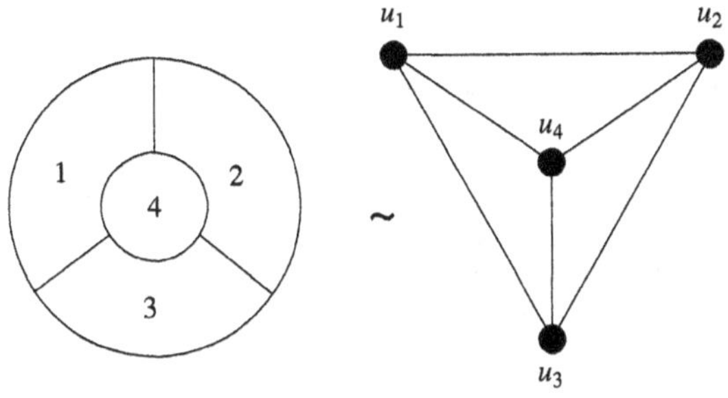

Figure 8.10: A geographical map and its plane dual graph.

graph, so is G. In order to confirm Guthrie's suspicion, it suffices to prove that $\chi(G) \leq 4$ for every planar graph G. (See Figure 8.10.)

Note 8.27 If we add a vertex or an edge to the plane graph G to obtain another plane graph G', then clearly we have $\chi(G) \leq \chi(G')$. Therefore, we can assume that G is a maximally planar graph, where all its faces are bounded by exactly three edges. Adding vertices to G to make it maximally planar corresponds to the process of forming a "small" cycle in G^* for every vertex of G^* of degree at least four, in order to make G^* a 3-regular plane graph. This process is shown in Figure 8.11.

This is precisely the way Cayley announced the four color problem to the London Mathematical Society in 1878, namely in terms of chromatic numbers of planar graphs. It has since been called the *Four Color Conjecture*. In 1879 Sir Alfred Kempe published a faulty proof, in which a flaw was discovered eleven years later. Although Kempe's proof was wrong, it contained fundamental ideas for addressing the problem. About a century later, in 1976, these ideas were used by Kenneth Appel, Wolfgang Haken, and John Koch to finally provide a lengthy but correct proof of the Four Color Conjecture. The Four Color Conjecture is therefore now justifiably called the *Four Color Theorem*.

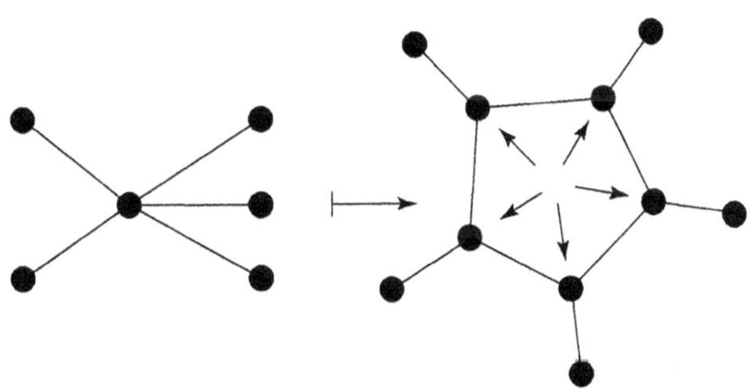

Figure 8.11: The process in G^* corresponding to the addition of a vertex in G.

Appel and Haken's proof depends on extensive computer calculation to check 1,476 nontrivial cases, so-called "unavoidable configurations." A new approach was given to the proof of the Four Color Theorem in 1997 by Neil Robertson, Daniel Sanders, Paul Seymour, and Robin Thomas, who reduced the number of cases of unavoidable configurations to 633. (See Fritsch and Fritsch [16].) Interestingly, they found the computer program worked faster when dealing with *edge colorings* instead of vertex colorings. We will discuss this point more fully in §8.5 and §8.6. This was a useful step forward, since a famous equivalence relation of Peter Guthrie Tait (1831–1901) from 1878 states that every planar graph satisfying certain conditions is edge 3-colorable if, and only if, it is vertex 4-colorable. The number of cases one needs to check is still too huge for one person to verify. Hence, the proof still depends extensively on computer computations. Many mathematicians and computer scientists are waiting hopefully for a self-contained and explanatory proof of the Four Color Theorem.

But going back to 1890, when a proof of the Four Color Conjecture was still a century away, Percy Heawood could "fix" Kempe's faulty proof, and he provided a correct proof of the *Five Color Theorem,* which we now present together with a polished version of Heawood's proof.

Theorem 8.28 **Five Color**

For a planar graph G, we have $\chi(G) \leq 5$.

Proof We can assume G is simple. We will use induction on n, the number of vertices of G.

Since the theorem is clearly true for any simple graph on five or fewer vertices, we can assume $n \geq 6$ and that the theorem is true for all planar graphs on $n-1$ and fewer vertices.

By Corollary 7.18 there is a vertex u of G with $d_G(u) \leq 5$. If $d_G(u) \leq 4$, then by the induction hypothesis the graph $G-u$ has a proper 5-coloring. Since we have five colors to choose from, at least one color is distinct from all of the colors of the neighbors of u in G. Therefore, this coloring can be extended to a proper 5-coloring of G. If $d_G(u) = 5$, then let $u_1, \ldots, u_5$ be the neighbors of u that are connected to u by the edges $e_1, \ldots, e_5$, respectively, as shown in Figure 8.12.

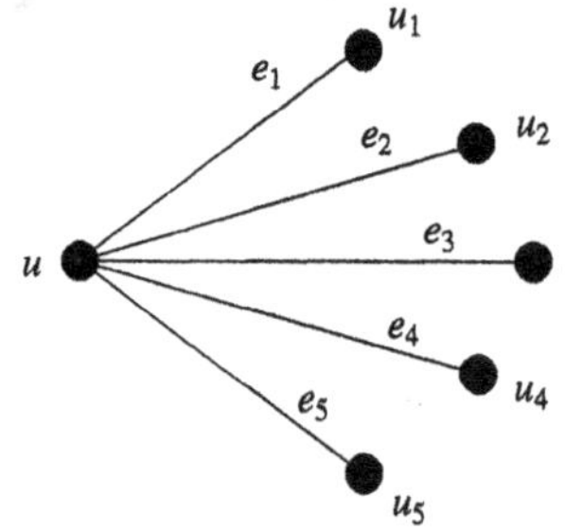

Figure 8.12: The five neighbors and incident edges of u.

Since G is planar and K_5 is not, there is a pair u_1 and u_2 of neighbors of u that are not adjacent. In this case the contracted graph $(G \cdot e_1) \cdot e_2$ is loopless and planar on $n-2$ vertices. By the induction hypothesis this graph has a proper 5-coloring. In this coloring, the vertices u_1, u_2, and u have all collapsed into one vertex receiving a single color. We use this coloring to color properly all the vertices from $V(G-u)$ in G. Note that in this coloring the vertices u_1 and u_2 receive the same color. Thus at most four colors have been used on all of the neighbors of u. We therefore have at least one color that we can assign to u to extend this coloring of $G-u$ to a proper 5-vertex coloring of G. This completes our inductive proof. ■

In 1976, as mentioned earlier, Appel, Haken, and Koch finally managed to prove the Four Color Conjecture with the aid of extensive computer calculations by dispatching 1,476 *unavoidable configurations* of planar graphs, which took close to 1,000 hours (about 42 days) of computer time. A *configuration* of a plane graph G is a cycle together with the induced subgraph inside it. They called it *unavoidable*, if a minimal counterexample to the Four Color Conjecture is forced to contain the configuration.

Theorem 8.29 **Four Color**

For a planar graph G, we have $\chi(G) \leq 4$.

The proof of this theorem is still beyond the scope of any textbook on general graph theory. Because of its reliance on such extensive computer calculations, this proof has not been immune to criticism. The authors have responded with a 741-page book (Appel and Haken [2]), which is an algorithmic version of their proof. The book addresses various criticisms and corrects many minor errors.

The great improvements to the proof of the Four Color Theorem by Robertson, Seymour, Sanders, and Thomas in 1997 managed to reduce the number of unavoidable configuration to 633, and they developed an efficient algorithm [in fact an $O(n^2)$ running-time algorithm; see Chapter 13 for definitions] to color planar graphs on n vertices using at most four colors. Their work is posted on the Internet at

`www.math.gatech.edu/~thomas/FC/fourcolor.html`

From this Web presentation, the computer code can be obtained, which now makes it possible for a desktop computer to "verify" the Four Color Theorem in a couple of hours.

The interesting part of this whole story is that the Four Color Theorem can be generalized to graphs of an arbitrary genus. What is even more interesting is that the proof of this generalization for graphs of an arbitrary genus of one or more is much easier and is completely within the scope of this book. We start with the chromatic numbers of toroidal graphs of genus one.

Theorem 8.30 For a graph G of genus $\gamma(G) \leq 1$, we have $\chi(G) \leq 7$.

Proof Let G be a graph on n vertices with $\gamma(G) \leq 1$. We may assume G is simple. If $\gamma(G) = 0$, then by Theorem 8.28, we have $\chi(G) \leq 5$. Hence, we may assume $\gamma(G) = 1$.

The theorem is clearly true if $n \leq 7$, so we assume $n \geq 8$ and proceed by induction on n.

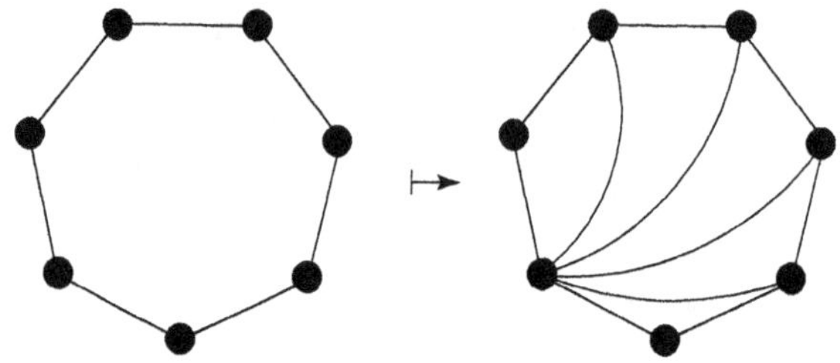

Figure 8.13: Since each face r is a 2-cell, we can subdivide the face into 2-cells that are each bounded by three edges.

Assume the statement is correct for all graphs of genus one with $n-1$ or fewer vertices. By Theorem 7.50 each face of a toroidal embedding of G is a 2-cell. If a face r is bounded by $k \geq 4$ edges, then we can pick one vertex u on its boundary, and add $k-3$ edges from u to the $k-3$ other bounding vertices of r, which are not the two neighbors of u on the cycle forming r. This subdivides r into faces that all are bounded by exactly three edges as shown in Figure 8.13.

This procedure does not alter the value of n, the genus, or reduce the chromatic number of G. Hence, we can assume that each face is bounded by exactly three edges, and further that each edge bounds exactly two faces. In that case, if m denotes the number of edges in G and f the number of faces, then in a similar way as in the proof of Corollary 7.15, we have $2m = 3f$. Thus, by Theorem 7.52

$$0 = 3n - 3m + 3f = 3n - m.$$

By the Hand-Shaking Theorem, we now have

$$\sum_{u \in V(G)} d(u) = 2m = 6n.$$

Hence, there is a vertex $u \in V(G)$ with $d_G(u) \leq 6$. The graph $G-u$ is toroidal, simple, and on $n-1$ vertices. By the induction hypothesis it has a proper 7-vertex coloring. Since we have at least one color to assign u, the 7-coloring of $G-u$ extends to a proper 7-vertex coloring of G. This completes our inductive proof. ■

Theorem 8.30 serves as the base case in our inductive proof of the next theorem—known as *Heawood's Formula* after Percy Heawood.

Theorem 8.31 **Heawood's Formula**

For a graph G of genus $\gamma(G) \geq 1$, we have

$$\chi(G) \leq \left\lfloor \frac{1}{2}\left(7 + \sqrt{1 + 48\gamma(G)}\right) \right\rfloor.$$

Proof Let $\kappa(\gamma) = 1/2(7 + \sqrt{1 + 48\gamma})$. We show that $\chi(G) \leq \lfloor \kappa(\gamma(G)) \rfloor$, proceeding by induction on $\gamma(G) + |V(G)|$.

Let $n = |V(G)|$ and $\gamma = \gamma(G)$. The theorem is clearly true for all $n \leq \lfloor \kappa(\gamma) \rfloor$, so we can assume that $n \geq \lfloor \kappa(\gamma) \rfloor + 1$. As in the proof of Theorem 8.30, we can assume that all of the faces are bounded by exactly three edges and that each edge bounds exactly two faces. Hence, if m denotes the number of edges in G and f the number of faces, then we have $2m = 3f$. Thus, by Theorem 7.52

$$6 - 6\gamma = 3n - 3m + 3f = 3n - m,$$

or $m = 3n + 6\gamma - 6$. Hence, by the Hand-Shaking Theorem, we obtain

$$\sum_{u \in V(G)} d(u) = 2e = 6n + 12(\gamma - 1).$$

Therefore, there is a vertex $u \in V(G)$ with

$$d_G(u) \leq 6 + \frac{12(\gamma - 1)}{n} \leq 6 + \frac{12(\gamma - 1)}{\kappa(\gamma)} = \kappa(\gamma) - 1, \tag{8.3}$$

by definition of $\kappa(\gamma)$. Since $d_G(u) \in \mathbb{N}$ is a natural number, we get

$$d_G(u) \leq \lfloor \kappa(\gamma) - 1 \rfloor = \lfloor \kappa(\gamma) \rfloor - 1.$$

Now, the graph $G - u$ has $n - 1$ vertices and genus at most γ. Hence, by the induction hypothesis, it has a proper $\lfloor \kappa(\gamma) \rfloor$-vertex coloring. Since we have at least one color available to assign to u, this coloring of $G - u$ extends to a proper $\lfloor \kappa(\gamma) \rfloor$-vertex coloring of G. Hence, we have completed the induction and therefore proved the theorem. ■

Note 8.32 For a planar graph G, we have $\gamma(G) = \gamma = 0$. Plugging this value into Heawood's Formula actually gives $\chi(G) \leq 4$, which is the Four Color Theorem! However, our base case $\gamma = 1$ in the proof of Theorem 8.31 implies that $\gamma \geq 1$. This is really needed in our arguments to conclude that

$$\frac{12(\gamma - 1)}{n} \leq \frac{12(\gamma - 1)}{\kappa(\gamma)}$$

in (8.3) in the proof of Heawood's Formula in Theorem 8.31. This deduction does not work for $\gamma(G) = 0$.

By Theorem 7.55 of Ringel, Youngs, and Mayer, we have

$$\gamma(K_n) = \left\lceil \frac{1}{12}(n - 3)(n - 4) \right\rceil;$$

and hence

$$\frac{1}{12}(n - 3)(n - 4) \leq \gamma(K_n) < \frac{1}{12}(n - 2)(n - 3).$$

Let $\gamma = \gamma(K_n)$ and $\kappa(\gamma) = 1/2(7 + \sqrt{1 + 48\gamma})$. By solving the previous inequalities for n, we obtain that

$$\kappa(\gamma) - 1 = \frac{1}{2}(5 + \sqrt{1 + 48\gamma}) < n \leq \frac{1}{2}(7 + \sqrt{1 + 48\gamma}) = \kappa(\gamma),$$

which yields $n = \lfloor \kappa(\gamma) \rfloor$. Since $n = \chi(K_n)$, each genus $\gamma \in \mathbb{N}$ has a graph, namely $G = K_{\lfloor \kappa(\gamma) \rfloor}$, of genus γ such that $\chi(G) = \lfloor \kappa(\gamma) \rfloor$.

We conclude this section by summarizing both the Four Color Theorem 8.29 and the Ringel, Youngs, and Mayer's Theorem 7.55.

Theorem 8.33 If $\gamma \in \mathbb{N} \cup \{0\} = \{0, 1, 2, \dots\}$ and χ_γ is the largest chromatic number possible for a graph of genus γ, then

$$\chi_\gamma = \lfloor 1/2(7 + \sqrt{1 + 48\gamma}) \rfloor.$$

Remark 8.34 Strictly speaking, "Heawood's Formula" is really the *equality* displayed in Theorem 8.33, even though Heawood proved only the "less than or equal to" part of it. He did claim that what Ringel, Youngs, and Mayer managed to prove in Theorem 7.55 was true, but provided no proof. Needless to say, Heawood did not prove the hardest case when $\gamma(G) = 0$, either—the Four Color Theorem.

For a more detailed discussion regarding the history of the Four Color Conjecture and the Four Color Theorem, together with the underlying topological foundations, we refer the reader to Fritsch and Fritsch [16].

8.5 Edge Coloring of a Graph

So far we have discussed proper vertex colorings of simple graphs, which are the most common colorings of simple graphs. Sometimes, however, one needs to consider colorings of the *edges* rather than the vertices of the graph. This happens when one is concerned about partitioning the binary relations among the objects, rather than partitioning the objects themselves. In this section our graphs need not be simple graphs.

Definition 8.35 Let $G = (V, E, \phi)$ be a graph. An *edge coloring* or a *proper edge coloring* of G is a function $c : E(G) \to \mathbb{N}$ satisfying

$$c(e) = c(f) \Rightarrow \phi(e) \cap \phi(f) = \varnothing \tag{8.4}$$

for all $e, f \in E(G)$; that is, no two adjacent edges receive the same color.

For $k \in \mathbb{N}$, a *k-edge coloring* or a *proper k-edge coloring* of G is an edge coloring $c : E(G) \to \{1, \dots, k\}$.

We say that G is *k-edge colc 'le* if G has a proper k-edge coloring.

Remark 8.36

- Unlike vertex colorings, it matters a great deal whether our graph G is simple or not, with loops or not.

- A map $c : V(E) \to \mathbb{N}$, where we do not necessarily assume (8.4) to hold, is called a *nonproper edge coloring*.
- As for vertex colorings, a graph G having m edges is m-edge colorable. Hence, any graph G has a proper k-edge coloring for some $k \le m = |E(G)|$.
- If G is k-edge colorable, then any subgraph G' of G is also k-edge colorable.

Definition 8.37 For a graph G, the least number $k \in \mathbb{N}$ such that G is k-edge colorable is called the *edge chromatic number* of G, or the *chromatic index* of G, and is denoted by $\chi'(G)$.

Note 8.38 For a graph G and any vertex $u \in V(G)$, all edges with u as an endvertex are adjacent and hence must receive different colors in a proper edge coloring of G. Hence, we note the obvious lower bound for the edge chromatic number of G

$$\chi'(G) \ge \Delta(G),$$

the maximum degree in G.

The next example illustrates the ideas just discussed.

Example 8.39 Consider the complete graph K_n on n vertices. From Note 8.38 we have in particular that

$$\chi'(K_n) \ge \Delta(K_n) = n - 1. \tag{8.5}$$

If n is odd, we can sharpen the inequality in (8.5) by one: For a proper k-edge coloring of K_n, each set of unicolored edges has at most $(n-1)/2$ edges in it. Hence, we have

$$k \cdot \frac{n-1}{2} \ge |E(K_n)| = \binom{n}{2},$$

and so $k \ge n$, which implies that $\chi'(K_n) \ge n$ in this case. By representing K_n as a regular n-gon, we can obtain a proper n-edge coloring of K_n by coloring each of the $(n-1)/2$ edges, which are parallel as straight line segments, the same color, thereby using at most n colors.

If $n \ge 4$ is even, the inequality of (8.5) is an equality: By the previous paragraph, we have a proper $(n-1)$-edge coloring of K_{n-1} since $n-1$ is odd. We form K_n from K_{n-1} by connecting a new vertex u to each of the vertices of K_{n-1}. For each vertex v in K_{n-1}, precisely $n-2$ colors are used to color all of the edges in K_{n-1} incident to v. So, we have exactly one color left to color the edge $\{u, v\}$, which is the color of the edge opposite the vertex v in K_{n-1}. Hence, all of the new edges from u receive different colors, and we therefore obtain a proper $(n-1)$-edge coloring of K_n in this case.

We summarize the result of Example 8.39 in the following observation.

Observation 8.40 For $n \in \mathbb{N}$, the edge chromatic number of K_n is given by

$$\chi'(K_n) = \begin{cases} n & \text{if } n \text{ is odd,} \\ n-1 & \text{if } n \text{ is even.} \end{cases}$$

Note that for odd n we have $\chi'(K_n) = \Delta(K_n) + 1$, and for even n we have $\chi'(K_n) = \Delta(K_n)$. In particular, we have

$$\Delta(K_n) \le \chi'(K_n) \le \Delta(K_n) + 1$$

for all $n \in \mathbb{N}$. As we will see later, this last inequality actually holds for all simple graphs. Before we discuss that result, we first prove the following theorem.

Theorem 8.41 For a bipartite graph G, we have

$$\chi'(G) = \Delta(G).$$

Proof We prove this by induction on $|E(G)|$, the number of edges of G. We can assume G has more than one edge.

By removing one edge $e = \{u, v\}$ from G, the graph $G' = G - e$ has a proper $\Delta(G)$-edge coloring by the induction hypothesis. Since each of the vertices u and v has degree strictly less than $\Delta(G)$ in G', there is a color $i \in \{1, \ldots, \Delta(G)\}$ that has not been used on any of the edges with u as an endvertex in G'. Also, there is a color j that has not been used on any of the edges with v as an endvertex in G'. If $i = j$, then we can color e by that color and we are done.

Otherwise if $i \neq j$, let $E(i, j) \subseteq E(G)$ be the set of edges that have the colors i or j in the edge coloring of G'. Let $C(i, j)$ be the component of $G'[E(i, j)]$ that contains the vertex u. Since the edges in $C(i, j)$ have only two colors, each vertex of $C(i, j)$ has degree one or two in $C(i, j)$, so $C(i, j)$ must be a connected path or a cycle, in which the edges are colored alternately by colors i and j. Since i is unused on edges with u as an endvertex, then $C(i, j)$ cannot be a cycle and therefore must be a path. Every vertex on $C(i, j)$ that is on the opposite part of u in the bipartition of G is an endvertex of an edge colored by j. This is shown in Figure 8.14.

Therefore, the vertex v is not contained in the component $C(i, j)$. In this case we can swap the colors i and j for each of the edges in $C(i, j)$ and still have a proper $\Delta(G)$-edge coloring of G', where j has not been used on either u or v. In this case we can now color e by the color j, and thereby obtain a $\Delta(G)$-edge coloring of G. This completes the induction. ■

From Theorem 8.41 we obtain the next corollary.

Corollary 8.42 For the complete bipartite graph $K_{m,n}$, we have

$$\chi'(K_{m,n}) = \max(\{m, n\}).$$

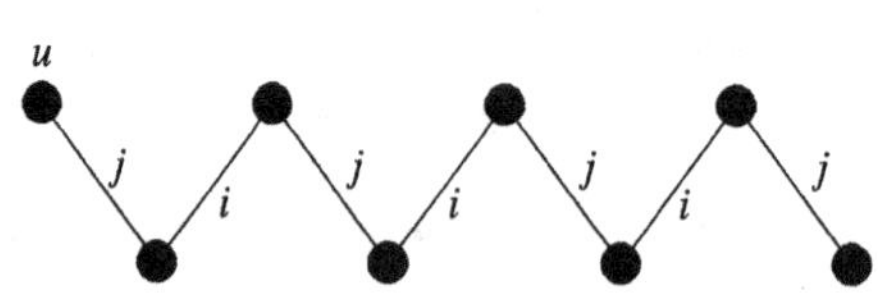

Figure 8.14: Every vertex on the opposite part from u is an endvertex of an edge of color j.

The following theorem gives us as tight bounds for $\chi'(G)$ as we can hope for when G is a simple graph. It was proved by Vizing in 1964 and independently by Gupta in 1966, although the latter proof never was published except as an abstract. It is usually referred to as Vizing's Theorem.

Theorem 8.43 **Vizing's**

If G is a simple graph, then

$$\Delta(G) \le \chi'(G) \le \Delta(G) + 1.$$

Proof We need to prove only the upper bound $\chi'(G) \le \Delta(G) + 1$. We use induction on $|E(G)|$, the number of edges of G. We assume that G has more than one edge and we let $e_0 = \{u, v_0\}$ be an edge in G. By the induction hypothesis, the graph $G' = G - e_0$ can be edge colored by $\Delta(G) + 1$ colors. Since the degree of each vertex $w \in V(G)$ is strictly less than $\Delta(G) + 1$, there is always a color in $\{1, \ldots, \Delta(G) + 1\}$ that has not been used to color edges incident to w. We refer to such a color as an *unused* color at w. We now apply the following procedure:

1. Let $k = 0$.
2. Seek a common unused color at both v_k and u.
   - If it exists, then we call that color c_{k+1} and we go to CASE ONE.
   - If it does not exist, then we seek an unused color at v_k (which, in particular, has not been used to color e_k) and has not been used to color any of the $e_1, \ldots, e_{k-1}$.
     - If such a color exists, then we call that color c_{k+1}. Since c_{k+1} in this case has been used at u, there is an edge $e_{k+1} = \{u, v_{k+1}\}$ of color c_{k+1}. We increment k by one, and go to step 2.
     - If such a color does not exist, then we go to CASE TWO.

We note that since u is adjacent to finitely many neighbors, the procedure must end in either CASE ONE or CASE TWO.

CASE ONE: We can recolor the edges $e_1, \ldots, e_k$ by coloring e_i with the color c_{i+1} for each $i \in \{1, \ldots, k\}$, and then color e_0 by c_1.

CASE TWO: We now have the edges $e_0, \ldots, e_k$ all with u as one endvertex and with $v_0, \ldots, v_k$ as the other endvertices. For each $i \in \{1, \ldots, k\}$ the edge e_i has the color c_i. We let c' be an unused color at u. The color c' is used at $v_0, \ldots, v_k$, since otherwise we would have ended up in CASE ONE. Let c'' be an unused color at v_k. This color has been used on one of the edges $e_1, \ldots, e_{k-1}$, so there is an $i \in \{1, \ldots, k-1\}$ with $c_i = c''$. Let $E(c', c'') \subseteq E(G)$ be the set of edges of colors either c' or c'', and let $C(c', c'')$ be the component of $E(c', c'')$ containing v_k. Since the edges in $C(c', c'')$ have two colors, the vertices of $C(c', c'')$ have degree one or two. Since the color c'' is unused at v_k, then $d_{C(c',c'')}(v_k) = 1$. Thus, $C(c', c'')$ is a path with v_k as an endvertex. Let w be the other endvertex of $C(c', c'')$. We note that $w \ne v_k$ and that $w \ne v_i$ since both $c'' = c_i$ and c' are used at v_i. Therefore, $d_{C(c',c'')}(v_i) = 2$.

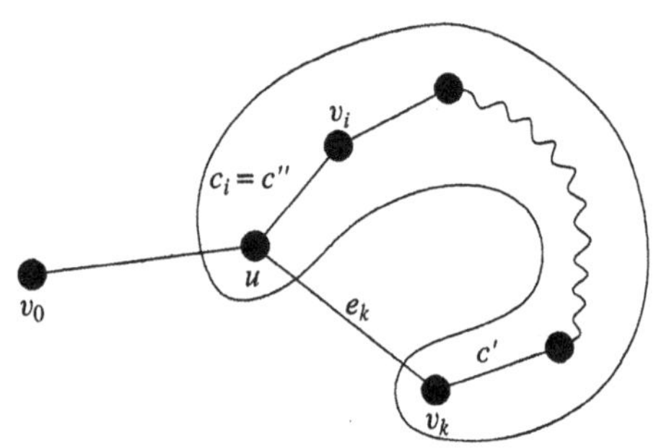

Figure 8.15: First the colors of $C(c', c'')$ are swapped, then we recolor the edges $e_1, \ldots, e_{i-1}$.

We now have three subcases to consider.

SUBCASE TWO-A ($w = u$): Since c' is unused at u, the vertex v_i must be adjacent to w in $C(c', c'')$. (See Figure 8.15.) We can recolor the edges of G' by first swapping the colors c' and c'' in the path $C(c', c'')$, after which the color $c'' = c_i$ is unused at u. Thus, we can recolor the edges $e_1, \ldots, e_{i-1}$ by assigning the color $c_{\ell+1}$ to e_ℓ for each $\ell \in \{1, \ldots, i-1\}$. Finally, we can assign the color c_1 to e_0, and thereby get a proper edge coloring of G.

SUBCASE TWO-B ($w = v_j \neq v_i$): Here, too, we recolor the edges of G' by swapping the colors c' and c'' in the path $C(c', c'')$, after which c' is unused at v_j (since e_j did not have the color c'' before). Now we can reassign c' to e_j and $c_{\ell+1}$ to e_ℓ for each $\ell \in \{0, \ldots, j-1\}$, and thereby obtain a proper edge coloring of G.

SUBCASE TWO-C ($w \neq u$ and $w \neq v_j$ for all $j \in \{1, \ldots, k\}$): We recolor the edges of G' by swapping the colors c' and c'' in the path $C(c', c'')$, after which c' is unused at v_k. Now we reassign c' to e_k and $c_{\ell+1}$ to e_ℓ for each $\ell \in \{0, \ldots, k-1\}$, and thereby obtain a proper edge coloring of G.

This completes our inductive step, and hence the proof. ■

Remark 8.44

- The proof of Vizing's Theorem yields an inductive algorithm to obtain a proper edge coloring of a given simple graph, and this algorithm can be implemented to run efficiently, contrary to what the actual proof indicates.
- The condition that G be simple is necessary. In Exercise 8.49 we ask for an example of a graph G with $\chi'(G) \geq \lceil 3\Delta/2 \rceil$ for all possible values of $\Delta \geq 2$.

From Vizing's Theorem we see that every simple graph G has either $\chi'(G) = \Delta(G)$ or $\chi'(G) = \Delta(G)+1$. Hence, the equivalence classes of simple graphs fall into two subclasses. This motivates the following definition.

Definition 8.45 Let G be a simple graph.

1. We say that G is of *class one* if $\chi'(G) = \Delta(G)$.
2. We say that G is of *class two* if $\chi'(G) = \Delta(G) + 1$.

If, in addition, G is regular and of class one, then we say that G is *one factorable*.

To determine which simple graphs are of class one and which are of class two is an unsolved problem. That is, nobody has been able to find easily checked conditions that determines whether a graph is of class one or of class two.

8.6 Tait's Theorem

Since graph coloring in general has been highly influenced by the Four Color Conjecture, it seems appropriate to conclude this chapter with a connection between edge coloring and the Four Color Theorem, a connection established by Tait in 1878. Tait thought, at the time, that he had found a proof of the Four Color Conjecture, but instead he really proved the equivalence stated in Theorem 8.46.

As noted earlier in §8.4, in 1997 Robertson, Sanders, Seymour, and Thomas developed a new and improved algorithm to check the cases of unavoidable configurations in the Four Color Theorem. Their result, which reduced the number of cases to 633 (as described in Fritsch and Fritsch [16]), actually used Theorem 8.46. It turned out that their computer program worked much faster when dealing with proper edge colorings rather than proper vertex colorings for the cases they considered.

Recall that a geographical map can be represented by a plane graph M, where the edges represent the proper borders between the regions, and the vertices represent border points where three or more regions meet (recall that we may assume M has no bridges). Clearly every vertex in M is of degree at least three. If, however, a vertex has degree $d \geq 4$, then we can form a d-cycle for that vertex, as shown in Figure 8.11 in the case for $d = 5$, and add a region within the d-cycle. Therefore, when considering coloring the regions or faces of the map M, we may assume that M is a 3-regular plane graph.

Theorem 8.46 **Tait's**

For a 3-regular plane graph M, the following are equivalent:

1. M is 4-face colorable; that is, using at most four colors we can assign colors to the faces of M such that no two faces of M having a common boundary edge receive the same color.
2. M is 3-edge colorable.

Theorem 8.46, which Tait proved in 1878 around the same time as Kempe published his faulty proof of the Four Color Conjecture, is referred to as *Tait's Theorem* or *Tait's Reformulation*. Strictly speaking, Tait's original proof was only a "sketch" of a rigorous proof, but it contained all of the main ideas that a precise reformulation would use. The sketchy parts are mainly of a topological nature. Since then, shorter and more elegant proofs have been found for Tait's Theorem 8.46, but in the following proof, we present Tait's original ideas.

Proof (Based on Tait's Sketch) We first prove the easier part, that a proper 4-face coloring of M implies a proper 3-edge coloring of M.

Assume that we have a proper 4-face coloring $c : F(M) \to \{1, 2, 3, 4\}$, where $F(M)$ denotes the set of faces of the plane graph M. We may assume that every edge borders exactly two faces of M. (If there are cut edges, we can remove them and deal with them as blocks, after we have colored the edges of each component of the remaining graph.) We define a proper 3-edge coloring $c' : E(M) \to \{1, 2, 3\}$ by the following two cases:

1. If e borders the two faces f_1 and f_2, where one of them is colored by 4, $c(f_i) = 4$, then we set $c'(e) = c(f_{3-i})$, the color of the other face.
2. If e borders the two faces f_1 and f_2, where $\{c(f_1), c(f_2)\} \subseteq \{1, 2, 3\}$, then we set $c'(e) = i$, where i is the color used on neither f_1 nor f_2—that is, $i \in \{1, 2, 3\} \setminus \{c(f_1), c(f_2)\}$.

In this way c' is a well-defined map, which in turn is a proper 3-edge coloring of M.

The other way is more involved. We now assume that we have a proper 3-edge coloring $c' : E(M) \to \{1, 2, 3\}$ from which we would like to obtain a proper 4-face coloring $c : F(M) \to \{1, 2, 3, 4\}$. Let M_1 and M_2 be the subplane graphs that we obtain from M by removing the edges from M with color 1 and color 2, respectively. Since M is 3-regular, both M_1 and M_2 are 2-regular graphs. Hence, they are both disjoint unions of cycles embedded in the plane.

We now assign the numbers 0 or 1 to each of the faces of M_i in the following way:

1. We label the infinite face by 0.
2. Label a face f of M_i by $[m_i(f) \bmod 2] \in \{0, 1\}$, where $m_i(f)$ is the least number of edges one needs to cross in order to get from the infinite face to f in M_i.

In this manner every pair of bounding faces of M_i receives different labelings. An example of such a labeling is shown in Figure 8.16.

Every face f in M is now an intersection of two unique faces f_1 of M_1 and f_2 of M_2.

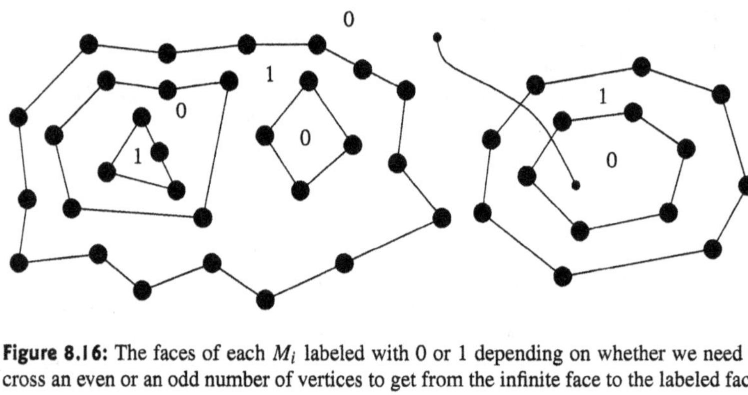

Figure 8.16: The faces of each M_i labeled with 0 or 1 depending on whether we need to cross an even or an odd number of vertices to get from the infinite face to the labeled face.

We define $c : F(M) \to \{1, 2, 3, 4\}$ by

$$c(f) = \begin{cases} 1 & \text{if } (m_1(f_1), m_2(f_2)) = (0, 0), \\ 2 & \text{if } (m_1(f_1), m_2(f_2)) = (0, 1), \\ 3 & \text{if } (m_1(f_1), m_2(f_2)) = (1, 0), \\ 4 & \text{if } (m_1(f_1), m_2(f_2)) = (1, 1). \end{cases}$$

To see that c is a proper 4-face coloring of M, we need to consider three cases for two bordering faces f and f' of M:

1. The common border edge has color 1. Here the labels from M_1 are the same—that is, $m_1(f_1) = m_1(f_1')$—but the labels from M_2 are different. Therefore, we have $c(f) \neq c(f')$ in this case.
2. The common border edge has color 2. Here the labels from M_2 are the same—that is, $m_2(f_2) = m_2(f_2')$—but the labels from M_1 are different. Therefore, we have $c(f) \neq c(f')$ in this case.
3. The common border edge has color 3. Here we have $m_i(f_i) \neq m_i(f_i')$ for both $i = 1, 2$. Therefore, we have $c(f) \neq c(f')$ in this final case as well.

We have now shown that c is a proper 4-face coloring of M, and hence we have completed the proof. ■

So far, we have seen that K_n is of class one when n is odd and of class two when n is even. Every bipartite graph is of class one. By the Four Color Theorem 8.29 and Tait's Theorem 8.46, we have the following corollary.

Corollary 8.47 Every 3-regular plane graph is of class one.

8.7 Exercises

8.1. Give an example of two nonisomorphic simple graphs of five vertices each that have a chromatic number of three.

8.2. Determine the chromatic number of the Petersen graph.

8.3. Prove Observation 8.6, that $k = \chi(G)$ is the smallest $k \in \mathbb{N}$ such that there is a homomorphism $\phi : G \to K_k$.

8.4. Determine the chromatic number of the Grötzsch graph shown in Figure 8.17.

8.5. Show that if G is a simple graph on n vertices and G is not the complete graph K_n, then $\chi(G) \leq n - 1$.

8.6. Show that if G is a simple graph on n vertices with $m \leq \frac{(n-1)(n-2)}{2} - 1$ edges, then $\chi(G) \leq n - 2$.

8.7. Consider the simple graph in Figure 8.8. How many distinct 3-colorings does it have? Up to permutation of the colors, how many 3-colorings does the graph have?

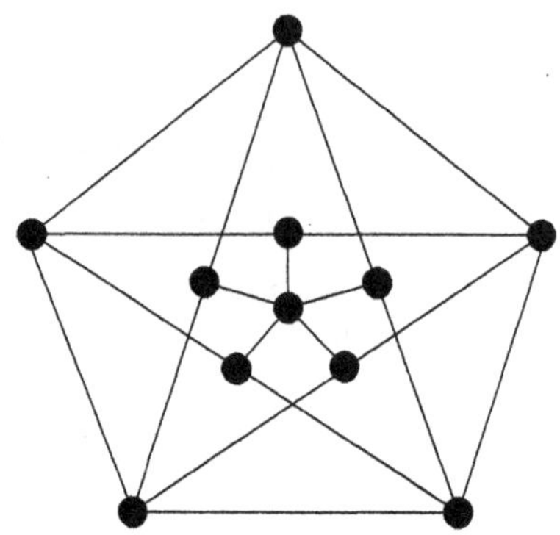

Figure 8.17: The Grötzsch graph.

8.8. Show that the *rigid ladder* graph in Figure 8.18 has chromatic number three.

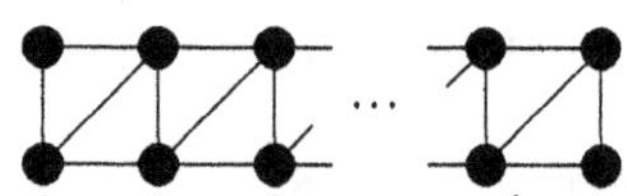

Figure 8.18: The rigid ladder graph on $2n$ vertices.

8.9. Let G and H be simple graphs. Define the *Cartesian product* $G \times H$ as the simple graph on the vertices $V(G) \times V(H)$ and edges given by the following.

$$E(G \times H) = \{\{(u, v), (u', v')\} : u = u' \text{ and } \{v, v'\} \in E(H), \text{ or } v = v' \text{ and } \{u, u'\} \in E(G)\}.$$

Show that $\chi(G \times H) = \max(\{\chi(G), \chi(H)\})$.

8.10. Show that for simple graphs G, H, and K we have the isomorphisms

$$G \times K_1 \cong K_1 \times G \cong G,$$
$$G \times H \cong H \times G,$$
$$(G \times H) \times K \cong G \times (H \times K).$$

(This shows that the isomorphism class of simple graphs form a commutative *monoid*.) Use this to show that $G_1 \times \cdots \times G_k$ is well defined for any $k \in \mathbb{N}$, and express $\chi(G_1 \times \cdots \times G_k)$ in terms of the $\chi(G_i)$'s.

8.11. For $k \in \mathbb{N}$, define the *k-cube* Q_k by letting the vertices be

$$V(Q_k) = \{(b_1, \ldots, b_k) : b_i \in \{0, 1\} \text{ for each } i \in \{1, \ldots, k\}\},$$

and connect two vertices $(b_1, \ldots, b_k)$ and $(b'_1, \ldots, b'_k)$ if, and only if,

$$\sum_{i=1}^{k} |b_i - b'_i| = 1.$$

(Compare this with the definition of a k-box graph in Exercise 5.19 in Chapter 5, which generalizes both the grid graph and the k-cube. Also, note that $Q_k = K_2 \times \cdots \times K_2$ as a Cartesian product from the preceding exercise.) By convention Q_0 is a single vertex. Observe that Q_1 is an edge together with its endvertices, Q_2 is the usual square C_4, and Q_3 is the usual three-dimensional cube. Show the following about Q_k:

(a) Q_k has 2^k vertices,
(b) Q_k is a regular graph,
(c) Q_k has $k2^{k-1}$ edges, and
(d) $\chi(Q_k) = 2$ for all $k \in \mathbb{N}$.

8.12. For $n \in \mathbb{N}$ let W_n be the wheel on $n+1$ vertices. Figure 8.19 lists the first five wheels $W_1, \ldots, W_5$. Show that

$$\chi(W_n) = \begin{cases} 3 & \text{if } n \geq 2 \text{ is even,} \\ 4 & \text{if } n \geq 3 \text{ is odd.} \end{cases}$$

What is $\chi(W_1)$?

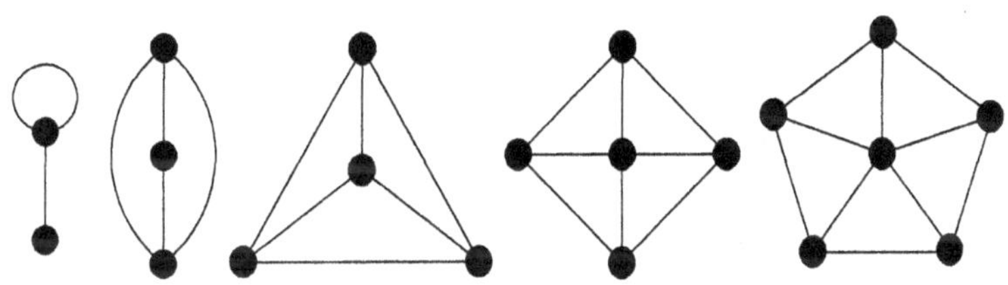

Figure 8.19: The first five wheels $W_1, \ldots, W_5$.

8.13. Show that if G and H are two simple 1-isomorphic graphs, then $\chi(G) = \chi(H)$.

8.14. Prove Observation 8.7 by induction on the number of components of the graph G.

8.15. Let G be a simple graph. Show that G has at least

$$m = \binom{\chi(G)}{2}$$

edges.

8.16. Let G be a simple graph with m edges. Show that

$$\chi(G) \leq \frac{1}{2} + \sqrt{2m + \frac{1}{4}}.$$

8.17. Let G_1 and G_2 be simple subgraphs of a given graph. Show that

$$\chi(G_1 \cup G_2) \leq \chi(G_1) \cdot \chi(G_2).$$

8.18. Let G be a simple graph on n vertices and $\overline{G}$ its complement. Show that

$$\chi(G) + \chi(\overline{G}) \geq 2\sqrt{n}.$$

8.19. Let G be a simple graph on n vertices and let $\omega(G)$ denote the number of vertices in the largest complete subgraph of G (this is called the *clique number* of G; see Chapter 9). Show that

$$\omega(G) + \omega(\overline{G}) \leq n + 1.$$

8.20. Let G be a simple graph on n vertices and $\overline{G}$ its complement. Show that

$$\chi(G) + \chi(\overline{G}) \leq n + 1. \tag{8.6}$$

[*Hint:* Let u be a vertex of G and $\overline{G}$. If $G' = G - u$, then $\chi(G) \leq \chi(G') + 1$ and $\chi(\overline{G}) \leq \chi(\overline{G'}) + 1$. If equality holds in both cases, then $d_G(u) \geq \chi(G)$ and $d_{\overline{G}}(u) \geq \chi(\overline{G})$. Use this to show that (8.6) holds. Otherwise, use induction on n.]

8.21. Let G be a simple graph and $\ell = \ell(G)$ be the length of the longest path in G.

(a) Let u be an endvertex of a longest path p in G. Show that all the neighbors of u in G must be on p.
(b) Show that u has at most $\ell - 2$ neighbors in G.
(c) Consider $G - u$ and use induction to show that $\chi(G) \leq \ell(G) - 1$.

8.22. Let G be a simple graph on n vertices with $\chi(G) = 2$. Show that G has at most $m = \lfloor n/2 \rfloor \lceil n/2 \rceil$ edges.

8.23. Let G be a simple bipartite graph and $\mathsf{A}(G)$ the adjacency matrix for G. Show that for all odd $k \in \mathbb{N}$ every diagonal entry of $\mathsf{A}(G)^k$ is zero. Does the converse hold as well?

8.24. Let G be a simple graph. Can the result in Exercise 8.23 be generalized to state that every diagonal entry of $\mathsf{A}(G)^k$ is zero for all k that are not divisible by $\chi(G) \geq 3$? [*Hint:* Consider 3-colorable graphs first.]

8.25. Prove that when we connect one pair of vertices to a complete k-partite graph G or any Turán-k graph, we obtain a complete subgraph on $k + 1$ vertices.

8.26. Let $k, m \in \mathbb{N}$. Show that for k nonnegative real numbers $x_1, \ldots, x_k$ summing up to a constant

$$\sum_{i=1}^{k} x_i = C,$$

we have

$$\sum_{i=1}^{k} x_i^2 \geq C^2/k,$$

where equality holds if, and only if, $x_i = C/k$ for each i. Use this fact to prove that a graph G on km vertices with $\chi(G) \leq k$ has the maximum number of edges if, and only if, $G = K_m^k$.

8.27. As $n \in \mathbb{N}$ gets large, which of the following classes of graphs on n vertices contains the graphs with the most edges?

- **(a)** Bipartite graphs.
- **(b)** Graphs G with $\chi(G) = 3$.
- **(c)** Graphs G with $\chi(G) = 4$, where the number of vertices of color i is i times the number of vertices of color 1, for $i \in \{2, 3, 4\}$.

8.28. Let G be a simple graph on n vertices. Show that if G has no coloring for which there are three vertices of the same color, then G contains two complete vertex disjoint subgraphs K_a and K_b covering the vertices of G—that is, $a + b = n$.

8.29. Give an example demonstrating that Exercise 8.28 does not hold in general. More precisely, if G is a simple graph on n vertices, with no coloring with k vertices of the same color, then G need not contain k complete vertex disjoint subgraphs $K_{a_1}, \ldots, K_{a_k}$ that cover all the n vertices—that is, $a_1 + \cdots + a_k = n$.

8.30. Compute the maximum number of edges a graph G on $n\binom{k}{2}$ vertices can have, if the vertices of G can be partitioned into k parts

$$V(G) = V_1 \cup \cdots \cup V_k$$

of sizes $m, 2m, \ldots, km$, respectively, such that no two vertices in the same part are adjacent.

8.31. For which $k \in \mathbb{N}$ and $a_1, \ldots, a_k \in \mathbb{N}$ is the k-partite graph $K_{a_1,\ldots,a_k}$ a planar graph?

8.32. Is it possible that the Delta Plus One Coloring Algorithm will use different numbers of colors to color the same graph? If so, provide a demonstration with a graph G and two such labelings.

8.33. Prove that Algorithm Greedy Coloring provides a proper vertex coloring for G using at most $\max(\{\min(\{i, d_i + 1\}) \mid 1 \leq i \leq n\})$ colors, where n is the number of vertices of G.

8.34. Let G be a simple graph on n vertices and let $\overline{G}$ be its complement. Use Exercise 8.33 to show that

$$\chi(G) + \chi(\overline{G}) \leq n + 1.$$

8.35. Let G be a plane graph.

(a) Show that if $\chi(G^*) = 2$, then $d_G(u)$ is even for all $u \in V(G)$.
(b) Show that an Eulerian graph cannot have an edge cut consisting of an odd number of vertices.
(c) Show that the plane dual G^* has no odd cycles.
(d) Show that $d_G(u)$ is even for all $u \in V(G)$ if, and only if, we have $\chi(G^*) = 2$.

8.36. Let G be an Eulerian graph of genus γ. If G^* is the geometric dual of a surface embedding of G in S_γ, is G^* then 2-colorable?

8.37. For a given $n \in \mathbb{N}$, how many edges can a graph G on n vertices have if $\chi(G) \leq 5$? Compare this to the maximum number of edges a planar graph on n vertices can have.

8.38. Let G be a simple Hamiltonian plane graph. Show that the geographical map formed by G in the plane P is 4-colorable. [*Hint:* Show that two colors suffice to color the regions inside the Hamilton cycle.]

8.39. Let G be a simple planar graph containing no triangles. Show that G has a vertex of degree at most three. Use this (and not the Four Color Theorem!) to show that $\chi(G) \leq 4$.

8.40. Consider a finite collection of straight lines in the plane P and the geographical map formed by these lines. Show that the regions of this map can be properly colored using two colors.

8.41. Consider a finite collection of straight lines in a *general position* in the plane P; that is, no three lines meet in a single point. Let G be the graph with vertices consisting of all the points of intersection of these lines, and edges consisting of all the straight line segments between the points. (See Figure 8.20.) Show that $\chi(G) \leq 3$.

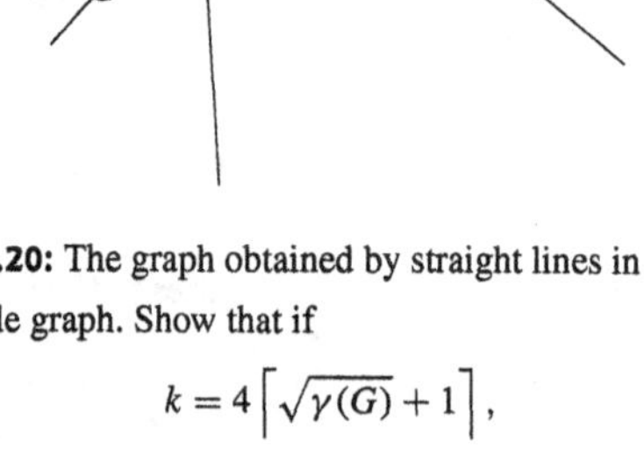

Figure 8.20: The graph obtained by straight lines in the plane.

8.42. Let G be a simple graph. Show that if

$$k = 4\left\lceil \sqrt{\gamma(G)} + 1 \right\rceil,$$

then G is a subgraph of the complete k-partite graph $K_{m_1,\ldots,m_k}$ for some $m_1, \ldots, m_k$.

8.43. Explicitly assign a proper edge coloring to $K_{3,4}$ using four colors. Do the same with $K_{4,5}$. Can you come up with a pattern for $K_{m,n}$?

8.44. In a certain local high school there are m teachers $t_1, \ldots, t_m$ and n classes $c_1, \ldots, c_n$, where $n, m \in \mathbb{N}$. Teacher t_i teaches the class c_j for h_{ij} hours, which do not have to be consecutive. The task is to make a schedule of classes such that the combined number of hours that all the m teachers use in the school itself is at minimum. Assuming that there is no classroom shortage, show how to convert this job into the task of finding an optimal proper edge coloring of a certain graph.

8.45. Let G be a (general) graph and let $G' \subseteq G$ be the graph obtained from G by removing all the loops from G. Prove that

$$\chi'(G) \leq \max\left(\left\{\chi'(G'), \frac{\Delta(G) + \Delta(G')}{2}\right\}\right).$$

8.46. (From the annual high school mathematics competition in Iceland, 1992.) In Mirkwood there live twelve dwarfs who bear the same names as the twelve months. They all live in cottages that have been painted either red or blue. Some of the dwarfs are mutual everlasting buddies, others are not. So it is, and so it will be unaltered through time to come. Each month the dwarf that bears the same name of that month pays each of his everlasting buddies a visit. If the majority of them have a different colored cottage, then he will repaint his cottage to have the same color as the majority of his buddies. Show that there will come a time when none of the dwarfs will need to repaint his house ever more. [*Hint:* Consider the simple graph G on twelve vertices representing the dwarfs, where two vertices are connected if, and only if, the corresponding dwarfs are everlasting buddies. Color the vertices the same color as their cottages—red or blue. Consider the nonproper edge coloring where we color an edge white if both its endvertices have the same color, and black if the endvertices have a different color. What can be said about the process that the set of the edges $E(G)$ undergoes each month when the dwarfs either repaint their cottages or not?]

8.47. Continuing with the twelve dwarfs from Mirkwood: What can we conclude if each dwarf decides to repaint his cottage if, and only if, the *minority* of his buddies have a different colored cottage?

8.48. This is a generalization of Exercises 8.46 and 8.47. Consider a simple graph G and a nonproper vertex coloring $c : V(G) \to \mathbb{N}$. Let $V(G) = \{u_0, \ldots, u_{n-1}\}$ be any labeling of the vertices. As i ranges through the positive integers, from one and onward, we alter the color of $u_{i \bmod n}$ to the color that the strict majority of its neighbors have. If there is no clear majority color, then we do not alter the color of $u_{i \bmod n}$. Show that this process will become stable, in the sense that there is an $N \in \mathbb{N}$ such that for $i \geq N$, no change of colors will take place.

8.49. By multiplying the edges of the triangle, or the 3-cycle C_3, show that for every $\Delta \in \mathbb{N}$ there is a graph G having $\Delta(G) = \Delta$ and with $\chi'(G) \geq \lceil 3\Delta/2 \rceil$. Shannon proved in 1949 that this lower bound is actually also an upper bound. Thus, it is the best lower bound possible.

8.50. For any graph G show by induction on $|E(G)|$, the number of edges of G, that

$$\chi'(G) \leq 2\Delta(G) - 1.$$

Use your proof to obtain a greedy algorithm to edge-color a graph G having at most $2\Delta(G) - 1$ colors.

8.51. For a graph G define the *line graph*, denoted by $L(G)$, as the graph with vertices consisting of the edges of G—that is, $V(L(G)) = E(G)$—and where $e, e' \in V(L(G))$ are adjacent in $L(G)$ if, and only if, they are adjacent in G. That is, $\phi(e) \cap \chi(e') \neq \varnothing$.

(a) Show that $\chi'(G) = \chi(L(G))$.
(b) Use this to prove Vizing's Theorem when $\Delta(G) = 3$.

8.52. Show that if a 3-regular plane graph M is Hamiltonian, then $\chi'(M) \leq 3$. [*Note!* Tait made this observation in the same paper in which he published his Theorem 8.46 in 1880. Also, there are 3-regular planar graphs that are not Hamiltonian, so this does not yield an alternative proof of Tait's Theorem 8.46.]

8.53. Show directly that a Hamiltonian plane graph M has a proper 4-face coloring. [*Hint:* Show that two colors suffice for the faces inside the Hamiltonian cycle of M.]

CHAPTER

9

Coloring Enumerations and Chordal Graphs

Chapter Goals

Define and study the chromatic polynomialof a graph.

Define and discuss interval and chordal graphs.

Present a characterization of chordal graphs.

Discuss colorings of chordal graphs.

Present an optimal coloring algorithm for chordal graphs.

Define and discuss d^{th} power of a graph and discuss its colorings.

9.1 The Chromatic Polynomial of a Graph

In Chapter 8 we discussed the chromatic number of a graph, which is the least number of colors one can use for a proper vertex coloring of the graph. In this section we will define and discuss the *chromatic polynomial* of a graph, which describes the number of different proper vertex colorings there are for a graph using a fixed number of colors. This is a more general problem than that of determining the chromatic number of the graph. It can often be of great value to consider a more general problem than the original one, even when we are interested only in solving the original problem of determining the chromatic number of the graph.

Clearly, for any graph G, there is no way to vertex-color G properly using zero colors (that is, the set of available colors is the empty set). Furthermore, if G has at least one edge, there is no way to vertex-color G properly using one available color. Continuing in this way, if G is not bipartite, then there is no way to vertex-color G properly using two available colors. In general, if $k < \chi(G)$, then there is no way to vertex-color G properly using k available colors—that is, from a set containing k colors. So, $\chi(G)$ is the least natural number k such that the number of ways to vertex-color G properly from a set of k colors is nonzero. Therefore, if for all natural numbers k we know the number of proper k-vertex colorings of G, then in particular, we can determine the chromatic number $\chi(G)$. Hence, for a graph

G and all $k \in \mathbb{N}$, we would like to know the number of possible proper k-vertex colorings of G. Since this is a more general problem than that of asking for the chromatic number of G, why even bother dealing with this more general question, when we cannot even solve a particular case of it, namely that of finding the chromatic number of G? Experience from many branches of mathematics has shown that asking a more general question can be very fruitful in answering the original question.

Note 9.1 The branch of mathematics called *algebraic geometry* deals with the study of sets of roots of polynomials in several variables over a given ring or field. Such sets of roots are called *algebraic varieties* and they possess a rich algebraic and topological structure. Knowledge of their properties is of great interest and importance for computational algebraists, number theorists, analysts, and theoretical physicists. The principle of asking more and more general questions in algebraic geometry is sometimes called the *Grothendieck Principle* after the French algebraic geometer, Alexander Grothendieck. In the late 1950s and the early 1960s, he was very successful in solving old unanswered questions by asking and then answering much more general questions than the original ones.

The idea of studying the number of k-vertex colorings for a graph is originally due to Birkhoff from 1912, where he hoped that this more general question would lead to a solution to the then Four Color Conjecture. As we now know, it did not turn out that way. Nonetheless, this idea is interesting from a mathematical point of view, since first of all, it yields a recursive way to explicitly compute the number of possible vertex colorings, and second, it provides a foundation for many inductive proofs for statements about vertex colorings and sometimes the actual structure of the graph itself.

Definition 9.2 Let G be a graph and $k \in \mathbb{N}$. Define $\chi_G(k)$ to be the number of proper k-vertex colorings $c : V(G) \to \{1, \ldots, k\}$.

Remark 9.3 By Definition 9.2 we immediately see the following:

- For each $k < \chi(G)$, we have $\chi_G(k) = 0$.
- For each $k \geq \chi(G)$, we have $\chi_G(k) > 0$.
- If G is a graph with loops, then $\chi_G(k) = 0$ for all $k \in \mathbb{N}$. Hence, Definition 9.2 is valid for all graphs, loopless or not.

For some simple graphs G, the number $\chi_G(k)$ is easy to compute, as we see in the next example.

Example 9.4 Let $k, n \in \mathbb{N}$. We assume there are k colors available to vertex-color the following graphs on n vertices.

- For the null graph N_n we have for each vertex $u \in V(N_n)$ exactly k choices with which to color u. The coloring of each vertex is completely independent from the coloring of the others. Hence,

$$\chi_{N_n}(k) = k^n.$$

  The least value $k \in \mathbb{N}$ for which $\chi_{N_n}(k) \neq 0$ is $k = 1$, and so $\chi(N_n) = 1$.

- For the complete graph K_n let $u_1, \ldots, u_n$ be any listing of its vertices. We have k choices with which to color u_1. Having colored u_1, we have $k-1$ choices with which to color u_2. In general, after having colored the first i vertices, we have $k-i$ choices with which to color u_{i+1}, as long as $k \geq i$. Hence, for $k \geq n-1$ we have

$$\chi_{K_n}(k) = k(k-1)\cdots(k-n+1).$$

The least value $k \in \mathbb{N}$ for which $\chi_{K_n}(k) \neq 0$ is $k = n$ and so $\chi(K_n) = n$.

- Note that any proper k-vertex coloring of K_n yields a proper k-vertex coloring of any simple graph G on n vertices, which again yields a proper k-vertex coloring of N_n. Hence, for any simple graph on n vertices and any $k \in \mathbb{N}$, we have $\chi_{K_n}(k) \leq \chi_G(k) \leq \chi_{N_n}(k)$. Therefore, for any simple graph G and for any $k \in \mathbb{N}$, we have that

$$k(k-1)\cdots(k-n+1) \leq \chi_G(k) \leq k^n,$$

since these expressions are, in fact, also valid for $k < n-1$.

Just as Theorem 4.4 gives us a recursive formula to compute $\tau(G)$, the number of spanning trees for a graph G, similarly we have a recursive formula to compute $\chi_G(k)$. Recall the notion of contraction of an edge in a graph from Definition 2.34.

Theorem 9.5 Let G be a graph and $e \in E(G)$ an edge which is not a loop. For any $k \in \mathbb{N}$, we have

$$\chi_G(k) = \chi_{G-e}(k) - \chi_{G \cdot e}(k).$$

Proof If the edge e is one of the multiple edges between its endvertices, then $\chi_G(k) = \chi_{G-e}(k)$ and $\chi_{G \cdot e}(k) = 0$, since $G \cdot e$ has a loop. So, the formula holds.

Otherwise, the edge e is the only edge connecting its endvertices u and v. Consider the vertex colorings of $G-e$ with k colors. These k-colorings are split into two sets of colorings: one set $\mathcal{C}_{1;k}$ in which u and v receive the same color, and the other set $\mathcal{C}_{2;k}$ in which u and v receive different colors. There is a one-to-one correspondence between $\mathcal{C}_{1;k}$ and colorings of $G \cdot e$ using k colors. Also, there is a one-to-one correspondence between $\mathcal{C}_{2;k}$ and colorings of G using k colors. Hence, we obtain

$$\chi_{G-e}(k) = |\mathcal{C}_{1;k}| + |\mathcal{C}_{2;k}| = \chi_{G \cdot e}(k) + \chi_G(k).$$

This completes the proof. ∎

Remark 9.6 If G is a graph other than the null graph, then for any edge e which is not a loop, both $G-e$ and $G \cdot e$ have strictly fewer edges than G. Therefore, we usually write $\chi_G(k)$ in terms of $\chi_{G-e}(k)$ and $\chi_{G \cdot e}(k)$, as shown in Theorem 9.5. This is a *recursive* formula, the continued use of which must eventually stop, since G has finitely many edges.

There are occasions, though, when it will be better to use the formula $\chi_{G-e}(k) = \chi_{G \cdot e}(k) + \chi_G(k)$, as we will see later.

The chromatic number for a loopless graph depends only on *whether* two vertices are connected, not on *how many* edges are between the two vertices. Thus, when examining

$\chi_G(k)$ for a given natural number k, it suffices to consider the simple graph obtained by replacing each collection of multiple edges between a pair of endvertices with a single edge between them. Recall the notion from Definition 2.36 of a simple contraction for simple graphs. Then, from Theorem 9.5, we immediately obtain the following observation.

Observation 9.7 Let G be a simple graph and $e \in E(G)$ an edge. For any $k \in \mathbb{N}$, we have

$$\chi_G(k) = \chi_{G-e}(k) - \chi_{G/e}(k) \tag{9.1}$$

As we did for the chromatic number $\chi(G)$ of a graph G, for $\chi_G(k)$ we can henceforth assume all graphs are simple, unless otherwise stated.

We know from Example 9.4 that $\chi_{N_n}(k) = k^n$. Hence we can, by repeated use of the recursive Formula 9.1, compute $\chi_G(k)$ for any simple graph G, since at each step the number of edges is reduced by at least one. Suppose $|E(G)| = m$. Notice that when writing each step out, if we do not keep track of duplicate calculations, we need to work out close to 2^m expressions. This is because we obtain two new formulas to evaluate for each edge that we remove from the graph. So, one should keep a close watch for repetitions, especially when considering simple graphs having a specific structure.

Example 9.8 Consider the graph G shown in Figure 9.1 and how we can evaluate $\chi_G(k)$ for any $k \in \mathbb{N}$.

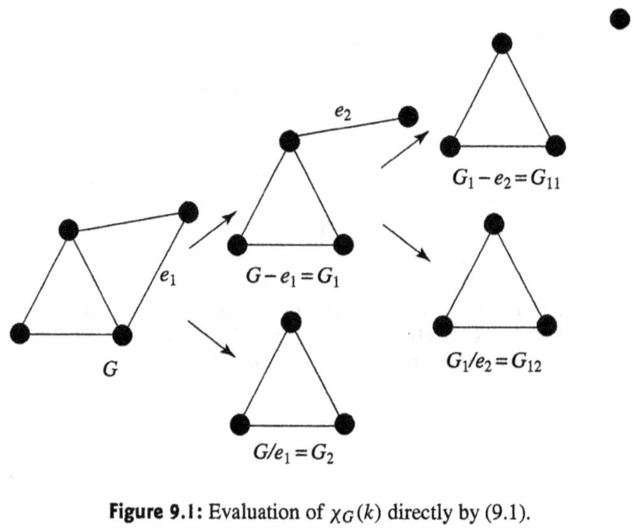

Figure 9.1: Evaluation of $\chi_G(k)$ directly by (9.1).

We obtain

$$\begin{aligned}
\chi_G(k) &= \chi_{G-e_1}(k) - \chi_{G/e_1}(k) \\
&= \chi_{G_1}(k) - \chi_{G_2}(k) \\
&= [\chi_{G_1-e_2}(k) - \chi_{G_1/e_2}(k)] - \chi_{G_2}(k) \\
&= [\chi_{G_{11}}(k) - \chi_{G_{12}}(k)] - \chi_{G_2}(k) \\
&= \chi_{G_{11}}(k) - [\chi_{G_{12}}(k) + \chi_{G_2}(k)].
\end{aligned}$$

Now, note that G_2 and G_{12} are both K_3, the complete graph on $n = 3$ vertices. Also note that G_{11} consists of the disjoint union of K_3 and K_1, a single vertex. Since the coloring of a single vertex is independent from the coloring of the rest of the graph, here we have that $\chi_{G_{11}}(k) = k \cdot \chi_{K_3}(k)$, and hence by Example 9.4 we get

$$\begin{aligned}\chi_G(k) &= \chi_{G_{11}}(k) - [\chi_{G_{12}}(k) + \chi_{G_2}(k)] \\ &= k \cdot \chi_{K_3}(k) - 2 \cdot \chi_{K_3}(k) \\ &= (k-2) \cdot \chi_{K_3}(k) \\ &= k(k-1)(k-2)^2.\end{aligned}$$

Note that here we have $\chi_G(1) = \chi_G(2) = 0$ and $\chi_G(3) = 6 \neq 0$, so $\chi(G) = 3$ is the least $k \in \mathbb{N}$ for which $\chi_G(k) \neq 0$.

The evaluation of $\chi_G(k)$ for any simple graph G and for any $k \in \mathbb{N}$ can always be done as in Example 9.8. In some other cases, however, it can be better to use the "reversed" version of (9.1)—that is to say,

$$\chi_{G-e}(k) = \chi_G(k) + \chi_{G/e}(k),$$

especially if our graph G in question is "close" to being a complete graph.

Example 9.9 Consider the graph G shown in Figure 9.1. We illustrate how to evaluate $\chi_G(k)$ for this graph for any $k \in \mathbb{N}$ in Figure 9.2.

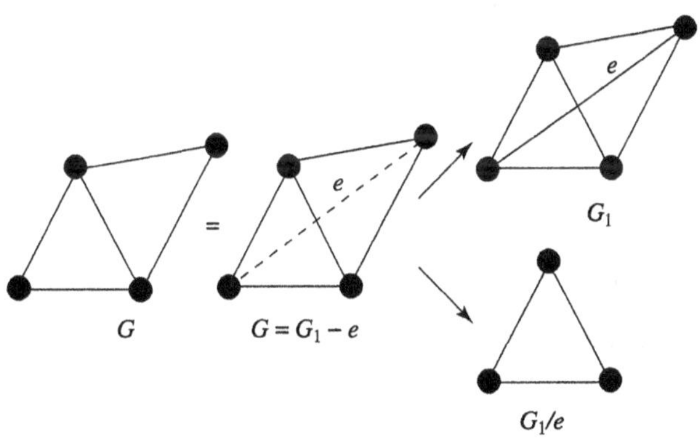

Figure 9.2: Evaluation of $\chi_G(k)$ by the reversed version of (9.1).

We obtain

$$\begin{aligned}\chi_G(k) &= \chi_{G_1-e}(k) \\ &= \chi_{G_1}(k) + \chi_{G_1/e}(k).\end{aligned}$$

Since G_1 is K_4, the complete graph on four vertices, and G_1/e is K_3, the complete graph on three vertices, here we have by Example 9.4 that

$$\begin{aligned}\chi_G(k) &= \chi_{G_1}(k) + \chi_{G_1/e}(k)\\ &= k(k-1)(k-2)(k-3) + k(k-1)(k-2)\\ &= k(k-1)(k-2)[(k-3)+1]\\ &= k(k-1)(k-2)^2,\end{aligned}$$

which is consistent with the result from Example 9.8.

In what we have seen so far, $\chi_G(k)$ is a polynomial expression in terms of k, for each $k \in \mathbb{N}$—that is, given by a polynomial in k which is *also* valid for $k < \chi(G)$. This is indeed always the case, for every simple graph G, as a direct consequence of Observation 9.7.

Theorem 9.10 For a simple graph G and $k \in \mathbb{N}$, the number $\chi_G(k)$ of proper k-vertex colorings of G is given by a polynomial expression in k.

Proof Since $\chi_{N_n}(k) = k^n$ is a polynomial in k for any $k, n \in \mathbb{N}$, the theorem follows from Observation 9.7 by induction on $|E(G)|$. ■

Theorem 9.10 gives rise to the following definition.

Definition 9.11 Let G be a simple graph. The *chromatic polynomial* $\chi_G(t)$ in the indeterminate t is the unique polynomial determined by $\chi_G(k)$ for all $k \in \mathbb{N}$.

Remark 9.12 The chromatic polynomial for the complete graph K_n, which from Example 9.4 is given by $t(t-1)\cdots(t-n+1)$, is sometimes called the *falling factorial* in t. It is denoted by either $(t)_n$ or $t^{\underline{n}}$. That is,

$$(t)_n = t^{\underline{n}} = t(t-1)\cdots(t-n+1).$$

9.2 Basic Properties of the Chromatic Polynomial

We start by presenting the following Theorem 9.13, in which we state the most commonly used basic properties of the chromatic polynomial of a graph. Its proof is easily obtained from Observation 9.7 by induction on $|E(G)|$, the number of edges of G. The result is left as Exercise 9.7.

Theorem 9.13 For a simple graph G, the chromatic polynomial $\chi_G(t)$ satisfies the following properties:

1. $\chi(G) = \min(\{k \in \mathbb{N} : \chi_G(k) \neq 0\})$.
2. If $\chi(G) = k$, then $\chi_G(t)$ is divisible by $(t)_k$, the falling factorial polynomial in t of degree k.
3. The degree of $\chi_G(t)$ is equal to the number of vertices of G.
4. The coefficient for the leading term of $\chi_G(t)$ is one.
5. The coefficient for the second-highest term of $\chi_G(t)$ is $-|E(G)|$.
6. The coefficient for the last term, the constant term of $\chi_G(t)$, is zero.

7. The least degree of a term of $\chi_G(t)$ is the number of components of G.
8. The coefficients for the powers of t in $\chi_G(t)$ alternate in sign.
9. If G has at least one edge, then $\chi_G(1) = 0$; that is, the sum of the coefficients of $\chi_G(t)$ is zero.

From Theorem 9.13 we can deduce the following corollary, whose proof we leave as Exercise 9.8.

Corollary 9.14 For a simple graph G on n vertices, we have the following:

$$G = N_n \Leftrightarrow \chi_G(t) = t^n,$$
$$G = K_n \Leftrightarrow \chi_G(t) = (t)_n.$$

When determining the chromatic polynomial for a particular graph, it is worthwhile to keep an eye open for various methods to simplify the task. If a simple graph G has K_m as a subgraph, then $\chi_G(k) = 0$ for all $k \in \{0, \ldots, m-1\}$. Therefore, $\chi_G(t)$ is divisible by the falling factorial $(t)_m$ and we can write $\chi_G(t)$ in the form $\chi_G(t) = (t)_m\alpha(t)$, where $\alpha(t)$ is some polynomial in t. If $k \geq \chi(G)$, then for each *fixed* proper k-vertex coloring of the subgraph K_m in G, we have that $\alpha(k)$ is the number of proper k-vertex colorings of G with the given coloring of K_m. With this in mind, we obtain the following useful "folklore" lemma.

Lemma 9.15 Suppose G is a simple graph such that $G = G' \cup G''$ and $G' \cap G'' = K_m$ for some $m \in \mathbb{N}$. Then we have

$$\chi_G(t) = \frac{\chi_{G'}(t) \cdot \chi_{G''}(t)}{(t)_m}.$$

Proof Let $k \geq \chi(G)$. For each of the $(k)_m$ proper k-vertex colorings of $G' \cap G'' = K_m$, there are $\alpha'(k)$ ways to color properly the rest of the vertices of G', and there are $\alpha''(k)$ ways to color properly the rest of the vertices of G''. Since the vertex colorings of $G' \setminus G''$ and $G'' \setminus G'$ are independent, we obtain that $\chi_G(k) = \alpha'(k) \cdot (k)_m \cdot \alpha''(k)$, which proves the lemma. (See Figure 9.3.) ■

As in Observation 8.7 and Theorem 8.9 for the chromatic number, we now obtain the following corollary for the chromatic polynomial.

Corollary 9.16 Let G be a simple graph.

1. If G has components $C_1, \ldots, C_h$, then we have

$$\chi_G(t) = \chi_{C_1}(t) \cdots \chi_{C_h}(t).$$

2. If G is connected and has blocks $B_1, \ldots, B_k$, then we have

$$\chi_G(t) = \frac{1}{t^{k-1}} \chi_{B_1}(t) \cdots \chi_{B_k}(t).$$

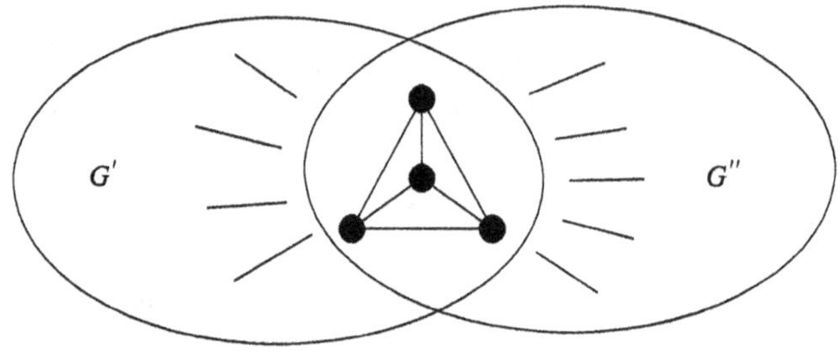

Figure 9.3: We can compute $\chi_{G' \cup G''}(t)$ in terms of $\chi_{G'}(t)$ and $\chi_{G''}(t)$ if $G' \cap G''$ is complete.

Proof The first part is clear, since the vertex colorings of each component are independent from those of any other component.

For the second part we proceed by induction on the number k of blocks of G. For $k = 1$ the statement is a tautology. Assume the statement is true for $k-1$ and that B_k corresponds to a leaf in the block-cutpoint graph BC(G). Let $H = B_1 \cup \cdots \cup B_{k-1}$. We have $G = H \cup B_k$, and $H \cap B_k$ consists of a single cut vertex. Hence, by Lemma 9.15 and then by the induction hypothesis, we obtain

$$\begin{aligned}\chi_G(t) &= \frac{\chi_H(t) \cdot \chi_{B_k}(t)}{t} \\ &= \frac{1}{t^{k-2}} \chi_{B_1}(t) \cdots \chi_{B_{k-1}}(t) \cdot \frac{\chi_{B_k}(t)}{t} \\ &= \frac{1}{t^{k-1}} \chi_{B_1}(t) \cdots \chi_{B_k}(t),\end{aligned}$$

which completes the proof. ■

Corollary 9.17 For a simple graph T, the following statements are equivalent:

1. T is a tree on n vertices.
2. $\chi_T(t) = t(t-1)^{n-1}$.

Proof Assume first that T is a tree on n vertices. Since T has exactly $n-1$ blocks, where each block consists of an edge together with its endvertices, we obtain by Corollary 9.16 that

$$\chi_T(t) = \frac{1}{t^{n-2}}(t(t-1))^{n-1} = t(t-1)^{n-1}.$$

Assume now that $\chi_T(t) = t(t-1)^{n-1}$. If T has $k \geq 2$ components, then t^k would divide $\chi_T(t)$, which is not the case here. Hence, T is connected and has n vertices. Since the coefficient for t^{n-1} is $n-1$, we have by Theorem 9.13 that T has precisely $n-1$ edges. By Theorem 3.7, T must be a tree. ■

Note that Lemma 9.15 can additionally be used to simplify the computation of the chromatic polynomials for graphs that contain vertex cuts consisting of complete subgraphs.

9.3 Interval and Intersection Graphs

As we saw earlier, vertex-coloring a graph G can be a computationally hard problem. To compute the chromatic number $\chi(G)$ is one task, but to actually vertex-color the graph G using $\chi(G)$ available colors is yet another. Increasing the number of available colors (perhaps just by one!) can sometimes allow efficient coloring of G. However, in a practical setting, this is not always possible, since we might have only a fixed number of colors available. Some classes of graphs are better suited for vertex colorings than others. Luckily, such classes often appear in practical settings. The class of *interval graphs* is an example of such an important class of graphs; it appears in many time-dependent scheduling problems. The interval graphs are a special subclass of *chordal graphs*, a class of graphs which are themselves well suited for optimal vertex coloring. In this section, we will define and discuss interval graphs as well as other *intersection graphs*. This will serve as a motivation for defining chordal graphs in §9.4. But first we need to motivate the study of chordal graphs. Note that a graph G with K_m as a subgraph must have $\chi(G) \geq m$.

Definition 9.18 Let G be a simple graph.

1. A *clique* in G is a subgraph of G that is complete. The *clique number* of G, denoted by $\omega(G)$, is the number of vertices in a maximum-sized clique of G.
2. An *independent set* in G is a subset of $V(G)$ that induces the null graph. The *independence number* of G, denoted by $\alpha(G)$, is the number of vertices in a maximum-sized independent set of G.

Remark 9.19 From Definition 9.18 we can make the following observations, where $\overline{G}$ denotes the complement of the simple graph G.

- We have $\omega(G) = \alpha(\overline{G})$ for any simple graph G.
- We have $\alpha(G) = \omega(\overline{G})$ for any simple graph G.
- We have $\chi(G) \geq \omega(G)$ for any simple graph G.

If G is a graph and G' is an induced subgraph of G, then by the last comment of the above remark we have $\chi(G') \geq \omega(G')$. The graphs where we have *equality* instead of a mere inequality have a special name.

Definition 9.20 A simple graph G is called *perfect* if $\chi(G') = \omega(G')$ for every induced subgraph G' of G.

Unfortunately, not all simple graphs are perfect. There are graphs where the chromatic number can be strictly greater than the clique number. We illustrate this with an example.

Example 9.21 For any odd $n \geq 5$, the cycle C_n on n vertices has no triangles. Hence, we have

$$\chi(G) = 3 > 2 = \omega(G).$$

In fact, as we now show, we can push this example further in order to make the difference between $\chi(G)$ and $\omega(G)$ as large as we want. Cyclically order the vertices $\{u_1, \ldots, u_n\}$

of the cycle C_n. Let $m \in \mathbb{N}$ and let $V_1 \cup \cdots \cup V_n$ be a disjoint union of n sets of vertices, where each set V_i has m vertices, constituting altogether mn vertices in the union. We form a graph G by first making each V_i into a clique (that is, by connecting all pairs of vertices within each V_i). Second, for distinct i and j we connect $u \in V_i$ and $v \in V_j$ if, and only if, u_i and u_j are connected in the cycle C_n. In this manner we obtain an n-"cycle," where each "vertex" is a clique of size m. An example where $n = 5$ and $m = 3$ is shown in Figure 9.4.

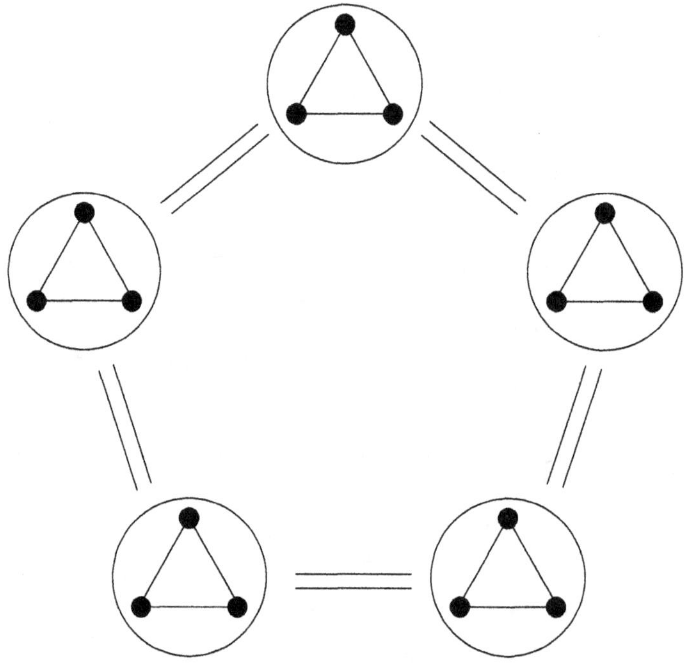

Figure 9.4: A 5-"cycle" on cliques of size three. A pair of parallel lines indicates all the possible edges between the 3-cliques—that is, a total of nine edges.

Clearly, each pair V_i and V_{i+1} forms a clique in G. If now three vertices are in distinct sets V_i, then at least two of them are not adjacent. Hence, it is easy to see that the clique number of G is $2m$.

To see that the chromatic number of G is given by $\chi(G) = \lceil \frac{2mn}{n-1} \rceil$, we first argue the lower bound.

Since n is odd, we can write $n = 2k + 1$ for some $k \in \mathbb{N}$. For a given proper vertex coloring of G, vertices that have the same color must be in different nonadjacent parts $V_1, \ldots, V_n$. Suppose there are $k + 1$ vertices having the same color. By the Pigeon-Hole Principle there must be a pair of parts containing vertices of the same color and whose corresponding vertices are adjacent in C_n. By definition of the n-"cycle" the vertices in the parts must also be adjacent. In this case we do not have a proper coloring. Hence, for each color, there are at most k vertices having that color. From this we obtain

$$\chi(G) \geq \left\lceil \frac{mn}{k} \right\rceil = \left\lceil \frac{2mn}{n-1} \right\rceil.$$

To see that $q = \lceil \frac{2mn}{n-1} \rceil$ colors suffice to vertex color G properly, we first list the mn vertices of G in such a way that $u \in V_i$ appears before $v \in V_j$ if $i < j$. There are many such listings. (See Exercise 9.11.) We need only one of them. We now consider two cases.

CASE ONE: If mn is divisible by k, then $mn = qk$, so we can cyclically color the mn vertices of G by the colors $1, 2, \ldots, q$ in a repeating cyclic manner:

$$1, 2, \ldots, q, \; 1, 2, \ldots, q, \; \ldots, \; 1, 2, \ldots, q \quad (k \text{ times}).$$

If two vertices u and v are adjacent, then either

1. u and v are contained in the same part V_i for some i, or
2. $u \in V_i$ and $v \in V_j$, where the corresponding vertices u_i and v_j are adjacent in cycle C_n.

In either case there are at most $2m - 2 \leq q - 3$ vertices between u and v in the listing. Hence, we have a proper vertex coloring of G in this case.

CASE TWO: If mn is not divisible by k, then $q = \lfloor \frac{2mn}{n-1} \rfloor + 1$. Letting $a = \lfloor \frac{2mn}{n-1} \rfloor$, we can write $mn = ka + b$, where $b \in \{1, 2, \ldots, k-1\}$. Therefore, we have

$$mn = b \cdot q + (k - b) \cdot (q - 1). \tag{9.2}$$

Hence, using the set $\{1, 2, \ldots, q\}$ of colors, we can by (9.2) color the vertices of G cyclically, first repeating the cycle $1, 2, \ldots, q$ of colors b times, and then repeating the cycle $1, 2, \ldots, q - 1$ of colors $k - b$ times, thereby completing the full cycle of all of the mn vertices of G in a cyclic manner. Note that there are at least $q - 2$ vertices between two vertices of the same color in the listing of $V(G)$. Since, however, two adjacent vertices in G have at most $2m - 2 \leq q - 3$ vertices between them in the listing, they cannot be of the same color. Hence, we have a proper vertex coloring of G in this case also.

Therefore, we have $\chi(G) = \lceil \frac{2mn}{n-1} \rceil$, and so

$$\chi(G) - \omega(G) = \left\lceil \frac{2mn}{n-1} \right\rceil - 2m \geq \frac{2m}{n-1}.$$

This quantity can be arbitrarily large for any fixed $n \geq 5$ when m tends toward infinity.

Fortunately, however, there is a large class of graphs for which the chromatic number and the clique number are the same, unlike the graphs in Example 9.21. Moreover, the structure of these graphs will often provide an efficient greedy coloring algorithm. The next example illustrates an application where such graphs arise naturally.

Example 9.22 Consider a project that requires a certain number of agents in order to be completed. The agents in this example can represent people in a company, processors in a computer, or genes in a biological network system carrying and passing genetic information to and from each

Table 9.1: Starting and ending times for ten tasks.

Task Number	t_{i1}	t_{i2}
T_1	1	4
T_2	2	6
T_3	3	7
T_4	5	10
T_5	8	12
T_6	9	15
T_7	11	18
T_8	13	16
T_9	14	20
T_{10}	17	19

other. Assume that the project in question involves ten tasks or subprojects that are labeled $T_1, T_2, \ldots, T_{10}$. These tasks depend on one another, some tasks need to be completed before other tasks can be started, and some tasks are more time consuming than others. We assume here that each task T_i takes place from the time t_{i1} to t_{i2} measured in some given timescale, as shown in Table 9.1.

If each agent is able to work on only one task at a time, what is the minimum number of agents needed in order to complete this project? Needless to say, an agent can work on one task, complete it, and move on to another task as long as the tasks do not overlap in time.

From Table 9.1, we form the corresponding *interval graph* G on ten vertices $V(G) = \{u_1, \ldots, u_{10}\}$ corresponding to the tasks $T_1, \ldots, T_{10}$ of the project. In the graph we connect two vertices u_i and u_j if, and only if, the corresponding tasks T_i and T_j overlap in time. Figure 9.5 provides a geometric interpretation of the tasks and the corresponding interval graph G.

Any assignment of the tasks to agents, in such a way that no agent operates on two tasks at the same time, yields a proper vertex coloring of G, where each agent corresponds to a unique color. Vice versa, any proper coloring of G yields a legitimate assignment of tasks to the agents with the corresponding color. Hence, the minimum number of agents that can possibly get the project done, under the constraints imposed, is given by $\chi(G)$. Looking at the geometric interpretation of the tasks $T_1, \ldots, T_{10}$, we can obtain a proper $\chi(G)$-coloring for this interval graph G (and for *any* interval graph for that matter!) in the following greedy fashion.

Start by listing the vertices $V(G) = \{u_1, \ldots, u_{10}\}$ according to when the corresponding tasks start. (We have done that already here, as shown in Figure 9.5.) For each $i \in \{1, \ldots, 10\}$ color the vertex u_i with the least color that has not been used on previous neighbors of u_i among $u_1, \ldots, u_{i-1}$. In this way we see that the number of colors used never exceeds the "thickness" of the intervals shown geometrically in Figure 9.5. This *thickness*, the maximum number of interval layers, is equal to four in this case. This is precisely the clique number of the graph G, so $\omega(G) = 4$ in this case. In fact, we obtain the following vertex coloring, $c : V(G) \to \{1, 2, 3, 4\}$, by employing this greedy procedure:

$$c(u_1) = 1,\ c(u_2) = 2,\ c(u_3) = 3,\ c(u_4) = 1,\ c(u_5) = 2,$$
$$c(u_6) = 3,\ c(u_7) = 1,\ c(u_8) = 2,\ c(u_9) = 4,\ c(u_{10}) = 2.$$

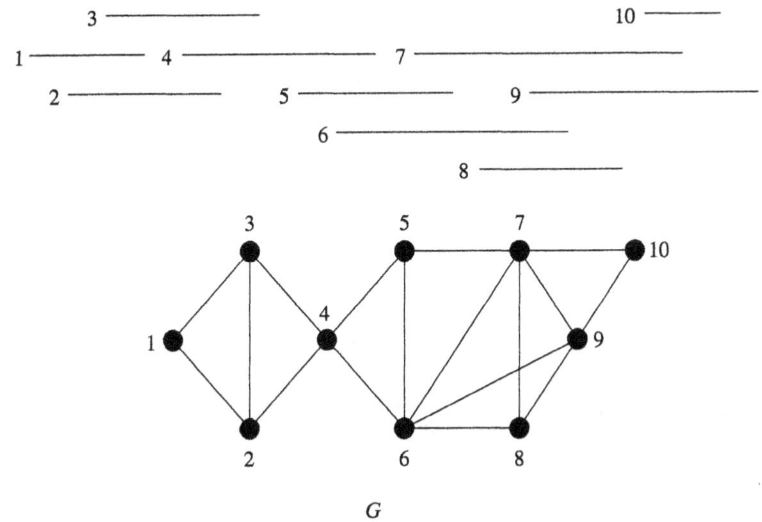

Figure 9.5: An interval graph corresponding to ten tasks.

Hence, our greedy coloring procedure properly vertex-colored G with $\omega(G)$ colors, a priori the least possible number of colors, and hence optimal. In particular, we have that

$$\chi(G) = \omega(G),$$

which, in fact, holds for every interval graph. Each interval graph can be properly vertex colored using this greedy coloring algorithm. The algorithm yields an optimal coloring.

Interval graphs are an important class of graphs, which in addition are well suited for various vertex coloring problems. As we saw in Example 9.22, the interval graph there was defined by a collection of intervals on the real line. As geometrical objects they either had a nonempty intersection as sets or not. The next definition is useful in formalizing these ideas.

Definition 9.23 Let $\widetilde{S} = (S_1, \ldots, S_n)$ be a list of subsets of a given fixed set or universe U. The *intersection graph* G of $\widetilde{S}$ is the simple graph having vertex set $V(G) = \{u_1, \ldots, u_n\}$, where each vertex u_i corresponds to the subset $S_i \subseteq U$, and edge set

$$E(G) = \{\{u_i, u_j\} : S_i \cap S_j \neq \varnothing\}.$$

A graph G is called an *intersection graph* if G is an intersection graph of some finite list $\widetilde{S}$ of subsets of some set U. In this case we say that G is *represented* by $\widetilde{S}$.

If all the sets $S_1, \ldots, S_n$ are distinct, then without ambiguity we can talk about the intersection graph of the *set* $\{S_1, \ldots, S_n\}$ rather than the ordered list $\widetilde{S}$. Also, in this case we say that G is *represented* by the set.

Many specific types of intersection graphs have special names in the graph theory literature. We describe a few of them below.

Interval graphs: If $\widetilde{S} = (I_1, \dots, I_n)$ is a list of closed intervals

$$I_i = [a_i; b_i] = \{x \in \mathbb{R} : a_i \leq x \leq b_i\},$$

where $a_i \leq b_i$ are real numbers, then the intersection graph G of $\widetilde{S}$ is called an *interval graph*.

If, in addition, no interval in the list $\widetilde{S}$ properly contains another interval in the list, then G is called a *proper interval graph*.

If each interval I_i of $\widetilde{S}$ has a unit length (that is, $b_i - a_i = 1$), then G is called a *unit interval graph*. Clearly each unit interval graph is a proper interval graph.

Many very special kinds of interval graphs are common in computer science. An example are the *threshold graphs*, which have been used in the study of which processes in computer programs can be done in parallel, to shorten time, and which cannot. A graph G is a threshold graph if G is represented by a list of intervals $(I_1, \dots, I_a, J_1, \dots, J_b)$ that satisfy the following conditions:

1. We have $I_1 \subseteq \cdots \subseteq I_a$, so the set $\{I_i : 1 \leq i \leq a\}$ represents a clique in G.
2. We have $J_i \cap J_j = \varnothing$ for $1 \leq i < j \leq b$, so the set $\{J_i : 1 \leq i \leq b\}$ represents an independent set in G.
3. For each i there is a unique j such that $J_i \subseteq I_{j+1} \setminus I_j$.

An example of a threshold graph is shown in Figure 9.6. There the short intervals represent the J_j's and the longer ones the I_i's. All threshold graphs have a representation like the one shown there. For some other presentations and properties of threshold graphs we refer to Merris [39].

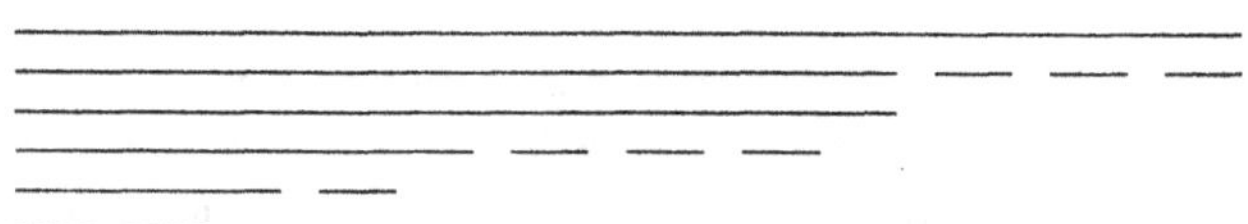

Figure 9.6: The long intervals represent the I_i's and the short ones represent the J_j's in a threshold graph.

Circular arc graphs: If $\widetilde{S} = (A_1, \dots, A_n)$ is a list of closed arcs on the unit circle

$$A_i = \{(\cos\theta, \sin\theta) \in \mathbb{R}^2 : a_i \leq \theta \leq b_i\},$$

where $a_i \leq b_i$ are real numbers, then the intersection graph G of $\widetilde{S}$ is a *circular arc graph*. An example of a circular arc graph, both the geometric configuration as arcs and also the corresponding graph, is given in Figure 9.7.

If, in addition, no arc in $\widetilde{S}$ properly contains another arc in the list, then G is called a *proper circular arc graph*. In fact, the circular arc representation on the left in Figure 9.7 is not proper, since arc 2 is contained in arc 3. However, there does exists a different circular arc representation of the graph on the right in Figure 9.7 as a proper circular arc graph. Hence, the graph on the right is a proper circular arc graph, although we cannot conclude that from Figure 9.7.

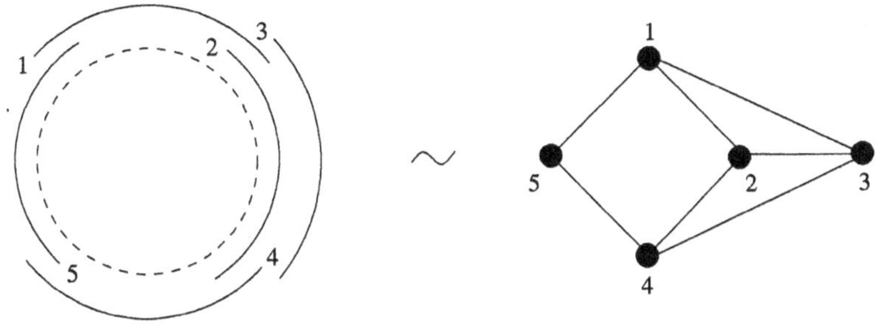

Figure 9.7: A circular arc graph. Arcs correspond to vertices.

Also, if all of the arcs have the same length (that is, $b_i - a_i$ is a constant), then G is a *unit circular arc graph*. Note that we could have demanded that each arc have unit arc length, but then we would have needed to consider arcs of circles of various radii. Again, each unit circular arc graph is a proper circular graph.

Disk graphs: If $\widetilde{S} = \{D_1, \ldots, D_n\}$ is a list of closed circular disks

$$D_i = \{(x, y) \in \mathbb{R}^2 : (x - a_i)^2 + (y - b_i)^2 \leq r_i\},$$

where $a_i, b_i, r_i \in \mathbb{R}$ and $r_i > 0$, then the intersection graph G of $\widetilde{S}$ is called a *disk graph*. An example of a disk graph, the geometric representation as disks in the Euclidean plane, and the corresponding graph is given in Figure 9.8.

If no disk properly contains another disk in the list, then G is called a *proper disk graph*. Note that the disk graph representation of the graph in Figure 9.8 is not proper, since disk 4 is contained in disk 5. However, the graph on the right in Figure 9.8 is a proper disk graph (see Exercise 9.16).

Also, if all of the disks have the same radius of one, $r_i = 1$ for all i, then G is a *unit disk graph*. Note that if G is an intersection graph of $\widetilde{S}$, a list of disks where all the disks have the same radius, then G is by scaling of the disks indeed a unit disk graph.

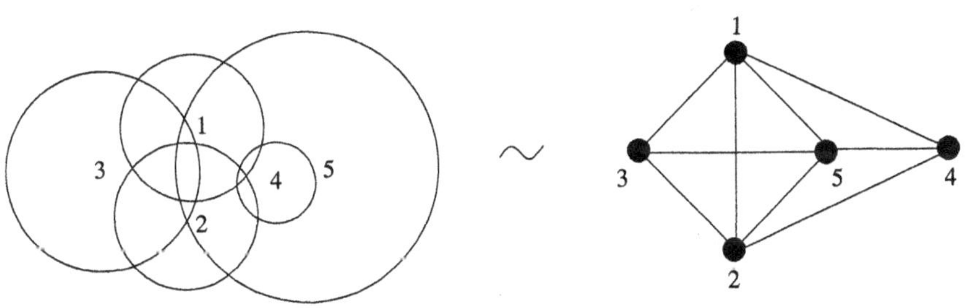

Figure 9.8: A disk graph. Each disk corresponds to a vertex.

Coin graphs: If $\widetilde{S}$ is a list of "coins"—that is, closed disks that only possibly intersect, or "touch," in their closed bounding circles

$$\sqrt{(a_i - a_j)^2 + (b_i - b_j)^2} \geq r_i + r_j$$

for all i and j—then G is a *coin graph*. Geometrically, we can represent every coin graph on n vertices by placing n coins of various sizes on a table, and connecting two vertices if, and only if, corresponding coins touch each other. An example of a coin graph, the geometric representation as "coins" in the Euclidean plane, and the corresponding graph is given in Figure 9.9. Note that there the circles have distinct radii, but there is another representation of the graph on the right in Figure 9.9 as a coin graph, where each circle has the same radius.

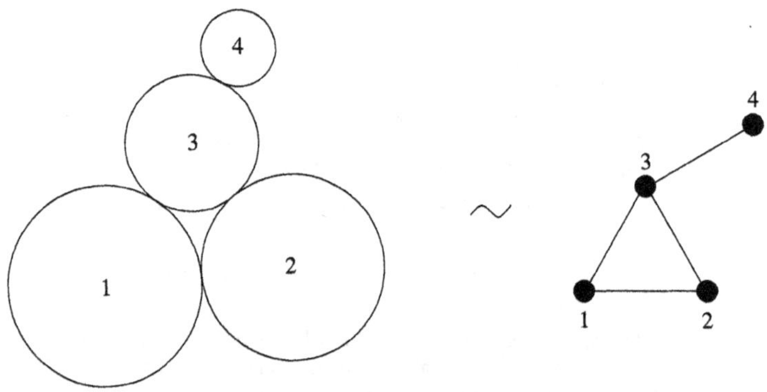

Figure 9.9: A coin graph. The coins (that is, the nonoverlapping circles) correspond to the vertices.

If every coin has the same radius of one, then G is a *unit coin graph*. By scaling, G is a unit coin graph if, and only if, it is an intersection graph of a list $\widetilde{S}$ of coins of same size. Hence, G is a unit coin graph if, and only if, we can represent G by a collection of coins of the same type on a table—be it a collection of pennies, nickels, dimes, or quarters.

If G is a coin graph, represented by a collection of coins, then we can obtain a planar embedding of G by letting the vertices be the center points of the coins and letting the edges be the straight line segments between the centers of touching coins. Therefore, every coin graph is a planar graph. In fact, we have the following theorem.

Theorem 9.24 **Thurston's**

For a simple graph G the following are equivalent:

1. G is a planar graph.
2. G is a coin graph.

The proof, which we omit, can be found in Thurston [53], written as course notes in 1978 at Princeton. The reading of these notes demands active participation by the reader. Numerous "details" are left to the reader to verify, many of them quite involved.

Thurston has since started to revise these notes, to write a more detailed version of them, and to make them more accessible to read. The first volume of these revisions, from 1997, appears in Thurston [54].

Remark 9.25 (i) Every simple graph G is an intersection graph of some list $\widetilde{S}$ of subsets of some fixed set U: If G has n vertices, let $V(G) = \{u_1, \ldots, u_n\}$ be a labeling of the vertices of G. Let $U = E(G)$ be the fixed underlying set or universe. For each $i \in \{1, \ldots, n\}$, define $S_i \subseteq U$ by

$$S_i = \{e \in U : u_i \in e\}.$$

That is, S_i is the set of all edges that have u_i as an endvertex. In this case G is an intersection graph of the list $\widetilde{S} = (S_1, \ldots, S_n)$ of subsets of U. In Exercise 9.14 we ask the reader to verify that this is indeed the case.

Hence, we see that intersection graphs are interesting only if the underlying universe U and its subsets possess an interesting mathematical structure.

(ii) A coin graph is an example of a *contact graph*, known also as a *tangency graph*, which is an intersection graph where the vertices can be represented by well-defined geometric bodies in some Euclidean space, like the plane $\mathbb{R}^2$ or the three-dimensional space $\mathbb{R}^3$, and where these bodies are either disjoint or intersect only possibly at their boundaries—hence the words "contact" or "tangency" for the bodies and "touch" for the coin graphs.

Needless to say, some intersection graphs appear more often in the scientific literature than others, and there are many more types than we have mentioned. It is, however, fair to state that interval graphs are one of the most important classes of intersection graphs that we have mentioned, since they appear most frequently in various applications. They were first introduced in connection with biology by Lekkerkerker and Boland in 1962 [35].

When dealing with and manipulating the actual geometric subsets of $\mathbb{R}^d$ of a list $\widetilde{S}$ of an intersection graph G, often we need to use the topological structure of $\mathbb{R}$: Consider an interval graph G of a list $\widetilde{S} = \{I_1, \ldots, I_n\}$ of real closed intervals. Let

$$x_1 < x_2 < \cdots < x_m \tag{9.3}$$

be the ordered list of *all* the distinct end points of the intervals of $\widetilde{S}$. If $m < 2n$, then there are two intervals with a common end point, say x_l. We can now find an $\epsilon > 0$, such that if we replace the x_l end point in one of these two intervals with either $x_l - \epsilon$ or $x_l + \epsilon$, then the relative order of $x_l - \epsilon$, x_l, and $x_l + \epsilon$, with respect to the other x_i's in (9.3), remains unchanged. Thus, we can replace either of the two identical end points of the intervals by $x_l - \epsilon$ or $x_l + \epsilon$ without changing the interval graph G.

Observation 9.26 Every interval graph G is an intersection graph of a list $\widetilde{S}$ of intervals, where all the end points are distinct.

By scaling and stretching the intervals of the list $\widetilde{S}$, we may further assume that all of the end points are positive integers. Hence, if $\widetilde{S}$ is a list of n intervals, we may assume that all of the end points are distinct and are among the integers $\{1, 2, \ldots, 2n\}$. Note this was indeed the case in Example 9.22.

Corollary 9.27 Every interval graph on n vertices is an intersection graph of a set of n distinct subpaths of P_{2n}, the path on $2n$ vertices.

We note that every path is, in particular, a tree. So every interval graph on n vertices is an intersection graph of distinct subtrees of a given tree. We will have more to say about this in the next section.

9.4 Chordal Graphs

In this section we will define chordal graphs and give a characterization that provides an efficient and optimal greedy coloring algorithm. In some sense, chordal graphs form precisely the largest class of intersection graphs that can be vertex colored optimally in a greedy fashion. As we saw in Example 9.22, the reason our greedy coloring algorithm uses precisely the a priori optimal number of $\omega(G)$ colors to color the vertices of G properly is the existence of an ordering $u_1, \ldots, u_n$ of the vertices of G such that the neighbors of each u_i among the previous ones $u_1, \ldots, u_{i-1}$ form a clique in G. By ordering the intervals of a list $\tilde{S}$ of a general interval graph by their left end points, then all of the previous neighbors of given interval $I_i = [a_i; b_i]$ have left end points less than a_i and right end points greater than a_i. Hence, all these intervals must in particular form a clique, since they all contain the real number a_i. We make a formal definition to capture these ideas.

Definition 9.28 Let G be a simple graph.

1. A vertex u is called a *simplicial* vertex of G if $N(u) = N_G(u)$ induces a clique in G. That is, the induced subgraph $G[N(u)]$ is a complete graph.
2. Assume $n = |V(G)|$. An ordering $V(G) = \{u_1, \ldots, u_n\}$ of the vertices such that for each $i \in \{2, \ldots, n\}$, the set $N(u_i) \cap \{u_1, \ldots, u_{i-1}\}$ of previous neighbors induces a clique in G, is called a *simplicial elimination ordering* of the vertices $V(G)$.

A simplicial elimination ordering of the vertices of a graph is sometimes also called a *perfect elimination ordering* in the graph theory literature.

We have seen that every interval graph G has a simplicial elimination ordering of its vertices and can therefore be properly vertex colored by at most $\omega(G)$ colors. This is optimal. Many other classes of graphs also have simplicial elimination orderings of their sets of vertices and can therefore also be properly vertex colored by an optimal clique number of colors. We present a couple of examples.

Example 9.29 Every tree T on n vertices has a simplicial eliminating ordering $V(T) = \{u_1, \ldots, u_n\}$ that we obtain in the following way: Let i run through the set $\{0, \ldots, n-1\}$ starting with $i = 0$, and let u_{n-i} be a vertex of degree one (that is, a leaf) in the tree $T - \{u_{n-i+1}, \ldots, u_n\}$. Here the set of vertices removed from T is empty if $n - i + 1 > n$—that is, when $i = 0$. Also, recall that when we remove a leaf from a tree, the resulting graph is a tree. In this case u_1 has no previous neighbors, and each u_i, where $i \in \{2, \ldots, n\}$, has exactly one previous neighbor, a clique consisting of one vertex. Hence, every tree T has a simplicial elimination ordering of its vertices.

Example 9.30 Let G be a *maximal outerplanar graph* on n vertices—that is, an outerplanar graph with a planar embedding with exactly one region bounded by n edges, and all other regions bounded by three edges. A maximal outerplanar graph on $n = 8$ vertices is shown in Figure 9.10.

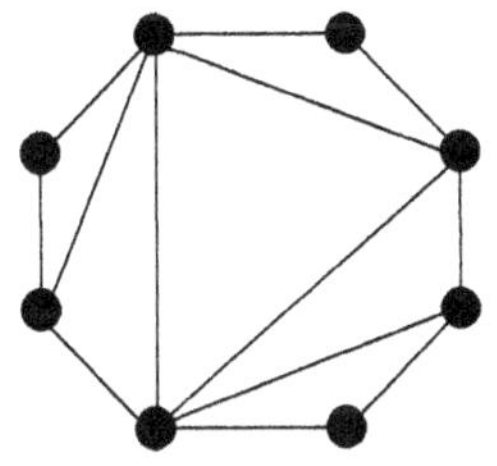

Figure 9.10: A maximal outerplanar graph on eight vertices.

By Euler's Formula one can deduce that G has precisely $2n-3$ edges. In fact, an outerplanar graph on n vertices is maximal if, and only if, it has $2n-3$ edges. (See Exercise 9.18.) One can show that each maximal outerplanar graph on three or more vertices has a vertex of degree two. We ask the reader to verify this in Exercises 9.19 and 9.20. We proceed as in Example 9.29 and establish a simplicial elimination ordering $V(G) = \{u_1, \dots, u_n\}$ of the vertices of G in the following way: Let i run through the set $\{0, \dots, n-1\}$ starting with $i = 0$, and let u_{n-i} be a vertex of degree two in the maximal outerplanar graph $G - \{u_{n-i+1}, \dots, u_n\}$. By Exercise 9.20, the removal of a degree-two vertex from a maximal outerplanar graph results in a maximal outerplanar graph. Here u_1 has no previous neighbors, u_2 has one previous neighbor, and each u_i where $i \geq 3$ has exactly two previous neighbors, which by the maximality of G are adjacent themselves, and hence form a clique K_2 consisting of two vertices. So, every maximal outerplanar graph G has a simplicial elimination ordering of its vertices.

We now present an algorithm to assign colors to the vertices of a simple graph G optimally, provided that we have a simplicial elimination ordering $V(G) = \{u_1, \dots, u_n\}$ of the vertices. The algorithm is greedy and nearly identical to ALGORITHM GREEDY COLORING. It is worth comparing the two algorithms.

ALGORITHM SIMPLICIAL COLORING assigns colors to the vertices of a graph G. For each $i \in \{1, \dots, n\}$, we let $N_i \subseteq V(G)$ denote the set of neighbors of u_i among the previous vertices $\{u_1, \dots, u_{i-1}\}$. Note that $N_1 = \varnothing$.

ALGORITHM SIMPLICIAL COLORING

`Input:` A list $(N_1, \dots, N_n)$ of lower-numbered neighbors of some simplicial elimination ordering $V(G) = \{u_1, \dots, u_n\}$ of G.

`Output:` A list $S = (c_1, \dots, c_n)$ of colors, where c_i is the color assigned to vertex u_i.

begin
    $m = \max(\{|N_1| + 1, \dots, |N_n| + 1\})$;
    $S_0 = ()$; // initial empty list of colors
    for $i = 1$ **to** n **do** {
        $C_i = \{c_j : u_j \in N_i\}$;
        $c_i = \min(\{1, \dots, m\} \setminus C_i)$;
        $S_i =$ append c_i to the $(i-1)$-tuple S_{i-1};
        }
    output S_n;
end.

Here ALGORITHM SIMPLICIAL COLORING yields a proper vertex coloring of G using at most m colors. Note that for each $i \in \{1, \ldots, n\}$, we have $|C_i| \leq |N_i| \leq m - 1$, and thus the color c_i is well defined for each i. ALGORITHM SIMPLICIAL COLORING uses an optimal number of colors for every graph with a simplicial elimination ordering of its vertices.

Observation 9.31 Let G be a simple graph and $V(G) = \{u_1, \ldots, u_n\}$ a simplicial elimination ordering of its vertices. ALGORITHM SIMPLICIAL COLORING yields a proper vertex coloring of G using the optimal number of $\omega(G)$ colors.

Proof If $V(G) = \{u_1, \ldots, u_n\}$ is a simplicial elimination ordering of the vertices of G, then clearly $|N_i| + 1 \leq \omega(G)$ for each i, and hence $m \leq \omega(G)$.

On the other hand, if i is the highest index of a vertex in a clique of size $k = \omega(G)$, then all the $k - 1$ neighbors of u are previous neighbors. So, $N_i = k - 1$; and hence, we get $m \geq \omega(G)$. Therefore, $m = \omega(G)$. ∎

As we saw in Figure 9.10, the maximal outerplanar graph on eight vertices is obtained by adding five internal edges, or chords, to the cycle graph C_8 on eight vertices. In other words, we have split the internal region of C_8, which is bounded by eight edges, into six triangles. A general maximal outerplanar graph on n vertices is also formed in this way from the cycle graph C_n on n vertices by adding chords and thereby splitting up the internal region bounded by n edges into $n - 2$ triangles. This geometric observation gives rise to the following definition.

Definition 9.32 A *chord* of a cycle C is an edge connecting two vertices of C that are not neighbors in C.

A simple graph G is said to be *chordal* or *triangulated* if for every cycle C in G with four or more vertices, the induced subgraph $G[C]$ of G has a chord in G.

Remark 9.33

- If G is a chordal graph and $V' \subseteq V(G)$, then the induced subgraph $G[V']$ of G is also chordal.
- A graph G is chordal if, and only if, every component of G is chordal.
- A graph G is chordal if, and only if, every block of G is chordal.

The interesting point is that chordal graphs are precisely the graphs that possess a simplicial elimination ordering on their vertices, and so can be colored in a greedy fashion with an optimal number of colors. Recall the notion of a vertex cut as a set of vertices, whose removal will disconnect the graph, or reduce the graph to a single vertex. Also, recall that a vertex cut is *minimal* if it contains no vertex cuts as proper subsets.

Theorem 9.34 Let G be a simple graph. The following are equivalent:

1. Every minimal vertex cut of G induces a clique in G.
2. G has a simplicial elimination ordering of its vertices.
3. G is a chordal graph.

Note 9.35 If every minimal vertex cut of G induces a clique in G, then the same holds for every induced subgraph G' of G, since any minimal vertex cut of G' is contained in a minimal vertex cut of G.

Before proving Theorem 9.34, we first present the following key fact, which provides the main step in our proof of Theorem 9.34. It is also of interest in its own right.

Lemma 9.36 Let G be a simple connected graph on $n \geq 3$ vertices. If $G \neq K_n$, and every minimal vertex cut induces a clique in G, then G has two nonadjacent simplicial vertices $u, v \in V(G)$.

Proof We use induction on n. If $n = 3$, then $G = P_3$, for which the statement clearly holds.

Assume $n \geq 4$. Suppose u' and u'' are two nonadjacent vertices. Let $W \subseteq V(G) \setminus \{u', u''\}$ be a minimal vertex cut that separates u' and u'' in G. Let $G[V']$ and $G[V'']$ be the connected components of $G - W$ that contain u' and u'', respectively. If $G[V' \cup W]$ is a complete graph, then $N_G(u') \subseteq V' \cup W$ induces a clique in G and we let $u = u'$. Otherwise, by Note 9.35, every minimal vertex cut of the induced subgraph $G[V' \cup W]$ of G induces a clique in $G[V' \cup W]$. Hence, by the induction hypothesis, there are two nonadjacent simplicial vertices u'_1 and u'_2 in $G[V' \cup W]$. By assumption W induces a clique in G, so u'_1 and u'_2 cannot both be contained in W. Hence, one of them, say u'_1, is contained in $V' \setminus W$. In this case we have that $N_G(u'_1) \subseteq V' \cup W$ induces a clique in G and we can let $u = u'_1$. In either case, there is a vertex $u \in V'$ that is simplicial in G.

By the same token, there is a vertex $v \in V''$ that is simplicial in G. Therefore, we have two simplicial vertices $u, v \in V(G)$ that are contained in two distinct components of $G - W$ and so cannot be adjacent in G. ■

Proof (Theorem 9.34) We prove the theorem by showing the following "cycle" of implications:

$$1 \Rightarrow 2 \Rightarrow 3 \Rightarrow 1.$$

$1 \Rightarrow 2$: Let G be a simple graph on $n \geq 3$ vertices in which every minimal vertex cut induces a clique. By Lemma 9.36, G has a simplicial vertex u_n. By Note 9.35 and induction on n, the induced subgraph $G - u_n$ has a simplicial elimination ordering $V(G - u_n) = \{u_1, \ldots, u_{n-1}\}$, which yields a simplicial elimination ordering $V(G) = \{u_1, \ldots, u_n\}$ of G.

$2 \Rightarrow 3$: Let C be a cycle of length four or more in G. Let $u \in V(C)$ be the vertex that is labeled with the largest index in some simplicial elimination ordering of $V(G)$. Since the two neighbors v and w of u in C are among the previously labeled neighbors of u, which induce a clique in G, they must be connected in G. Hence, $\{u, w\}$ is a chord of C. This holds for all cycles of length four or more in G, so G is chordal.

$3 \Rightarrow 1$: Let G be a chordal graph and W a minimal vertex cut of G. If G is complete, we are done, so we may assume there are two distinct components $G[V']$ and $G[V'']$ of $G - W$, containing two nonadjacent vertices u' and u'', respectively. Assume that W does not induce a clique in G, so there are two vertices u and v in W that are not connected in G.

If $N_G(u) \cap V' = \emptyset$, then there are no edges with one endvertex in $V'' \cup \{u\}$ and the other in V'. Hence, $W \setminus \{u\}$ would be a vertex cut of G. This contradicts our assumption that W is minimal. The same holds true for v. Thus, $G[V' \cup \{u, v\}]$ is connected. Likewise, $G[V'' \cup \{u, v\}]$ is also connected.

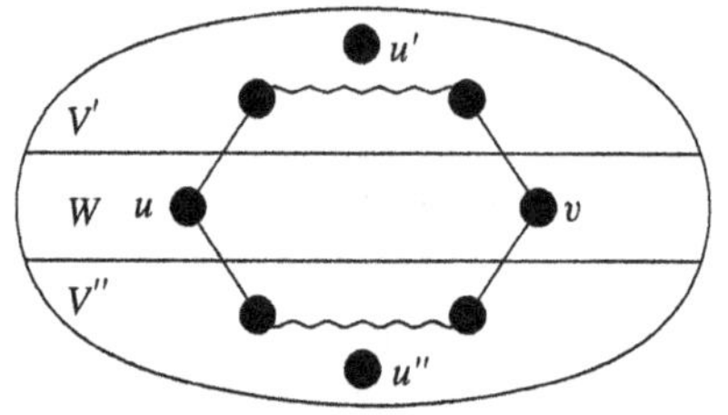

Figure 9.11: The closed composite walk $C = p'p''^{-1}$ is a cycle of length at least four.

Let p' and p'' be the shortest paths from u to v in $G[V' \cup \{u, v\}]$ and $G[V'' \cup \{u, v\}]$, respectively. The closed composite walk $C = p'p''^{-1}$ is clearly a cycle, and it has length at least four, since u and v are not connected in G, as depicted in Figure 9.11.

By assumption G is chordal and so C has a chord. Since there is no edge with one endvertex in V' and the other in V'', there are no chords between $V(p') \setminus \{u, v\}$ and $V(p'')\setminus\{u, v\}$. Also, since both p' and p'' are of shortest length, the only possibility left is for $\{u, v\}$ to be a chord. This contradicts our initial assumption that W does not induce a clique.

This concludes the proof. ∎

By Remark 9.33, every induced subgraph G' of a chordal graph G is also chordal, and hence has a simplicial elimination ordering of its vertices. By ALGORITHM SIMPLICIAL COLORING, we have that $\chi(G') = \omega(G')$. Therefore Remark 9.33, Theorem 9.34, and Observation 9.31 imply the following result that was first proved by Berge in 1960.

Corollary 9.37 Every chordal graph is perfect.

Note 9.38 Claude Berge also made the following famous *Strong Perfect Graph Conjecture* in 1961: A simple graph G is perfect if, and only if, G has no induced subgraph which is an odd cycle of length ≥ 5 or a complement of an odd cycle of length ≥ 5. In 2003 Chudnovsky, Robertson, Seymour, and Thomas managed to prove this outstanding conjecture, thereby giving a complete characterization of perfect graphs. For an excellent survey of the main aspects of perfect graphs and outlines of the proof, see [9]. Their proof is a landmark in the area of perfect graphs.

Because of Theorem 9.34, a chordal graph is sometimes *defined* as a graph that possesses a simplicial elimination ordering of its vertices. Here we have only touched upon the study of chordal graphs, and there are many different other types of elimination orderings of vertices of a graph. In Jamison and Laskar [30] several other elimination orderings are studied, and the graphs atmitting them are then completely characterized.

As we initially saw in Example 9.22, the underlying property that allowed us to use a greedy algorithm to color properly the vertices of an interval graph G was that every interval graph is chordal. Clearly, interval graphs are very special kinds of intersection graphs. As we saw in Corollary 9.27, every interval graph is an intersection graph of a list of subpaths of a given fixed path. The following result, whose proof we omit, can be found in (Gavril [18]).

Theorem 9.39 For a simple graph G on n vertices, the following are equivalent:

1. G is a chordal graph.
2. G is an intersection graph of a list $\widetilde{S} = (T_1, \ldots, T_n)$ of subtrees of some fixed tree T.

In fact, the above Theorem 9.39 can also be found in McKee [38], where in addition some interesting applications of chordal graphs in statistics, matrix theory, and computer science are studied and discussed. Also, this article is rich in references on topics on chordal graphs and related topics in the sciences.

So far, we have discussed chordal graphs and some of their properties and have given an algorithm for an optimal vertex coloring from an arbitrary simplicial elimination ordering of their vertices.

Consider a chordal graph G on n vertices, and let $(N_1, \ldots, N_n)$ be a list of previous neighbors from a given simplicial elimination ordering $V(G) = \{u_1, \ldots, u_n\}$ of the vertices. Let $d_i = |N_i|$ be the number of previously labeled neighbors of u_i. For an arbitrary $k \geq \chi(G)$, we can explicitly count the number of proper k-colorings $c : V(G) \to \{1, \ldots, k\}$ of G in the following way: There are $k - d_1 = k$ choices for $c(u_1) \in \{1, \ldots, k\}$. For each choice of legitimate values for $c(u_1), \ldots, c(u_{i-1})$, there are $k - d_i$ choices for $c(u_i)$ for every $i \in \{2, \ldots, n\}$. By the product rule, we have the following.

Theorem 9.40 Let $(N_1, \ldots, N_n)$ be the list of previous neighbors in $V(G)$ from some simplicial elimination ordering of the vertices of a chordal graph G. If $d_i = |N_i|$ for each i, then the chromatic polynomial of G is given by

$$\chi_G(t) = \prod_{i=1}^{n}(t - d_i) = t\prod_{i=2}^{n}(t - d_i).$$

Remark 9.41 Since u_1 has no previous neighbors, then $d_1 = 0$. Moreover, if $m = \omega(G)$, then $\chi_G(r) = 0$ for each $r \in \{0, 1, \ldots, m-1\}$. This implies that

$$\{0, \ldots, m-1\} \subseteq \{d_1, \ldots, d_n\}$$

for every chordal graph G, so the values of the d_i's cannot be arbitrary for $i \in \{2, \ldots, n\}$. In other words, the falling factorial $(t)_m$ divides the chromatic polynomial $\chi_G(t)$. This is, of course, consistent with the fact that for *any* simple graph G, the falling factorial factorial $(t)_{\chi(G)}$ always divides $\chi(G)$.

Theorem 9.40 yields an easy way to compute the chromatic polynomial for any chordal graph, provided that we have the simplicial elimination ordering of the vertices. Often when we have a chordal graph it is not too hard to "spot" a simplicial elimination ordering of its vertices. Hence, we can sometimes use Theorem 9.40 together with Theorem 9.5 to quickly compute the chromatic polynomial of a given graph that is "close" to being chordal.

Example 9.42 Consider the wheel W_4 on five vertices, made from the 4-cycle C_4, by connecting an additional center vertex to the four vertices of C_4. Clearly W_4 is not chordal, since C_4 has no chord. However, both removal of an edge from C_4 and contraction of an edge of C_4 result in chordal graphs, $W_4 - e$ and $W_4/e = K_4$, respectively. This is shown in Figure 9.12.

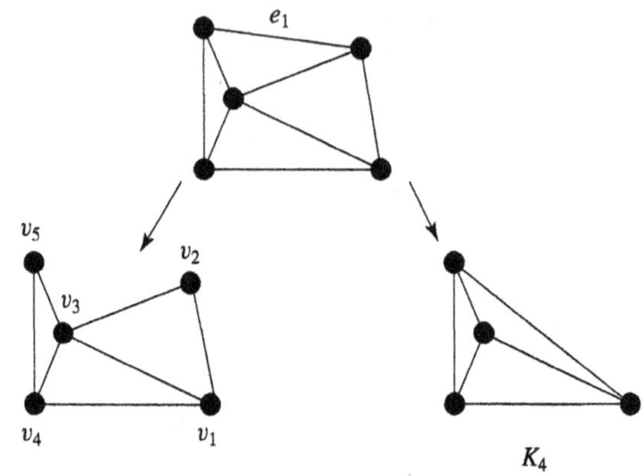

Figure 9.12: Removal and contraction of one edge e of the cycle of W_4 yields two chordal graphs.

Hence, the chromatic polynomial of W_4 is a difference of two polynomials, both of which split into linear factors.

$$\begin{aligned}\chi_{W_4}(t) &= \chi_{W_4-e}(t) - \chi_{K_4}(t)\\ &= t(t-1)(t-2)^3 - (t)_4\\ &= t(t-1)(t-2)(t^2-5t+7).\end{aligned}$$

We have seen that if G is a chordal graph, then $\chi_G(t)$ factors completely into linear factors of the form $t - d_i$, where the d_i are nonnegative integers. The other way does not hold, since there are graphs that are not chordal whose chromatic polynomials split into linear factors of the form $t - d$, where d is a nonnegative integer. (See Exercise 9.22.)

For a more detailed discussion on the chromatic polynomials of chordal graphs and some subclasses of them, such as unit interval graphs and threshold graphs, we refer to Chandrasekharan, Laskar, and Veni Madhavan [8], where their forms are obtained.

9.5 Powers of Graphs

We previously discussed proper vertex colorings of simple graphs. These are vertex colorings where adjacent vertices receive different colors for all pairs of adjacent vertices of a graph. Often, however, a proper vertex coloring of a graph is not sufficient, and we need a vertex coloring in which not only neighboring vertices receive different colors, but also vertices that are even farther apart. The following example provides some motivation.

Example 9.43 Consider a collection of radio stations that communicate by sending out signals with a certain frequency. Assuming that all of the stations have the same fixed geographical range of transmission, we can form a graph G by letting $V(G)$ represent the stations, where we

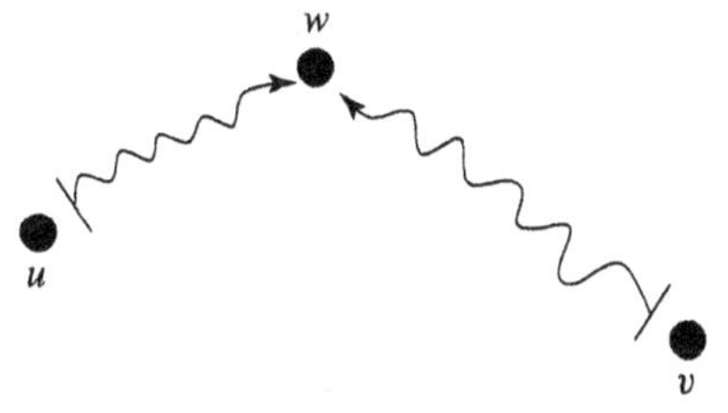

Figure 9.13: Signals of the same frequency from two stations result in a clashing of signals at an adjacent station.

connect a pair of vertices if, and only if, the corresponding stations are within each other's reach—that is, they are able to communicate directly by sending signals from one station to another. Assume further that a limited number of frequencies are available for the stations. We would like to find out the least number of distinct frequencies needed for the stations to communicate without difficulties.

The problem we need to worry about here is that of clashing or interfering signals. Every station must be able to distinguish among all signals within its reach, including its own. This means, first of all, that two stations with corresponding neighboring vertices u and v cannot use the same frequency for their signals. If that were the case, then an outgoing signal from u to v would clash with, and scramble, an incoming signal from v to u. In addition, if two vertices u and v are not adjacent, but are both adjacent to a common vertex w, then the station corresponding to w will not be able to distinguish between signals from the station corresponding to u and the station corresponding to v, as depicted in Figure 9.13.

These are the necessary and sufficient conditions that the signals from the stations must satisfy. Hence, the least number of signals needed for the stations is the least number of colors needed to color the vertices of G in such a way that

- Adjacent vertices of G must receive distinct colors.
- Vertices adjacent to a third vertex must receive distinct colors.

We are therefore seeking a vertex coloring of G such that $\partial_G(u, v) \leq 2$ implies that the vertices u and v of G will receive different colors. The least number of colors that allows such a vertex coloring of G is thus the desired minimum number of frequencies of the radio stations.

Example 9.43 motivates the following definition, which is a generalization of Definition 8.1.

Definition 9.44 Let G be a simple graph and $d \in \mathbb{N}$. A *distance-d vertex coloring* of G is a map $c : V(G) \to \mathbb{N}$ such that

$$c(u) = c(v) \Rightarrow \partial_G(u, v) > d \tag{9.4}$$

for all $u, v \in V(G)$. We drop the word "vertex" when no ambiguity results.

We say that G is *distance-d k-vertex colorable* or *distance-d k-colorable* if G has a distance-d coloring $c : V(G) \to \{1, \ldots, k\}$.

Remark 9.45 If G is a simple graph, then the following clearly hold:

- Definition 9.44 is a generalization of Definition 8.1, since a simple graph G has a distance-1 k-coloring if, and only if, G has a proper k-vertex coloring.
- For any simple graph G on n vertices and for any $d > 0$, G has a distance-d n-coloring simply by assigning the color i to the vertex u_i in any listing $V(G) = \{u_1, \ldots, u_n\}$ of the vertices of G. In particular, for any given d every simple graph has a distance-d k-coloring for some $k \in \mathbb{N}$.
- If G is a simple connected graph on n vertices and $d \geq \text{diam}(G)$, then G has no distance-d n'-coloring, for any $n' < n$, since any two vertices in G have a distance of d or less and therefore cannot be assigned the same color.

We illustrate Definition 9.44 next.

Example 9.46 Consider the simple graph G shown in Figure 9.14. Since $\partial_G(u, v) \leq 2$ for every pair of vertices u and v in $V(G)$, no two vertices of G can receive the same color in a distance-2 coloring of G. So, in a distance-2 coloring of G, each vertex must have its own color. By Example 8.2 we see here that G has a distance-1 3-coloring and a distance-d 5-coloring for any $d \geq 2$.

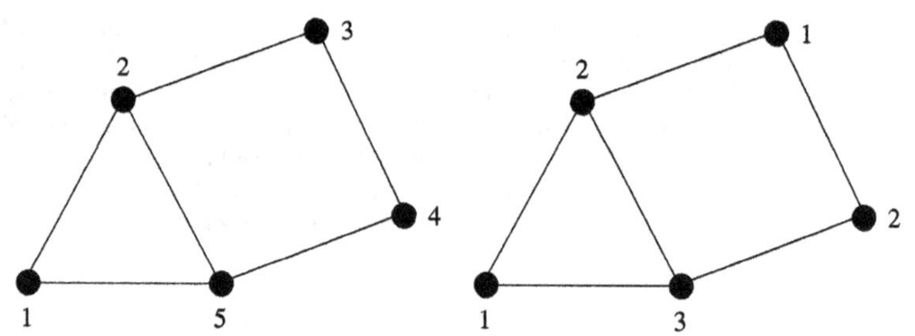

Figure 9.14: A simple graph G, a distance-2 coloring and a distance-1 coloring of G.

For a simple graph G and $d \in \mathbb{N}$, one would ideally like to obtain a distance-d coloring of G using a minimum number of colors, as was the case for the usual proper vertex coloring of G. By Remark 9.45, the following is a valid definition.

Definition 9.47 Let G be a simple graph and $d \in \mathbb{N}$. The least $k \in \mathbb{N}$ such that G has a distance-d k-vertex coloring is called the *distance-d chromatic number* of G and is denoted by $\chi_d(G)$.

Remark 9.48 For any simple graph G, we have the following:

- $\chi_1(G) = \chi(G)$, the chromatic number of G.
- If $d' \leq d$, then $\chi_{d'}(G) \leq \chi_d(G)$. In particular, $\chi(G) \leq \chi_d(G)$ for any $d \in \mathbb{N}$.

An example helps to clarify Definition 9.47.

Example 9.49 Continuing from Example 9.46, we revisit the graph G from Figure 9.14. In Example 8.5 we established that $\chi(G) = 3$. Hence, from Example 9.46, we have that

$$\chi_d(G) = \begin{cases} 3 & \text{if } d = 1, \\ 5 & \text{if } d \geq 2. \end{cases}$$

Closely related to a distance-d coloring of a graph is the notion of a *power of a graph*. We define this notion in Definition 9.50. Although a distance-d coloring is more general than the usual vertex coloring, it allows us to apply all of the tools we have developed for a proper vertex coloring to a distance-d coloring.

Definition 9.50 Let G be a simple graph and $d \in \mathbb{N}$. The *dth power* of G, denoted by G^d, is the simple graph on the same set of vertices as G, where we connect every pair of vertices having distance d or less in G. That is,

$$V(G^d) = V(G),$$
$$E(G^d) = \{\{u, v\} \mid \partial_G(u, v) \leq d\}.$$

Remark 9.51 For any simple graph G, we have the following:

- Clearly, $G^1 = G$ by definition.
- If $d' \leq d$, then $G^{d'} \subseteq G^d$.
- For any $d \in \mathbb{N}$, the graph G^d has the same number of components as G. In fact, two vertices in G^d are in the same component if, and only if, they are in the same component of G.
- If G is connected on n vertices and $d \geq \text{diam}(G)$, then $G^d = K_n$.

The dth power of a given graph G is directly related to distance-d coloring of G. If one is not concerned about keeping track of the original graph G, then one can eliminate the notion of distance-d colorings of G by considering the usual proper vertex colorings of G^d instead, forgetting about G itself. The proof of the following observation is left as Exercise 9.25.

Observation 9.52 For any simple graph G and $d \in \mathbb{N}$, we have

$$\chi_d(G) = \chi(G^d).$$

We now consider an example illustrating the squaring of a tree.

Example 9.53 Figure 9.15 depicts a tree T on eight vertices and its square T^2, in which every pair of vertices of distance two or less in T have been connected with an edge. To the original seven edges we added ten more edges to T to obtain T^2. Note that for every internal vertex, it plus its neighbors form a clique in T^2.

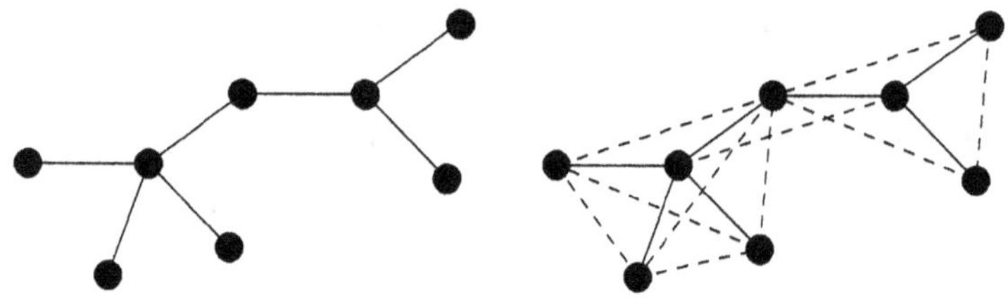

Figure 9.15: The tree T and its square T^2. The additional edges are shown as dashed lines.

With Example 9.53 in mind, consider a tree T on $n \geq 3$ vertices. The removal of a single internal vertex $u \in V(T)$ will disconnect T but not T^2, since every two distinct neighbors of u are connected in T^2. Hence, no vertex cut of T^2 consists of a single vertex. However, if T is not a star, the removal of a pair of adjacent internal vertices of T will disconnect the square T^2, since there is no edge in T^2 with its endvertices in each of the components of $T - \{u, v\}$.

Let $S \subseteq V(T)$ be an independent set of vertices of T; that is, no two vertices in S are adjacent in T. If $u, v \in V(T) \setminus S$ are two distinct vertices, then there is a unique path p from u to v in T. On p, no two consecutive vertices are from S, so one can traverse from u to v via the vertices of p by edges of T^2. Hence, there is a path from u to v in $T^2 - S$. Therefore, S is not a vertex cut of T^2. From this we have that any vertex cut of T^2 must contain a pair of adjacent vertices of T. From these remarks, we have the following.

Observation 9.54 For a tree T on three or more vertices that is not a star, any minimal vertex cut of T^2 consists of a pair of adjacent internal vertices of T.

From Observation 9.54 we have in particular that a minimal vertex cut of T^2 induces a clique in T^2. Thus, by Theorem 9.34, T^2 is a chordal graph. By generalizing these arguments for T^2, one can show the following (see Agnarsson, Greenlaw, and Halldórsson [1]).

Theorem 9.55 If T is a tree and $d \in \mathbb{N}$, then T^d is a chordal graph.

Note 9.56 Theorem 9.55 cannot directly be generalized to any chordal graph G, since there is a chordal graph G where G^2 is not chordal. (See Exercise 9.31.)

By definition every tree is a chordal graph, and we know that any power of a tree is again a chordal graph. Also, we know that if G is a chordal graph, then G^d is not necessarily chordal. So a quite natural question is: Are there other classes of chordal graphs (except for trees) all of whose powers are also chordal?

Consider an interval graph G on n vertices $V(G) = \{u_1, \ldots, u_n\}$. Let each u_i correspond to the interval $I_i = [a_i; b_i]$, where $a_i < b_i$. By Observation 9.26 we may assume that all of the end points of the intervals are distinct. We may further assume that $a_1 < a_2 < \cdots < a_n$—that is, this listing of the vertices is a simplicial elimination ordering of the vertices of G.

Let $d \in \mathbb{N}$. For each $i \in \{1, \ldots, n\}$ define a new interval $I_i(d) = [a_i; b_i(d)]$ having the same left end point as I_i, and where the right end point $b_i(d)$ is defined by

$$b_i(d) = \max(\{b_\alpha : \partial_G(u_i, u_\alpha) \leq d - 1\}).$$

Clearly, $b_i(d) \geq b_i$ for each i. We can obtain an interval graph as the intersection graph of the set $\{I_1(d), \ldots, I_n(d)\}$ of real intervals. With this framework, we have the following from Agnarsson, Greenlaw, and Halldórsson [1].

Theorem 9.57 Suppose G is an interval graph on n vertices represented by the set $\{I_1, \ldots, I_n\}$ of real intervals sorted by their distinct left end points. Then for any $d \in \mathbb{N}$, the graph G^d is also an interval graph represented by the set $\{I_1(d), \ldots, I_n(d)\}$ of real intervals.

Theorem 9.57 is constructive in the sense that we know not only that an arbitrary power of an interval graph is again an interval graph—that is, G^d is an intersection graph of *some* set of real intervals—but that it is an intersection graph of a set of intervals that have been *directly constructed* from the intervals representing G.

Proof (Theorem 9.57) Using the notation just introduced, we need to prove that

$$\partial_G(u_i, u_j) \leq d \Leftrightarrow I_i(d) \cap I_j(d) \neq \emptyset.$$

Throughout, we may assume that $i < j$, and hence $a_i < a_j$, since the intervals have been ordered by their left end points.

Assume that $\partial_G(u_i, u_j) \leq d$. By assumption we know there exists a sequence of intervals $I_{i_1}, \ldots, I_{i_d}$ such that $I_i = I_{i_1}$ and $I_j = I_{i_d}$, where $I_{i_\ell} \cap I_{i_{\ell+1}} \neq \emptyset$ for all $\ell \in \{1, \ldots, d-1\}$. Since, in particular, $I_{i_{d-1}} \cap I_j \neq \emptyset$, we have that the right end point of $I_{i_{d-1}}$ satisfies $a_j < b_{i_{d-1}}$. Since now $\partial_G(u_i, u_{i_{d-1}}) \leq d - 1$, we have by definition of $b_i(d)$ that

$$a_i < a_j < b_{i_{d-1}} \leq b_i(d),$$

which implies that $I_i(d) \cap I_j(d) \neq \emptyset$.

The other implication works similarly. Assuming $I_i(d) \cap I_j(d) \neq \emptyset$, then by definition of $I_i(d)$, there is an $\alpha \in \{1, \ldots, n\}$ with $b_i(d) = b_\alpha$, where $\partial_G(u_i, u_\alpha) \leq d - 1$ and $a_j < b_\alpha$. If $I_\alpha \cap I_j = \emptyset$, then by the last inequality, we must have

$$a_j < b_j < a_\alpha < b_\alpha,$$

in which case $I_j \subseteq I_i(d)$, and hence

$$\partial_G(u_i, u_j) \leq \partial_G(u_i, u_\alpha) \leq d - 1 < d.$$

Otherwise, $I_\alpha \cap I_j \neq \emptyset$ implies that u_α and u_j are adjacent in G, and hence

$$\partial_G(u_i, u_j) \leq \partial_G(u_i, u_\alpha) + \partial_G(u_\alpha, u_j) \leq (d - 1) + 1 = d,$$

which completes our proof. ■

Since every interval graph is chordal, then by Theorems 9.34 and 9.57, we immediately obtain the following corollary.

Corollary 9.58 For every interval graph G and $d \in \mathbb{N}$, the power graph G^d is also an interval graph and hence also chordal.

For a general chordal graph G it can be shown that if $d \in \mathbb{N}$ is an odd number, then G^d is indeed also chordal, as was first shown explicitly by Balakrishnan and Paulraja [4]. Moreover, this result, together with a complete characterization of the chordal graphs G whose squares G^2 are also chordal from Laskar and Shier [34], yields a complete characterization of those chordal graphs G whose power G^d is also chordal for an arbitrary $d \in \mathbb{N}$.

9.6 Exercises

9.1. Compute the chromatic polynomial of the graph shown in Figure 8.1 recursively using Observation 9.7.

9.2. Compute the chromatic polynomials of C_3 and C_4, the cycles on $n = 3$ and $n = 4$ vertices. Use these results and Lemma 9.15 to compute the chromatic polynomial of the graph shown in Figure 8.1.

9.3. Prove that the chromatic polynomial for a tree T on n vertices is given by $\chi_T(t) = t(t-1)^{n-1}$ in the two following ways:

(a) Use Theorem 3.4 and induction on n.

(b) Choose a root of T and directly count all possible k-vertex colorings.

9.4. As we saw in Example 9.4, the chromatic polynomial $\chi_G(t)$ of a simple graph G on n vertices satisfies $\chi_G(k) \le k^n$ for all $n \in \mathbb{N}$. Show that if G is connected, then we have further that

$$\chi_G(k) \le k(k-1)^{n-1}$$

for all $k \in \mathbb{N}$.

9.5. Use induction on n to show that for the cycle C_n on n vertices, we have

$$\chi_{C_n}(t) = (t-1)[(t-1)^{n-1} + (-1)^n].$$

9.6. Use the result of Exercise 9.5 to show directly that for odd $n \in \mathbb{N}$, the chromatic polynomial $\chi_{C_n}(t)$ has $t-2$ as a polynomial factor. Also show that for even $n \in \mathbb{N}$, the chromatic polynomial $\chi_{C_n}(t)$ does not have $t-2$ as a polynomial factor. Interpret the results in terms of the chromatic number of C_n.

9.7. Use induction on $|E(G)|$ to prove Theorem 9.13.

9.8. Use Theorem 9.13 to prove Corollary 9.14.

9.9. Show that for any simple graph G, its chromatic polynomial $\chi_G(t)$ has no real roots x where $0 < x < 1$. [*Hint:* Use induction on $|E(G)|$, the number of edges of G. Note that it suffices to consider connected graphs and that the statement holds for trees.]

9.10. Prove that two isomorphic graphs G and G' have the same chromatic polynomial, $\chi_G(t) = \chi_{G'}(t)$. Does the converse hold for two graphs G and G' on n vertices, for an arbitrary $n \in \mathbb{N}$? That is, does $\chi_G(t) = \chi_{G'}(t)$ imply that G and G' are isomorphic, if both graphs have n vertices?

9.11. Consider the graph G from Example 9.21, where the mn vertices are listed in such a way that $u \in V_i$ appears before $v \in V_j$ if $i < j$. Find the total number of such listings that are possible.

9.12. Consider the graph G from Example 9.21. If the parts do not have the same number of vertices, say V_i has m_i vertices for each $i \in \{1, \ldots, n\}$, what are $\omega(G)$ and $\chi(G)$?

9.13. Show that the graph G from Example 9.21 is a circular arc graph and not an interval graph. Is it a proper circular arc graph? Is it a unit circular arc graph?

9.14. Complete the argument from Remark 9.25, and prove that all simple graphs are intersection graphs for some list $\widetilde{S}$ of subsets of some set U.

9.15. Show that if G is a threshold graph, then both G and $\overline{G}$ are chordal graphs.

9.16. Show that the graph on the right in Figure 9.8 is a proper disk graph; that is, it does have a disk representation where no disk contains another.

9.17. Show that every coin graph is a planar graph. Show further that every unit coin graph on $n > 3$ vertices has at most $3n - 7$ edges.

9.18. This and the next two exercises are to verify the properties about maximal outerplanar graphs mentioned in Example 9.30.

Show that every outerplanar graph on n vertices has at most $2n - 3$ edges. Additionally, show that an outerplanar graph on n vertices is a maximal outerplanar graph (that is, all regions except one are bounded by exactly three edges) if, and only if, it has $2n - 3$ edges.

9.19. Let G be a maximal outerplanar graph and consider a plane embedding of G where all vertices are bordering the infinite region. Show that if $e \in E(G)$ is any edge in this embedding that borders the infinite region, then the simple contraction G/e is also a maximal outerplanar graph. [*Hint:* Show that G/e is an outerplanar graph. Then count the vertices and edges, and use Exercise 9.18.]

9.20. Let G be a maximal outerplanar graph on $n \geq 3$ vertices. Use induction on n and Exercise 9.19 to show that G has a vertex u of degree two. Also show that $G - u$ is a maximal outerplanar graph on $n - 1$ vertices. [*Note!* This is what is essential to establish the simplicial elimination ordering of the vertices of a maximal outerplanar graph as in Example 9.30.]

9.21. For $n \geq 3$, use Observation 9.7 to compute $\chi_{K_n - e}(t)$, where e is some edge of K_n.

9.22. For $n \geq 3$, let G be the graph obtained from K_n by subdividing an edge of K_n into a path of length two. Show that G is not chordal. Use Exercise 9.21 to show that $\chi_G(t)$ splits into linear factors if, and only if, $n \in \{2, 6\}$. [*Hint:* An expression of the form $x^2 - 4$, where x is an integer, can be a square of an integer, say $x^2 - 4 = y^2$ for some integer y, if, and only if, $4 = (x - y)(x + y)$. Note that if both x and y are integers, then so are $x - y$

and $x + y$. Since the only positive factors of 4 are 1, 2, and 4, there are only finitely many possibilities for the values of $x - y$ and $x + y$, and hence also for x and y.]

This shows that for $n = 6$, the graph G is not chordal, although $\chi_G(t)$ splits into linear factors.

9.23. For a simple graph G and $p, q \in \mathbb{N}$, show that

$$(G^p)^q = G^{pq}.$$

9.24. For a simple graph G and $d \in \mathbb{N}$, show that the distance function in G^d satisfies

$$\partial_{G^d}(u, v) = \lceil \partial_G(u, v)/d \rceil,$$

for all $u, v \in V(G)$.

9.25. Prove Observation 9.52, which shows that all distance-d coloring problems can be formulated in terms of proper vertex colorings of some power graphs.

9.26. Let G be the graph from Problem 1.4 shown in Figure 1.5.

(a) Show that $\chi(G) = 4$.
(b) Show that G is isomorphic to C_2^7, the square of the cycle C_7.
(c) Show that both C_7 and C_7^2 are circular arc graphs.
(d) What is C_7^3?

9.27. For $n \in \mathbb{N}$, let S_n be a star on $n + 1$ vertices—that is, a tree on $n + 1$ vertices with n leaves. Compute $\chi_{S_n}(t)$ and $\chi_{S_n^2}(t)$.

9.28. Let $n \geq 3$. Let W_n be the wheel on $n + 1$ vertices. Compute $\chi_{W_n}(t)$ and $\chi_{W_n^2}(t)$.

9.29. Let $m, n \in \mathbb{N}$. Form a graph G by connecting the center vertices of the stars S_m and S_n by an edge. Show that G^2 is chordal. Also, compute $\chi_G(t)$, $\chi_{G^2}(t)$, and $\chi_{G^3}(t)$.

9.30. Let $n, m \geq 3$ be natural numbers. Form a graph G by connecting with an edge any vertex of the wheel W_m to any vertex of the wheel W_n. Compute $\chi_G(t)$. Does $\chi_G(t)$ depend on which two vertices we connect?

9.31. Let C_8 be the cycle on eight vertices, which we label cyclically $u_1, \ldots, u_8$. Form the simple graph G from C_8 by adding the following five edges: $\{u_1, u_3\}$, $\{u_3, u_5\}$, $\{u_5, u_7\}$, $\{u_7, u_1\}$, and $\{u_1, u_5\}$. Show that G is a chordal graph, and that its square G^2 is not chordal.

9.32. Let T be a tree and $\Delta(T)$ the maximum degree of T. Show that

$$\chi(T^2) = \Delta(T) + 1.$$

9.33. Use Theorems 9.34 and 9.55 to express $\chi(T^3)$ in terms of some parameters of T.

9.34. Consider the interval graph G from Example 9.22. Find interval representations of G^2 directly from the interval representation of G shown in Figure 9.5. Use that to draw the graph G^2. Finally, draw the graph G^2 directly from G as shown in that figure. Find the least $d \in \mathbb{N}$ such that $G^d = K_{10}$.

CHAPTER

10

Independence, Dominance, and Matchings

Chapter Goals

Discuss independence of vertices in a graph G as well as G's independence number $\alpha(G)$.

Discuss dominance of vertices in a graph G as well as G's dominance number $\gamma(G)$.

Study matchings in graphs, especially matchings in bipartite graphs.

Characterize every maximum matching M in terms of M-augmenting paths.

Present Hall's Marriage Theorem for bipartite graphs.

Discuss the deficiency of matchings of bipartite graphs.

10.1 Independence of Vertices

In Chapters 8 and 9 we discussed vertex and edge colorings of graphs. They relate naturally to the study of *independence* among vertices and edges of a graph: A set of vertices is independent if no two of its vertices are adjacent. So, for a fixed vertex coloring of a graph G, each set of vertices having the same color is independent in G. Similarly, for a fixed edge coloring of G, each set of edges of the same color is independent, in the sense that no two edges of the same color are adjacent. Further, in Chapters 8 and 9 on graph colorings, the overall goal was to color either the vertices or the edges, using the *minimum* number of colors, in such a way that each color class of either vertices or edges was independent. There the object was to find a minimum size *partition.*

In this chapter we focus on one *color class*, so to speak, and our objective here is to determine the *maximum* possible size of a color class, among all proper vertex colorings of the graph.

Definition 10.1 Let G be a graph and $S \subseteq V(G)$. A set S is an *independent set* if the induced subgraph $G[S]$ is a null graph.

1. The independent set S is *maximal* if S is not a proper subset of any independent set of G.
2. The independent set S is *maximum* if no other independent set has more vertices than S.

As was the case for vertex coloring graphs, when dealing with independent sets, what matters is whether vertices are adjacent or not, not the extent of their adjacencies. Clearly, a vertex that is an endvertex of a loop is not contained in any independent set. Also, a set of vertices in a graph G is independent in G if, and only if, the set is independent in the graph obtained from G by replacing all multiple edges with a single edge.

Recall Definition 9.18, where we defined the independence number, $\alpha(G)$. The following statement is clear.

Observation 10.2 For a graph G having a maximum independent set $S_{\max} \subseteq V(G)$, the independence number $\alpha(G)$ is given by

$$\begin{aligned}\alpha(G) &= \max(\{|S| : S \subseteq V(G) \text{ is an independent set}\})\\ &= |S_{\max}|.\end{aligned}$$

Note 10.3 For a bipartite graph G with bipartition $V(G) = X \cup Y$, both parts X and Y are independent sets, and hence we have

$$\alpha(G) \geq \max(\{|X|, |Y|\}).$$

As shown in Figure 10.1, we see that $\alpha(G)$ can be much larger than $\max(\{|X|, |Y|\})$. In fact, we see that $\alpha(G) = 6$, whereas both X and Y have four vertices each. This construction can, of course, be pushed further, to construct a bipartite graph on $2k$ vertices, each part containing k vertices, and where the independence number is $2(k-1)$, or almost twice the number of vertices of each part.

Example 10.4 A certain county wants to determine how many distinct radio stations it can have broadcasting within its borders. The terrain is hilly and even mountainous. Each potential radio station, then, must be assigned a frequency that is robust and differs substantially from the frequency assigned to any other station in the county. Nine different frequencies are available, some of them close or similar enough to be confused with each other, so that they should not be used by stations within the same county.

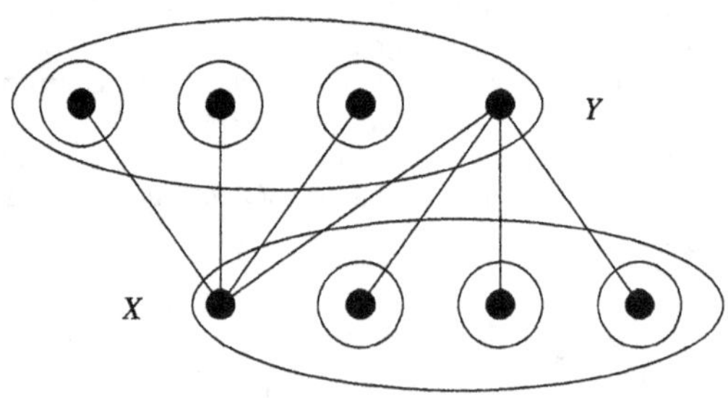

Figure 10.1: Note that $\alpha(G)$ can be much larger than the size of either part of a bipartite graph G.

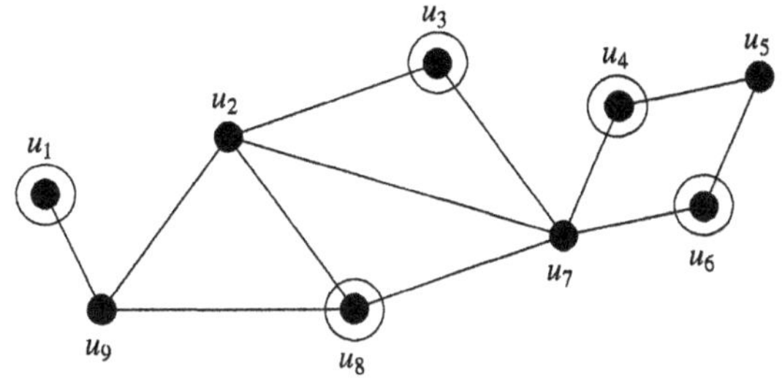

Figure 10.2: Of the possible nine frequencies, the ones connected by an edge are similar and can be confused with one another.

Consider the simple graph G shown in Figure 10.2, where the vertices $V(G) = \{u_1, \ldots, u_9\}$ represent the available frequencies for the county, and an edge $\{u_i, u_j\} \in E(G)$ indicates that the frequencies corresponding to u_i and u_j are similar.

The maximum number of radio stations that can possibly be built in the county is the same as the maximum number of available frequencies no two of which are similar or close. This value is the same as the maximum size of an independent set of G. Hence, the maximum number of possible stations is given here by $\alpha(G)$. In Figure 10.2 we have depicted an independent set $S = \{u_1, u_3, u_4, u_6, u_8\}$, which clearly is maximal, since any other vertex is a neighbor of at least one of the vertices in S. Furthermore, S is a maximum independent set. We can see this directly by noting that each of the blocks of G has either one or two independent vertices. Hence, $\alpha(G) = 5$ in this case, so the county can build at most five noninterfering radio stations.

Example 10.5 Let $k \in \mathbb{N}$ and consider $P_{6k} = (u_1, e_2, u_2, \ldots, e_{6k}, u_{6k})$, the simple path on the vertices $u_1, \ldots, u_{6k}$, and the edges $e_i = \{u_{i-1}, u_i\}$ for $2 \leq i \leq 6k$. Clearly, the sets

$$\begin{aligned} S_1 &= \{u_1, u_4, u_7, \ldots, u_{6k-2}, u_{6k}\}, \\ S_2 &= \{u_2, u_5, u_8, \ldots, u_{6k-1}\}, \end{aligned}$$

are both maximal independent sets of P_{6k} containing $2k + 1$ and $2k$ vertices, respectively. Hence, a graph can have different-sized maximal independent sets.

Also, the sets

$$\begin{aligned} S_3 &= \{u_1, u_3, \ldots, u_{6k-1}\}, \\ S_4 &= \{u_2, u_4, \ldots, u_{6k}\}, \end{aligned}$$

are, by the Pigeon-Hole Principle, both maximum independent sets. We see $\alpha(S_3) = \alpha(S_4) = 3k$. Therefore, a maximum independent set in a graph is not necessarily unique.

For a vertex coloring of a graph G, each color class is an independent set (maximum or not). If the vertex coloring is optimal with $\chi(G)$ colors, then the largest color class must contain at least $|V(G)|/\chi(G)$ vertices. Hence, a maximum independent set of G must contain at least $|V(G)|/\chi(G)$ vertices. This was indeed the case in Example 10.5, where $\chi(P_{6k}) = 2$ and each maximum independent set (S_3 or S_4) contained $3k = |P_{6k}|/\chi(P_{6k})$ vertices. Since the independence number of any graph is an integer, we have the following observation.

Observation 10.6 For a graph G, the independence number $\alpha(G)$ satisfies

$$\alpha(G) \geq \left\lceil \frac{|V(G)|}{\chi(G)} \right\rceil.$$

Example 10.7 The lower bound given in Observation 10.6 is not a good bound in general. In fact, the ratio $(\alpha(G)\chi(G))/|V(G)|$ can be arbitrarily large, as the next argument illustrates: For $m, n \in \mathbb{N}$, let G be the graph on $m+n$ vertices, where m of the vertices are isolated and the remaining n vertices form a complete graph K_n. Clearly, we have $\alpha(G) = m + 1$, and the chromatic number is $\chi(G) = n + 1$. Hence, we obtain

$$\frac{\alpha(G)\chi(G)}{|V(G)|} \geq \frac{mn}{m+n} + 1,$$

which can be arbitrarily large, depending on our choices of m and n.

In general, there is no easy way to find or "spot" a maximum independent set in a graph G. However, there are some facts worth mentioning when determining $\alpha(G)$ and a corresponding maximum independent set $S \subseteq V(G)$ with $|S| = \alpha(G)$. The following observation provides a couple of such facts; we leave the proof as Exercise 10.2.

Observation 10.8 For a graph G we have the following:

1. If G has components $C_1, \ldots, C_k$, then

$$\alpha(G) = \sum_{i=1}^{k} \alpha(C_i).$$

2. If G is connected with blocks $B_1, \ldots, B_k$, then

$$\alpha(G) \leq \sum_{i=1}^{k} \alpha(B_i). \tag{10.1}$$

Furthermore, equality holds in (10.1) if, and only if, each block B_i has a maximum independent set containing no cut-vertices of G.

For a fixed graph G, a natural question is whether there is a systematic way to determine $\alpha(G)$ or, further, to find a corresponding maximum independent set of the graph. Consider a graph G having $u \in V(G)$ as a vertex. Clearly, a maximum independent set of G either contains u or does not. If G has a maximum independent set that does not contain u, then $\alpha(G) = \alpha(G - u)$. If, however, all maximum independent sets contain u, then each maximum independent set of G cannot contain any of u's neighbors in G. Recall from Definition 1.22 that the closed neighborhood $N_G[u] \subseteq V(G)$ is the set of the neighbors of u together with u itself. With this in mind, we have the following theorem. The result is depicted in Figure 10.3.

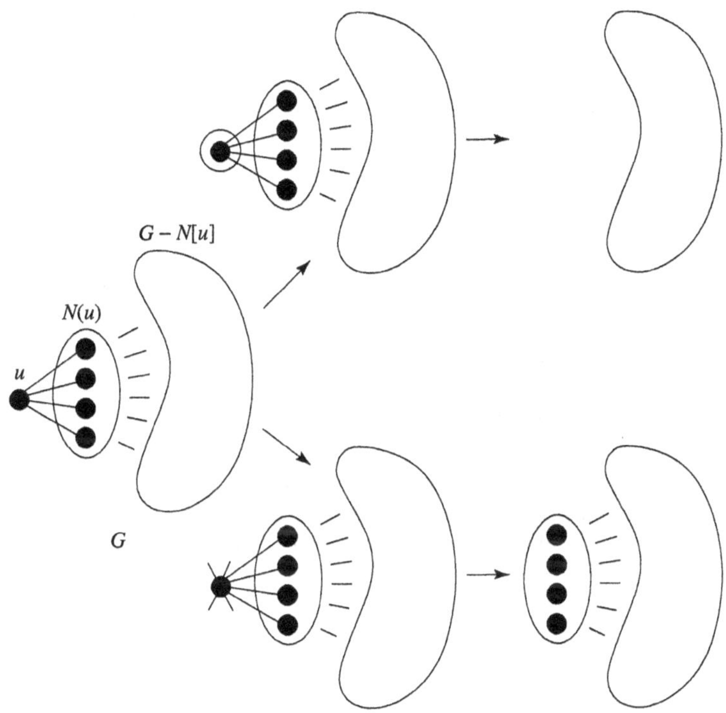

Figure 10.3: A recursive way to determine $\alpha(G)$ from a fixed vertex u of G.

Theorem 10.9 Let G be a graph and $u \in V(G)$ a vertex of G. The independence number $\alpha(G)$ satisfies

$$\alpha(G) = \max(\{\alpha(G - u), 1 + \alpha(G - N_G[u])\}). \tag{10.2}$$

Proof Clearly, every independent set of $G - u$ yields an independent set of G. Also, an independent set of $G - N_G[u]$ with u added to it is an independent set of G as well.

If G has a maximum independent set that does not contain the vertex u, then $\alpha(G) = \alpha(G-u)$. Otherwise, every maximum independent set of G contains u and hence contains no vertex in $N_G(u)$. Therefore, in this case

$$S \leftrightarrow S \setminus \{u\}$$

yields a bijective correspondence between maximum independent sets of G and maximum independent sets of $G - N_G[u]$. Hence, in this case we have $\alpha(G) = 1 + \alpha(G - N_G[u])$, which completes the proof. ∎

Whether or not the recursive process of determining $\alpha(G)$ by (10.2) is efficient or not depends completely on the graph G at hand. As for the recursive method of computing the chromatic polynomial $\chi_G(t)$ of G, given by (9.1), where the number of steps in our computation depends on a careful choice of the edge to contract, sometimes choosing the vertex $u \in V(G)$ with foresight can also (in this case) reduce the number of computations quite a bit.

In contrast to many inductive proofs on trees, where we are concerned with choosing a leaf, here it is a good rule of thumb to choose a vertex $u \in V(G)$ that is of high degree or is a cut-vertex which splits the graph G up into many parts of much smaller size!

Example 10.10 Consider the graph G on nine vertices $V(G) = \{u_1, \ldots, u_9\}$ from Figure 10.2. The vertex $u_7 \in V(G)$ is a cut-vertex and also a vertex of maximum degree—five! Hence, u_7 should a priori be a vertex that we use in our computation of $\alpha(G)$.

By Theorem 10.9 we obtain the following:

$$\begin{aligned}\alpha(G) &= \max(\{\alpha(G-u_7), 1+\alpha(G-N_G[u_7])\}) \\ &= \max(\{\alpha(G_1), 1+\alpha(G_2)\}),\end{aligned}$$

where G_1 and G_2 are the graphs shown in Figure 10.4. Since G_2 is the disjoint union of K_1 and K_2, the single vertex u_5 and the edge $\{u_1, u_9\}$, we have $1 + \alpha(G_2) = 3$.

To compute $\alpha(G_1)$, we note that G_1 is a disjoint union of two connected graphs G_1' and G_1'', the second of which is P_3, the path of length three; hence, $\alpha(G_1'') = 2$. For G_1' we can use Theorem 10.9 once more on the vertex u_2 and obtain

$$\begin{aligned}\alpha(G_1') &= \max(\{\alpha(G_1'-u_2), 1+\alpha(G_1'-N_{G_1'}[u_2])\}) \\ &= \max(\{\alpha(G_{11}'), 1+\alpha(G_{12}')\}),\end{aligned}$$

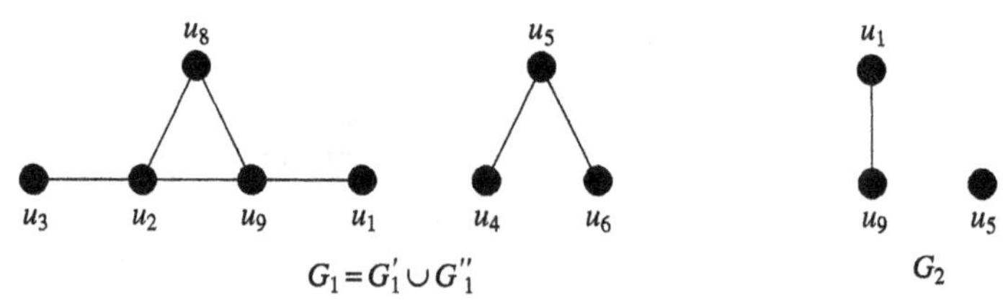

Figure 10.4: By computing $\alpha(G_1)$ and $\alpha(G_2)$, we determine $\alpha(G)$.

where G'_{11} is a disjoint union of K_1 and P_3, and G'_{12} is just the single vertex u_1. Hence, $\alpha(G'_{11}) = 3$, and $1 + \alpha(G'_{12}) = 2$; and therefore, $\alpha(G'_1) = \max(\{3, 2\}) = 3$.

From the previous calculations, we finally obtain

$$\begin{aligned}\alpha(G) &= \max(\{\alpha(G_1), 1 + \alpha(G_2)\}) \\ &= \max(\{\alpha(G'_1) + \alpha(G''_1), 3\}) \\ &= \max(\{3 + 2, 3\}) \\ &= 5.\end{aligned}$$

Note that by backtracking through the computation to see which graphs actually yielded the maximum independence number $\alpha(G) = 5$, we can obtain a maximum independent set of G (in fact, all of them!). (See Exercise 10.5.) For example, the subgraph $G_1 = G'_1 \cup G''_1$ yielded the independence number $\alpha(G) = 5$ of G. Further, the subgraph G'_{11} yielded the independence number of G'_1. So, all in all, the independent set of G that gave rise to the independence number of G is the union of the maximum independent sets of G'_{11} and G''_1, which is given by

$$S = \{u_1, u_3, u_8\} \cup \{u_4, u_6\} = \{u_1, u_3, u_4, u_6, u_8\}.$$

10.2 Domination of Vertices

Every maximal independent set S of vertices in a simple graph G has the property that every vertex of the graph either is contained in S or is adjacent to a vertex in S. Such a property is called *domination* of the graph G. Disregarding whether a subset of vertices of G is independent or not, one can focus on this domination property. Clearly, every vertex in a graph G is contained in (or is adjacent to) a vertex in $V(G)$, so in this sense $V(G)$ is a *dominating set* of the vertices of G. It is therefore of interest to find a *small* set of vertices with the domination property, unlike the problem of finding a set of independent vertices, where the object was to have the set be as large as possible.

Recall Definition 1.22 and the following Lemma 1.23 of the open and closed neighborhood of an arbitrary subset $D \subseteq V(G)$ of a graph G. The open neighborhood and the closed neighborhood of D can be given respectively by

$$\begin{aligned}N_G(D) &= \bigcup_{v \in D} N_G(v), \\ N_G[D] &= \bigcup_{v \in D} N_G[v].\end{aligned}$$

In what follows, when there is no danger of ambiguity, we write $N(D)$ and $N[D]$ for $N_G(D)$ and $N_G[D]$, respectively.

Definition 10.11 For a graph G, a set $D \subseteq V(G)$ of vertices is a *dominating set* of G if $N_G[D] = V(G)$—that is, if every vertex in $V(G)$ is either in D or adjacent to a vertex in D.

1. The dominating set D is *minimal* if D does not properly contain a dominating set of vertices.

2. The dominating set D is a *minimum* if no other dominating set has fewer vertices than D.
3. The cardinality of a minimum dominating set of G is called the *domination number* of G, and is denoted by $\gamma(G)$.

Note that by an unfortunate convention in the graph-theory literature, both the genus of a graph G and the domination number of G are denoted by $\gamma(G)$. However, since these two parameters rarely are used in the same context, there should be no (or at least little!) danger of confusion. In the rare case where we need to discuss both the genus and the domination number of a graph simultaneously, we can temporarily denote one of these parameters by γ_G and the other by $\gamma(G)$.

As for the independence number $\alpha(G)$, we have the following observation.

Observation 10.12 For a graph G having a minimum dominating set $D_{\min} \subseteq V(G)$, the domination number $\gamma(G)$ is given by

$$\begin{aligned}\gamma(G) &= \min(\{|D| : D \subseteq V(G) \text{ is a dominating set}\}) \\ &= |D_{\min}|.\end{aligned}$$

Example 10.13 In a certain state there are nine cities, each of which needs a police force to patrol the streets. A police station in one city is able to patrol the streets of that city and every city in the vicinity that has a good road system.

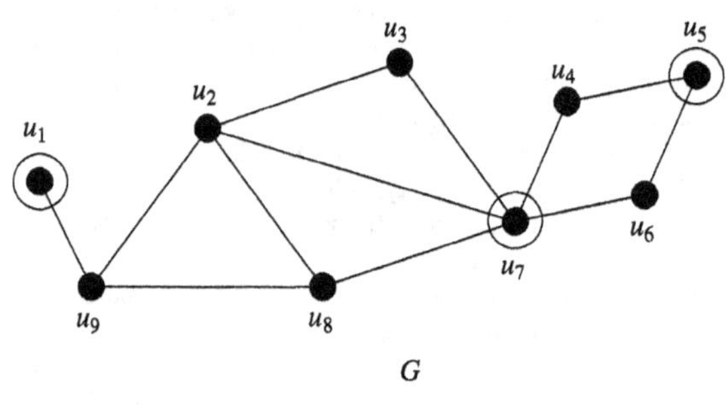

Figure 10.5: Cities that are close to one another and linked via a good road system are connected by an edge.

Consider the simple graph G shown in Figure 10.5, where the vertices $V(G) = \{u_1, \ldots, u_9\}$ represent the cities and an edge $\{u_i, u_j\} \in E(G)$ indicates that the corresponding two cities are close to one another and connected by a good road system. The minimum number of police stations needed to be built to ensure that all streets are patrolled is the the same as the minimum number of vertices that dominate each of the nine vertices of the graph G; therefore, this value is given by $\gamma(G)$. In Figure 10.5 we have shown a dominating set $D = \{u_1, u_5, u_8\}$, which is clearly a minimal dominating set, since the removal of any of these three vertices will leave a set consisting of two vertices that does not dominate G. The set D is in fact a minimum dominating set, since the combined neighborhood of any pair of vertices in G does not dominate G. (See Example 10.19.) The set D is not a unique minimum dominating set, since $D' = \{u_2, u_5, u_9\}$ and $D'' = \{u_4, u_7, u_9\}$ are also dominating sets of G with the same number of vertices, and there are still more! (See Exercise 10.11.)

In this example, we have that $\gamma(G) = 3$, which means that the state in question needs to set up police stations only in three chosen cities.

Example 10.14 Let $k \in \mathbb{N}$ and consider P_{3k}, the simple path on the vertices $u_1, \ldots, u_{3k}$ and the edges $e_i = \{u_{i-1}, u_i\}$ for $2 \leq i \leq 3k$. The set $D = \{u_2, u_5, \ldots, u_{3k-1}\}$ on k vertices is a dominating set of P_{3k}, since for each $i \in \{1, \ldots, n\}$ the vertices subscripted $3i - 1$ are clearly dominated by D (they are in D), and each vertex subscripted $3i - 2$ or $3i$ is adjacent to the vertex subscripted $3i - 1$ in P_{3k}. Also, since each vertex in P_{3n} can dominate at most three vertices, namely the ones in its closed neighborhood, any set of $k - 1$ vertices can dominate at most $3(k - 1) < 3k$ vertices of P_{3k}. Hence, D is a minimum dominating set of P_{3k} and therefore $\gamma(P_{3k}) = k$.

Furthermore, in this case D is the unique minimum dominating set. (See Exercise 10.12.)

Remark 10.15 For a graph G we see that the following hold:

- Every minimum dominating set is a minimal dominating set.
- A graph can have many minimum dominating sets.
- A dominating set may or may not be an independent set.
- Loops and multiple edges do not affect domination.

By Remark 10.15, we may assume that our graphs are simple when dealing with their dominating sets.

Consider a simple graph G and let $S \subseteq V(G)$ be a maximal independent set of G. By the maximality of S, every vertex $u \in V(G) \setminus S$ is adjacent to some vertex in S. Thus, S is by definition a dominating set of G that, in addition, is independent. Conversely, if $D \subseteq V(G)$ is a dominating set of G that is independent, then D is a maximal independent set, since every vertex in $V(G) \setminus D$ is adjacent to some vertex in D. We summarize in the following.

Observation 10.16 For a graph G we have

$$\gamma(G) \leq \alpha(G).$$

In fact, a set of vertices $S \subseteq V(G)$ is a maximal independent set if, and only if, it is dominating and independent.

Example 10.17 The lower bound given in Observation 10.16 is not a good bound in the general case. In fact, the ratio $\alpha(G)/\gamma(G)$ can be arbitrarily large. For $n \in \mathbb{N}$, let G be the star on $n + 1$ vertices. Clearly, $\alpha(G) = n$ and $\gamma(G) = 1$, so the ratio $\alpha(G)/\gamma(G) = n$ can be made arbitrarily large.

As was the case for finding a maximum independent set, there is no easy way to find or "spot" a minimum dominating set in a graph G. There are, though, some observations worth mentioning, when determining $\gamma(G)$ and a corresponding dominating set D with $|D| = \gamma(G)$. The proof of the following observation is left as Exercise 10.17.

Observation 10.18 For a graph G we have the following:

1. If G has components $C_1, \ldots, C_k$, then

$$\gamma(G) = \sum_{i=1}^{k} \gamma(C_i).$$

2. If G is connected with blocks $B_1, \ldots, B_k$, then

$$\gamma(G) \leq \sum_{i=1}^{k} \gamma(B_i). \tag{10.3}$$

Furthermore, equality holds in (10.3) if, and only if, G has a minimum dominating set D such that (i) each $D_i = D \cap V(B_i)$ is a minimum dominating set of the block B_i, and (ii) the D_i's are pairwise disjoint.

Given a graph G, an obvious question is how one determines $\gamma(G)$. As with many mathematical problems, where one has to determine the number of elements in a largest or a smallest set among many feasible subsets having a certain property, this problem of determining the domination number is in general a hard one. In many particular cases one can use logical arguments, applicable only to the graph G in question, to determine $\gamma(G)$.

In the following example, note that for every vertex $u \in V(G)$, the closed neighborhood $N_G[u]$ contains at least one vertex contained in every dominating set of G—in particular, in every minimum dominating set.

Example 10.19 Continuing from Example 10.13, to determine $\gamma(G)$ of the graph G shown in Figure 10.5, one could argue as follows: Since $N[u_1] = \{u_1, u_9\}$ and $N[u_5] = \{u_4, u_5, u_6\}$, then each of these two sets must contain at least one vertex of every dominating set D of G—in particular, of every minimum dominating set of G. However, these two closed neighborhoods together dominate all but one vertex u_3 of G, as shown in Figure 10.6, where the sets $N[u_1]$ and $N[u_5]$ are indicated with closed curves, and the sets that they dominate are indicated with dotted curves. Hence, the domination number of G cannot possibly be two. Since we know from Example 10.13 that there is a dominating set containing three vertices, we have that $\gamma(G) = 3$.

The expanded statement of the idea used here is set forth more precisely in Exercise 10.22.

When the graph G in question is connected but "sparse," the following lemma can be of value to yield a lower bound quickly; often this bound turns out to be the correct value of $\gamma(G)$.

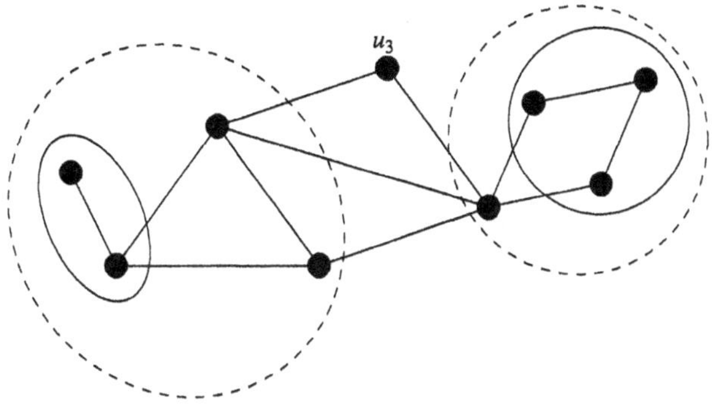

Figure 10.6: The closed neighborhoods $N[u_1]$ and $N[u_5]$ together dominate all but one vertex u_3 of G.

Lemma 10.20 For a connected graph G, we have

$$\gamma(G) \geq \left\lceil \frac{\text{diam}(G) + 1}{3} \right\rceil.$$

Proof Let u_0 and u_k be vertices of G of distance $\partial_G(u_0, u_k) = \text{diam}(G)$ from each other, and let $p = (u_0, u_1, \ldots, u_k)$ be a path of length $k = \text{diam}(G)$. Note, without loss of generality, we can assume that G is simple, so we have omitted the edges in the specification of p.

If D is a dominating set of G, then D must, in particular, dominate each vertex u_i on the path p. Clearly, every vertex u_i in D that lies on p can dominate at most three vertices on p: itself and its (at most) two neighbors on p, namely u_{i-1} and u_{i+1}. Also, every vertex in D that is not on p can dominate at most three vertices on p, which in addition must be consecutive on the path p, since otherwise we have a shorter path between u_0 and u_k, contradicting the definition of diam(G).

Therefore, we have that each vertex in D can dominate at most three vertices on p. Since the number of vertices of p is $k + 1$, we have that

$$3|D| \geq k + 1.$$

Since this inequality holds for all dominating sets D, and $|D|$ is an integer, the lemma follows. ■

We now want to describe a systematic way to determine the domination number $\gamma(G)$ of a graph G and to find a minimum dominating set D with $|D| = \gamma(G)$. In general, there is no efficient way to compute $\gamma(G)$, as there is no efficient way to compute the independence number $\alpha(G)$, but when the graph in question does not have too many vertices and edges, the following method can be of value.

Let G be a graph. We would like to compute the domination number $\gamma(G)$ of G, and without loss of generality we can assume that G is simple. Let $e = \{u, v\} \in E(G)$ be an edge. Let

$$E_v(u) = \{e' \in E(G) : u \in e', e' \neq \{u, v\}\},$$
$$E_u(v) = \{e' \in E(G) : v \in e', e' \neq \{u, v\}\};$$

that is, $E_v(u)$ is the set of edges incident to u not including the edge e, and $E_u(v)$ is the set of edges incident to v other than e, as depicted in Figure 10.7. With this setup, we have the following theorem.

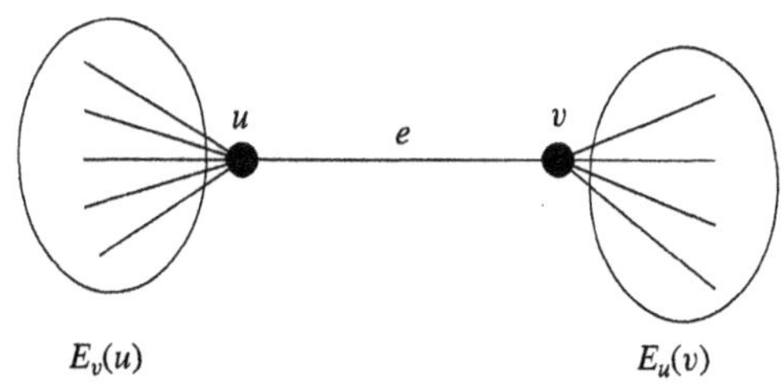

Figure 10.7: For the edge $e = \{u, v\}$ we have the edge sets $E_v(u)$ and $E_u(v)$.

Theorem 10.21 Let G be a simple graph and $e = \{u, v\} \in E(G)$ an edge. Let $E_v(u)$ and $E_u(v)$ be the sets of edges as described previously. In this case the domination number $\gamma(G)$ satisfies

$$\gamma(G) = \min(\{\gamma(G - e), \gamma(G - E_v(u)), \gamma(G - E_u(v))\}). \tag{10.4}$$

Proof Clearly, by removing any set of edges from G, we can only increase the domination number of G. Therefore, we have

$$\gamma(G) \leq \min(\{\gamma(G - e), \gamma(G - E_v(u)), \gamma(G - E_u(v))\}).$$

We now show that equality holds in (10.4).

The removal of the edge $e = \{u, v\}$ from G is not going to increase the domination number in the following cases:

1. G has a minimum dominating set that does not contain either u or v.
2. G has a minimum dominating set containing both u and v.
3. G has a minimum dominating set containing u and not v, and contains a neighbor of v other than u.
4. G has a minimum dominating set containing v and not u, and contains a neighbor of u other than v.

Since the removal of e in any of these four cases is not going to affect the current domination, the domination number will not be increased. Hence, in any of these four cases, we have that $\gamma(G) = \gamma(G - e)$.

The remaining cases are the following two:

1. G has a minimum dominating set D containing exactly one vertex from $N_G[u]$, namely v.
2. G has a minimum dominating set containing exactly one vertex from $N_G[v]$, namely u.

In the former case, none of the edges in $E_v(u)$ have endvertices contained in the dominating set D; hence, their removal will not affect the current domination of G. So here we have that $\gamma(G) = \gamma(G - E_v(u))$. Similarly, in the latter case, the removal of the edges in $E_u(v)$ will not affect the domination of G; therefore, we have that $\gamma(G) = \gamma(G - E_u(v))$.

We can conclude that the smallest number among $\gamma(G - e)$, $\gamma(G - E_v(u))$, and $\gamma(G - E_u(v))$ is precisely $\gamma(G)$, depending on the corresponding case. This completes the proof. ■

In general, using (10.4) recursively is not very efficient for large graphs, especially when using it blindly, without any consideration of the choice of the edge $e = \{u, v\}$. The structure of G determines which edge e provides the best choice. In this case it is a good rule of thumb to choose an edge e that is a bridge, or an edge whose removal has a large impact on destroying the connectivity of G. As a second rule, we should try to have $d_G(u) + d_G(v)$ as large as possible, so that we can remove as many edges as possible at each phase of our computation.

Example 10.22 Consider the graph G on nine vertices $V(G) = \{u_1, \ldots, u_9\}$ that is shown in Figure 10.8. Here G has a bridge $\{u_3, u_8\}$, the removal of which yields two components that are easier to process. Using (10.4) on the edge $\{u_3, u_8\}$, we obtain

$$\begin{aligned}\gamma(G) &= \min(\{\gamma(G - \{u_3, u_8\}), \gamma(G - E_{u_8}(u_3)), \gamma(G - E_{u_3}(u_8))\}) \\ &= \min(\{\gamma(G_1), \gamma(G_2), \gamma(G_3)\}),\end{aligned}$$

where $G_1 = G - \{u_3, u_8\}$, $G_2 = G - \{\{u_2, u_3\}, \{u_3, u_9\}\}$, and $G_3 = G - \{\{u_4, u_8\}, \{u_7, u_8\}\}$ are the graphs shown in Figure 10.9.

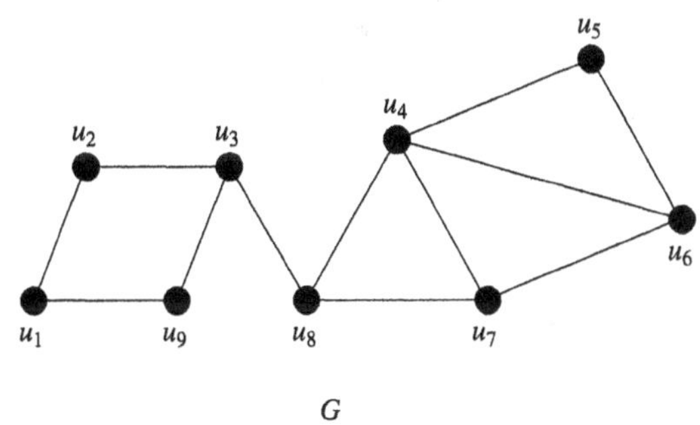

Figure 10.8: Finding the domination number of a graph G in a systematic way.

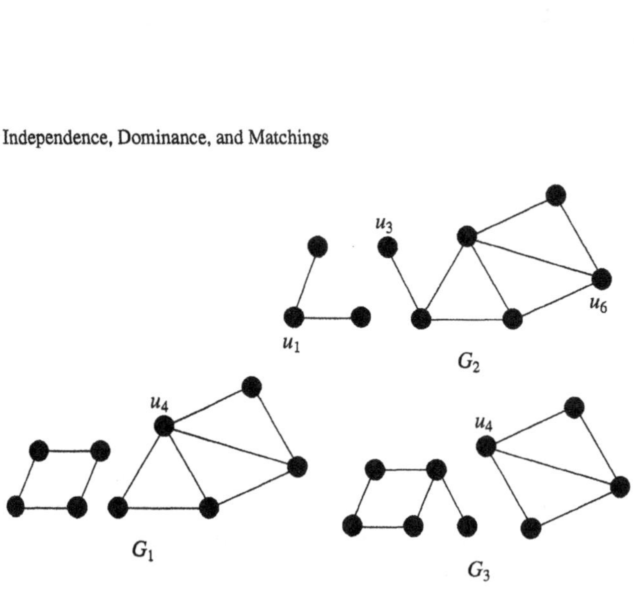

Figure 10.9: By computing $\gamma(G_1)$, $\gamma(G_2)$, and $\gamma(G_3)$, we can determine $\gamma(G)$.

To determine $\gamma(G_1)$, we note that the smaller component of G_1 has a domination number of two, and the larger component has a domination number of one, since u_4 dominates every vertex in that component. Hence, we have $\gamma(G_1) = 3$.

To determine $\gamma(G_2)$, we note that the smaller component of G_2 is dominated by u_1, and the larger component has a diameter of three. By Lemma 10.20, the larger component has a domination number of at least two. Therefore, we have that $\gamma(G_2) = 3$.

Finally, the component of G_3 on four vertices is dominated by u_4. The other component on five vertices has a diameter of three, and hence has a domination number of at least two. Thus, we have $\gamma(G_3) = 3$. All in all, we have

$$\gamma(G) = \min(\{\gamma(G_1), \gamma(G_2), \gamma(G_3)\}) = 3.$$

Note 10.23 As for the systematic way of determining the independence number $\alpha(G)$ recursively by (10.2), by keeping track of the graphs with domination number $\gamma(G)$ when using (10.4) recursively, and ignoring those with higher domination numbers, we can in fact determine *all* the minimum dominating sets of G as well!

Domination of vertices has been studied extensively, and we have only scratched the surface of it. For those who want to study domination in depth we refer to two excellent books on the matter, [27] and [26], both by Haynes, Hedetniemi, and Slater. The first book is the first (long-awaited) comprehensive treatment of domination in graphs and the other deals with some collected advanced material.

10.3 Matchings in a Graph

When dealing with an independent set of edges, one is often more concerned with the objects represented by the endvertices than with the edges themselves. Each edge unites or *matches* a pair of vertices of G. In fact, an independent set of edges in G is called a *matching*. Many assignment problems can be formulated as finding a special matching in a bipartite graph, where the vertices in one part represent a given set of agents and the vertices in the other part represent projects to be assigned to the agents.

Obviously, an empty set of edges, and every set containing a single edge, is an independent set of edges, and hence a matching. Thus, we are interested in matchings that are *large* in cardinality, especially the largest or the *maximum* matchings in a given graph.

This section is devoted to matchings in a graph G and a description of a maximum matching M in G in terms of M*-augmenting paths* in G. A matching of G is related to an edge coloring of G, since each color class of a proper edge coloring of G yields an independent set of edges. As mentioned, we are usually interested in interpreting the meaning of an independent set of edges in terms of the endvertices of the edges. The vertices of a graph often represent two *different* kinds of objects, so that each edge in a set of independent edges matches two different types of vertices. Therefore, we are mostly interested in matchings in bipartite graphs, where the vertices in one part represent objects of one kind, and the vertices of the other part represent objects of another (related) kind.

At first glance, one might think that limiting one's attention to bipartite graphs, instead of discussing matchings in general graphs, is overly restrictive. It turns out, however, that Tutte's Theorem from 1947, which completely characterizes when a graph G has a matching covering all of its vertices (that is, a matching of G, which is also an edge cover of G), is, in fact, a *consequence* of Hall's Marriage Theorem from 1935, which describes matchings in bipartite graphs. We will address this point further in §10.4.

Example 10.24 Consider Problem 1.3, again, where we have seven applicants $A_1, \ldots, A_7$ for a job at a company EX that has five jobs $J_1, \ldots, J_5$ to offer. For each specific job J_i certain qualifications are needed, which a given applicant A_j either does or does not possess, and where an edge between applicant A_j and job J_i indicates that applicant A_j is qualified for job J_i. For example, the jobs J_1 and J_4 might involve managing employees at an international level, for which an applicant must have good people skills and must be able to speak several languages well for smooth communications, whereas jobs J_2 and J_5 might involve computer programming, for which an applicant needs some basic computer programming skills but few people skills. The ultimate goal for company EX is to hire as many qualified applicants as possible. The applicants and the jobs they are qualified for are as follows:

$$\begin{aligned} A_1&: J_5,\\ A_2&: J_1,\ J_2,\ J_4,\\ A_3&: J_2,\ j_3,\\ A_4&: J_5,\\ A_5&: J_2,\ J_3,\\ A_6&: J_5,\\ A_7&: J_5. \end{aligned}$$

The situation is completely described by the bipartite graph shown in Figure 1.4. The question becomes: Is it possible for EX to hire qualified applicants for all of its current openings? If not, for how many jobs can EX hire qualified applicants? These questions are equivalent to finding a matching of maximum size (that is, with the maximum number of edges) for the graph depicted in Figure 1.4.

EX might even have a more specific scaling or rating system for the applicants in their hiring process, where the company EX scales the qualifications that applicant A_j has for the job J_i on a scale from 0 to 100. In this case, we can consider a weighted bipartite graph (G, W), where

$$W(\{A_j, J_i\}) = w_{ij} \in \{0, 1, 2, \ldots, 100\}$$

for each $1 \le i \le 5$ and $1 \le j \le 7$. For example, a low weight, $w_{ij} = 26$, indicates that applicant A_j has some qualifications for the job I_j, but is not a stellar choice for the position. A larger weight, say $w_{ij} = 83$, indicates that the applicant A_j is highly qualified for the job J_i. Note that a weight of zero, $w_{ij} = 0$, indicates that applicant A_j is not qualified for the job J_i at all, something that is represented by omitting the edge between A_j and J_i in the graph G in Figure 1.4. Hence, for this weighted problem, it makes no difference whether we consider the complete bipartite graph $K_{5,7}$ and the edges which have nonnegative weight, or the graph G in Figure 1.4, with edges of positive weights as shown in Figure 10.10. Here, the ultimate goal for company EX is to hire applicants with the maximum total qualifications score. This result is the same as finding a matching of maximum weight (the weight of the matching being the sum of the weights of the edges in the matching) for the graph G in Figure 10.10.

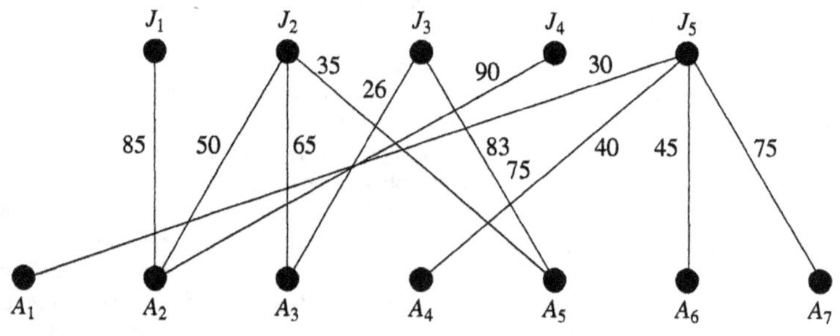

Figure 10.10: Bipartite graph describing job qualification.

In order to solve matching problems in general, we need to formalize the intuitive ideas just presented by introducing precise definitions and theorems. Our first definition is related to the notion of domination discussed in §10.2.

Definition 10.25 Let G be a graph, $L \subseteq V(G)$ a set of vertices of G, and $N \subseteq E(G)$ a set of edges of G.

1. The set L *covers* N if $\phi(e) \cap L \ne \varnothing$ for all $e \in N$—that is, if every edge in N has at least one endvertex in L. The set L is called a *vertex cover* if L covers all of $E(G)$.
2. The set N *covers* L if $L \subseteq \bigcup_{e \in N} \phi(e)$—that is, if every vertex in L is an endvertex of some edge in N. The set N is called an *edge cover* if N covers all of $V(G)$.

Figure 10.11 shows an example of a vertex cover that consists of the vertices contained in the oval region. There, every edge has at least one endvertex located inside the oval.

Definition 10.26 Let G be a graph. A subset $M \subseteq E(G)$ containing no loops is a *matching* of G if $\phi(e) \cap \phi(e') = \varnothing$ for all distinct pairs of edges $e, e' \in M$. That is, no two edges in M can be adjacent.

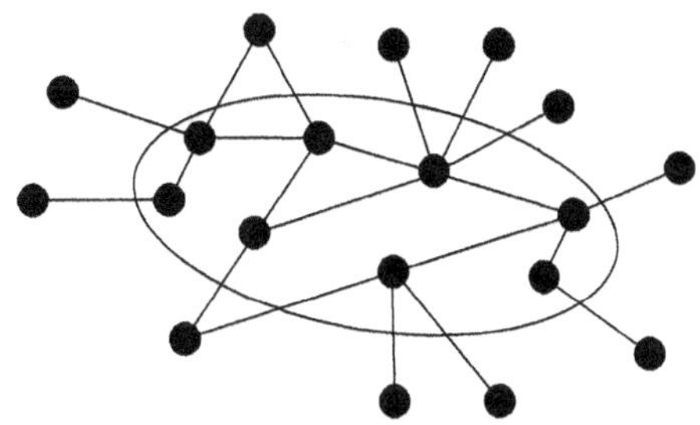

Figure 10.11: Every edge has one endvertex contained inside the oval region.

1. A matching M is *maximal* if M is not a proper subset of any matching of G.
2. A matching M is *maximum* if no other matching covers a greater number of vertices than M.
3. A matching M is *perfect* if M is an edge cover of G—that is, M covers all of $V(G)$.

Example 10.27 In Figure 10.12 we see two sets of nonadjacent edges of a graph G. Clearly, both are maximal, since it is impossible to add an edge to either one of them without obtaining a pair of adjacent vertices. Also, both sets are of size three. The set on the right is a perfect matching, since it covers all the vertices of G; hence, it is also maximum. The set on the left is not a matching, since it contains a loop.

Although matchings are well defined for graphs in general, it is somewhat counterintuitive to consider matchings in graphs with loops. Hence, in our discussion on matchings, we can assume that G has no loops. With the main definitions of this section at hand, we make some easy observations in the following remark.

Remark 10.28 For a graph G we have the following:

- A maximum matching in G is a maximal matching.
- A perfect matching in G is a maximum matching.
- A maximum matching in G is not necessarily perfect.

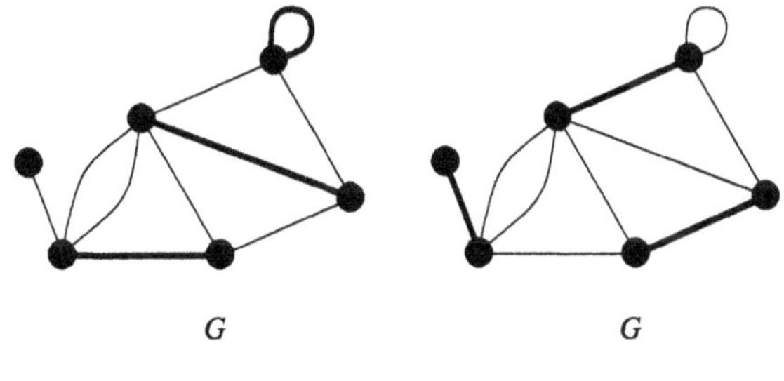

Figure 10.12: Two sets of nonadjacent edges of G are depicted. The one on the right is a perfect matching; the one on the left is not a matching.

- The graph G can have many distinct perfect matchings.
- A matching in G can have at most $\lfloor n/2 \rfloor$ edges.
- A matching is perfect if, and only if, it has $n/2$ edges. In particular, if G has a perfect matching, then G has an even number of vertices.
- If G is bipartite with bipartition $V(G) = X \cup Y$, then the size of a maximum matching of G is at most $\min(\{|X|, |Y|\})$. In particular, if $|X| \neq |Y|$, then G cannot have a perfect matching.

To check whether a given matching in a graph is maximal or not is an easy task. We simply process the remaining edges and see if one of them can be added to the matching to obtain a strictly larger matching. However, given a maximal matching M that is not perfect, it is a priori not easy to determine whether the matching is maximum or not. As we saw in Example 10.27, a maximal matching can be either maximum or not.

Example 10.29 Consider maximal matchings of the simple path P_{2n}. Assuming that $V(P_{2n}) = \{u_1, \ldots, u_{2n}\}$ and $E(P_{2n}) = \{\{u_i, u_{i+1}\} : i \in \{1, \ldots, 2n-1\}\}$, the following are two maximal matchings for P_{2n}:

$$\begin{aligned} M_1 &= \{\{u_{2k}, u_{2k+1}\} : 1 \leq k \leq n-1\}, \\ M_2 &= \{\{u_{2k-1}, u_{2k}\} : 1 \leq k \leq n\}. \end{aligned}$$

In both of these maximal matchings for P_{2n}, the edges along the path are alternately contained in the matching or not.

This simple example is a fundamental one and plays a key role in a characterization of maximum matchings for any graph.

Example 10.29 gives rise to the following definition.

Definition 10.30 Let M be a matching in a graph G and $k \in \mathbb{N}$.

1. A simple path $p = (u_0, e_1, u_1, \ldots, e_k, u_k)$ in G is called an *M-alternating path* if the edges $e_1, e_2, \ldots, e_k$ are alternately in M and $E(G) \setminus M$. (Note, the alternation can begin with e_1 being in either set.)
2. A simple path $p = (u_0, e_1, u_1, \ldots, e_k, u_k)$ in G is called an *M-augmenting path* if p is an M-alternating path in G and neither u_0 nor u_k is covered by M. That is, they are not endvertices of any edges in M.

Note 10.31 If $p = (u_0, e_1, u_1, \ldots, e_k, u_k)$ is M-augmenting, then the edges $e_2, e_4, \ldots, e_{k-1}$ are in M. Thus $k-1$ must be even. That is, k, the length of p, must be odd, so $k = 2\ell + 1$ for some $\ell \in \mathbb{N}$.

In Figure 10.13 we see the path graph P_8 with a matching $M = \{e_1, e_2, e_3\}$. The path is M-augmenting, since its endvertices are not covered by M. The following theorem, which ties together maximum matchings and augmenting paths, is due to the French graph theorist Claude Berge from 1957.

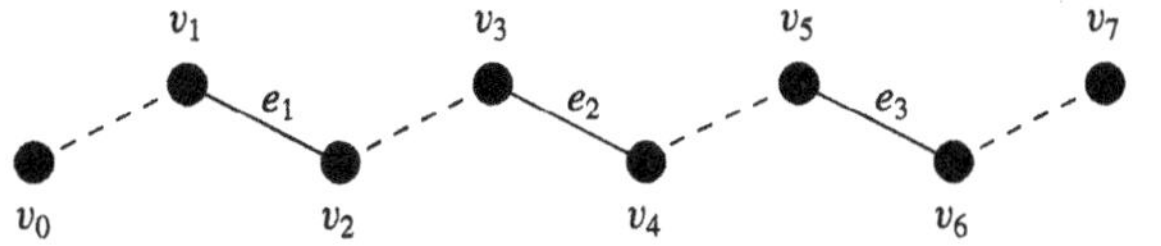

Figure 10.13: An augmenting path of the graph P_8.

Theorem 10.32 Let G be a graph and M a matching in G. The following are equivalent:

1. M is a maximum matching of G.
2. G has no M-augmenting path.

Proof Assume that there is an M-augmenting path, which by Note 10.31 has the form $p = (u_0, e_1, u_1, \ldots, u_{2k}, e_{2k+1}, u_{2k+1})$, where the edges $e_2, e_4, \ldots, e_{2k}$ are in M. If $E(p)$ is the set of edges in p, then we obtain a new matching $M \mathbin{\triangle} E(p)$ given by

$$M \mathbin{\triangle} E(p) = (M \setminus \{e_2, e_4, \ldots, e_{2k}\}) \cup \{e_1, e_3, \ldots, e_{2k+1}\},$$

which has strictly more edges than M. Hence, M is not a maximum matching.

The other part is more involved. Assume that M is not a maximum matching. We need to show that there is an M-augmenting path in G. Let M' be a maximum matching in G, and consider the symmetric difference

$$M \mathbin{\triangle} M' = (M \setminus M') \cup (M' \setminus M)$$

of these two edge sets. Let $H = G[M \mathbin{\triangle} M']$ be the corresponding induced subgraph in G. Each vertex in H is an endvertex of an edge in $M \setminus M'$ or an edge in $M' \setminus M$. So, $d_H(u) \geq 1$ for every vertex $u \in V(H)$. Also, each vertex $u \in V(H)$ can be an endvertex of at most two edges, one from M and the other from M'. Thus, $d_H(u) \leq 2$ for all $u \in V(H)$. Therefore, H is a subgraph of G in which every vertex has degree of one or two in H.

Any graph in which every vertex has degree one or two must be a disjoint union of simple paths and cycles, as depicted in Figure 10.14. Clearly, no loops in G are contained in either M or M'. Each component of H consisting of a cycle has alternating edges in M and M', and so each cycle in H must have an even length. Hence, the number of edges from M and M' in each cycle of H, and in each path of even length in H, is the same.

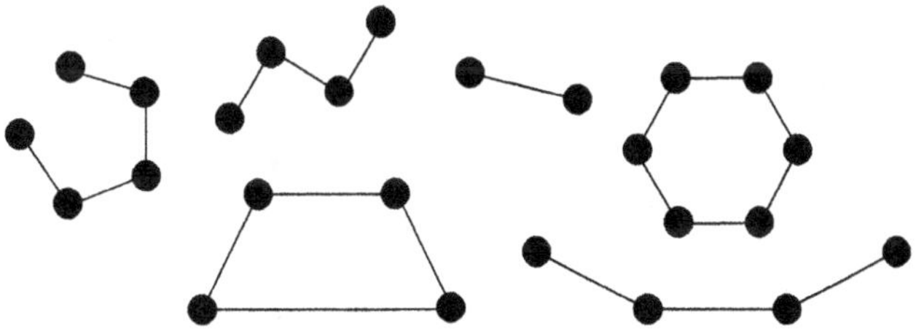

Figure 10.14: The subgraph H induced by the symmetric difference of the two matchings M and M' is a disjoint union of paths and cycles.

By assumption M is not a maximum matching, but since M' is maximum, we have $|M'| > |M|$. Hence, H must have at least one component consisting of a path $p = (u_0, e_1, u_1, \ldots, u_{2k}, e_{2k+1}, u_{2k+1})$ of an odd length, in which the evenly labeled edges are in M and the odd ones in M'. In this case neither u_0 nor u_{2k+1} is covered by M. So, p is an M-augmenting path. This completes the proof. ■

10.4 Hall's Marriage Theorem

Hall's Marriage Theorem, which we discuss in this section, provides a necessary and sufficient condition for a bipartite graph to have a matching that covers all of the vertices in the smaller part of the vertex partition.

Any graph has a maximum matching but not necessarily a perfect matching—the "best" matching possible. As we noted in Remark 10.28, bipartite graphs with an uneven partition of vertices never have a perfect matching. Nevertheless, we can hope for a matching that covers the *smaller* part of the partition of vertices.

Recall that for $S \subseteq V(G)$, the set $N(S) = N_G(S)$ of neighbors of S denotes the set of vertices adjacent to at least one vertex of S. Consider a bipartite graph G with a partition $V(G) = X \cup Y$, where $|X| \leq |Y|$. If G has a matching covering all of X, then clearly for every $S \subseteq X$, the set $N_G(S) \subseteq Y$ must contain at least as many vertices as S, since in that case M provides a one-to-one correspondence between any $S \subseteq X$ and a subset of $N_G(S) \subseteq Y$.

The interesting part, however, is that this is sufficient for G to have a matching covering all of X, which is precisely what Hall's Marriage Theorem states. This theorem from 1935 is due to the English mathematician Philip Hall.

Theorem 10.33 **Hall's Marriage**

Let G be a bipartite graph with bipartition $V(G) = X \cup Y$. The following are equivalent:

1. The graph G has a maximum matching M that covers all of X.
2. For all $S \subseteq X$, we have $|S| \leq |N(S)|$.

Proof We need to show only that the second condition implies the first one, so we assume that $|S| \leq |N(S)|$ for any $S \subseteq X$. We want to show that if M is a maximum matching of G, then M covers all of X. Assume this not to be the case. Then there is an $x \in X$ that is not covered by M. Let $A \subseteq V(G)$ be the set of vertices that can be connected to x with an M-alternating path in G. Let $S = A \cap X$ and $T = A \cap Y$. In Figure 10.15 we see a portion of G, where the thick lines indicate edges in the matching M and the dotted lines indicate edges in G that are not in M. Figure 10.15 shows, for instance, that $y \in S$ and $z, w \in T$. Since M is a maximum matching, using Theorem 10.32, each $a \in A \setminus \{x\}$ is covered by M. Therefore, M yields a one-to-one correspondence between $S \setminus \{x\}$ and T. Thus, $|T| = |S| - 1$.

Each vertex in T is an endvertex of an edge in M, whose other endvertex is in S. So, $T \subseteq N(S)$. Also, by definition we have that $N(S) \subseteq Y$. Furthermore, from x to each element in S there is an M-alternating path, the last edge of which is from M. We see that

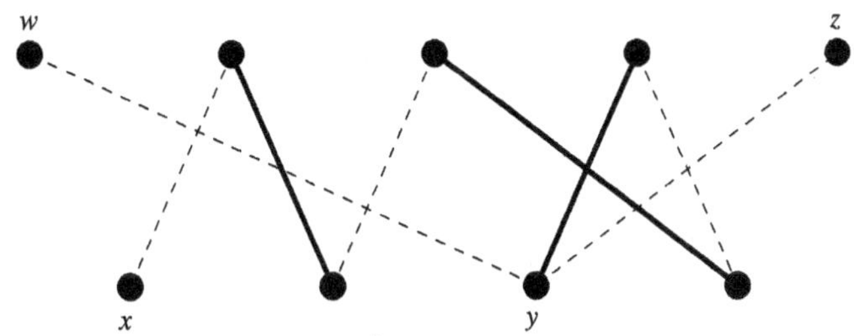

Figure 10.15: Here $y \in A \cap X = S$ and $z, w \in A \cap Y = T$.

there is also an M-alternating path from x to any neighbor of S. Hence, by definition of T, we have that $N(S) \subseteq T$. This fact means that

$$|N(S)| = |T| = |S| - 1 < |S|,$$

which contradicts the second condition. This completes the proof. ■

Remark 10.34

- The name "Marriage Theorem" comes from the following presentation: Suppose we have two groups of people, one of women and the other of men. Then each woman can marry a man that she likes if, and only if, each group of women collectively likes at least as many men.
- The case of Hall's Marriage Theorem, when $|X| = |Y|$, was first proved by Frobenius in 1917.
- The second condition in Hall's Marriage Theorem is called *Hall's Condition.*
- As mentioned in West [57], Hall's Marriage Theorem is, as the English graph theorist Nash-Williams and later other researchers called it, a *TONCAS* theorem: "These Obvious Necessary Conditions Are Sufficient." Nash-Williams (who died in January 2001) was a true champion in graph theory, with many strong and beautiful results in the field to his credit. He was also a genuine English gentleman, not in the posh "snobbish" sense, but rather in the true meaning of the phrase.

We illustrate the use of Hall's Marriage Theorem in the next example.

Example 10.35 Consider the graph G shown in Figure 1.4 and used in Example 10.24. To find out whether the company EX is able to hire for all of its jobs, we note that for $S = \{J_1, J_4\} \subseteq \{J_1, \ldots, J_5\}$, we have

$$|N_G(S)| = |N_G(\{J_1, J_4\})| = |\{A_2\}| = 1 < 2 = |\{J_1, J_4\}| = |S|.$$

Hence, by Hall's Marriage Theorem, EX is not able to hire for all the available jobs.

Example 10.36 In a certain high school in the Shenandoah Valley in Virginia, there is a $k \in \mathbb{N}$ such that each senior girl would like to be accompanied by any one of exactly k senior boys to the senior promenade (that is, "the prom"), and each boy would like to take any one of exactly k girls to the prom as well. Show that every girl and boy in this senior class can happily go to the prom accompanied by a member of the opposite sex whom they like.

We form a bipartite graph G with the vertex set consisting of the senior girls, A, and the senior boys, B. An edge between a girl a_i and a boy b_j indicates that the girl a_i and the boy b_j would like to accompany each other to the senior prom. We want to show that G has a perfect matching: The fact that each girl likes k senior boys means that each vertex $a_i \in A$ is adjacent to k vertices $b_j \in B$—that is, $d_G(a_i) = k$ for all i. Likewise, $d_G(b_j) = k$ for all j. Here we have that

$$k|A| = |E(G)| = k|B|;$$

hence, $|A| = |B|$—that is, the numbers of senior girls and boys are equal. Therefore, it suffices to show that there is a matching that covers all of A, since such a matching automatically covers all of B as well, and hence is perfect. So, if $S \subseteq A$, then there are exactly $k|S|$ edges going from S to $N_G(S) \subseteq B$. Since each vertex $b_j \in N_G(S)$ has exactly k edges coming into it, then at most k of those edges are coming from S. Therefore, we have

$$|N_G(S)| \geq \frac{k|S|}{k} = |S|.$$

By Hall's Marriage Theorem 10.33 the bipartite graph G on $A \cup B$ has a matching covering all of A; hence, G has a perfect matching. Therefore, all the seniors can go happily to the prom with a member of the opposite sex whom they like.

By Example 10.36, we have the following corollary.

Corollary 10.37 If $k \in \mathbb{N}$, then every k-regular bipartite graph has a perfect matching.

From this result we obtain (Exercise 10.32) the following corollary, which is also a direct consequence of Theorem 8.41.

Corollary 10.38 If $k \in \mathbb{N}$, then every k-regular bipartite graph is k-edge colorable.

So far, we have discussed conditions to determine whether or not there is a matching that covers all of the vertices in one part of a bipartite graph. In the case where we can conclude that there is no such matching, we would still like to know *how many* vertices we can cover, even though we cannot cover them all. The following definition captures this notion formally.

Definition 10.39 Let G be a bipartite graph with bipartition $V(G) = X \cup Y$. Define the *deficiency* δ_X of G at X by

$$\delta_X = \max(\{|S| - |N(S)| : S \subseteq X\}).$$

Note 10.40

- Since $\varnothing \subseteq X$, we have $\delta_X \geq |\varnothing| - |N(\varnothing)| = 0$.
- G has a matching covering all of X if, and only if, $\delta_X = 0$.

The following theorem is due to the Norwegian graph theorist Oeystein Ore.

Theorem 10.41 Let G be a bipartite graph with bipartition $V(G) = X \cup Y$. Then the maximum number of vertices of X that can be covered by a matching of G is $|X| - \delta_X$.

Proof Add δ_X new vertices to Y and connect each of them to all of the vertices of X. By Hall's Marriage Theorem 10.33, this new bipartite graph has a matching that covers all of the vertices in X. Since at most δ_X of the edges in this matching are not in G, we have the theorem. ■

We illustrate Theorem 10.41.

Example 10.42 Consider the graph G shown in Figure 1.4 and used in Example 10.24. As demonstrated in Example 10.24, the graph G does not have a matching that covers all of the J_i vertices, representing the jobs that the company EX has to offer. That is, EX is not able to hire for all of its job openings. Hence, a natural question is: *How* many open positions can EX fill?

Here, as seen in Figure 1.4, the deficiency is given by

$$\delta = \delta_{\{J_1,\dots,J_5\}} = 1,$$

which is obtained with $S = \{J_1, J_4\}$—that is,

$$\delta_X = \max(\{|S| - |N(S)| : S \subseteq X\}) = |\{J_1, J_4\}| - |\{A_2\}| = 2 - 1 = 1.$$

Hence, by Ore's Theorem 10.41, a maximum matching of G covers at most $|\{J_1, \dots, J_5\}| - \delta = 5 - 1 = 4$ vertices. Therefore, the matching M given by

$$M = \{\{A_2, J_1\}, \{A_3, J_2\}, \{A_5, J_3\}, \{A_6, J_5\}\}$$

is a maximum matching of G.

Remark 10.43 Let G be a bipartite graph with bipartition $V(G) = X \cup Y$ and $m \in \mathbb{N}$. Any matching of G covering m vertices in X also covers m vertices of Y, and vice versa. Hence,

$$|X| - \delta_X = |Y| - \delta_Y.$$

Moreover, this value is the largest number of edges that a matching of G can have. We will have more to say about this in Chapter 11.

We finish this section by briefly discussing matchings in graphs that are not necessarily bipartite.

Consider a graph G that has a perfect matching $M \subseteq E(G)$. Let $S \subseteq V(G)$ and let $C_1, \dots, C_k$ be all of the components of $G - S$. Clearly, the collection of components depends on the set S of vertices. Let $o(G - S)$ be the number of components with an odd number of vertices. Clearly, we have $\ell = o(G - S) \leq k$. Also, since M is a perfect matching, then each of the ℓ odd components must contain a vertex that is matched to another vertex, which must lie in S. All of these vertices in S must be distinct, so we have, in particular, that S contains at least ℓ vertices, since M is a matching. That is,

$$o(G - S) \leq |S|,$$

for all $S \subseteq V(G)$. In Figure 10.16 we depict the components of $G - S$ and show how the odd ones must have a vertex matched to distinct vertices in S.

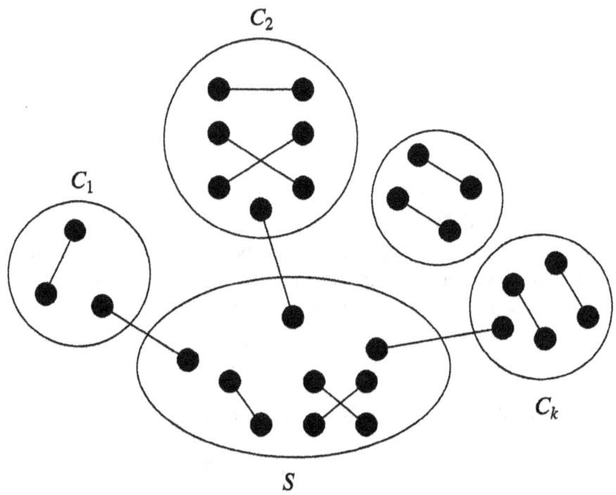

Figure 10.16: A graph G, a set $S \subseteq V(G)$, and odd components of the graph $G - S$.

Also, here it is interesting that this condition is sufficient for the graph G to have a perfect matching. This is precisely what is stated by Tutte's Theorem, which is due to the Canadian graph theorist William Tutte and is from 1947. Tutte (who died in May 2002) was a true heavyweight champion of graph theory.

Theorem 10.44 **Tutte's**

For a graph G, the following are equivalent:

1. The graph G has a perfect matching M.
2. For all $S \subseteq X$, we have $o(G - S) \leq |S|$.

Remark 10.45

- The second condition in Tutte's Theorem is called *Tutte's Condition*.
- Like Hall's Marriage Theorem 10.33, Tutte's Theorem is yet another *TONCAS* theorem.

Many proofs of Tutte's Theorem 10.44 have appeared since he presented his original proof in 1947. For some readable, well-presented proofs, we refer the reader to Bollobás [5], Diestel [12], and West [57].

It is interesting to note that Hall's Theorem and Tutte's Theorem are, in some sense, equivalent: Hall's Theorem 10.33 can be used to prove Tutte's Theorem 10.44 and vice versa, as described in West [57]. Hence, our decision to restrict our attention mostly to matchings in bipartite graphs is not such a big restriction after all.

10.5 Exercises

10.1. Show that the graph depicted in Figure 10.2, which was discussed in Example 10.4, has a *unique* maximum independent set S given by $S = \{u_1, u_3, u_4, u_6, u_8\}$.

10.2. Prove Observation 10.8.

10.3. For a graph G and a vertex $u \in V(G)$, let G'_u be the contraction of G obtained by contracting the closed neighborhood $N_G[u]$ into one vertex u' of G'_u. Show that

$$\alpha(G) = \max(\{\alpha(G - u), 1 + \alpha(G'_u - u')\}).$$

10.4. If G is a graph that has a maximum independent set of vertices, which contains no cut-vertex of G, must the equality in (10.1) hold?

10.5. Show that for a graph G, the backtracking method described in Example 10.10 yields all of the maximum independent sets of G. [*Hint:* Use induction on $n = |V(G)|$.]

10.6. How many maximum independent sets of vertices can a path on n vertices have?

10.7. Show that a tree in which every maximal path has an even length has a unique maximum independent set. Can you describe it?

10.8. Let T be a tree in which every leaf is of distance 2 from another leaf. Show that every maximum independent set of T must contain all the leaves of T.

10.9. Show that a tree in which every maximal path has an even length has a unique maximum independent set. Can you describe the vertices of this unique set?

10.10. Let T be a tree in which every leaf is of distance two from another leaf. Show that all the leaves must be contained in every maximum independent set of T.

10.11. Find a dominating set of the graph G shown in Figure 10.5. Your dominating set should contain three elements and be a different dominating set from those mentioned in Example 10.13.

10.12. Let $k \in \mathbb{N}$. Show that the set $D = \{u_2, u_5, \ldots, u_{3k-1}\}$ is the unique minimum dominating set of P_{3k}.

10.13. Let $n \in \mathbb{N}$. Show that $\gamma(P_n)$ is given by

$$\gamma(P_n) = \left\lceil \frac{n}{3} \right\rceil.$$

10.14. Let $k \in \mathbb{N}$. Show that P_{3k+1}, the path on $3k+1$ vertices, has at least two minimum dominating sets, each containing a pair of adjacent vertices.

10.15. Let $k \in \mathbb{N}$. Show that P_{3k+2} has no minimum dominating sets that contain two adjacent vertices of P_{3k+2}.

10.16. For $n \in \mathbb{N}$, what is the maximum size of a minimal dominating set of P_n?

10.17. Prove Observation 10.18. Further, give an example of a graph G having blocks $B_1, \ldots, B_k$, where for each pair $i < j$ there *exists* a disjoint pair D_i and D_j of minimum dominating sets of B_i and B_j respectively, but where the inequality in (10.3) is strict.

10.18. Give an example of a graph G with blocks $B_1, \ldots, B_k$ such that for every distinct i and j, every pair of minimum dominating sets D_i and D_j of the blocks B_i and B_j, respectively, are disjoint, and where the inquality of (10.3) is strict.

10.19. Give an example of a graph G with blocks $B_1, \ldots, B_k$ such that every minimum dominating set D_i contains no cut-vertices of G, and where the inquality of (10.3) is strict.

10.20. Give an example of a graph G having an induced path of arbitrary length ℓ, with $\gamma(G) = 2$. [*Note!* This result is in contrast to the lower bound of $\gamma(G)$ given in Lemma 10.20.]

10.21. Let G be a graph and $u \in V(G)$. Let $N_G[u]$ denote the closed neighborhood of u in G, as in Definition 1.22. Show that if G has vertices $u_1, \ldots, u_k$ such that $N_G[u_i] \cap N_G[u_j] = \varnothing$ for all $i \neq j$, then $\gamma(G) \geq k$.

10.22. Let G be a graph and $u \in V(G)$. Let $N_G^2[u]$ denote the closed neighborhood of the closed neighborhood $N_G[u]$ of u in G, as in Definition 1.22. That is,

$$N_G^2[u] = N_G(N_G[u]).$$

Show that if G has vertices $u_1, \ldots, u_k$ such that

$$N_G^2[u_i] \cap N_G^2[u_j] = \varnothing \qquad \text{for all } i \neq j,$$
$$V(G) \setminus \bigcup_{i=1}^{k} N_G^2[u_i] \neq \varnothing,$$

then $\gamma(G) > k$. [*Hint:* This result is the idea of Example 10.19 expanded.]

10.23. Let G be a graph on n vertices and of maximum degree Δ. Show that the number of minimum dominating sets of G is not more than

$$\binom{n+1}{n+1-\lceil n/(\Delta+1)\rceil}.$$

10.24. Let G be a graph and let $U \subseteq V(G)$ be the set of vertices of G that are endvertices of loops in G. If $|U| = \ell$ and $G - U$ has a matching of size m, show that G has $m + \ell$ edges, no two of which are adjacent.

10.25. Let G be a graph with no loops. Let G' be the graph obtained from G by replacing all multiple edges with a single edge. Show that G has a matching of size m if, and only if, G' has a matching of size m.

10.26. Let $m \geq n \geq 1$. Directly determine the number of matchings in the complete bipartite graph $K_{m,n}$ such that all n vertices are covered in one part. How many perfect matchings does $K_{n,n}$ have? [*Hint:* Note that there is a one-to-one correspondence between matchings of $K_{m,n}$ covering all of the n vertices, and the set of all injective functions from $\{1, \ldots, n\}$ to $\{1, \ldots, m\}$.]

10.27. Find the number of perfect matchings in K_{2n} by using direct combinatorial methods. [*Hint:* A perfect matching in K_{2n} can be obtained by a list $u_1, u_2, \ldots, u_{2n}$, where each u_{2i-1} is matched with u_{2i}, for each $i \in \{1, \ldots, n\}$. Count the number of such listings, and then count the number of those listings yielding the same matching.]

10.28. Find the number of perfect matchings for the graph on $2n$ vertices that is shown in Figure 10.17.

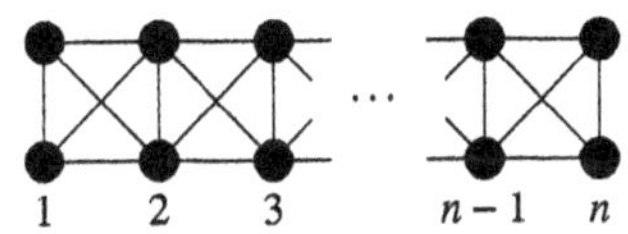

Figure 10.17: How many perfect matchings does this graph have?

10.29. For $n \in \mathbb{N}$, determine the number of perfect matchings in the complete 3-partite graph $K_{2n}^3 = K_{2n,2n,2n}$.

10.30. Let G be a graph with an even number of vertices such that

$$d_G(u) \geq |V(G)|/2$$

for every vertex $u \in V(G)$. Show that G has a perfect matching.

10.31. Let G be a bipartite graph with bipartition $V(G) = X \cup Y$. If there is a $k \in \mathbb{N}$ such that

$$d_G(x) \geq k \geq d_G(y)$$

for all $x \in X$ and $y \in Y$, then show that G has a matching which covers all of X. [*Hint:* Apply the idea used in Example 10.35.]

10.32. Prove Corollary 10.38. [*Hint:* Use induction on k.]

10.33. Prove the following *Theorem of Distinct Representatives.*

Theorem 10.46 For sets $A_1, \ldots, A_n$, the following are equivalent:

(a) For each $i \in \{1, \ldots, n\}$ one can pick an $x_i \in A_i$ such that the $x_1, \ldots, x_n$ are all distinct.

(b) For each $k \in \{1, \ldots, n\}$ and distinct $i_1, \ldots, i_k$, we have

$$|A_{i_1} \cup \cdots \cup A_{i_k}| \geq k.$$

[*Hint:* Construct a bipartite graph with bipartition $V(G) = X \cup Y$, where $X = \{A_1, \ldots, A_n\}$ and $Y = A_1 \cup \cdots \cup A_n$.]

10.34. *The Harem Problem:* Let X be a set of sultans, and let Y be the set of potential wives. Find necessary and sufficient conditions such that every sultan in X can marry at least k women that he likes. [*Hint:* Form k copies of each sultan and use Hall's Marriage Theorem. (*Note!* This problem could be phrased as *The Flirting Problem:* For a given set of X cheerleaders and a set of Y football players, find necessary and sufficient conditions such that every cheerleader can flirt with at least k football players that she likes, in such a way that no two cheerleaders are flirting with the same football player.)]

10.35. Let Δ be the maximum degree in the bipartite graph G. Show that G has a matching of size of at least $|E(G)|/\Delta$.

10.36. Let M be a maximal matching in a graph G, and let M' be a maximum matching in G. Show that $|M'| \leq 2|M|$.

10.37. Without the use of Theorem 10.44, show directly that if a graph G satisfies Tutte's Condition, then $|V(G)|$ must be even.

CHAPTER

11

Cover Parameters and Matching Polynomials

Chapter Goals

Discuss vertex covers of a graph G and the corresponding vertex covering number $\beta(G)$.

Discuss edge covers of a graph G and the corresponding edge covering number $\beta'(G)$.

Demonstrate how the vertex and edge covering numbers relate to the independence number and the matching number of a graph.

Discuss the rook polynomial of a general chessboard and how it relates to matchings in a bipartite graph.

Discuss the matching defect polynomial of a graph and how it relates to the characteristic polynomial.

Present two matching algorithms, the AUGMENTING PATH ALGORITHM and the KUHN-MUNKRES ALGORITHM.

11.1 Covers and Related Parameters

This section will discuss further and in more detail how independence and covers relate to one another via their corresponding parameters. Recall Definition 10.25 of a vertex and edge cover. As for independent and dominating sets, the smallest covers are usually of the most interest. We make this notion precise.

Definition 11.1 Let G be a graph.

1. A vertex cover $L \subseteq V(G)$ is *minimal* if no vertex cover is properly contained in L.
2. A vertex cover $L \subseteq V(G)$ is *minimum* if no other vertex cover exists which has a smaller number of vertices.
3. An edge cover $N \subseteq E(G)$ is *minimal* if no edge cover is properly contained in N.
4. An edge cover $N \subseteq E(G)$ is *minimum* if no other edge cover exists which has a smaller number of edges.

Example 11.2 Let $k \in \mathbb{N}$ and consider P_{2k+1}, the simple path on the vertices $u_1, \ldots, u_{2k+1}$ and the edges $e_i = \{u_{i-1}, u_i\}$ for $2 \leq i \leq 2k+1$.

The vertex cover $L_1 \subseteq \{u_1, u_2, u_3, \ldots, u_{2k+1}\}$ given by the odd-labeled vertices

$$L_1 = \{u_1, u_3, u_5, \ldots, u_{2k+1}\}$$

is a minimal vertex cover, since each edge e_i has at least one endvertex in L_1, but the removal of one vertex from L_1, say the vertex u_{2i+1}, will result in the edge $e_{2i+1} = \{u_{2i}, u_{2i+1}\}$ not having any of its two endvertices in L_1.

The vertex cover $L_2 \subseteq \{u_2, u_4, \ldots, u_{2k}\}$ given by the even-labeled vertices

$$L_2 = \{u_2, u_4, \ldots, u_{2k}\}$$

is a minimum vertex cover, since each edge has at least one endvertex in L_2, but since P_{2k+1} has precisely $2k$ edges and each vertex is of degree at most two, each vertex can cover at most two edges. Hence, any vertex cover of P_{2k+1} must have at least $|E(P_{2k+1})|/2 = k$ vertices.

Example 11.3 Let $k \in \mathbb{N}$ and consider P_{6k}, the simple path on the vertices $u_1, \ldots, u_{6k}$ and the edges $e_i = \{u_{i-1}, u_i\}$ for $2 \leq i \leq 6k$.

The edge cover $N_1 \subseteq E(P_{6k})$ given by

$$N_1 = \bigcup_{i=1}^{2k} \{\{u_{3i-2}, u_{3i-1}\}, \{u_{3i-1}, u_{3i}\}\}$$

is a minimal edge cover, since every vertex is an endvertex of at least one edge in N_1, and the removal of one edge will destroy the coverage: If an edge $\{u_{3i-2}, u_{3i-1}\}$ is removed, then the vertex u_{3i-2} will not be covered by N_1, and if an edge $\{u_{3i-1}, u_{3i}\}$ is removed, then the vertex u_{3i} will not be covered by N_1.

The edge cover $N_2 \subseteq E(P_{6k})$ given by

$$N_2 = \bigcup_{i=1}^{3k} \{\{u_{2i-1}, u_{2i}\}\} = \{\{u_1, u_2\}, \{u_3, u_4\}, \ldots, \{u_{6k-1}, u_{6k}\}\}$$

is a minimum edge cover, since it is a perfect matching of P_{6k}.

Remark 11.4 A matching in a graph is perfect if, and only if, it is an edge cover, in which case it is a minimum edge cover. Likewise, an edge cover of a graph is a matching if, and only if, it is perfect, in which case it is a maximum matching.

Note that $V(G)$ is always a vertex cover of G. Also, if G has no isolated vertices, then $E(G)$ is always an edge cover.

Observation 11.5

1. Any graph G has a vertex cover. The graph G has the empty set as a vertex cover if, and only if, G is a null graph.
2. A graph G has an edge cover if, and only if, it has no isolated vertices.

Remark 11.6 Let G be a graph without isolated vertices. Note that every edge with a leaf as an endvertex must be contained in every edge cover of G. More generally, at least one of the $d_G(u)$ edges with $u \in V(G)$ as an endvertex must be contained in an edge cover of G. Hence, $N \subseteq E(G)$ is an edge cover of G if, and only if, for all $u \in V(G)$, we have

$$d_{G-N}(u) \leq d_G(u) - 1.$$

That is, every vertex $u \in V(G)$ has a degree in $G - N$ strictly less than that in G.

We conclude this section with a discussion of the cardinality of minimum covers (both vertex and edge) and the cardinality of maximum independent sets (both vertex and edge), and we see how these two important concepts relate to one another. We now define the parameters corresponding to Definition 11.1.

Definition 11.7 Let G be a graph.

1. The *matching number* $\alpha'(G)$ of G is the number of edges in a maximum matching of G.
2. The *vertex cover number* $\beta(G)$ of G is the number of vertices in a minimum vertex cover of G.
3. The *edge cover number* $\beta'(G)$ of G is the number of edges in a minimum edge cover of G.

An edge in a graph has either one or two endvertices. Therefore, an edge cover of a graph G on n vertices must have at least $\lceil n/2 \rceil$ edges. Noting that the edges of a spanning forest of G also form an edge cover, we have the following.

Observation 11.8 Let $n \in \mathbb{N}$ and G be a graph on n vertices having no isolated vertices. Then the edge cover number satisfies

$$\lceil n/2 \rceil \leq \beta'(G) \leq n - 1.$$

Both the lower and upper bound can be obtained, and therefore are tight.

Example 11.9 We demonstrate that both the lower and the upper bound given in Observation 11.8 can be reached.

- Consider the complete graph K_n. If $n = 2m$ is even, then a perfect matching in K_n has precisely $m = \lceil n/2 \rceil$ edges. If $n = 2m + 1$ is odd, let $u \in V(K_n)$ be a fixed vertex. In this case a perfect matching in $K_n - u$, together with one edge having u as an endvertex in K_n, has $m + 1 = \lceil n/2 \rceil$ edges. Hence, $\beta'(K_n) = \lceil n/2 \rceil$.

- Consider the star $K_{1,n-1}$ on n vertices, which is a tree with $n - 1$ leaves. By Remark 11.6 every edge with a leaf as an endvertex must be contained in every edge cover. In particular, an edge cover of $K_{1,n-1}$ must contain all the $n - 1$ edges of $K_{1,n-1}$. Hence, $\beta'(K_{1,n-1}) = n - 1$.

Like the independence number $\alpha(G)$, which we can compute recursively by (10.2), the matching number $\alpha'(G)$ can similarly be computed recursively.

Let G be a graph and $e \in E(G)$ an edge of G. If e is not in any maximum matching of G, then clearly any maximum matching in G is also a maximum matching in $G - e$, and so $\alpha(G) = \alpha(G - e)$ in this case. If $\phi(e) = \{u, v\}$, then here let $G - \{u, v\}$ denote the graph obtained from G by removing both vertices u and v from G, along with all edges having one of them as an endvertex in G. With this notation in mind, we have the following theorem.

Theorem 11.10 Let G be a graph and $e \in E(G)$ an edge of G with $\phi(e) = \{u, v\}$. The matching number $\alpha'(G)$ satisfies

$$\alpha'(G) = \max(\{\alpha'(G - e), 1 + \alpha'(G - \{u, v\})\}). \tag{11.1}$$

Proof We observe that every matching in $G - e$ is also a matching in G. Likewise, a matching M in $G - \{u, v\}$ yields a matching $M \cup \{e\}$ in G.

If there is any maximum matching of G that does not contain the edge e, then clearly $\alpha'(G) = \alpha'(G - e)$.

Otherwise, suppose every maximum matching in G contains the edge $e = \{u, v\}$. Such maximum matchings contain no edges adjacent to e. In this case we have that

$$M \leftrightarrow M \cup \{e\}$$

yields a bijective correspondence between a maximum matching M of $G - \{u, v\}$ and a maximum matching $M \cup \{e\}$ of G. Therefore, we have that $\alpha'(G) = 1 + \alpha'(G - \{u, v\})$, which completes our proof. ■

In a similar way, we can count the number of *perfect* matchings $\mu(G)$ of G.

Theorem 11.11 Let G be a graph and $e \in E(G)$ an edge of G with $\phi(e) = \{u, v\}$. In this case the number of perfect matchings of G, $\mu(G)$ satisfies

$$\mu(G) = \mu(G - e) + \mu(G - \{u, v\}). \tag{11.2}$$

Proof The total number of perfect matchings of G is the number of perfect matchings of G that do not include the edge e, plus the number of perfect matchings of G that do include e. The number of those perfect matchings not including e is given by $\mu(G - e)$, and the number of perfect matchings including e is given by $\mu(G - \{u, v\})$, since none of those perfect matchings include edges adjacent to e. ■

Example 11.12 We argue that a perfect matching in a forest is unique, if it exists. This fact is clearly true for the forest on a single vertex. Let F be a forest and let $e = \{u, v\} \in E(G)$ be an edge having a leaf $u \in V(F)$ as an endvertex. Since $d(u) = 1$, the edge e must be included in any perfect matching of F, and so $\mu(F - e) = 0$. By Theorem 11.11 we have that

$$\mu(F) = \mu(F - \{u, v\}),$$

which, by induction, equals either zero or one.

The idea applied to the leaves in Example 11.12 can be pushed further to any collection of degree-one vertices of a graph G, as stated in the following lemma, whose proof we leave as Exercise 11.10.

Lemma 11.13 Let G be a graph and $D_1 \subseteq V(G)$ the leaves of G; that is, $d_G(u) = 1$ for all $u \in D_1$. If $N_G[D_1]$ is the closed neighborhood of the leaves of G, then the number of perfect matchings $\mu(G)$ of G satisfies

$$\mu(G) = \mu(G - N_G[D_1]).$$

Example 11.14 Consider the "4-ladder" G on eight vertices as shown in Figure 11.1. We apply Theorem 11.11 on the edge $e = \{u, v\}$ and obtain that

$$\mu(G) = \mu(G - e) + \mu(G - \{u, v\}).$$

Note that $G - e$ has two leaves, namely u and v. Thus by Lemma 11.13, we obtain that $\mu(G - e) = \mu(C_4)$, the cycle on four vertices. Since C_4 has two perfect matchings, we have that $\mu(G - e) = 2$ as well.

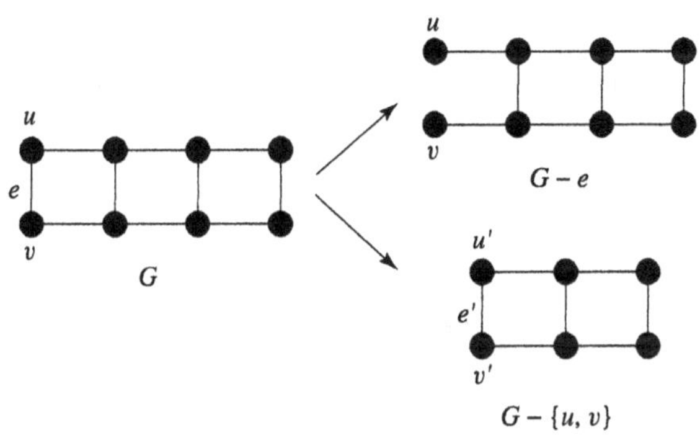

Figure 11.1: The "4-ladder" G together with $G - e$ and $G - \{u, v\}$, respectively.

Also note that $G' = G - \{u, v\}$ is the "3-ladder," so in the same manner we obtain that

$$\mu(G') = \mu(G' - e') - \mu(G' - \{u', v'\}),$$

where $G' - e'$ has two leaves; and therefore, $\mu(G - e') = 1$. Also, $G' - \{u', v'\}$ is C_4, and so $\mu(G' - \{u', v'\}) = 2$ and $\mu(G') = 3$. From this series of steps we have finally that $\mu(G) = 2 + 3 = 5$.

If G is a graph, then for any vertex cover L of G it must be the case that $|L| \geq \alpha'(G)$, since every edge in a matching of size $\alpha'(G)$ must have at least one endvertex in L. Therefore, we see that $\beta(G) \geq \alpha'(G)$.

Assume further that G is a bipartite graph G with a bipartition $V(G) = X \cup Y$. By Theorem 10.41 we have that $\alpha'(G) = |X| - \delta_X$. If $S \subseteq X$ is a subset with $\delta_X = |S| - |N(S)|$, then $L = N(S) \cup (X \setminus S)$ is a vertex cover of size $|X| - \delta_X = \alpha'(G)$. Therefore, we obtain the following theorem, which was first proved by the Hungarian mathematicians Kőnig and Egerváry in 1931.

Theorem 11.15 For a bipartite graph G, the following holds:

$$\alpha'(G) = \beta(G).$$

Theorem 11.15 is our first theorem relating an independent set of edges and a cover of vertices. Our next results relate the independence number of a graph to the vertex cover number, both quantities involving numbers of vertices. If $S \subseteq V(G)$ is an independent set of vertices, then no edge has both its endvertices in S. Therefore, each edge has at least one endvertex in the complement $\overline{S}$, so $\overline{S}$ is a vertex cover. Conversely, if $\overline{S}$ is a vertex cover, then every edge must have at least one endvertex in $\overline{S}$; therefore, S is independent. We summarize in the following.

Lemma 11.16 In any graph G the following are equivalent for $S \subseteq V(G)$:

1. The set $S \subseteq V(G)$ is independent.
2. The set $\overline{S} \subseteq V(G)$ is a vertex cover.

From Lemma 11.16 we obtain a corollary relating the independence number and vertex cover number.

Corollary 11.17 For any graph G on n vertices, we have the following:

$$\alpha(G) + \beta(G) = n.$$

In light of Theorem 11.15 and Corollary 11.17, a natural question to pose is whether similar relations hold for edge covers and matchings. This is indeed the case. Gallai proved the following theorem in 1959 (see Gardner [17]).

Theorem 11.18 For a graph G with n nonisolated vertices, we have the following:

$$\alpha'(G) + \beta'(G) = n.$$

Proof If $M \subseteq E(G)$ is a maximum matching of G and $N \subseteq E(G)$ is an edge cover of size $n - |M|$, then that implies $\beta'(G) \leq n - \alpha'(G)$. Likewise, if N is a minimum edge cover and M is a matching of size $n - |N|$, then that implies $\alpha'(G) \geq n - \beta'(G)$. With these two inequalities, we have the theorem. We conclude our proof by providing such an edge cover and matching, respectively.

Let M be a maximum matching of G. For each vertex $u \in V(G)$ that is not covered by M, we add to M one edge with u as an endvertex and obtain an edge cover N. Since the number of vertices in G that are not covered by M is precisely $n - 2|M|$, the total number of edges in N is

$$|M| + (n - 2|M|) = n - |M|.$$

Let N be a minimum edge cover of G. As a minimum and hence also minimal edge cover of G, the edges of N form a spanning forest of G, in which every tree is a star. (See

Exercise 11.4.) Since G has no isolated vertices, each of these stars contains at least one edge. Let k be the number of these stars. Since each star has one fewer edges than vertices, we have that $n = |N| + k$. By choosing one edge from each star, we obtain a matching M of size $k = n - |N|$. This completes the proof. ∎

We obtain the following corollary that summarizes the relations of $\alpha(G)$, $\alpha'(G)$, $\beta(G)$, and $\beta'(G)$ for a bipartite graph G. Note that some of the statements in Corollary 11.19 do hold even if G is not bipartite, or if G has isolated vertices, as stated in Theorems 11.15 and 11.18 together with Corollary 11.17.

Corollary 11.19 For a bipartite graph G on n nonisolated vertices, we have the following:

$$\begin{aligned} \alpha(G) &= \beta'(G), \\ \alpha'(G) &= \beta(G), \\ \alpha(G) + \beta(G) &= n, \\ \alpha'(G) + \beta'(G) &= n. \end{aligned}$$

Proof The only part left to prove is the last equality, which is a direct consequence of the other three. By appropriately substituting the first two equations into the third one, we get the last equality. ∎

Note 11.20 Kőnig first proved the relation $\alpha(G) = \beta'(G)$ in 1916 [31]. Needless to say, his proof was more involved than our proof, since neither Theorem 11.15 nor 11.18 was available at that time.

If G is a bipartite graph and we know one of the parameters $\alpha(G)$, $\alpha'(G)$, $\beta(G)$, or $\beta'(G)$, then by Corollary 11.19, we know them all. For example, if we have successfully computed the independence number $\alpha(G)$, then we automatically have the following:

$$\begin{aligned} \alpha'(G) &= n - \alpha(G), \\ \beta(G) &= n - \alpha(G), \\ \beta'(G) &= \alpha(G). \end{aligned}$$

Similar formulas can be derived for each of the other parameters. (See Exercise 11.7.)

11.2 Rook Polynomials and Bipartite Graphs

In a game of chess, played on an 8×8 board, the *castle* or *rook* is a piece that can move in a straight line along its row or column, over any number of empty squares. The rook can capture any single chess piece of the opponent in its path on these straight lines, up, down, to the left, or to the right, as depicted in Figure 11.2.

Assume that we have not two, but k mutually fiendish rooks that we want to place on the chessboard. Assume further that our chessboard C is not necessarily a regular 8×8 square chessboard on 64 squares, but rather a chessboard of *any* shape on $m \geq k$ squares, where

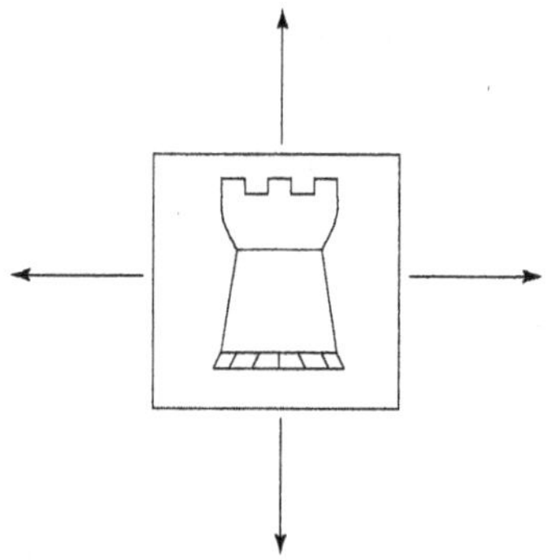

Figure 11.2: The rook can move along empty squares in its row or column on the chessboard.

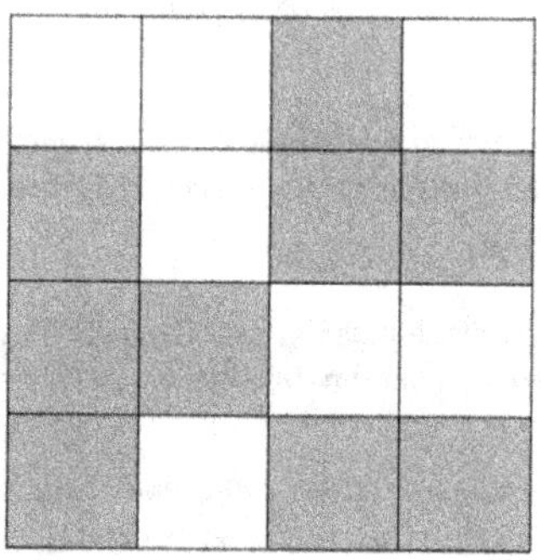

Figure 11.3: A general chessboard C on 9 squares, as indicated by the shaded squares.

$m \in \mathbb{N}$—that is, a board composed of a subset of m squares of an $n \times n$ chessboard, for some $n \in \mathbb{N}$. In fact, any board composed of m squares is a subset of an $m \times m$ chessboard. (See Exercise 11.13.) In Figure 11.3 we see an example of one possible "general" chessboard C on nine squares. There C is the subset consisting of the shaded squares of the 4×4 chessboard.

A natural question to ask is: In how many ways we can place these k rooks on the chessboard C, in such a way that no rook can capture another rook? This statement means precisely that no two rooks can be located in the same row or column.

Example 11.21 Consider a regular 8×8 chessboard. If $k > 8$, then, by the Pigeon-Hole Principle, when these k rooks are placed on the chessboard, at least two rooks must be located in the same row. Hence, if r_k denotes the number of ways to place k rooks on the chessboard, then $r_k = 0$ if $k \geq 9$.

Assume now that $k \leq 8$. To place the k rooks on the chessboard, then, first of all, no two rooks can be in the same row. Hence, there are $\binom{8}{k}$ ways of choosing the k rows where the rooks are going to be placed. In the first row of these k chosen ones, there are eight spots to place the rook. In the second row, there are seven spots to place the rook, since it cannot be in the same column as the first rook. Continuing in this manner, we see there are $8 - i + 1$ spots to place the ith rook in the ith chosen row, since it cannot be in the same column as the previous $i - 1$ rooks. Hence, when $k \leq 8$, we obtain that

$$r_k = \binom{8}{k} \cdot 8 \cdot 7 \cdots (8 - k + 1) = \frac{P(8, k)^2}{k!}. \tag{11.3}$$

For a general chessboard C, let $r_k(C)$ denote the number of possible locations of placing k rooks on C such that no rook can capture another rook. In Example 11.21, we obtained that for the regular 8×8 chessboard C, $r_k(C) = r_k$, where r_k is given by (11.3). By convention we let $r_0(C) = 1$ for all chessboards C. With this fact in mind, we state the following definition.

Definition 11.22 Let C be a general chessboard, and let $r_k(C)$ be the number of possible locations of placing k rooks on C, such that none of the k rooks can capture another rook. The *rook polynomial* $\mu_C(t)$ of the chessboard C is defined by

$$\mu_C(t) = r_0(C) + r_1(C)t + r_2(C)t^2 + \cdots = \sum_{k \geq 0} r_k(C)t^k.$$

Example 11.23 Consider the 3×3 chessboard C. By the same methodology as in Example 11.21, we have that $r_k(C) = 0$ for $k \geq 4$, and that

$$r_k(C) = \frac{P(3, k)^2}{k!}$$

for $k \in \{0, 1, 2, 3\}$. Therefore, we see that

$$\mu_C(t) = r_0(C) + r_1(C)t + r_2(C)t^2 + r_3(C)t^3 = 1 + 9t + 18t^2 + 6t^3.$$

Note that $r_1(C)$ is the number of squares of C, and that the leading term of $r_C(t)$ is $3! = 6$, the number of permutations of three elements.

Remark 11.24 Let C be a general chessboard.

- The rook polynomial $\mu_C(t)$ is indeed a polynomial, since $r_k(C) = 0$ if k is greater than the number of squares of C.
- The rook polynomial $\mu_C(t)$ is the generating function of the sequence $(r_k(C))_{k \geq 0}$.
- The rook polynomial $\mu_C(t)$ is sometimes called the *matching generating polynomial* of C in the graph-theory literature. The "generating" part comes from the previous point. The "matching" part of the expression will be described soon.

There are techniques and observations to make the computation of the rook polynomial a bit easier. The first such method is the following.

Theorem 11.25 Let C be a general chessboard that is a union of two chessboards $C = C_1 \cup C_2$, where both the rows and the columns of C_1 and C_2 are disjoint. In this case we have that

$$\mu_C(t) = \mu_{C_1}(t)\mu_{C_2}(t).$$

Proof Since the rows and the columns of C_1 and C_2 are disjoint, the placement of rooks on C_1 is independent from the placement of rooks on C_2. Therefore, the number of ways to place k rooks on C is given by

$$\begin{aligned} r_k(C) &= r_k(C_1 \cup C_2) \\ &= r_0(C_1)r_k(C_2) + r_1(C_1)r_{k-1}(C_2) + \cdots + r_k(C_1)r_0(C_2) \\ &= \sum_{i=0}^{k} r_i(C_1)r_{k-i}(C_2). \end{aligned} \tag{11.4}$$

Hence, by (11.4) we obtain

$$\begin{aligned} \mu_{C_1}(t)\mu_{C_2}(t) &= \left(\sum_{i\geq 0} r_i(C_1)t^i\right)\left(\sum_{j\geq 0} r_j(C_2)t^j\right) \\ &= \sum_{k\geq 0}\left(\sum_{i=0}^{k} r_i(C_1)r_{k-i}(C_2)\right)t^k \\ &= \sum_{k\geq 0} r_k(C)t^k \\ &= \mu_C(t), \end{aligned}$$

which completes the proof. ■

Example 11.26 Consider the chessboard C marked with shaded squares in Figure 11.4. There C is a row/column disjoint union of three subchessboards C_1, C_2, and C_3, where the chessboard C_1 consists of the shaded squares in columns 1, 2, and 4 and rows 3, 4, and 6, the chessboard C_2 consists of the single shaded square in column 3 and row 5, and the chessboard C_3 consists of the three shaded squares in columns 5 and 6 and rows 1 and 2. Hence, by Theorem 11.25, we have that

$$\mu_C(t) = \mu_{C_1}(t)\mu_{C_2}(t)\mu_{C_3}(t).$$

Note that C_1 is equivalent to the 3×3 chessboard, by moving the columns and rows together. Hence, by Example 11.23, we have that $\mu_{C_1}(t) = 1 + 9t + 18t^2 + 6t^3$. Clearly, we have that $\mu_{C_2}(t) = 1 + t$. For the chessboard C_3, we immediately obtain that $\mu_{C_3}(t) = 1 + 3t + t^2$. Hence, we have that

$$\begin{aligned} \mu_C(t) &= (1 + 9t + 18t^2 + 6t^3)(1 + t)(1 + 3t + t^2) \\ &= 1 + 13t + 58t^2 + 115t^3 + 105t^4 + 42t^5 + 6t^6. \end{aligned}$$

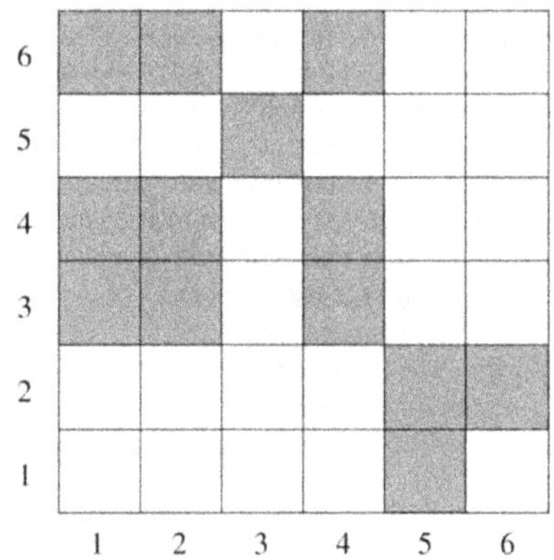

Figure 11.4: The chessboard is the union of three row/column-disjoint subchessboards.

For a general chessboard to be such a union of smaller subchessboards, where all of the rows are disjoint and all of the columns are also disjoint, is a pretty strong condition. However, there is a recursive way to systematically compute the rook polynomial for any chessboard C.

Theorem 11.27 Let C be a general chessboard. For a square s in C, let C' be the chessboard obtained by removing the square s from C, and let C'' be the chessboard obtained by removing all of the squares in the same row and same column as s from C. In this case we have

$$\mu_C(t) = \mu_{C'}(t) + t\mu_{C''}(t).$$

Proof The possibilities of placing r rooks on C are split into two classes: one class where no rook is placed on the square s in C and the second class where a rook is placed on s. The number of possibilities of the first class is $r_k(C')$ and of the second class is $r_{k-1}(C'')$. By the definition of the rook polynomial, the theorem follows. ■

Although the rook polynomial can be computed using Theorem 11.27, this computation can be very time consuming for large chessboards. Also, keeping an eye open for which squares s in C yield the most reduction in the computation of $\mu_C(t)$ is a good rule of thumb.

Consider a chessboard C that has m columns and n rows. In that case C is a subchessboard contained in the $m \times n$ rectangular chessboard. In fact, the $m \times n$ square chessboard is the smallest rectangular chessboard that can contain C. We form a bipartite graph G_C with bipartition $V(G_C) = X \cup Y$, where

$$X = \{u_1, \ldots, u_m\}$$
$$Y = \{v_1, \ldots, v_n\},$$

that is, X corresponds to the columns of C, and Y corresponds to the rows of C. For every square s in C, we connect the column of s in X to the row of s in Y with an edge. In this way, the simple bipartite graph G_C is completely determined by C. Permuting the columns and rows of C, and obtaining an equivalent chessboard C', corresponds to relabeling the vertices of X and Y; hence, the corresponding bipartite graph $G_{C'}$ is isomorphic to G_C. Note that by the minimality of m and n, no vertex in G is isolated. In this way the squares in C correspond to the edges of G_C.

Conversely, let G be a simple bipartite graph with a bipartition $V(G) = X \cup Y$, where $|X| = m$ and $|Y| = n$. In this case G determines a unique chessboard C_G within the rectangular $m \times n$ chessboard, made up by the squares s, where the vertex in X corresponding to the column of s is adjacent to the vertex in Y corresponding to the row of s. So, the edges of G correspond to the squares of C_G.

But placing a rook on a square s on C is the same as matching the column of s with the row of s, since no other rook can be in the same column or row as s. Hence, a placement of k rooks on C yields a matching of G_C of size k, and vice versa. We summarize in the following.

Observation 11.28 For chessboards and simple bipartite graphs we have the following:

1. There is a one-to-one correspondence between chessboards and simple bipartite graphs, given by $C \mapsto G_C$ or $G \mapsto G_C$, such that

$$C_{G_C} = C,$$
$$G_{C_G} = G.$$

2. For each $k \in \mathbb{N}$, there is a one-to-one correspondence between placements of k rooks on C and matchings of G_C of size k. In particular, we have that

$$r_k(C) = \mu_k(G_C),$$

the number of matchings of G_C of size k.

By Observation 11.28, it makes no difference whether we count matchings of size k of a simple bipartite graph, or count the number of ways to place k rooks on the corresponding chessboard. In that sense, we can view Theorem 11.27 as a direct consequence of Exercise 11.9, when applied to a bipartite graph. Moreover, this observation allows us to define the rook polynomial for any graph G, bipartite or not.

Definition 11.29 For a graph G, let $\mu_k(G)$ be the number of matchings of G with k edges. Then the *rook polynomial* $\mu_G(t)$ of G is defined by

$$\mu_G(t) = \mu_0(G) + \mu_1(G)t + \mu_2(G)t^2 + \cdots = \sum_{k \geq 0} \mu_k(G)t^k.$$

Remark 11.30 For a graph G on n vertices, we have the following:

- If G is bipartite, then the rook polynomial for G is identical to the rook polynomial for the chessboard C_G—that is, $\mu_G(t) = \mu_{C_G}(t)$.
- When G is not bipartite, then the rook polynomial $\mu_G(t)$ of G is usually called the *matching generating polynomial* of G.
- Note that the rook polynomial $\mu_G(t)$ for any graph G is indeed a polynomial, since $\mu_k(G) = 0$ for all $k > \lceil n/2 \rceil$.

Analogous to Theorems 11.25 and 11.27, we have the following Observation, whose proof we leave as Exercise 11.23.

Observation 11.31 For a graph G we have the following:

1. If G has components $G_1, \ldots, G_k$, then

$$\mu_G(t) = \mu_{G_1}(t)\mu_{G_2}(t)\cdots\mu_{G_k}(t).$$

2. If $e \in E(G)$ has endvertices $u, v \in V(G)$, then

$$\mu_G(t) = \mu_{G-e}(t) + t\mu_{G-\{u,v\}}(t).$$

We conclude this section with an example.

Example 11.32 Let G be a simple separable graph with a cut-vertex $u \in V(G)$. Let G_1 and G_2 be the corresponding subgraphs with $V(G_1) \cap V(G_2) = \{u\}$. We would like to express the rook polynomial of G in terms of G_1 and G_2.

For $k \in \mathbb{N}$ the matchings of G consisting of k edges are of three classes. The first class consists of those matchings that do not cover the cut-vertex u. The second class consists of those matchings where u is matched with a vertex v of G_1. The third class consists of those matchings where u is matched with a vertex w in G_2. For $i \in \{1, 2\}$, let $N_i = N_{G_i}(u)$ be the neighbors of u in G_i. Hence, $d_G(u) = |N_1| + |N_2|$. If μ' is the number of matchings in class one, then we have that

$$\mu' = \sum_{i=0}^{k} \mu_i(G_1 - u)\mu_{k-i}(G_2 - u).$$

If μ'' is the number of matchings in class two, then for each $v \in N_1$ that u is matched with, we count the remaining of the $k-1$ edges in the matching. Therefore,

$$\mu'' = \sum_{v \in N_1} \left(\sum_{i=0}^{k-1} \mu_i(G_1 - \{u, v\})\mu_{k-1-i}(G_2 - u) \right).$$

Similarly, we obtain

$$\mu''' = \sum_{w \in N_2} \left(\sum_{i=0}^{k-1} \mu_i(G_1 - u)\mu_{k-1-i}(G_2 - \{u, w\}) \right).$$

From this we have that $\mu_k(G) = \mu' + \mu'' + \mu'''$, and we can deduce that the rook polynomial of G is given by

$$\begin{aligned}\mu_G(t) &= \mu_{G_1-u}(t)\mu_{G_2-u}(t) \\ &\quad + t \sum_{v \in N_1} \mu_{G_1-\{u,v\}}(t)\mu_{G_2-u}(t) \\ &\quad + t \sum_{w \in N_2} \mu_{G_1-u}(t)\mu_{G_2-\{u,w\}}(t).\end{aligned}$$

11.3 The Matching Defect Polynomial

In addition to the rook polynomial $\mu_G(t)$ of a graph G, there is another polynomial obtained from matchings of G. It is similar to the rook polynomial but perhaps not quite as natural. Nonetheless, it has some nice properties that we will discuss in this section. But first we need the definition.

Definition 11.33 For a graph G on n vertices, let $\mu_k(G)$ be the number of matchings of G with k edges. The *matching defect polynomial* $\delta_G(t)$ of G is defined by

$$\delta_G(t) = \sum_{k \geq 0} (-1)^k \mu_k(G) t^{n-2k}.$$

Example 11.34 Consider P_6, the simple path on six vertices. For $k \in \{1, 2, 3\}$ we have

$$\begin{aligned} \mu_0(P_6) &= 1, \\ \mu_1(P_6) &= 5, \\ \mu_2(P_6) &= 6, \\ \mu_3(P_3) &= 1. \end{aligned}$$

By definition we now obtain that the rook polynomial for P_6 is given by

$$\mu_{P_6}(t) = \mu_0(P_6) + \mu_1(P_6)t + \mu_2(P_6)t^2 + \mu_3(P_6)t^3 = 1 + 5t + 6t^2 + t^3,$$

and the matching defect polynomial for P_6 is given by

$$\delta_{P_6}(t) = \mu_0(P_6)t^6 - \mu_1(P_6)t^4 + \mu_2(P_6)t^2 - \mu_3(P_6) = t^6 - 5t^4 + 6t^2 - 1.$$

Note the similarity between $\mu_{P_6}(t)$ and $\delta_{P_6}(t)$ reflected in the equation $\delta_{P_6}(t) = t^6 \mu_{P_6} \times (-t^{-2})$, something that holds in general. (See the remark below.)

Remark 11.35 For a graph G on n vertices we have the following:

- The matching defect polynomial is indeed a polynomial, since we have that $\mu_k(G) = 0$, when $k > \lceil n/2 \rceil$.
- If G has n vertices, then the leading term of $\delta_G(t)$ is t^n. Also, $\delta_G(t)$ has no t^{n-1} term.
- If G has m edges, then the coefficient for the t^{n-2}-th term is $-m$.
- The name "matching defect polynomial" is due to Lovász and Plummer, from their book [36] on matchings. This name is descriptive, since for each matching on k edges, the power $n - 2k$ denotes the defect—how far the matching is from being perfect.
- There are other names for this polynomial in the graph-theory literature, such as the *acyclic polynomial* or simply the *matching polynomial*.

- The matching defect polynomial of a graph G is very much related to the rook polynomial of G, since we have
$$\delta_G(t) = t^n \mu_G(-t^{-2}).$$

As in Observation 11.31, we have similarly for the matching defect polynomial that

$$\delta_G(t) = \delta_{G-e}(t) - \delta_{G-\{u,v\}}(t). \tag{11.5}$$

Note the minus sign, since the summands in $\delta_G(t)$ alternate in sign. Further, if G has components $G_1, \ldots, G_k$, then it is easy to show that

$$\delta_G(t) = \delta_{G_1}(t) \cdots \delta_{G_k}(t). \tag{11.6}$$

The proof of (11.6) is left as Exercise 11.24.

Consider now all the edges having a vertex u as an endvertex. By repeatedly using Observation 11.31 on all these edges, we obtain the following corollary, whose complete proof we leave as Exercise 11.25.

Corollary 11.36 For a graph G and a vertex $u \in V(G)$, we have the following:

$$\delta_G(t) = t\delta_{G-u}(t) - \sum_{v \in N_G(u)} \delta_{G-\{u,v\}}(t).$$

Example 11.37 Consider the tree T shown in Figure 11.5. Let $u \in V(T)$ be the unique vertex of degree three. By Corollary 11.36, we now obtain for T that

$$\begin{aligned}\delta_T(t) &= t\delta_{T-u}(t) - \sum_{v \in N_T(u)} \delta_{T-\{u,v\}}(t)\\ &= t\delta_{2\times P_2 \cup P_3}(t) - 2\delta_{P_1 \cup P_2 \cup P_3}(t) - \delta_{3\times P_2}(t).\end{aligned}$$

Now, it is immediate by the definition of the matching defect polynomial that $\delta_{P_1}(t) = t$, $\delta_{P_2}(t) = t^2 - 1$, and $\delta_{P_3}(t) = t^3 - 2t$. By (11.6), we therefore obtain that

$$\begin{aligned}\delta_T(t) &= t\delta_{P_2}(t)^2\delta_{P_3}(t) - 2\delta_{P_1}(t)\delta_{P_2}(t)\delta_{P_3}(t) - \delta_{P_2}(t)^3\\ &= t(t^2-1)^2(t^3-2t) - 2t(t^2-1)(t^3-2t) - (t^2-1)^3\\ &= t^8 - 7t^6 + 14t^4 - 9t^2 + 1.\end{aligned}$$

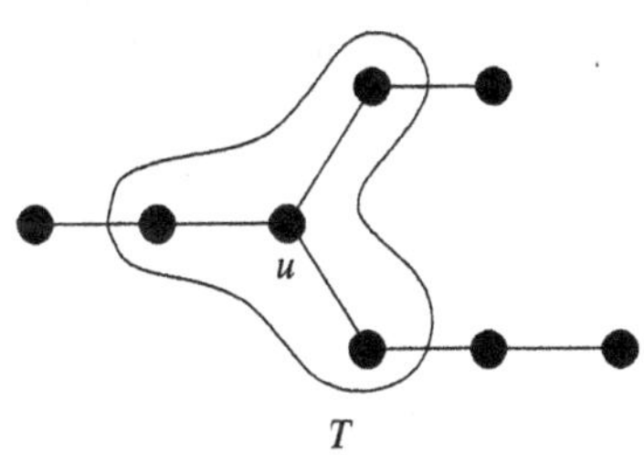

Figure 11.5: For this tree it is best to consider the vertex u of degree three.

Note that the leading term is t^8, since T has eight vertices, and that the coefficient for the t^6 term is -7, since T, being a tree, has seven edges. We further see that since $\delta_G(0) = 1 \neq 0$, G has a perfect matching—in fact, a unique one.

We dedicate the rest of this section to a result that relates the matching defect polynomial of a graph G to the adjacency matrix $\mathsf{A}(G)$ of G. Recall that a *characteristic polynomial* in t of a square $n \times n$ matrix A is given by $\det(t\mathsf{I}_n - \mathsf{A})$, where I_n is the identity matrix of the same size as A. With this fact in mind, we have the following lemma, whose proof we leave as Exercise 11.26.

Lemma 11.38 Any two orthogonally equivalent square matrices have the same characteristic polynomial.

By Lemma 11.38 we have that the characteristic polynomial of the adjacency matrix $\mathsf{A}(G)$ of a graph G is independent of the labeling of the vertices of G. In particular, two adjacency matrices of two isomorphic graphs are identical. Therefore, the following definition is completely sensible.

Definition 11.39 Let G be graph on n vertices and $\mathsf{A}(G)$ an adjacency matrix of G. The *characteristic polynomial* $\phi_G(t)$ of G is defined by

$$\phi_G(t) = \det(t\mathsf{I}_n - \mathsf{A}(G)),$$

where I_n is the $n \times n$ identity matrix.

We are now able to interpret the coefficients of $\phi_G(t)$ in the case when G is a forest. The following result is due to Mowshowitz from 1972 (see Lovász and Plummer [36] and Mowshowitz [41]).

Theorem 11.40 If G is a forest, then $\delta_G(t) = \phi_G(t)$.

Proof The theorem clearly holds if G is a null graph with no edges, since in that case both polynomials are t^n, where n is the number of vertices of G.

So assume that G is a forest with at least one edge. We proceed by induction on n. Since G has an edge, there is a labeling $V(G) = \{u_1, u_2, \ldots, u_n\}$ of the vertices of G, such that u_1 is a leaf, with u_2 as its unique neighbor. By Corollary 11.36 we obtain that

$$\delta_G(t) = t\delta_{G-u_1}(t) - \delta_{G-\{u_1,u_2\}}(t). \tag{11.7}$$

From the given labeling of the vertices, the characteristic polynomial is given by

$$\phi_G(t) = \det(t\mathsf{I}_n - \mathsf{A}(G)) = \begin{vmatrix} t & -1 & 0 & \cdots & 0 \\ -1 & t & * & \cdots & * \\ 0 & * & * & \cdots & * \\ \vdots & \vdots & \vdots & \ddots & \vdots \\ 0 & * & * & \cdots & t \end{vmatrix},$$

where the first row has the indeterminate t in the first entry, -1 in the second entry, and zero in all of the other entries in the matrix. Expanding this determinant along the first row, we obtain that

$$\begin{aligned}\phi_G(t) &= \det(t\mathsf{I}_n - \mathsf{A}(G)) \\ &= t\det(t\mathsf{I}_{n-1} - \mathsf{A}(G-u_1)) - \det(t\mathsf{I}_{n-2} - \mathsf{A}(G - \{u_1, u_2\})), \qquad (11.8)\end{aligned}$$

where I_k denotes the $k \times k$ identity matrix, for each $k = n, n-1, n-2$. By the induction hypothesis, we have that $\delta_{G-u_1}(t) = \det(t\mathsf{I}_{n-1} - \mathsf{A}(G-u_1))$ and $\delta_{G-\{u_1,u_2\}}(t) = \det(t\mathsf{I}_{n-2} - \mathsf{A}(G - \{u_1, u_2\}))$, so the theorem follows from (11.7) and (11.8). ■

Remark 11.41 Theorem 11.40 provides a more convenient and often quicker way of computing the number, $\mu_k(G)$, of matchings having k edges in G in the case when G is a forest, since by using linear algebra in the computation of the characteristic polynomial, the determinant can be computed in a fast manner.

11.4 Matching Algorithms

So far, we have discussed independent sets, covers, and polynomials obtained from matchings but have presented no specific algorithms. In this section we describe two algorithms: the AUGMENTING PATH ALGORITHM, which is used to obtain a maximum matching for any bipartite graph, and the KUHN-MUNKRES ALGORITHM, which is used to determine a perfect matching of maximum weight in a weighted graph which is isomorphic to the complete bipartite graph $K_{n,n}$. We begin with a definition.

Definition 11.42 Let G be a graph, M a matching in G, and $r \in V(G)$ a vertex in G that is not covered by M. A rooted tree $(T, r) = ((V(T), E(T)), r)$ is called an *M-alternating subtree of G* if the root r is not covered by M and T is a subtree of G in which each simple path in T with r as an endvertex is an M-alternating path in G.

Remark 11.43 Rooting T at r, an edge of T is contained in M if, and only if, it connects vertices on level $2i-1$ and $2i$, where $i \in \{1, \ldots, \lfloor \text{ht}(T)/2 \rfloor\}$. Note that no edges from level 0 to level 1 are in M.

Recall that for a bipartite graph G with bipartition $V(G) = X \cup Y$, we obtain a maximum matching by establishing a matching that covers $|X| - \delta_X$ vertices of X. Thus, we concentrate on finding a matching that covers the most vertices of X.

The following algorithm takes any matching of a bipartite graph and enlarges it to a maximum matching. To make the algorithm more readable, we introduce the following notation and conventions:

1. For a matching M and an M-augmenting path p in G, let $M \vartriangle E(p)$ be the corresponding augmented matching of G, as in the proof of Theorem 10.32.
2. We say that a set $\widetilde{T}$ of vertex disjoint and M-alternating subtrees of G, rooted at vertices in X, is *M-saturated* if no vertex of G can be added as a leaf to any tree in $\widetilde{T}$ in such a way that $\widetilde{T}$ remains a set of vertex disjoint M-alternating subtrees of G.

3. If $\widetilde{T}$ is not M-saturated and $u \in V(G)$ can be added to $T \in \widetilde{T}$ as a leaf, we let T_u denote the rooted tree obtained by adding u as a leaf to T. Also, we let

$$\widetilde{T}_{u,T} = (\widetilde{T} \setminus \{T\}) \cup \{T_u\}$$

denote the resulting set of vertex disjoint and M-alternating subtrees of G.

4. We say that a set $\widetilde{T}$ of vertex disjoint and M-alternating subtrees of G is *M-augmented* if there is an M-alternating subtree $T \in \widetilde{T}$ and a leaf $y \in V(T) \cap Y$ such that y is not covered by M.

The Augmenting Path Algorithm takes as input a bipartite graph G with bipartition $V(G) = X \cup Y$ together with a matching M of G, and yields a maximum matching $M_{\max}$ of G.

```
AUGMENTING PATH ALGORITHM
Input: A bipartite graph G with bipartition V(G) = X ∪ Y, a matching M of
G, and a set U ⊆ X of vertices that are not endvertices of any edge in M.
Output: A maximum matching M_max of G.
begin
    M_0 = M;                                  // initial matching of G
    T~_00 = {(({u}, ∅), u) : u ∈ U};   // initial set of rooted subtrees
    while T~_ij is not M_i-saturated do {
            pick z ∈ X ∪ Y and T ∈ T~_ij
                to which z can be added as a leaf, such that (T~_ij)_z,T
                consists of vertex disjoint M_i-alternating subtrees;
            if (T~_ij)_z,T is M_i-augmented then {
                p = M_i-alternating path from root(T) to z;
                M_i+1 = M_i Δ E(p);
                T~_i+1 j = T~_ij \ {T};
            }
            else
                T~_i j+1 = (T~_ij)_z,T;
    }
    output M_max = M_i;
end.
```

Consider the following example of how the Augmenting Path Algorithm runs on a given bipartite graph.

Example 11.44 Let G be the complete bipartite graph $K_{3,3}$ with one edge removed. Assume the bipartition $V(G) = X \cup Y$ is given by

$$X = \{x_1, x_2, x_3\} \quad \text{and} \quad Y = \{y_1, y_2, y_3\},$$

where all pairs $\{x_i, y_j\}$ for $1 \leq i, j \leq 3$ are connected, except for $\{x_1, y_3\}$, as shown in Figure 11.6. Assume that G has a matching M given by

$$M = \{\{x_2, y_1\}, \{x_3, y_2\}\},$$

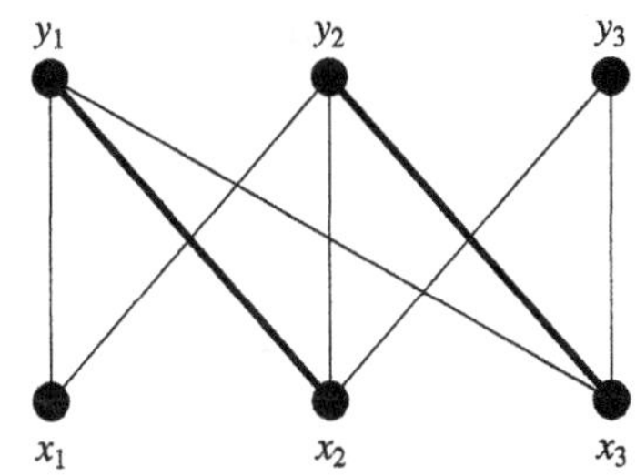

Figure 11.6: A bipartite graph G provided with a matching M indicated by thick edges.

the edges of which are indicated by thick edges in Figure 11.6. In this case we clearly have $U = \{x_1\}$, since $x_1 \in X$ is the only vertex of X that is not covered by M. Hence, for our initial input here for the Augmenting Path Algorithm, we have

$$\begin{aligned} M_0 &= M, \\ \widetilde{T}_{0\,0} &= \{(x_1; \varnothing)\}. \end{aligned}$$

[To minimize notation in this example we will indicate the root trees contained in $\widetilde{T}_{i\,j}$ by their root (listed first) and edges, which we will write as "uv" instead of the more precise "$\{u, v\}$".]

The set $\widetilde{T}_{0\,0}$ is not M_0-saturated, since $z = y_1$ can be added as a leaf to the vertex x_1 of $T = T_0 = (x_1, \varnothing) \in \widetilde{T}_{0\,0}$ to obtain a new set of vertex disjoint M_0-alternating subtrees of G. Since $(\widetilde{T}_{0\,0})_{T,z} = (\widetilde{T}_{0\,0})_{T_0,y_1}$ is not M_0-augmented, we let this new set be

$$\widetilde{T}_{0\,1} = (\widetilde{T}_{0\,0})_{T,z} = (\widetilde{T}_{0\,0})_{T_0,y_1} = \{(x_1; x_1y_1)\}.$$

The set $\widetilde{T}_{0\,1}$ is not M_0-saturated, since $z = y_2$ can be added as a leaf to the vertex x_1 of $T = T_1 = (x_1; x_1y_1) \in \widetilde{T}_{0\,1}$ to obtain a new set of vertex disjoint M_0-alternating subtrees of G. Since $(\widetilde{T}_{0\,1})_{T,z} = (\widetilde{T}_{0\,1})_{T_1,y_2}$ is not M_0-augmented, we let this new set be

$$\widetilde{T}_{0\,2} = (\widetilde{T}_{0\,1})_{T,z} = (\widetilde{T}_{0\,1})_{T_1,y_2} = \{(x_1; x_1y_1, x_1y_2)\}.$$

The set $\widetilde{T}_{0\,2}$ is not M_0-saturated, since $z = x_2$ can be added as a leaf to the vertex y_1 of $T = T_2 = (x_1; x_1y_1, x_1y_1) \in \widetilde{T}_{0\,2}$ to obtain a new set of vertex disjoint M_0-alternating subtrees of G. Since $(\widetilde{T}_{0\,2})_{T,z} = (\widetilde{T}_{0\,2})_{T_2,x_2}$ is not M_0-augmented, we let this new set be

$$\widetilde{T}_{0\,3} = (\widetilde{T}_{0\,2})_{T,z} = (\widetilde{T}_{0\,2})_{T_2,x_2} = \{(x_1; x_1y_1, x_1y_2, y_1x_2)\}.$$

The set $\widetilde{T}_{0\,3}$ is not M_0-saturated, since $z = x_3$ can be added as a leaf to the vertex y_2 of $T = T_3 = (x_1; x_1y_1, x_1y_1, y_1x_2) \in \widetilde{T}_{0\,3}$ to obtain a new set of vertex disjoint M_0-alternating subtrees of G. Since $(\widetilde{T}_{0\,3})_{T,z} = (\widetilde{T}_{0\,3})_{T_3,x_3}$ is not M_0-augmented, we let this new set be

$$\widetilde{T}_{0\,4} = (\widetilde{T}_{0\,3})_{T,z} = (\widetilde{T}_{0\,3})_{T_3,x_3} = \{(x_1; x_1y_1, x_1y_2, y_1x_2, y_2x_3)\}.$$

The set $\widetilde{T}_{0\,4}$ is not M_0-saturated, since $z = y_3$ can be added as a leaf to the vertex x_2 of $T = T_4 = (x_1; x_1y_1, x_1y_1, y_1x_2, y_2x_2) \in \widetilde{T}_{0\,4}$ to obtain a new set of vertex disjoint M_0-alternating subtrees of G. Since $(\widetilde{T}_{0\,4})_{T,z} = (\widetilde{T}_{0\,4})_{T_4,y_3}$ is not M_0-augmented, we let this new set be

$$\widetilde{T}_{0\,5} = (\widetilde{T}_{0\,4})_{T,z} = (\widetilde{T}_{0\,4})_{T_4,y_3} = \{(x_1; x_1y_1, x_1y_2, y_1x_2, y_2x_3, x_2y_3)\}.$$

At this point we note that $\widetilde{T}_{0\,5}$ is M_0-augmented, since the simple path $p = (x_1, y_1, x_2, y_3)$ in the tree $T = T_5 = (x_1; x_1y_1, x_1y_2, y_1x_2, y_2x_3, x_2y_3) \in \widetilde{T}_{0\,5}$ from the root x_1 to y_3 is an M_0-augmenting path of G. Hence, we let

$$\begin{aligned} M_1 &= M_0 \,\Delta\, E(p), \\ \widetilde{T}_{1\,5} &= \widetilde{T}_{0\,5} \setminus \{T_5\}. \end{aligned}$$

At this point we have $\widetilde{T}_{1\,5} = \varnothing$, and hence $\widetilde{T}_{1\,5}$ is M_1-saturated, so we stop and obtain our maximum matching:

$$M_{\max} = M_1 = \{\{x_1, y_1\}\{x_2, y_3\}, \{x_3, y_2\}\}.$$

In Figure 11.7 we see the trees T_1, T_2, T_3, T_4, and T_5. Note that since both y_1 and y_2 are both covered by M_0, then neither of the one-edge-paths x_1y_1 or x_1y_2 is M_0-augmenting.

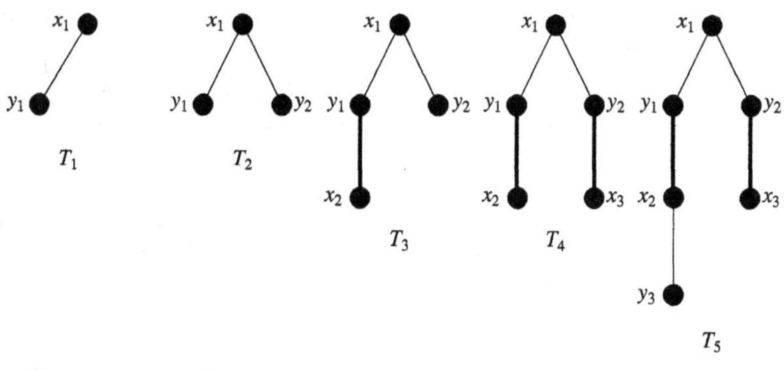

Figure 11.7: The trees $T_1, \ldots, T_5$ created by the AUGMENTING PATH ALGORITHM on the bipartite graph G from Example 11.44.

In Figure 11.8 we see the matching M_1, which is in fact a maximum matching $M_1 = M_{\max}$, since the corresponding $\widetilde{T}_{1\,5}$ is an empty set of M_1-alternating subtrees of our bipartite graph G.

Note that if we had chosen the vertex y_3 right after x_2, instead of y_2, then the AUGMENTING PATH ALGORITHM would have taken four steps instead of five.

Note, as indicated by Example 11.44, that the idea of the AUGMENTING PATH ALGORITHM is simple. By Theorem 10.32, a matching M of G is maximum if, and only if, G has no M-augmenting path. The set of M-alternating trees and the related terminology

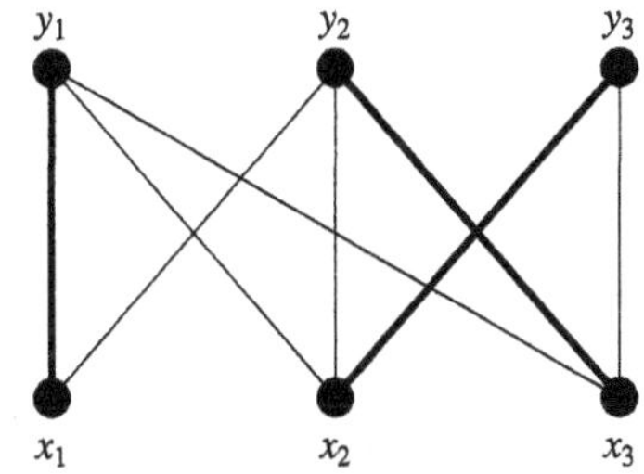

Figure 11.8: The matching M_1 created from $M = M_0$ by the AUGMENTING PATH ALGORITHM on G from Example 11.44 is a maximum one.

is simply a bookkeeping device to help us keep track of what paths we have attempted as augmenting paths, which ones can possibly be extended to M-augmenting paths, and which paths cannot. In all simplicity, the augmenting path algorithm is a systematic check of all M-alternating paths to see whether one exists that is M-augmenting.

Theorem 11.45 Let G be a bipartite graph. The AUGMENTING PATH ALGORITHM yields a maximum matching $M_{\max}$ of G.

Proof Since G is a bipartite graph with a finite number of vertices and edges, the AUGMENTING PATH ALGORITHM eventually halts.

We only need to show that when the algorithm halts, if $\widetilde{T}_{i\,j}$ is M_i-saturated, then M_i is a maximum matching of G. Let $R \subseteq X$ be the set of vertices in X that are not included in any of the trees in $\widetilde{T}_{i\,j}$ when the algorithm stops. Let $S \subseteq Y$ be the set of vertices in Y contained in the trees in $\widetilde{T}_{i\,j}$. First we show that $L = R \cup S$ is a vertex cover. If L is not a vertex cover, then there is an edge e with one endvertex $x \in X \setminus R$ and the other $y \in Y \setminus S$. By the definition of R, we have that $x \in V(T)$ for some $T \in \widetilde{T}_{i\,j}$. Hence, either $x = \text{root}(T)$, in which case $e \notin M_i$, or x is the endvertex of some $e' \in M_i$, in which case $e \notin M_i$. Since y is not a vertex in any tree in $\widetilde{T}_{i\,j}$, then in both cases we can add the edge e and the vertex y as a leaf to T in such a way that $\widetilde{T}_{i\,j}$ remains M_i-disjoint. This statement contradicts the fact that the AUGMENTING PATH ALGORITHM has stopped, so $L = R \cup S$ is a vertex cover.

Next we show that $|L| \leq |M|$, which by Theorem 11.15 implies that M_i is of maximum size. We show that each vertex in L is covered by M and that every edge in M can have at most one endvertex in L.

Let $y \in S$. Since y is contained in a tree $T \in \widetilde{T}_{i\,j}$, which is not M_i-augmented, vertex y must be covered by an edge $e \in M_i$. If x is the other endvertex of e, then x cannot be contained in R, since then we could add the edge e and its endpoint x to T, contradicting the fact that $\widetilde{T}_{i\,j}$ is M_i-saturated. Likewise, a vertex $x \in R$ must be covered by an edge $e \in M_i$, since otherwise it would be the root of some tree in $\widetilde{T}_{i\,j}$. If y is the other endvertex of e, then $y \notin S$, since in that case y would be contained in some tree $T \in \widetilde{T}_{i\,j}$ to which we could add y as a leaf, again contradicting that $\widetilde{T}_{i\,j}$ is M_i-saturated.

Therefore, we have that each vertex in $L = R \cup S$ is covered by an edge in M_i, where the other endvertex is not in L. This shows that $|L| \leq |M|$ and completes the argument. ∎

For the rest of this section we briefly discuss the bipartite graph $K_{n,n}$ with weights attached to its edges. We present the idea for the KUHN-MUNKRES ALGORITHM, which yields a perfect matching of maximum weight.

Recall the definition of a weighted graph. Consider the case where G is a bipartite graph with bipartition $V(G) = X \cup Y$. Assume n is large enough for G to be a subgraph of $K_{n,n}$. We can view G as a weighted graph $(K_{n,n}, W)$ by defining for each $e \in E(K_{n,n})$

$$W(e) = \begin{cases} 0 & \text{if } e \notin E(G), \\ 1 & \text{if } e \in E(G). \end{cases}$$

In this case a maximum matching M of G is really a matching in $(K_{n,n}, W)$ of maximum weight.

Note that if $W(e) \geq 0$ for all edges in a weighted bipartite graph (G, W) with bipartition $V(G) = X \cup Y$, then we can assume that all vertices between X and Y are adjacent by letting the weight of the new edges be zero.

Suppose the cardinalities of X and Y differ, say $|X| < |Y|$. By adding vertices to X and weight zero edges from all those vertices to all vertices in Y, we may also assume that $|X| = |Y| = n$, in which case our underlying graph G is isomorphic to $K_{n,n}$. Although we restrict our attention to the weighted graph $(K_{n,n}, W)$, this result shows that we can solve the weighted maximum matching problem for *all* weighted bipartite graphs (G, W), where the weights are nonnegative. So we are seeking the perfect matching of $(K_{n,n}, W)$ with maximum weight.

A weighted bipartite graph $(K_{n,n}, W)$ with a bipartition $V(K_{n,n}) = X \cup Y$, where $X = \{x_1, \ldots, x_n\}$ and $Y = \{y_1, \ldots, y_n\}$, is completely determined by an $n \times n$ matrix $\mathsf{W} = [w_{i,j}]_{i,j \in \{1,\ldots,n\}}$, where $w_{i,j} = W(\{x_i, y_j\})$ denotes the weight of the edge between x_i and y_j (in Exercise 11.27 you can see how this relates to the adjacency matrix).

Definition 11.46 Let $\mathsf{W} = [w_{i,j}]_{i,j \in \{1,\ldots,n\}}$ be the matrix of weights for the graph $(K_{n,n}, W)$ with bipartition $X = \{x_1, \ldots, x_n\}$ and $Y = \{y_1, \ldots, y_n\}$. We call a function $f : X \cup Y \to \mathbb{R}$ a *weighted cover* of W if

$$f(x_i) + f(y_j) \geq w_{i,j}$$

for all $i, j \in \{1, \ldots, n\}$. For such a weighted cover f, we further let

$$E_f = \{\{x_i, y_j\} : f(x_i) + f(y_j) = w_{i,j}\},$$

and $G_f = (X \cup Y, E_f)$ be the subgraph of $K_{n,n}$ on the same set of vertices as $K_{n,n}$ and edges E_f.

Example 11.47 Let $n \in \mathbb{N}$, and consider the weighted graph $(K_{n,n}, W)$ with bipartition $X = \{x_1, \ldots, x_n\}$ and $Y = \{y_1, \ldots, y_n\}$ and weights $w_{i,j}$ given by

$$W(\{x_i, y_j\}) = w_{i,j} := 2ij.$$

Let the function $f : X \cup Y \to \mathbb{R}$ be defined by

$$f(x_i) = f(y_i) := i^2,$$

for each $i \in \{1, \ldots, n\}$. Since $(i - j)^2 \geq 0$ for all $i, j \in \{1, \ldots n\}$, we have

$$f(x_i) + f(y_j) = i^2 + j^2 \geq 2ij = w_{i,j},$$

showing that f is indeed a weighted cover of W. Also, since the equality $i^2 + j^2 = 2ij$ holds if, and only if, $i = j$, we see that

$$E_f = \{\{x_i, y_j\} : f(x_i) + f(y_j) = w_{i,j}\} = \{\{x_i, y_i\} : 1 \leq i \leq n\},$$

and the corresponding subgraph G_f therefore consists of a collection of n copies of K_2, which constitutes a complete matching of $K_{n,n}$, where each x_i is matched with y_i.

With the terminology from Definition 11.46, we have the following useful lemma.

Lemma 11.48 If $f : X \cup Y \to \mathbb{R}$ is a weighted cover of W and G_f has a perfect matching M, then M is a perfect matching of maximum weight for $(K_{n,n}, W)$.

Proof Clearly a perfect matching in G_f is also a perfect matching for $(K_{n,n}, W)$. By the definitions of E_f and M, we have that

$$W(M) = \sum_{e \in M} W(e) = \sum_{i=1}^{n} f(x_i) + \sum_{j=1}^{n} f(y_j) = \sum_{u \in X \cup Y} f(u).$$

For any perfect matching M' of $(K_{n,n}, W)$, we have, however, that

$$W(M') = \sum_{e \in M'} W(e) \leq \sum_{i=1}^{n} f(x_i) + \sum_{j=1}^{n} f(y_j) = \sum_{u \in X \cup Y} f(u),$$

and so $W(M') \leq W(M)$. ■

Example 11.49 Continuing from Example 11.47, we note that G_f has a perfect matching M—in fact, $M = E(G_f)$ is a perfect matching. By Lemma 11.48, M is therefore a matching of G of maximum weight $M = M_{\max}$ given by

$$W(M_{\max}) = \sum_{i=1}^{n} w_{i,i} = \sum_{i+1}^{n} 2i^2 = \frac{n(n+1)(2n+1)}{3}.$$

Without the use of a weighted cover f of $(K_{n,n}, W)$, a direct proof that the matching M is of maximum weight—that is, $M = M_{\max}$—is a bit more tricky, since we need to prove that for any permutation $\sigma : \{1, \ldots, n\} \to \{1, \ldots, n\}$ we have that

$$\upsilon_1 + 2\sigma_2 + 3\sigma_3 + \cdots + n\sigma_n \leq \frac{n(n+1)(2n+1)}{6}.$$

In Exercise 11.34 we ask for such a direct proof.

We note that for any weight matrix $\mathsf{W} = [w_{i,j}]_{i,j\in\{1,\ldots,n\}}$ determining the weighted graph $(K_{n,n}, W)$ as in Definition 11.46, a weighted cover with $E_f \neq \varnothing$ can be given by defining f as follows:

$$f(x_i) = \max(\{w_{i,j} : j \in \{1,\ldots,n\}\})$$
$$f(y_j) = 0.$$

The following procedure is an outline of the KUHN-MUNKRES ALGORITHM, which yields a perfect matching of $(K_{n,n}, W)$ of maximum weight.

Let $\mathsf{W} = [w_{i,j}]_{i,j\in\{1,\ldots,n\}}$ be as in Definition 11.46, f a weighted cover of $(K_{n,n}, W)$ with $E_f \neq \varnothing$, and M some matching of G_f. We obtain a perfect matching of $(K_{n,n}, W)$ of maximum weight by the following steps:

1. **(a)** If each vertex in X is covered by M, then M is a perfect matching of $(K_{n,n}, W)$ of maximum weight, and stop.

   (b) Otherwise, choose a vertex $r \in X$ that is not covered by M and let $T = (\{r\}, \varnothing)$ be the tree with r as a root.
2. Let $A = V(T) \cap X$ and $B = V(T) \cap Y$.
3. **(a)** If $N_{G_f}(A) \nsubseteq B$, then go to step 4.

   (b) Otherwise, define a_f by

   $$a_f = \min(\{f(x) + f(y) - W(\{x, y\}) : x \in A \text{ and } y \in Y \setminus N_{G_f}(A)\})$$

   and define a new weighted cover f' by

   $$f'(u) = \begin{cases} f(u) - a_f & \text{if } u \in A, \\ f(u) + a_f & \text{if } u \in N_{G_f}(A), \\ f(u) & \text{otherwise.} \end{cases}$$

   In this case $N_{G_{f'}}(A) \nsubseteq B$. Replace f by f' and G_f by $G_{f'}$.
4. Choose $u \in N_{G_f}(A) \setminus B$ and add u to T by joining u with an edge from E_f to a vertex in A.
5. **(a)** If u is covered by $e \in M$, then let w be the other endvertex of e. Add both w and e to T and go to step 2.

   (b) Otherwise, T is an M-alternating rooted tree with r as a root. The path p from r to u in T is an M-alternating path. Replace M by $M \vartriangle E(p)$ and go to step 1.

The KUHN-MUNKRES ALGORITHM and Theorem 11.50 are due to Kuhn and Munkres, who independently obtained the algorithm in 1955 and 1957, respectively. Kuhn called the procedure the HUNGARIAN ALGORITHM in honor of the Hungarians Kőnig and Egerváry.

Theorem 11.50 **Kuhn-Munkres**

The KUHN-MUNKRES ALGORITHM yields a perfect matching M of $(K_{n,n}, W)$ of maximum weight.

Proof By the definition of a weighted cover, we see that if f is a weighted cover, then so is f' in step 3. Also, note that $B \subseteq N_{G_f}(A)$ always holds since we always add edges of G_f. So, if in step 3, $N_{G_f}(A) \nsubseteq B$ does not hold, then we must have $N_{G_f}(A) = B$. This condition implies that $N_{G_{f'}}(A) \nsubseteq B$ by the definition of f'.

Clearly the procedure does not halt until every vertex $r \in X$ is included in M. By Lemma 11.48 the result must then be a matching of maximum weight in $(K_{n,n}, W)$. This proves the theorem. ■

Remark 11.51 As with the AUGMENTING PATH ALGORITHM, the M-alternating tree T is for bookkeeping to maintain information in obtaining an M-alternating path in G_f.

11.5 Exercises

11.1. Show that the vertex cover L_2 from Example 11.2 is the unique minimum vertex cover for the simple path P_{2k+1}.

11.2. Let G be a connected graph. Show that the edges of a spanning tree of G are an edge cover of G.

11.3. Let G be a graph with no isolated vertices. Show that if N is a minimal edge cover of G, then the edges of N form a spanning forest of G.

11.4. Let G be a graph with no isolated vertices. Show that N is a minimal edge cover of G if, and only if, the edges of N are a disjoint union of stars—that is, graphs of the form $K_{1,k}$ for some $k \in \mathbb{N}$. [*Hint:* Show first that if $e_1, e_2, e_3 \in N$ form a path in this order in G, then $N \setminus \{e_2\}$ is also an edge cover.]

11.5. Show that a Hamiltonian graph has an edge cover consisting of a disjoint union of edges and 2-paths.

11.6. Show that a Hamiltonian graph on an even number of vertices always has a perfect matching.

11.7. Let G be a bipartite graph with n nonisolated vertices. For each of the parameters $\alpha(G)$, $\alpha'(G)$, $\beta(G)$, and $\beta'(G)$ write the others in terms of the one chosen by using Corollary 11.19.

11.8. Let G be a simple planar graph on $n \geq 3$ vertices. Show that the number of perfect matchings of G is at most $\frac{3}{25}(\sqrt{5})^n$. [*Hint:* Every simple planar graph has a vertex of degree five or less. Use induction.]

11.9. Let G be a graph and $e = \{u, v\} \in E(G)$ an edge of G. Denote by $\mu_r(G)$ the number of matchings of G that contain exactly k edges. Similarly to (11.2) in Theorem 11.11, show that

$$\mu_k(G) = \mu_k(G - e) + \mu_{k-1}(G - \{u, v\}).$$

11.10. Prove Lemma 11.13.

11.11. List all of the perfect matchings of the 4-ladder from Example 11.14. How many matchings of size two does the 4-ladder have?

11.12. As in Example 11.14, let μ_n denote the number of perfect matchings in the n-ladder on $2n$ vertices, as shown in Figure 11.9. Show that μ_n satisfies the following:

$$\begin{aligned} \mu_1 &= 1, \\ \mu_2 &= 2, \\ \mu_n &= \mu_{n-1} + \mu_{n-2}, \end{aligned}$$

for all $n \geq 3$. Use this framework to find a formula for μ_n. [*Note!* The numbers F_1, F_2, $F_3, \ldots,$ given by $F_1 = 0$, $F_1 = 1$, and $F_{n+1} = \mu_n$ for $n \in \mathbb{N}$ are called the *Fibonacci numbers*, and are due to the Italian mathematician Leonardo Pisano Fibonacci (1170–1250 (probably)). The Fibonacci number sequence $(F_n)_{n\geq 1}$ has a wide range of interesting properties that are still to this day being discovered. The journal *The Fibonacci Quarterly* is completely devoted to this number sequence.]

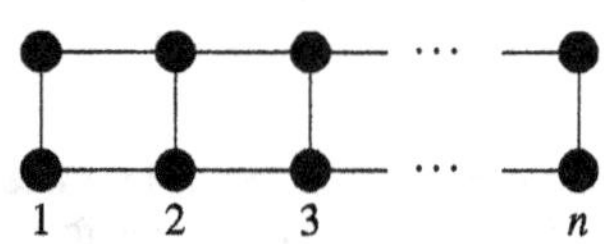

Figure 11.9: The n-ladder on $2n$ vertices.

11.13. Show that any general chessboard C on m squares is equivalent to a general chessboard C' on m squares inside the $m \times m$ chessboard, in the sense that we have a one-to-one correspondence between the squares of C and C' in such a way that any two squares are in the same column (or row) in C if, and only if, the corresponding squares are in the same column (or row) in C'.

11.14. For $k, n \in \mathbb{N}$ with $k \leq n$, show that the number r_k of possible ways to place k rooks on the $n \times n$ chessboard is given by

$$r_k = \frac{P(n,k)^2}{k!}.$$

For a fixed k, what is the asymptotic behavior of r_k, when n tends to infinity?

11.15. Let C be the $2n \times 2n$ chessboard, where the squares have been colored black and white in the same pattern as on a regular 8×8 chessboard. Show that the number r'_k of possibilities of placing k rooks on black squares is given by

$$r'_k = \sum_{i=0}^{k} \frac{P(n,i)^2}{i!} \frac{P(n,k-i)^2}{(k-i)!}.$$

[*Hint:* Start by permuting the rows and columns such that you have two "noninterfering" $n \times n$ black squares.]

11.16. Compute the rook polynomial of the chessboard C shown in Figure 11.3.

11.17. Compute the rook polynomial for an $n \times n$ chessboard. What is the degree of this polynomial, and what does the leading coefficient describe?

11.18. Compute the rook polynomial $\mu_C(t)$ for the chessboard C consisting of the black squares of the $2n \times 2n$ chessboard in Exercise 11.15. Show that

$$\mu_C(t) = (\mu_{C'}(t))^2,$$

where C' is the $n \times n$ chessboard.

11.19. Consider the $2n \times 2n$ chessboard from Exercise 11.15. Assume that each black square is split into $2m \times 2m$ smaller boards. Color these smaller squares black and white as on a usual 8×8 chessboard. Show that the number of small black squares is $4m^2n^2$. Compute the rook polynomial of the chessboard C consisting of these $4m^2n^2$ small black squares.

11.20. Let C be a chessboard with m columns and n rows. If $N \geq n$ and $M \geq n$, show that the number $e(m, n)$ of ways to embed C into the rectangular $M \times N$ chessboard is given by

$$e(m, n) = P(M, m)P(N, n).$$

[*Note!* By an "embedding" we mean a subchessboard of the $M \times N$ chessboard that is equivalent to C.]

11.21. For a graph G, what is the degree of the rook polynomial $\mu_G(t)$ of G?

11.22. Show that a chessboard C is a union of two chessboards C_1 and C_2, the columns of which are disjoint and the rows also, if, and only if, the corresponding graph G_C has G_{C_1} and G_{C_2} as distinct components.

11.23. Prove Observation 11.31, the first part by induction on k.

11.24. Prove that the matching defect polynomial $\delta_G(t)$ of a graph G satisfies (11.6).

11.25. Use Observation 11.31 to prove Corollary 11.36 by induction on $d_G(u)$.

11.26. Prove Lemma 11.38 directly by using that the determinant function is a multiplicative—that is, $\det(\mathsf{AB}) = \det(\mathsf{A})\det(\mathsf{B})$ for any square matrices A and B of the same size.

11.27. Let G be a bipartite graph with bipartition $V(G) = X \cup Y$, where $X = \{x_1, \ldots, x_m\}$ and $Y = \{y_1, \ldots, y_n\}$. Show that with respect to the labeling $V(G) = \{x_1, \ldots, x_m, y_1, \ldots, y_n\}$, G has an adjacency matrix of the form

$$\mathsf{A}(G) = \begin{bmatrix} 0 & \mathsf{A}^t \\ \mathsf{A} & 0 \end{bmatrix},$$

where A is an $n \times m$ matrix. In terms of $\mathsf{A}(G)$ and its entries, interpret what a matching of G is. What does it mean in terms of $\mathsf{A}(G)$ for a matching to be maximal, maximum, or perfect?

11.28. Let $\mathsf{N} = [n_{ij}]$ be a square matrix with nonnegative integer entries n_{ij}. Show that the following are equivalent:

(a) The sum of all entries in each row and each column is k.
(b) $\mathsf{N} = \mathsf{P}_1 + \cdots + \mathsf{P}_k$, where each P_i is a permutation matrix.

[*Hint:* Use Hall's Marriage Theorem and induction on k.]

11.29. A square matrix $\mathsf{X} = [x_{ij}]$ with nonnegative real entries x_{ij} is called a *doubly stochastic matrix*, if the sum of the entries in each row and in each column is one. Show that the following are equivalent:

(a) X is a doubly stochastic matrix.

(b) X has the form $\mathsf{X} = c_1\mathsf{P}_1 + \cdots + c_k\mathsf{P}_k$, where $c_i \geq 0$, $\sum_{i=1}^{k} c_i = 1$, and each P_i is a permutation matrix.

[*Hint:* For each stochastic matrix X form a bipartite graph G with bipartition $V(G) =$ (rows of X) $\cup$ (columns of X). Connect the ith row to the jth column if, and only if, $x_{ij} > 0$. Show that G has a perfect matching, and use induction on the number of positive entries of X.]

11.30. An $m \times n$ matrix $\mathsf{M} = [m_{ij}]$ of natural numbers m_{ij} that satisfies

(a) $m_{ij} \in \{1, \ldots, n\}$ for all i, j,
(b) no two entries in any row or in any column are equal,

is called a *Latin rectangle*. If, in addition, $m = n$, then M is called a *Latin square*. Prove the following theorem.

Theorem 11.52 Suppose $n > m$ are natural numbers. Let M be an $m \times n$ Latin rectangle. Then M can be extended to a Latin square by adding $n - m$ new rows.

[*Hint:* For each $i \in \{1, \ldots, n\}$, let A_i be the set of those natural numbers from $1, \ldots, n$ that do not occur in the ith column. Now show that M can be extended to an $(m+1) \times n$ Latin rectangle by using Theorem 10.46 from Exercise 10.33 to show that $\{A_1, \ldots, A_n\}$ has n distinct representatives.]

11.31. Let M be an $n \times n$ Latin square. Show that M can be represented by an n-edge coloring

$$c : E(K_{n,n}) \to \{1, \ldots, n\}.$$

Show further that each color class $M_i = \{e \in E(K_{n,n}) : c(e) = i\}$ is a perfect matching of $K_{n,n}$, and hence that M can be represented by an n-tuple $(M_1, \ldots, M_n)$ of edge-disjoint perfect matchings of $K_{n,n}$. Use this result to show that $n!$ divides the number of $n \times n$ Latin squares.

11.32. Let $B = [b_{ij}]$ be an $m \times n$ binary matrix. The *term rank* r is the minimum number of rows and columns that together contain all of the 1's of B. Prove the binary matrix version of Theorem 11.15.

Theorem 11.53 The term rank of a binary matrix B is equal to the largest number μ of 1's of B, no two of which sit in the same row or column.

[*Hint:* First note that the term rank of B cannot exceed μ. To prove the equality, assume that all the 1's are contained in r rows and s columns, where $r + s = \mu$. By rearranging the rows and columns one can further assume that B has the form

$$B = \begin{bmatrix} R & X \\ 0 & S \end{bmatrix},$$

where 0 is an $(m - r) \times (n - s)$ matrix containing only 0's as entries. For each $i \leq r$, let A_i be the set of those $j \leq n - s$ such that $b_{ij} = 1$. Use Theorem 10.46 to prove that $\{A_1, \ldots, A_r\}$ has distinct representatives, and in turn to show that R contains a set of r 1's, no two of which lie in the same row or column. Repeat this for S to obtain a total of $r + s = \mu$ 1's, no two of which lie in the same row or column. Note this result shows that one can use Hall's Marriage Theorem to prove Theorem 11.15.]

11.33. Use Theorem 11.15 to prove Hall's Marriage Theorem 10.33. [*Hint:* Regard an $m \times n$ binary matrix $B = [b_{ij}]$ as the incidence matrix of a family of n sets $\{A_1, \ldots, A_n\}$ from $\{1, \ldots, m\}$. That is, b_{ij} is 1 if A_i contains j, and b_{ij} is 0 otherwise. Note that this exercise, together with Exercise 11.32, shows us that in some sense Theorem 10.33 and Theorem 11.15 are equivalent.]

11.34. Show directly that the matching M from Example 11.34 is of maximum weight, by proving the following:

Let $n \in \mathbb{N}$ and $\sigma : \{1, \ldots, n\} \to \{1, \ldots, n\}$ be a permutation. Show that

$$\sigma_1 + 2\sigma_2 + 3\sigma_3 + \cdots + n\sigma_n \leq \frac{n(n+1)(2n+1)}{6}.$$

[*Hint:* Assume $\sigma_k = n$. Show that the sum on the left increases if we replace $k\sigma_k + n\sigma_n$ by $k\sigma_n + n\sigma_k$. Continue inductively.]

CHAPTER

12

Graph Counting

Chapter Goals

Define what is meant by a labeled and an unlabeled counting of graphs.

Discuss generating functions and exponential generating functions for graph counting.

Present two kinds of set partitions and discuss relating Stirling and Bell numbers.

Present the Labeled Counting Lemma and discuss some applications.

Present the Exponential Formula and discuss some applications.

Count some types of simple graphs that relate to the number two.

12.1 Introduction

This chapter presents the basic definitions related to counting, gives some examples, and provides useful counting techniques. These methods are generally used when counting or estimating the number of graphs on a given set of vertices, or when counting the number of nonisomorphic graphs on a given number of vertices. Several results from this chapter (and much more!) can be found in Harary and Palmer [24], Wilf [58], and the two volumes Stanley [50] and [51]. The classic text (Harary and Palmer [24]) on graph enumeration is still considered by many to be the "bible" on how to count graphs. Although the book was published in 1973, the counting techniques introduced there are still in wide use. Wilf [58] provides an excellent concise introduction to various combinatorial counting techniques. Wilf [58] uses generating functions to their fullest to obtain nice and closed expressions or formulas for the number of items being counted. Last but not least, it is impossible to discuss the use of generating functions in combinatorics without mentioning the two-volume treatise by (Stanley [50] and [51]). These volumes are the true professional's handbook on the use of generating functions in various combinatorial settings, in addition to graphs.

Before we start counting graphs, it is necessary to specify precisely what we want to count. The question, "What is the number of graphs having property P?" is natural and sounds reasonable enough, and could for that matter have been addressed right after the definition of a graph. However, this question can be ambiguous: Do we mean the number of graphs with a given fixed set of vertices? Do we mean the number of different shapes of graphs on a given fixed number of vertices? In general, during counting, when do we consider two or more graphs to be the "same" (so they contribute only $+1$ to the counting), and when do we consider them distinct?

Clearly, any question concerning counting is interesting only when both the number of vertices and the number of edges are bounded. If we allow an unbounded number of vertices or edges, it is easy to generate an infinite number of distinct graphs, regardless of what we mean by "distinct graphs." Therefore, we always consider a collection of graphs having a bounded number of vertices and edges. In particular, a collection of *simple* graphs on n vertices automatically has a bounded number of edges, since any simple graph on n vertices is a subgraph of the complete graph K_n, and so has at most $\binom{n}{2}$ edges. However, when determining the number of possible nonsimple graphs in a collection of graphs on n vertices, we must put a bound on the number of edges as well. This difference between simple and nonsimple graphs is fundamental when considering graph counting. Although for other graph-theoretic properties, such as connectivity and graph coloring, we can reduce to the study of simple graphs, when counting nonsimple graphs we must pay special attention to the edge-multiplicity between every pair of vertices in the graph. Hence, we will consider both the simple case and the nonsimple one simultaneously so we can compare the difference.

In previous chapters we have dealt with graph counting in a sporadic manner. We have slipped in a result here and there on the number of graphs satisfying certain conditions. In what follows we review a few of those results.

In Figure 3.2 we listed the six possible *shapes* of trees on six vertices. This result is the same as saying that there are precisely six nonisomorphic trees on six vertices. If, on the other hand, we have a set of six vertices $\{u_1, \dots, u_6\}$ and ask how many distinct trees we can form on them, then some of these trees are bound to be isomorphic. For instance, there are $6!/2 = 360$ simple paths on $\{u_1, \dots, u_6\}$, all of them are isomorphic to the leftmost tree shown in Figure 3.2. By Cayley's Theorem 4.7 there are exactly $6^{6-2} = 1{,}296$ distinct trees on $\{u_1, \dots, u_6\}$. All these trees, however, form only six nonisomorphic shapes.

In Theorem 3.36 the number of ordered rooted trees on $n + 1$ vertices and the number of regular binary trees on $2n + 1$ vertices are given in terms of the Catalan number C_n. There, the left/right layout matters, in the sense that two distinct rooted trees can have the same shape, and hence be the same when viewed as regular trees, and even be the same when viewed as trees on a fixed set of vertices. To clarify this point further we consider an example.

Example 12.1 The upper half of Figure 12.1 shows three trees T_1, T_2, and T_3 on the set of vertices $V = \{u_1, u_2, u_3, u_4, u_5\}$. The same trees, where the labeling of the vertices is ignored, are depicted in the lower half of that Figure 12.1.

The trees T_1, T_2, and T_3 are clearly all distinct when viewed as ordered rooted trees on the vertices V, since the left-to-right order of each set of siblings matters.

If we remove the labelings of the vertices, the trees T_1 and T_2 are the same when viewed as ordered rooted trees, since they have exactly the same left/right layout. This fact is easy to see in the lower half of Figure 12.1. Since u_2 does not have the same degree in T_1 as in T_2, they are distinct as trees on the vertices of V.

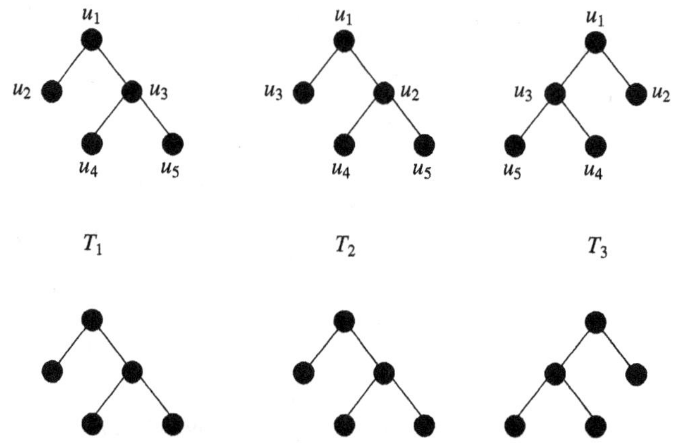

Figure 12.1: Three labeled trees and the same three trees with their labels removed.

The trees T_1 and T_3 are the same as trees on the vertices V. However, they are distinct as ordered rooted trees, since their left/right layouts are not the same, as we see in the lower half of Figure 12.1.

The trees T_2 and T_3 are distinct both as ordered rooted trees and as trees on V. They are, however, the same as rooted (unordered) trees!

Finally, it is clear that T_1, T_2, and T_3 are the same as (unrooted) trees, since they are all isomorphic.

Example 12.1 shows that we must state in a precise manner the objects that we are counting. If two graphs are the "same," they contribute only $+1$ to the counting. If two graphs are "different," they contribute $+2$ to the counting.

12.2 Basic Counting Results

To make matters precise, we introduce the following definition.

Definition 12.2 Let $n \in \mathbb{N}$ and P be a graph-theoretic property. A *labeled counting* or *labeled enumeration* is a counting of the number of graphs on the vertices $\{u_1, \ldots, u_n\}$ that satisfy the property P.

Let G and G' be graphs on the vertices $\{u_1, \ldots, u_n\}$ that both satisfy the property P. In a labeled counting G and G' are the same if, and only if, every two vertices u_i and u_j are joined by the same number of edges in both graphs.

Remark 12.3

- The property P mentioned in Definition 12.2 may be any property that can be stated in graph-theoretic terms: being simple, connected, Eulerian, Hamiltonian, 5-colorable, planar, acyclic, having a perfect matching, containing exactly 2,004 edges, and so on, or any combination of such graph-theoretic properties.

- Note that two graphs G and G' on vertices $\{u_1, \ldots, u_n\}$ are the same in a labeled counting, if the edge-multiplicity of each pair of vertices in G is the same as in G'. Needless to say, the edges are not labeled at all.

Example 12.4 Let us count the number of labeled simple graphs on three vertices. This problem is the same as counting the number of simple graphs on the vertices $V = \{u_1, u_2, u_3\}$. Here the property P from Definition 12.2 is "being simple." Note that in a simple graph with vertex set V, for every pair u_i and u_j of distinct vertices from V, either they are adjacent or not. There are $\binom{3}{2} = 3$ distinct pairs of vertices in V, namely $\{u_1, u_2\}$, $\{u_2, u_3\}$, and $\{u_1, u_3\}$. Hence, there are precisely $2^{\binom{3}{2}} = 8$ possibilities of forming a simple graph on $V = \{u_1, u_2, u_3\}$. So, there are eight *labeled* simple graphs on three vertices.

Figure 12.2 shows these simple graphs. In general, note that two graphs are distinct if, and only if, there is at least one pair of vertices $u_i, u_j \in V$ such that they are adjacent in one graph but not in the other. Also, note that the three graphs that have exactly one edge each, are isomorphic, and the same holds for the three graphs that have exactly two edges each.

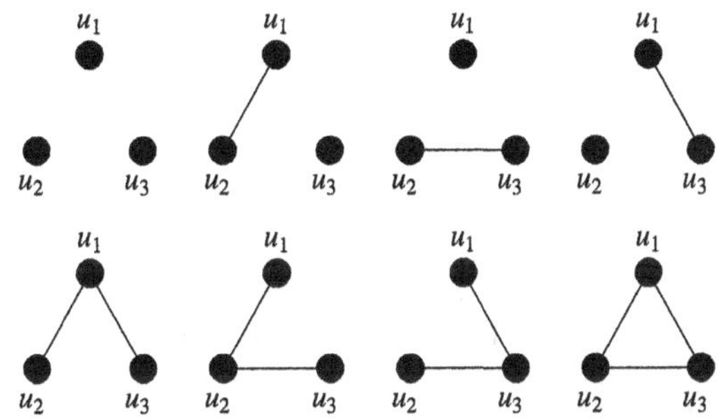

Figure 12.2: The eight labeled simple graphs on three vertices.

For a given set of n vertices $V = \{u_1, \ldots, u_n\}$, a simple graph with V as a vertex set has either u_i and u_j adjacent or not. Since there are a total of $\binom{n}{2}$ pairs of vertices, we have the following observation.

Observation 12.5 Let $m, n \in \mathbb{N}$, where $0 \leq m \leq \binom{n}{2}$.

1. The number g_n of simple graphs on n labeled vertices is given by
$$g_n = 2^{\binom{n}{2}}.$$
2. The number $g_n(m)$ of simple graphs on n labeled vertices having exactly m edges is given by
$$g_n(m) = \binom{\binom{n}{2}}{m}.$$

We leave the proof of the second part as Exercise 12.1.

Remark 12.6 The function $n \mapsto g_n = 2^{\binom{n}{2}}$ grows very fast, faster than any exponential function, faster than the factorial function, and faster than n^n as well. Some sample values illustrate this fact.

$$\begin{aligned} g_4 &= 64, \\ g_5 &= 1{,}024, \\ &\vdots \\ g_{100} &> 10^{1490}; \end{aligned}$$

that is, g_{100} has more than 1,490 digits when it is written out in decimal notation.

What is the number of graphs, which are not necessarily simple, on n labeled vertices $V = \{u_1, \ldots, u_n\}$? This counting problem can be solved in a manner similar to that for simple graphs. First we prove a useful combinatorial lemma.

Lemma 12.7 For $k, N \in \mathbb{N}$, the number of solutions to the equation

$$x_1 + x_2 + \cdots + x_k = N \tag{12.1}$$

in nonnegative integers $x_1, \ldots, x_k$ is $\binom{N+k-1}{k-1}$.

Proof Every solution to (12.1) in nonnegative integers $x_1, \ldots, x_k$ yields a solution in positive integers to

$$y_1 + y_2 + \cdots + y_k = N + k, \tag{12.2}$$

by letting $y_i = x_i + 1$, and also vice versa. Hence, the number of solutions to (12.2) in positive integers $y_1, \ldots, y_k$ is the same as the number of solutions to (12.1) in nonnegative integers $x_1, \ldots, x_k$. However, the number of solutions to (12.2) in positive integers $y_1, \ldots, y_k$ can be obtained by lining up $N + k$ identical marbles and then counting the number of ways that one can put $k - 1$ "separators" into the $N + k - 1$ slots between the marbles. For example, if $k = 5$ and $N = 10$, the configuration

$$\bullet\bullet \mid \bullet \mid \bullet\bullet\bullet \mid \bullet\bullet\bullet\bullet\bullet \mid \bullet\bullet\bullet\bullet$$

corresponds to the solution $2 + 1 + 3 + 5 + 4 = N + k = 15$ in positive integers to (12.2). Since, in general, there are precisely $\binom{N+k-1}{k-1}$ ways of choosing $k - 1$ separators among $N + k - 1$ possible slots, we have the lemma. ∎

As we mentioned earlier, a nonsimple graph on n vertices does not necessarily have finitely many edges. Hence, we need a bound on the number of edges as well in this case.

Theorem 12.8 Let $m, n \in \mathbb{N}$.

1. The number $\dot{g}_n(m)$ of graphs on n labeled vertices, m edges, and having no loops is given by

$$\dot{g}_n(m) = \binom{m + \binom{n}{2} - 1}{m}.$$

2. The number $\dot{g}_n(\leq m)$ of graphs on n labeled vertices, at most m edges, and having no loops is given by

$$\dot{g}_n(\leq m) = \binom{m + \binom{n}{2}}{m}.$$

3. The number $\ddot{g}_n(m)$ of graphs on n labeled vertices and m edges is given by

$$\ddot{g}_n(m) = \binom{m + \binom{n+1}{2} - 1}{m}.$$

4. The number $\ddot{g}_n(\leq m)$ of graphs on n labeled vertices and at most m edges is given by

$$\ddot{g}_n(\leq m) = \binom{m + \binom{n+1}{2}}{m}.$$

Proof Let $V = \{u_1, \ldots, u_n\}$ be a given set of n labeled vertices. For $i \neq j$, let $x_{i,j}$ denote the number of edges between u_i and u_j. In this way the number $\dot{g}_n(m)$ of graphs on the n labeled vertices V, having m edges and no loops, is equal to the number of solutions to the equation

$$\sum_{1 \leq i < j \leq n} x_{i,j} = m$$

in the $k = \binom{n}{2}$ nonnegative integers $x_{i,j}$. By Lemma 12.7 we have $\dot{g}_n(m) = \binom{N+k-1}{k-1}$, where $k = \binom{n}{2}$ and $N = m$. Hence, we have the first part of the theorem.

For the third part of the theorem, let x_i be the number of loops at the vertex u_i. In this way the number $\ddot{g}_n(m)$ of graphs on the n labeled vertices V having m edges is equal to the number of solutions to the equation

$$\left(\sum_{1 \leq i \leq n} x_i\right) + \left(\sum_{1 \leq i < j \leq n} x_{i,j}\right) = m$$

in the $k = n + \binom{n}{2} = \binom{n+1}{2}$ nonnegative integers x_i and $x_{i,j}$. By Lemma 12.7 we have $\ddot{g}_n(m) = \binom{N+k-1}{k-1}$, where $k = \binom{n+1}{2}$ and $N = m$. Hence, we have the third part.

The second and the fourth parts of the theorem are direct consequences of the first and the third parts, respectively. (See Exercise 12.5.) ∎

Example 12.9 By Theorem 12.8 there are precisely

$$\binom{3 + \binom{3}{2} - 1}{3} = \binom{5}{3} = \binom{5}{2} = 10$$

labeled graphs on three vertices, having three edges, and no loops. They are depicted in Figure 12.3.

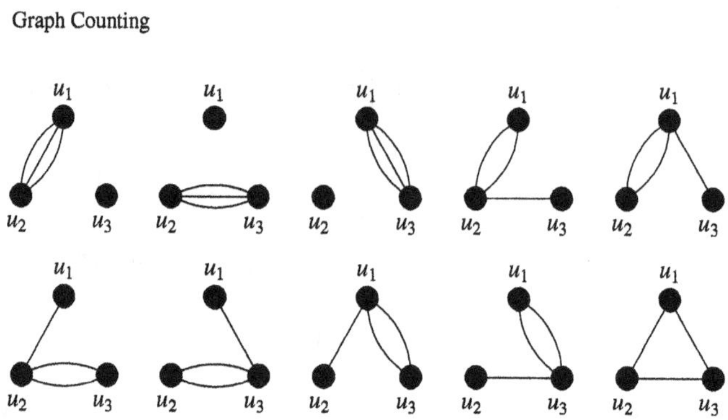

Figure 12.3: All loopless labeled graphs on three vertices and three edges.

Each of these graphs corresponds uniquely to a solution $(x_{1,2}, x_{2,3}, x_{3,1})$ of $x_{1,2} + x_{2,3} + x_{3,1} = m = 3$ in nonnegative integers $x_{i,j}$. These solutions, in the same order as shown in Figure 12.3, are given by the following ten possible solutions $(x_{1,2},\ x_{2,3},\ x_{3,1})$:

$$(3,0,0),\ (0,3,0),\ (0,0,3),\ (2,1,0),\ (2,0,1),$$
$$(1,2,0),\ (0,2,1),\ (1,0,2),\ (0,1,2),\ (1,1,1).$$

Note that only one of these ten graphs is a simple graph, namely the last one. It is the only graph that corresponds to a solution with each $x_{i,j} \in \{0, 1\}$.

Example 12.10 As in the Example 12.9, by Theorem 12.8 part 3, there are precisely

$$\binom{3 + \binom{3+1}{2} - 1}{3} = \binom{8}{3} = 56$$

graphs on the three labeled vertices $V = \{u_1, u_2, u_3\}$ that have three edges. This number 56 includes all the graphs on V that have no loops. Instead of simply listing all these 56 simple graphs, we describe how to list them in a systematic manner.

Note that a graph with vertex set V on three edges has m_e nonloop edges and m_l loops, where $m_e + m_l = m = 3$. We consider each case for $m_l \in \{0, 1, 2, 3\}$.

Example 12.9 presented the 10 graphs where $m_l = 0$. They are shown in Figure 12.3.

The graphs with exactly one loop have two nonloop vertices. There are three possibilities to place a loop on each of the vertices u_1, u_2, and u_3, and by Theorem 12.8 part 1, there are $\dot{g}_3(2) = 6$ graphs with vertex set V having two edges and no loops. Hence, there are $3 \cdot 6 = 18$ labeled graphs on three vertices having one loop and two nonloop edges, and they are shown in Figure 12.4. (For the general case for $m_e + m_l = m$, see Exercise 12.10.) Note that, so far, we have $10 + 18 = 28$ graphs.

To obtain the remainder of the graphs having two or more loops and three edges, we note that for graphs on $n = 3$ labeled vertices, we have a special dual-bijection between graphs with m_l loops and m_e nonloop edges, and graphs with m_e loops and m_l nonloop edges. This bijection, or duality, is given by transforming a graph on m_l loops and m_e nonloops to its dual-graph on m_e loops and m_l nonloops, by changing an edge between u_i

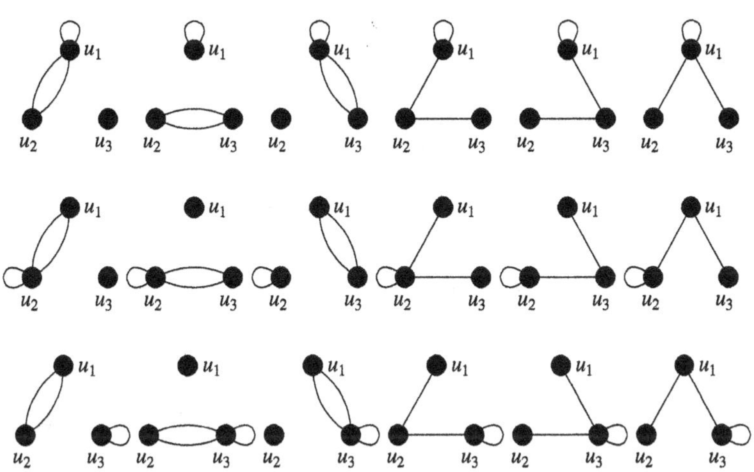

Figure 12.4: All labeled graphs on three vertices having one loop and two nonloop edges.

and u_j to a loop at u_k, where $k \in \{1, 2, 3\} \setminus \{i, j\}$, and changing a loop at u_i to an edge between u_j and u_k, where $\{j, k\} = \{1, 2, 3\} \setminus \{i\}$. In Figure 12.5 we see a graph and its dual-graph on its right. Clearly, the dual of the dual is the original graph.

In this way we obtain the remaining 28 graphs on 3 labeled vertices with m_l loops and m_e nonloop vertices, where $m_l \in \{2, 3\}$, altogether $28 + 28 = 56$ graphs, which is consistent with what we obtained from Theorem 12.8 part 3.

This alternative way of counting, or listing, all the graphs on n labeled vertices and m edges can be generalized, as we ask for in Exercise 12.11.

In this section we have defined what we mean by a labeled counting of a graph and given some basic examples of such countings. The other model for counting is a harder one (and sometimes more interesting from a theoretical point of view!). There we count the isomorphism classes and consider two graphs to be the "same" if they are isomorphic. To be precise we make the following definition.

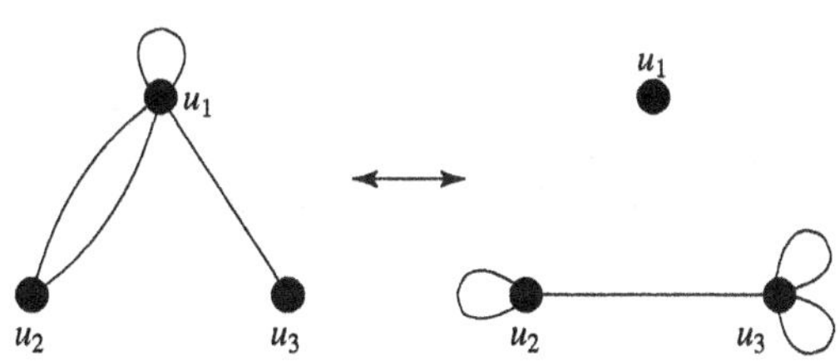

Figure 12.5: The special duality between graphs with m_l loops and m_e nonloop edges, and graphs with m_e loops and m_l nonloop edges.

Definition 12.11 Let $n \in \mathbb{N}$ and P be a graph-theoretic property. An *unlabeled counting* or an *unlabeled enumeration* is a counting of the number of graphs on the vertices $\{u_1, \ldots, u_n\}$ that satisfy the property P and are not isomorphic.

Let G and G' be graphs on the vertices $\{u_1, \ldots, u_n\}$ that both satisfy the property P. In an unlabeled counting, G and G' are considered the same if, and only if, they are isomorphic—that is, $G \cong G'$.

We present an example to illustrate the definition.

Example 12.12 Let us count the number of unlabeled simple graphs on three vertices. From Figure 12.2 we see that all simple graphs with one edge are isomorphic, as are all simple graphs with two edges. No other pairs of labeled simple graphs on 3 labeled vertices $V = \{u_1, u_2, u_3\}$ are isomorphic. Hence, the number of unlabeled simple graphs on $n = 3$ vertices is four. They are shown in Figure 12.6.

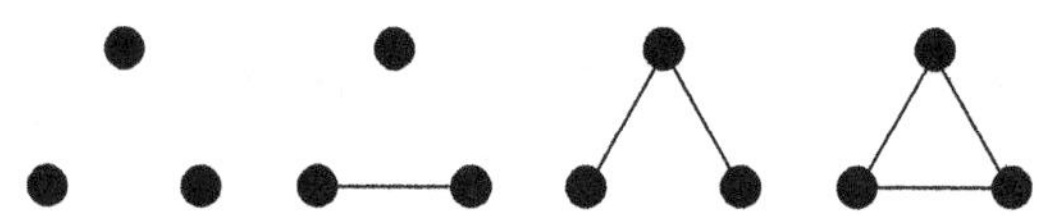

Figure 12.6: There are only four nonisomorphic simple graphs on three vertices.

Note that either exactly one or exactly three labeled graphs are isomorphic to each unlabeled graph.

To count the number of unlabeled graphs on three vertices and three edges, we could simply go through the list of all 56 labeled graphs from Example 12.10 and pick those that are not isomorphic to any of the ones that we have already chosen. This process is quite cumbersome, even in the small case where we consider only $n = 3$ vertices and $m = 3$ edges! Instead, in Example 12.13, we consider the cases where there are m_l loops and m_e nonloop edges, where $m_l + m_e = m = 3$ separately, as we did in Example 12.10.

Example 12.13 In Figure 12.3 we see all the labeled graphs on three vertices and three edges, with no loops—that is, $m_l = 0$ and $m_e = 3$. We see that there are exactly three nonisomorphic graphs. Now consider the unlabeled graphs on three vertices with $m_l = 1$ loops and $m_e = 2$ nonloop edges. From Figure 12.4 we see that there are precisely four nonisomorphic graphs on three vertices with one loop and two nonloop edges. All seven unlabeled graphs on three vertices, three edges, and $m_l \leq 1$ loops are shown in Figure 12.7.

In order to obtain all the other unlabeled graphs, we note that the duality described in Example 12.10 and depicted in Figure 12.5 preserves isomorphism. That is to say, two graphs are isomorphic if, and only if, their transformed dual graphs are isomorphic. Hence, taking the transformed duals of the graphs shown in Figure 12.7, we obtain the rest of the unlabeled graphs on three vertices and three edges, namely those having at least two loops. These graphs are shown in Figure 12.8. Therefore, we see from Figures 12.7 and 12.8 that there are precisely 14 unlabeled graphs on three vertices and three edges.

As we have seen in the last two examples, unlabeled counting is more difficult than labeled counting. However, using the right tools, we can perform unlabeled counting in a nice systematic fashion. Harary and Palmer [24] devote eight out of nine chapters to

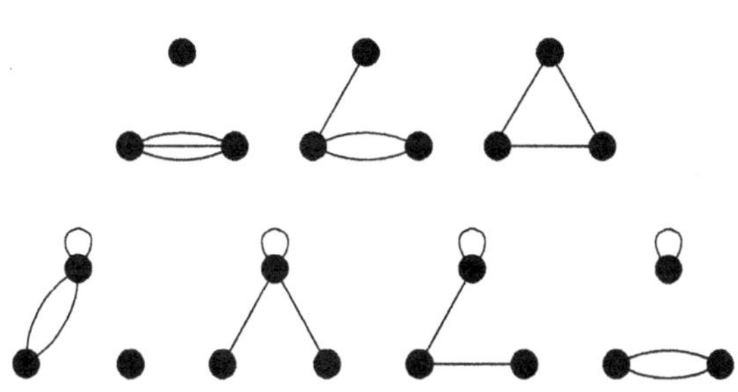

Figure 12.7: The seven unlabeled graphs on three vertices, three edges, and having at most one loop.

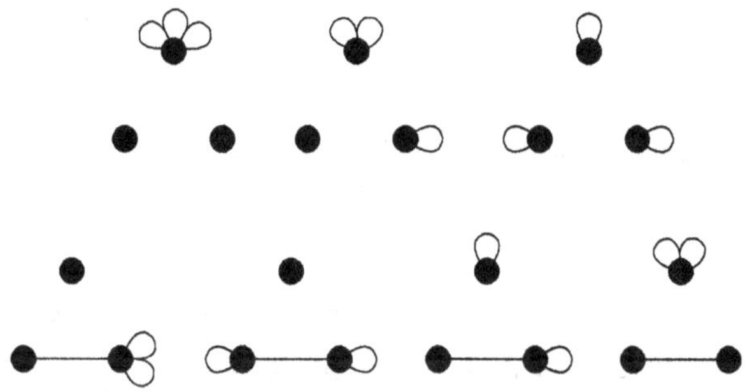

Figure 12.8: The seven unlabeled graphs on three vertices, three edges, and having at least two loops.

unlabeled graph counting! Only one chapter there is devoted to labeled counting. Another classic work is Riordan [45], which presents a nice discussion on how to count unlabeled trees on a given number of vertices. There, however, since the text was written in 1958, the current terminology was not used, so that book can be slightly hard to read at times. A more modern treatment of unlabeled counting of trees, where much of the current terminology is used, can be found in Deo [11].

For the remainder of this chapter we will concentrate on the easier of the two countings, namely labeled counting. In order to treat that subject systematically, we will need a short introduction to the notion of generating functions.

12.3 Generating Functions

Generating functions are tools that connect discrete mathematics to continuous analytic mathematics. In this way we include additional tools in our toolbox to treat discrete objects, namely the known methods and techniques from analytic mathematics. This connection has proved invaluable for solving many discrete combinatorial problems.

Although we will not dwell on the foundation of the analytic part, some minimal knowledge is needed in order to understand generating functions fully. We will assume some knowledge in the standard power series presentations (that is, a Taylor series expansion around $x = 0$) of some common real (respectively, complex) functions, such as $(1 + x)^\alpha$, $\log(1 - x)$, and e^x, to name a few. The power series presentations of these functions, and of many more, can be found in any comprehensive college calculus text.

The purpose of generating functions is twofold:

1. To use known analytic methods on a function, $f(x)$ or $f(x, y)$, to obtain the corresponding closed expression or formula for the number of possibilities of a certain finite combinatorial structure.
2. To encode the number of possibilities of a certain finite combinatorial structure in terms of a simple analytic function, $f(x)$ or $f(x, y)$, since often it is possible to state the combinatorial result *precisely* in terms of functions, whereas it is impossible to do so in a discrete fashion. By decoding the function, many times one can obtain just enough information for the desired purpose.

What the second statement means is that often when we have no chance of finding an exact formula for the number of possible finite combinatorial structures from the generating function, we can still obtain previously hidden information about our discrete structure—for example, a nontrivial recursive formula or an asymptotic result.

Here we will be concentrating on generating functions of one indeterminate or variable x.

Definition 12.14 Let $(a_0, a_1, a_2, \ldots) = (a_n)_{n\geq 0}$ be a given sequence of real (respectively, complex) numbers.

1. The *generating function* $a(x)$ of the sequence $(a_n)_{n\geq 0}$ is given by the series

$$a(x) = a_0 + a_1 x + a_2 x^2 + \cdots = \sum_{n\geq 0} a_n x^n.$$

2. The *exponential generating function* $\bar{a}(x)$ of the sequence $(a_n)_{n\geq 0}$ is given by the series

$$\bar{a}(x) = a_0 \frac{1}{0!} + a_1 \frac{x}{1!} + a_2 \frac{x^2}{2!} + \cdots = \sum_{n\geq 0} a_n \frac{x^n}{n!}.$$

We present an example to illustrate this definition. Remark 12.16 discusses the values of x for which these expressions hold.

Example 12.15

- The generating function for the constant sequence $(1, 1, 1, \ldots)$, consisting entirely of 1's, is given by $1 + x + x^2 + \cdots = \sum_{n\geq 0} x^n = \frac{1}{1-x}$.
- The generating function for the sequence $(1, 1, \frac{1}{2}, \frac{1}{6}, \ldots, \frac{1}{n!}, \ldots)$ is given by $1 + x + \frac{x^2}{2} + \cdots = \sum_{n\geq 0} \frac{x^n}{n!} = e^x$, the usual exponential function.
- The generating function for the sequence $(1, 1, 2, 6, \ldots, n!, \ldots)$ is given by $1 + x + 2x^2 + 6x^3 + \cdots = \sum_{n\geq 0} n! x^n$.

The usefulness of the exponential generating function will be discussed shortly. For now we will concentrate on the ordinary generating function for a number sequence, since, after all, the analytical foundation of both types is the same.

Remark 12.16 When substituting x in a generating function for a real (respectively, complex) number, we note the following:

- The first generating function in Example 12.15 is valid only for real (respectively, complex) numbers x satisfying $|x| < 1$.
- The second generating function is valid for all real (respectively, complex) numbers x.
- Viewed as a genuine function from a subset of $\mathbb{R}$ (respectively, $\mathbb{C}$) into $\mathbb{R}$ (respectively, $\mathbb{C}$), the third generating function is valid only for $x = 0$. That is to say, only for $x = 0$ can we evaluate the sum of the series and obtain a real (or complex) number (which is zero in this case). Nonetheless, it is still useful from a combinatorial point of view to consider the actual series, although it never yields a real (respectively, complex) number (except for $x = 0$)!

As we see in Remark 12.16, it is important to view generating functions not just as an infinite sum of real (respectively, complex) numbers, but rather as their own entities, disregarding whether they will converge to a given real (respectively, complex) value. When viewed as such entities, these generating functions are *formal power series*, which can be manipulated algebraically.

Before stating the properties of formal power series in more detail, we look at some examples that better explain the practical use of generating functions.

Example 12.17 For $m, n \in \mathbb{N}$, as in Observation 12.5, let $g_n(m)$ denote the number of simple graphs on n labeled vertices having m edges. So, $g_n(m) = \binom{\binom{n}{2}}{m}$. For fixed n, let $g_n(x)$ be the generating function for the labeled simple graphs on n vertices—that is, for the sequence $(g_n(m))_{m\geq 0}$. In this case we have

$$g_n(x) = \sum_{m\geq 0} g_n(m)x^m = \sum_{m=0}^{\binom{n}{2}} \binom{\binom{n}{2}}{m} x^m = (1+x)^{\binom{n}{2}}.$$

Example 12.18 Consider the sequence of trees $T_1, T_2, \ldots$ shown in Figure 3.7 relating to Example 3.8. There t_k denotes the number of vertices of the kth tree T_k, and it satisfies

$$\begin{aligned} t_1 &= 1, \\ t_2 &= 4, \\ t_k &= 3t_{k-1} - 2t_{k-2} \text{ for } k \geq 3. \end{aligned}$$

Letting $a_k = t_{k-1}$, we derive a closed formula for a_k, and hence for t_k as well, by considering the following generating function:

$$a(x) = a_0 + a_1 x + a_2 x^2 + \cdots = \sum_{k\geq 0} a_k x^k. \tag{12.3}$$

Since $a_0 = 1$, $a_1 = 4$, and $a_k = 3a_{k-1} - 2a_{k-2}$ for $k \geq 2$, for the generating function $a(x)$ we obtain that

$$\begin{aligned} a(x) &= a_0 + a_1 x + \sum_{k\geq 2} a_k x^k \\ &= 1 + 4x + \sum_{k\geq 2} (3a_{k-1} - 2a_{k-2})x^k \\ &= 1 + 4x + 3x\left(\sum_{k\geq 2} a_{k-1}x^{k-1}\right) + 2x^2\left(\sum_{k\geq 2} a_{k-2}x^{k-2}\right) \\ &= 1 + 4x + 3x[a(x) - 1] + 2x^2 a(x). \end{aligned}$$

Solving for $a(x)$ and using partial fractions, we obtain that

$$a(x) = \frac{1+x}{1-3x+2x^2} = \frac{3}{1-2x} - \frac{2}{1-x}.$$

Expanding the last two functions into their geometric series, we get

$$a(x) = 3\sum_{k\geq 0} 2^k x^k - 2\sum_{k\geq 0} x^k.$$

Since a_k is the coefficient of x^k in $a(x)$, it must be the coefficient of x^k on the right-hand side as well. Therefore, we obtain that $a_k = 3 \cdot 2^k - 2$, and hence $t_k = 3 \cdot 2^{k-1} - 2$, which is the desired formula for t_k.

Example 12.19 Recall the Catalan numbers $C_0, C_1, C_2, \ldots$ from Definition 3.34. They are defined recursively by

$$\begin{aligned} C_0 &= 1, \\ C_n &= C_0 C_{n-1} + C_1 C_{n-2} + \cdots + C_{n-1} C_0. \end{aligned}$$

We now derive the formula for C_n, which was given ad hoc in (3.3), by means of generating functions. Let $C(x)$ be the generating function for the Catalan numbers

$$C(x) = C_0 + C_1 x + C_2 x^2 + \cdots = \sum_{n\geq 0} C_n x^n.$$

From this equation we see that $C(x) \cdot C(x) = C(x)^2$ is given by

$$\begin{aligned} C(x)^2 &= C_0^2 + (C_0C_1 + C_1C_0)x + (C_0C_2 + C_1C_1 + C_0C_2)x^2 + \cdots \\ &= \sum_{n\geq 1}\left(\sum_{i=0}^{n-1} C_i C_{n-i-1}\right)x^{n-1}. \end{aligned}$$

Multiplying through by the variable x and using the recursive definition of C_n yields

$$xC(x)^2 = \sum_{n\geq 1}\left(\sum_{i=0}^{n-1} C_i C_{n-i-1}\right)x^n = \sum_{n\geq 1} C_n x^n = C(x) - 1.$$

Solving for $C(x)$, and keeping in mind that $C(x)$ evaluated at $x = 0$ gives $C(0) = C_0 = 1$, we obtain that $C(x) = \frac{1}{2x}\left(1 - \sqrt{1-4x}\right)$. Using the Generalized Binomial Theorem, $(1+x)^\alpha = \sum_{n\geq 0}\binom{\alpha}{n}x^n$ for any real α, where $\binom{\alpha}{n} = \alpha(\alpha-1)\cdots(\alpha-n+1)/n!$, we obtain for $C(x)$ (see Exercise 12.17) that

$$\begin{aligned} C(x) &= \frac{1}{2x}\left(1-(1-4x)^{\frac{1}{2}}\right) \\ &= \frac{-1}{2x}\sum_{n\geq 1}\binom{\frac{1}{2}}{n}(-4x)^n \\ &= \sum_{n\geq 1}\frac{1}{n}\binom{2n-2}{n-1}x^{n-1}. \end{aligned}$$

Comparing coefficients for x^{n-1}, we get $C_{n-1} = \frac{1}{n}\binom{2n-2}{n-1}$ for all $n \geq 1$. Therefore, $C_n = \frac{1}{n+1}\binom{2n}{n}$ for all $n \geq 0$, which agrees with what is given in (3.3).

As we have seen in the previous two examples, we manipulated the generating functions themselves as power series to obtain closed formulas for what we were counting. The manipulations were done *formally* in the sense that we did not take proper care as to whether or not the manipulations were legitimate or sensible operations when x was replaced by a real (respectively, complex) number. We conclude this section with some formal rules, which should make clearer which kinds of manipulations are legitimate and which ones are not.

By viewing a generating function $a(x) = \sum_{n\geq 0} a_n x^n$ as a formal object, where the a_n's are some real (respectively, complex) numbers, we have the following:

- Two formal power series, $\sum_{n\geq 0} a_n x^n$ and $\sum_{n\geq 0} b_n x^n$, are *equal* if, and only if, $a_n = b_n$ holds for every $n \geq 0$.
- There is a one-to-one bijection between the set of all infinite sequences $(a_n)_{n\geq 0}$ of real (respectively, complex) numbers and the set of all formal power series $\sum_{n\geq 0} a_n x^n$ over the indeterminate x. The bijection is given by the formation of the generating function of the sequence.

The set of all formal power series of sequences of real (respectively, complex) numbers over the indeterminate x is denoted by $\mathbb{R}[[x]]$ (respectively, $\mathbb{C}[[x]]$). The nth coefficient—that is, the coefficient for x^n—of an element $a(x) \in \mathbb{R}[[x]]$ is denoted by $[x^n]a(x)$. So, if $a(x) = \sum_{n\geq 0} a_n x^n$, then $[x^n]a(x) = a_n$. With this notation in mind, the following theorem explains some of the basic operations we can perform on formal power series, and hence on any generating function of a number sequence. We state the theorem in terms of the real numbers $\mathbb{R}$, but it also holds for the complex numbers $\mathbb{C}$, or any number field for that matter. The proof of the theorem is left as Exercise 12.20.

Theorem 12.20 Let $a(x), b(x) \in \mathbb{R}[[x]]$ be elements with $[x^n]a(x) = a_n$ and $[x^n]b(x) = b_n$. In this case we have the following:

1. Their sum is a formal power series $a(x) + b(x) \in \mathbb{R}[[x]]$, and for all $n \geq 0$ we have

$$[x^n](a(x) + b(x)) = a_n + b_n.$$

2. Their product is a formal power series $a(x)b(x) \in \mathbb{R}[[x]]$, and for all $n \geq 0$ we have
$$[x^n](a(x)b(x)) = a_0b_n + a_1b_{n-1} + \cdots + a_nb_0 = \sum_{i=0}^{n} a_ib_{n-i}.$$
In particular, for a fixed nonnegative integer m we have
$$[x^n](x^m a(x)) = \begin{cases} 0 & \text{if } n < m, \\ a_{n-m} & \text{if } n \geq m. \end{cases}$$
3. If $a_0 \neq 0$, then $a(x)$ has a unique multiplicative inverse $a(x)^{-1} \in \mathbb{R}[[x]]$ as a formal power series, satisfying $a(x)a(x)^{-1} = a(x)^{-1}a(x) = 1$.
4. The formal derivative $a'(x) = \frac{d}{dx}a(x) \in \mathbb{R}[[x]]$ of $a(x)$ has
$$[x^n]a'(x) = (n+1)a_{n+1}$$
for all $n \geq 0$. Moreover, the formal derivative is identical to the derivative of the real function $x \mapsto a(x)$ which is represented by the formal power series $a(x)$.
5. The formal derivative obeys the usual additive rule, product rule, and the quotient rule for differentiation.
6. If $a_0 = 0$, then the formal composition $(b \circ a)(x) = b(a(x)) \in \mathbb{R}[[x]]$ is a well-defined formal power series.
7. The formal derivative of a formal composition obeys the usual chain rule for differentiation.

The rules from Theorem 12.20 say that it is legitimate to manipulate the generating functions for a number sequence algebraically, just as if they were well-defined functions for all real (respectively, complex) values of x. In short, every formal manipulation makes perfect sense when dealing with generating functions.

Example 12.21 We would like to find the generating function of the number of all directed graphs on n labeled vertices $V = \{u_1, \ldots, u_n\}$ that have just one directed edge. In Figure 12.9 we show all nine digraphs on three vertices and one directed edge.

Clearly there are n possibilities for choosing the tail of the lone directed edge, and there are again n possibilities for choosing the head. Hence, there are precisely n^2 digraphs on n labeled vertices with one directed edge. Our generating function is given formally by
$$f(x) = x + 4x^2 + 9x^3 + \cdots = \sum_{n\geq 0} n^2x^n.$$
To encode $f(x)$ as a compact function, we first note that
$$\begin{aligned}\left(x \cdot \frac{d}{dx}\right)^2 \left(\sum_{n\geq 0} x^n\right) &= \sum_{n\geq 0} \left(x \cdot \frac{d}{dx}\right)^2 x^n \\ &= \sum_{n\geq 0} x \cdot \frac{d}{dx}\left(x \cdot \frac{d}{dx}(x^n)\right) \\ &= \sum_{n\geq 0} n^2x^n \\ &= f(x).\end{aligned}$$

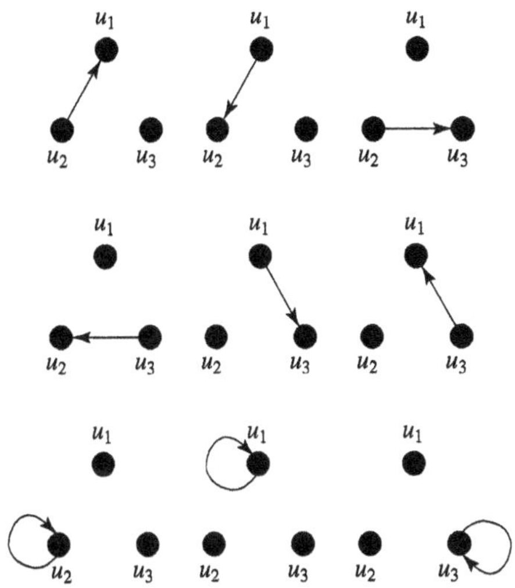

Figure 12.9: The nine labeled digraphs on three vertices and one directed edge.

Since we already have that $\sum_{n\geq 0} x^n = 1/(1-x)$, we obtain a compact form of $f(x)$ given by

$$f(x) = \left(x \cdot \frac{d}{dx}\right)^2 \left(\frac{1}{1-x}\right) = \frac{x(1+x)}{(1-x)^3},$$

which then is the generating function for the number of labeled directed graphs on n vertices and one directed edge—that is, for the number sequence $(0, 1, 4, 9, \ldots, n^2, \ldots)$.

Remark 12.22 The usefulness of generating functions in Example 12.21 was only for the actual encoding of the sequence $(0, 1, 4, 9, \ldots, n^2, \ldots)$ into a single function $x(1+x)/(1-x)^3$. For more complex situations the general scenario is not to know the actual number sequence $(a_0, a_1, a_2, \ldots, a_n, \ldots)$ for the number of graphs on n vertices satisfying a certain property, although the generating function is given by a nice closed formula. Such situations come up more frequently when we consider the exponential generating functions for number sequences, which are something that we discuss in more detail in the next two sections.

12.4 Partitions of a Finite Set

The presentation of a fundamental Labeled Counting Lemma for graphs depends on the use of exponential generating functions and the use of *weak ordered partitions* of a set. Hence, this section is dedicated to describing some properties of such partitions. The terminology of Definition 12.23 is similar to that used in Stanley's books [50] and [51].

We start by describing what we mean by weak ordered partitions of a given finite set S. Then we discuss some interesting numbers associated with these partitions of S.

Definition 12.23 Let $k \in \mathbb{N}$ and let S be a fixed finite set containing $n \geq k$ elements.

1. A *weak ordered partition* of S into k parts is an ordered k-tuple $(S_1, \ldots, S_k)$ of subsets of S satisfying:

$$S_i \cap S_j = \varnothing, \text{ for all } i \neq j,$$
$$\bigcup_{i=1}^{k} S_i = S.$$

2. A *proper partition* of S into k parts is a k-element set $\{S_1, \ldots, S_k\}$ of nonempty subsets of S satisfying the above two conditions.

The next examples illustrate the definition.

Example 12.24 Consider the set $S = \{1, 2, 3, 4\}$. The following are all the weak ordered partitions of S into two parts (S_1, S_2):

$$\begin{array}{llll}(\varnothing, \{1,2,3,4\}), & (\{1\}, \{2,3,4\}), & (\{2\}, \{1,3,4\}), & (\{3\}, \{1,2,4\}), \\ (\{4\}, \{1,2,3\}), & (\{1,2\}, \{3,4\}), & (\{2,3\}, \{1,4\}), & (\{3,4\}, \{1,2\}), \\ (\{1,3\}, \{2,4\}), & (\{2,4\}, \{1,3\}), & (\{1,4\}, \{2,3\}), & (\{1,2,3\}, \{4\}), \\ (\{1,2,4\}, \{3\}), & (\{1,3,4\}, \{2\}), & (\{2,3,4\}, \{1\}), & (\{1,2,3,4\}, \varnothing).\end{array}$$

Note that $(\{1\}, \{2, 3, 4\})$ and $(\{2, 3, 4\}, \{1\})$ are distinct weak ordered partitions. There are precisely $2^4 = 16$ weak ordered partitions of $S = \{1, 2, 3, 4\}$ into two parts, since there are precisely 2^4 subsets of $\{1, 2, 3, 4\}$.

However, when we consider proper partitions of S into two nonempty parts $\{S_1, S_2\}$, there is no distinction between $\{S_1, S_2\}$ and $\{S_2, S_1\}$. Hence, there are precisely seven proper partitions of S into two nonempty parts $\{S_1, S_2\}$:

$$\begin{array}{llll}\{\{1\}, \{2,3,4\}\}, & \{\{2\}, \{1,3,4\}\}, & \{\{3\}, \{1,2,4\}\}, & \{\{4\}, \{1,2,3\}\}, \\ \{\{1,2\}, \{3,4\}\}, & \{\{2,3\}, \{1,4\}\}, & \{\{1,3\}, \{2,4\}\}. & \end{array}$$

Note that the number of weak ordered partitions of S into two proper parts is given by $2!S(4, 2) + 2$, where $S(4, 2) = 7$ is the number of proper partitions of S into two nonempty parts.

Example 12.25 For nonnegative integers k and n with $k \leq n$, we denote by $S(n, k)$ the number of ways to partition the n-element set $S = \{1, \ldots, n\}$ properly into k nonempty parts. As we saw in Example 12.24, we have $S(4, 2) = 7$.

By convention $S(n, k) = 0$ if $k > n$ and $S(0, 0) = 1$. This convention makes sense, since there is one way of not partitioning the empty set. However, since there is no way to partition a nonempty set S properly into zero parts, we have $S(n, 0) = 0$ for all $n \in \mathbb{N}$. Likewise, $S(n, 1) = 1$, for all $n \in \mathbb{N}$. To partition $S = \{1, \ldots, n\}$ into two nonempty parts, $\{S_1, S_2\}$, one set is uniquely determined by the other by being its complement $S_2 = \overline{S_1}$ and vice versa. The number of ways to choose a subset $S_1 \subseteq S$ in such a way that neither S_1 nor S_2 is empty is $2^n - 2$, since S_1 can be any subset of S, except the empty set and the whole set S. In this way we have obtained all the weak ordered partitions (S_1, S_2) of S into two nonempty parts, and therefore we have the following:

$$S(n, 2) = \frac{2^n - 2}{2!} = 2^{n-1} - 1,$$

which is consistent with Example 12.24, since $S(4, 2) = 7 = 2^{4-1} - 1$.

The numbers $S(n, k)$ are called *Stirling numbers of the second kind.* We ask for some additional properties of these numbers in Exercises 12.28, 12.29, 12.30, and 12.31.

The properties of the $S(n, k)$'s described in the following theorem are defining properties of the Stirling numbers of the second kind.

Theorem 12.26 The Stirling numbers of the second kind satisfy the following recursive equation:

$$S(n, k) = kS(n - 1, k) + S(n - 1, k - 1) \tag{12.4}$$

for all $k, n \in \mathbb{N}$.

Proof There are two types of partitions of the set $\{1, \dots, n\}$ into k parts.

The first type consists of those partitions where the set containing n is the singleton set $\{n\}$. In this case the remaining $k - 1$ sets form a nonempty partition of $\{1, \dots, n - 1\}$. Hence, there are $S(n - 1, k - 1)$ partitions of the first type.

The second type consists of those nonempty partitions where the set containing n has two or more elements in it. Removing n from that set yields one of the $S(n - 1, k)$ partitions of $\{1, \dots, n\}$ into k parts. Since n could have been in any of the k possible sets, we see that there are $k \cdot S(n - 1, k)$ partitions of the second type. This completes the proof. ∎

The boundary conditions for $S(n, k)$ discussed in Example 12.25, together with (12.4), can be used to compute $S(n, k)$ for any nonnegative pair of integers $k, n \in \mathbb{N}$.

Note that if it hadn't been for the factor k, the recursive formula (12.4) would have been that of the binomial coefficients $\binom{n}{k}$.

Remark 12.27 The Stirling numbers of the second kind are named after the Scottish mathematician James Stirling (1692–1770). As the name $S(n, k)$ indicates, there are also *Stirling numbers of the first kind.* It is amusing to note that Stirling himself considered the Stirling numbers of the second kind *first*, before he considered the ones of the first kind. For definitions and a more detailed discussion of Stirling numbers of both kinds, we refer the reader to Graham, Knuth, and Patashnik [22].

Example 12.28 Let $S = \{1, \dots, n\}$, where $n \in \mathbb{N}$. The number $B(n)$ of unrestricted nonempty partitions of S, regardless of size, depends only on n. The unrestricted partitions of S with arbitrary sizes come in n types, where for each $k \in \{1, \dots, n\}$, the kth type is a proper partition of size k. We know that there are $S(n, k)$ proper partitions of size k. Hence, we have

$$B(n) = S(n, 1) + S(n, 2) + \cdots + S(n, n) = \sum_{k=1}^{n} S(n, k). \tag{12.5}$$

By convention $B(0) = 1$, and by (12.5) we have $B(1) = S(1, 1) = 1$ and $B(2) = S(2, 1) + S(2, 2) = 1 + 1 = 2$. The numbers $B(n)$ for nonnegative integers n are called *Bell numbers* after the Scottish/American mathematician Eric Temple Bell (1883–1960).

Let us compute the next two Bell numbers $B(3)$ and $B(4)$. By Exercise 12.29 we have $S(3, 2) = \binom{3}{2} = 3$ and hence $B(3) = S(3, 1) + S(3, 2) + S(3, 3) = 1 + 3 + 1 = 5$. The following is the complete list of all the unrestricted nonempty partitions of $\{1, 2, 3\}$.

$$\{\{1, 2, 3\}\}, \ \{\{1\}, \{2, 3\}\}, \ \{\{2\}, \{1, 3\}\}, \ \{\{3\}, \{1, 2\}\}, \ \{\{1\}, \{2\}, \{3\}\}.$$

For $B(4)$, we note from Example 12.25 that $S(4, 2) = 7$ and from Exercise 12.29 that $S(4, 3) = \binom{4}{3} = 4$. Hence, by (12.5), we have $B(4) = 1 + 7 + 4 + 1 = 13$. Further properties of the Bell numbers $B(n)$ are discussed in Exercises 12.32, 12.33, and 12.34.

The Bell numbers also satisfy a recursive formula that can be used to define them, as was the case in Theorem 12.26.

Theorem 12.29 The Bell numbers $B(n)$, where n is a nonnegative integer, satisfy the recursive formula

$$\begin{aligned} B(n+1) &= \binom{n}{0}B(0) + \binom{n}{1}B(1) + \cdots + \binom{n}{n}B(n) \\ &= \sum_{k=0}^{n} \binom{n}{k} B(k). \end{aligned} \tag{12.6}$$

Proof For $n \in \mathbb{N}$, the unrestricted nonempty partitions of $\{1, \ldots, n, n + 1\}$ are of exactly $n + 1$ types, where for $k \in \{0, 1, \ldots, n\}$ the kth type contains partitions where $n + 1$ is contained in a set with k elements from $\{1, \ldots, n\}$. There are $\binom{n}{k}$ ways of choosing k elements to join $n+1$ in a set, and there are $B(n-k)$ unrestricted partitions of the remaining $n - k$ elements. Since $\binom{n}{k} = \binom{n}{n-k}$, we have

$$B(n+1) = \sum_{k=0}^{n} \binom{n}{k} B(n-k) = \sum_{k=0}^{n} \binom{n}{k} B(k),$$

which completes our proof. ∎

12.5 The Labeled Counting Lemma

In this section we delve into the significance of the exponential generating function in further detail in regard to counting labeled graphs of a more difficult type than we have done so far. In particular, we will see that one can express certain combinatorial results neatly in terms of functions, although it is very messy to do this explicitly.

We use the partitions discussed in §12.4 to obtain our results. The next lemma, called the *Labeled Counting Lemma*, is fundamental for counting labeled graphs. Its proof is easy. When we are counting various labeled graphs, what is important is the mere statement of this lemma, as we will see shortly.

Lemma 12.30 **Labeled Counting**

For a nonnegative n, let a_n be the number of labeled graphs on n vertices satisfying the graph property $P(a)$, and let b_n be the number of labeled graphs on n vertices satisfying the graph property $P(b)$. Let $\bar{a}(x)$ and $\bar{b}(x)$ be the corresponding exponential generating functions for the sequences $(a_n)_{n\geq 0}$ and $(b_n)_{n\geq 0}$, respectively. If c_n is defined by

$$c_n = \sum_{(S_1,S_2)} a_{|S_1|}b_{|S_2|},$$

where the sum is taken over all weak ordered partitions (S_1, S_2) of $\{1, \ldots, n\}$, then the exponential generating function $\bar{c}(x)$ of the sequence $(c_n)_{n\geq 0}$ is given by

$$\bar{c}(x) = \bar{a}(x)\bar{b}(x).$$

Proof For each $n \in \mathbb{N}$ and $i \in \{0, 1, \ldots, n\}$, there are precisely $\binom{n}{i}$ weak ordered partitions (S_1, S_2) of $\{1, \ldots, n\}$, where $|S_1| = i$ and $|S_2| = n - i$. Hence, we have $c_n = \sum_{i=0}^{n} \binom{n}{i} a_i b_{n-i}$, which is the coefficient for $x^n/n!$ in $\bar{a}(x)\bar{b}(x)$. ∎

The next example demonstrates how to apply the Labeled Counting Lemma 12.30.

Example 12.31 For $n \in \mathbb{N}$, let $V = \{u_1, \ldots, u_n\}$ be a set of vertices. Let c_n be the number of simple graphs on n vertices that are a union of a simple cycle and a star, which is rooted at its center. We would like to find the exponential generating function $\bar{c}(x)$ for the c_n's. In Figure 12.10 we show one of the possible labeled graphs contributing to the number c_{11}, where one part is a cycle and the other part is a rooted star.

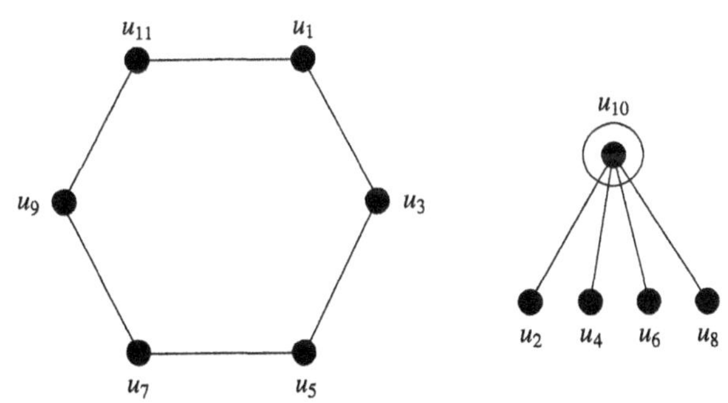

Figure 12.10: An 11-vertex labeled graph consisting of C_6 and a rooted star on five vertices, where the root is at the center.

Let $P(a)$ be the graph property of being a simple cycle, and let $P(b)$ be the graph property of being a rooted star, where the root is in the center of the star. For each $n \geq 3$, we note that there are $a_n = (n-1)!/2$ simple cycles on n labeled vertices, and none for $0 \leq n \leq 2$, so $a_n = 0$ for $0 \leq n \leq 2$. (See Exercise 5.22 in Chapter 5.) Also, for each

$n \geq 0$, there are exactly $b_n = n$ rooted stars on n labeled vertices, since there are n possible ways to choose the root and then all the other vertices automatically become the leaves. In this way we have that $c_n = \sum_{(S_1,S_2)} a_{|S_1|} b_{|S_2|}$, where the sum is over all the weak ordered partitions (S_1, S_2) of $\{1, \ldots, n\}$. By the Labeled Counting Lemma 12.30, we have that

$$\begin{aligned}
\bar{c}(x) &= \left(\sum_{n\geq 0} a_n \frac{x^n}{n!}\right)\left(\sum_{n\geq 0} b_n \frac{x^n}{n!}\right)\\
&= \left(\sum_{n\geq 3} \frac{(n-1)!}{2n!} x^n\right)\left(\sum_{n\geq 0} \frac{n}{n!} x^n\right)\\
&= \frac{x}{2}\left(\sum_{n\geq 3} \frac{x^n}{n}\right)\left(\sum_{n\geq 1} \frac{x^{n-1}}{(n-1)!}\right)\\
&= \frac{x}{2}\left(\int \frac{x^2}{1-x} dx\right) e^x\\
&= \frac{xe^x}{2}\left(\log\left(\frac{1}{1-x}\right) - x - \frac{x^2}{2}\right),
\end{aligned}$$

where "log" is the natural logarithm. This result shows that the c_n are compactly stored in the generating function $\bar{c}(x)$. Moreover, we can expand both e^x and $\log(1-x)$ into their series and obtain from those expansions an explicit formula for c_n, although we cannot hope for it to look very nice.

The Labeled Counting Lemma 12.30 can, of course, be generalized to a weak ordered partition of an arbitrary size k. In that case we call it the *General Labeled Counting Lemma*.

Lemma 12.32 **General Labeled Counting**

Let $k \in \mathbb{N}$. For each $i \in \{1, \ldots, k\}$, let $a_{i;n}$ be the number of labeled graphs on $n \geq 0$ vertices satisfying the graph property $P(a_i)$, and let $\bar{a}_i(x)$ be the exponential generating function of the corresponding number sequence $(a_{i;n})_{n\geq 0}$. If c_n is defined by

$$c_n = \sum_{(S_1,\ldots,S_k)} a_{1;|S_1|} \cdots a_{k;|S_k|},$$

where the sum is taken over all weak ordered partitions $(S_1, \ldots, S_k)$ of $\{1, \ldots, n\}$ of size k, then the exponential generating function $\bar{c}(x)$ of the sequence $(c_n)_{n\geq 0}$ is given by

$$\bar{c}(x) = \bar{a}_1(x) \cdots \bar{a}_k(x) = \prod_{i=1}^{k} \bar{a}_i(x).$$

The proof of Lemma 12.32 is left as Exercise 12.36. The next example shows how to apply the General Labeled Counting Lemma 12.32.

Example 12.33 For $n \in \mathbb{N}$, let $V = \{u_1, \ldots, u_n\}$ be a set of uncolored vertices. Let d_n be the number of ways that a simple plane graph on V can be formed by the disjoint union of a simple cycle, a nontrivial path, and isolated vertices, where some of the isolated vertices have been placed inside the cycle. We would like to find the exponential generating function $\bar{d}(x)$ for the d_n's.

In Figure 12.11 we see two of the possible labeled plane graphs contributing to the number d_{12}, where some of the isolated vertices have been placed inside the cycle. Note that the plane cycles are distinct, since the one on the right is a plane reflection of the one on the left, and hence the left cycle cannot be transformed by topological maneuvering to the one on the right. Also, the right-hand cycle has different isolated vertices inside than the left one. We do not care whether or not the path is inside or is outside the cycle.

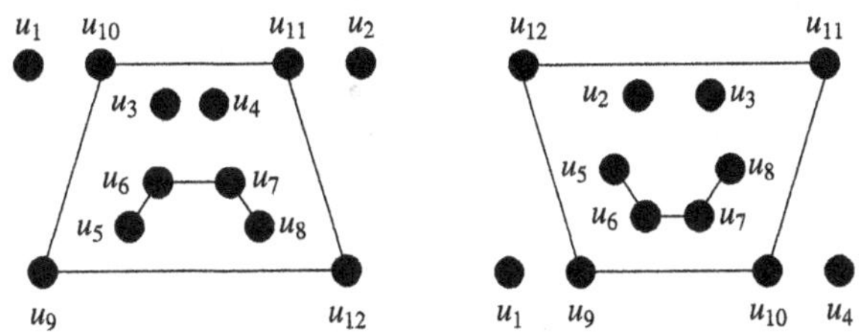

Figure 12.11: Two of the labeled graphs contributing to the value of d_{12} from Example 12.33.

Let $P(c)$ be the graph property of being a simple cycle, let $P(p)$ be the graph property of being a nontrivial path, and let $P(s)$ be the property that some isolated vertices are inside the cycle. There are no simple cycles on $n \le 2$ vertices, and there are precisely $c_n = (n-1)!$ plane cycles for $n \ge 3$, since each labeled cycle has two embeddings in the plane, one clockwise and another counterclockwise. There are no nontrivial simple paths on $n \le 1$ vertices, and exactly $p_n = n!/2$ plane paths on $n \ge 2$ vertices. Finally, for n isolated vertices there are $s_n = 2^n$ ways of choosing a subset of them to put inside the cycle. In this way we have that $d_n = \sum_{(S_1,S_2,S_3)} c_{|S_1|} p_{|S_2|} s_{|S_3|}$, where the sum is over all the weak ordered partitions (S_1, S_2, S_3) of $\{1, \ldots, n\}$. By Lemma 12.32, we obtain that

$$\begin{aligned}
\bar{d}(x) &= \left(\sum_{n\ge 0} c_n \frac{x^n}{n!}\right)\left(\sum_{n\ge 0} p_n \frac{x^n}{n!}\right)\left(\sum_{n\ge 0} s_n \frac{x^n}{n!}\right) \\
&= \left(\sum_{n\ge 3} \frac{(n-1)!}{n!} x^n\right)\left(\sum_{n\ge 2} \frac{n!}{2n!} x^n\right)\left(\sum_{n\ge 0} \frac{2^n}{n!} x^n\right) \\
&= \frac{x^2 e^{2x}}{2(1-x)}\left(\log\left(\frac{1}{1-x}\right) - x - \frac{x^2}{2}\right).
\end{aligned}$$

As in previous Example 12.31, we can retrieve the actual coefficients d_n from the series expansion of the exponential generating function $\bar{d}(x)$.

Note 12.34 The conditions on the graphs in Examples 12.31 and 12.33 were all mutually exclusive! That is, a graph can never be both a cycle and a rooted star, and so on. This is the reason why we could directly consider weak ordered partitions of $\{1, \dots, n\}$, and use the Labeled Counting Lemmas 12.30 and 12.32.

What about graph-theoretic conditions or properties that overlap, or more explicitly, are the same? In that case we cannot consider weak ordered partitions, but rather, we need general proper partitions—a topic that we discuss in the following section.

12.6 The Exponential Formula

The *Exponential Theorem* is similar to the General Labeled Counting Lemma for counting labeled graphs, except that we consider unrestricted proper partitions of a set of labeled vertices, and the graph-theoretic property is the *same* on all parts, instead of mutually exclusive, as was the case for the Labeled Counting Lemmas.

We first present the Exponential Theorem, often simply called the *Exponential Formula*, and then explain its use by some easy examples.

Theorem 12.35 **Exponential**

Let $(a_n)_{n\geq 0}$ be a given number sequence with $a_0 = 0$, and $\bar{a}(x)$ its exponential generating function. If $c_0 = 1$, and for $n \geq 1$ we define c_n by

$$c_n = \sum_{\{S_1,\dots,S_k\}} a_{|S_1|} \cdots a_{|S_k|},$$

where the sum is taken over all proper partitions $\{S_1, \dots, S_k\}$ of $\{1, \dots, n\}$ and where k runs through $\{1, \dots, n\}$, then the exponential generating function $\bar{c}(x)$ of the sequence $(c_n)_{n\geq 0}$ is given by

$$\bar{c}(x) = e^{\bar{a}(x)}.$$

Remark 12.36 Since $a_0 = 0$ is not used in the definition of c_n for $n \geq 1$, we could also have considered the sequence $(a_n)_{n\geq 1}$, which starts at $n = 1$. Note, however, if $\bar{a}(0) = a_0 \neq 0$, then $e^{\bar{a}(x)}$ is not defined as a formal power series.

Example 12.37 Let c_n be the number of ways that a labeled simple graph can be formed by the union of simple paths of positive length. For each $n \geq 0$, let a_n be the number of such paths on n labeled vertices. In this case we have that $a_0 = a_1 = 0$ and $a_n = n!/2$ for $n \geq 2$. Therefore, we have that the exponential generating function $\bar{a}(x)$ is given by

$$\bar{a}(x) = \sum_{n\geq 2} \frac{n!}{2} \frac{x^n}{n!} = \frac{x^2}{2} \sum_{n\geq 0} x^n = \frac{x^2}{2(1-x)}.$$

Since $c_n = \sum_{\{S_1,\dots,S_k\}} a_{|S_1|} \cdots a_{|S_k|}$, where the sum is taken over all proper partitions $\{S_1, \dots, S_k\}$ of $\{1, \dots, n\}$ for $1 \leq k \leq n$, then by the Exponential Theorem 12.35, we have that the exponential generating function $\bar{c}(x)$ is given by

$$\bar{c}(x) = e^{\frac{x^2}{2(1-x)}}.$$

What is interesting about the Exponential Formula is that we can set up an equation for an unknown sequence $(c_n)_{n\geq 0}$ in terms of something that we know already. This process is reversed from Example 12.37, where we already knew how many ways we could form a path on n labeled vertices, but did not know how many ways we could form a union of paths of positive length.

Recall that by Observation 12.5, the number of simple graphs on n labeled vertices is given by

$$g_n = 2^{\binom{n}{2}}.$$

Here the value g_n counts all simple graphs, both connected and disconnected. So, how do we count only the simple connected graphs? In this case we can let c_n be the number of simple labeled connected graphs, act as if we knew the c_n's, and then obtain g_n from them. Since every simple graph is a union of one or more connected simple graphs, then g_n, the number of all simple graphs, is given by

$$g_n = \sum_{\{S_1,\ldots,S_k\}} c_{|S_1|}\cdots c_{|S_k|},$$

where the sum is taken over all proper partitions $\{S_1, \ldots, S_k\}$ of $\{1, \ldots, n\}$, where $1 \leq k \leq n$. Here $c_{|S_i|}$ counts the number of connected graphs on vertices with indices in $S_i \subseteq \{1, \ldots, n\}$ for each $i \in \{1, \ldots, k\}$. By the Exponential Formula (Theorem 12.35) we obtain the following result—first obtained by Riddell in 1951 in his Ph.D. dissertation.

Theorem 12.38 If $\bar{g}(x)$ is the exponential generating function of the number of simple graphs on $n \geq 0$ labeled vertices, and $\bar{c}(x)$ is the exponential generating function of the number of simple connected graphs on $n \geq 1$ labeled vertices, then $\bar{g}(x) = e^{\bar{c}(x)}$. Hence, we have that

$$\bar{c}(x) = \log(\bar{g}(x)) = \log\left(\sum_{n\geq 0} 2^{\binom{n}{2}}\frac{x^n}{n!}\right).$$

We now present an application of this result.

Example 12.39 Recall that $\log(1 + x) = x - x^2/2 + x^3/3 - x^4/4 + \cdots = \sum_{n\geq 0}(-x)^n/n$, and thus by Theorem 12.38, we have that

$$\begin{aligned}\bar{c}(x) &= \log\left(\sum_{n\geq 0} 2^{\binom{n}{2}}\frac{x^n}{n!}\right)\\ &= \log\left(1 + x + x^2 + \frac{4}{3}x^3 + \cdots\right)\\ &= 1\cdot\frac{x}{1!} + 1\cdot\frac{x^2}{2!} + 4\cdot\frac{x^3}{3!} + \cdots.\end{aligned}$$

Hence, $c_1 = c_2 = 1$ and $c_3 = 4$. A list of all the simple connected graphs on $n \leq 3$ vertices is shown in Figure 12.12.

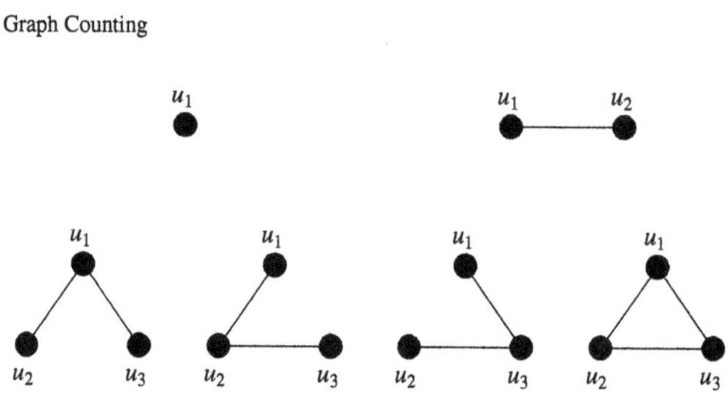

Figure 12.12: The simple connected graphs on at most three labeled vertices.

Instead of computing the c_n directly, as we did in Example 12.39, a recursive method sometimes works better, especially when we know the value of each c_i, where $i < n$. The following lemma is useful in this regard.

Lemma 12.40 Let $a(x) = \sum_{n\geq 0} a_n x^n$ and $b(x) = \sum_{n\geq 0} b_n x^n$ be two formal power series with $a_0 = 0$ and $b_0 = 1$ that satisfy $b(x) = e^{a(x)}$. If the b_n's are known, then the a_n's satisfy the recursive formula

$$\begin{aligned} a_n &= b_n - \frac{1}{n}((n-1)b_1 a_{n-1} + (n-2)b_2 a_{n-2} + \cdots + b_{n-1}a_1) \\ &= b_n - \frac{1}{n}\left(\sum_{i=1}^{n-1}(n-i)b_i a_{n-i}\right). \end{aligned}$$

Proof By taking the formal derivative of both sides of the equation $b(x) = e^{a(x)}$, and using the fact from Theorem 12.20 that the formal derivative is identical to the functional one, we obtain that

$$b'(x) = e^{a(x)} \cdot a'(x) = b(x)a'(x). \tag{12.7}$$

Comparing the coefficient of x^{n-1} on both sides of (12.7), we get $nb_n = \sum_{i=0}^{n}(n-i)b_i a_{n-i}$. Keeping in mind that $a_0 = 0$ and $b_0 = 1$ and solving for a_n, we get the lemma. ∎

By Theorem 12.38 and Lemma 12.40, we obtain the following recursive formula for the number of connected simple graphs.

Corollary 12.41 If c_n is the number of simple connected graphs on $n \geq 1$ labeled vertices, then $c_0 = 0$, $c_1 = 1$, and for $n \geq 2$ we have that

$$c_n = 2^{\binom{n}{2}} - \frac{1}{n}\left(\sum_{i=1}^{n-1}(n-i)\binom{n}{i}2^{\binom{i}{2}}c_{n-i}\right).$$

In Exercise 12.48 we ask for a combinatorial proof of the recursive formula presented in Corollary 12.41. Note, however, that this case is an exception to the general rule. Usually, one is unable to find a simple and nice combinatorial argument for an involved counting formula. More commonly one is able to arrive at an exact counting formula by the use of generating functions, without utilizing any slick combinatorial arguments. In fact, many simple combinatorial proofs appear later, when the methods of generating functions have revealed a simple answer! And a simple answer often implies a simple combinatorial proof.

12.7 The Number Two and Related Graphs

In concluding this chapter we consider various types of simple graphs that relate to the number two. This section could have been entitled "Counting X Simple Graphs," replacing "X" by any of the following: "Two-Regular," "Two-Colorable," "Even," or "Eulerian." All of these concepts are related by the number two, since by Theorem 5.7, a graph is Eulerian if, and only if, it is even and connected. This section provides further applications of the Exponential Formula for counting these two-related types of simple graphs on n labeled vertices. Although such countings all make use of the Exponential Formula, some of them, as we shall see, are harder than others.

We start with the easiest problem of all, to count the two-regular graphs on n labeled vertices.

12.7.1 Two-Regular Graphs

If G is a simple graph where each vertex has degree two, then each component must be a simple cycle on $n \geq 3$ vertices. This result can easily be seen by induction on $n = |V(G)|$. In fact, if we view a single vertex with a loop on it as a 1-cycle and a pair of edges connected with a double edge as a 2-cycle, we obtain the next observation.

Observation 12.42 For a graph G, we have the following:

1. If G is simple, then G is two-regular if, and only if, G is a union of cycles, each on at least three vertices.
2. If G is general, then G is two-regular if, and only if, G is a union of cycles, each on at least one vertex.

In Figure 12.13 we depict the first five cycles—that is, cycles on $k \in \{1, 2, 3, 4, 5\}$ vertices. A cycle is simple if, and only if, it has three or more vertices, so the first two cycles are considered only when looking at general graphs. Combining Observation 12.42 together with Exercises 12.44 and 12.45, we have the following.

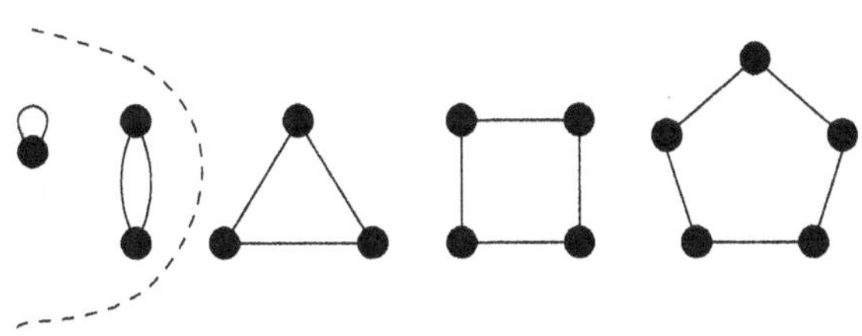

Figure 12.13: The first five cycles. A cycle is simple only if it has three or more vertices.

Theorem 12.43

1. The exponential generating function $\bar{r}_2(x)$ of two-regular simple graphs is given by

$$\bar{r}_2(x) = \sqrt{\frac{e^{-x(\frac{x}{2}+1)}}{1-x}}.$$

2. The exponential generating function $\bar{\bar{r}}_2(x)$ of two-regular graphs is given by

$$\bar{\bar{r}}_2(x) = \sqrt{\frac{e^{x(\frac{x}{2}+1)}}{1-x}}.$$

In the next subsection we examine two-colorable graphs.

12.7.2 Two-Colorable Graphs

Another word for two-colorable is bipartite. In this subsection we would like to count simple bipartite graphs. However, there is a benefit to viewing bipartite graphs as two-colorable graphs. The word "two-colorable" indicates that we can impose a certain ordered partition of the vertices into two distinct parts, which is nothing more than a two-coloring of the graph, say with the colors red and blue. This partition itself will help us in our counting endeavors here.

Recall that by Theorem 5.3 the bipartition of a *connected* bipartite graph is unique, and hence there are two ways of vertex coloring the graph with two colors, red and blue. However, if our bipartite graph has k components, then there are 2^k vertex colorings using the colors red and blue. In Figure 12.14 we show a bipartite graph having two components. All $2^2 = 4$ possible ways of vertex coloring the graph with red (1) and blue (2) are illustrated.

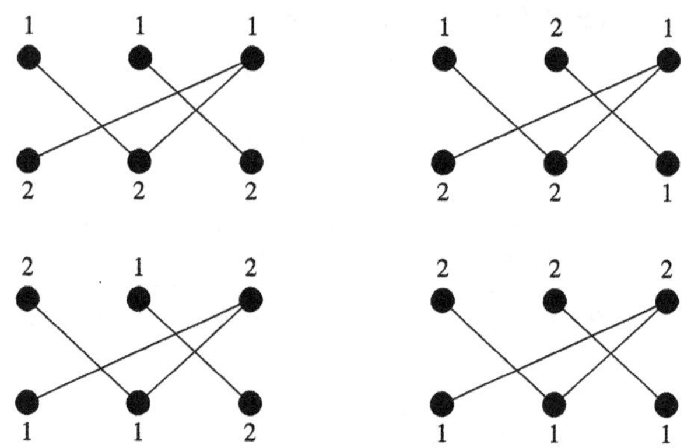

Figure 12.14: A bipartite graph with two components has $2^2 = 4$ distinct two-colorings.

Our ultimate goal is to obtain a formula for the number of simple bipartite graphs on n vertices or a generating function $\bar{b}(x)$ of the sequence $(b_n)_{n\geq 0}$, where b_n is the number of simple bipartite graphs on n labeled vertices. If a_n denotes the number of simple *connected* bipartite graphs on n labeled vertices, then clearly we have $b_n = \sum_{\{S_1,\ldots,S_k\}} a_{|S_1|} \cdots a_{|S_k|}$, where the sum is taken over all proper partitions $\{S_1, \ldots, S_k\}$ of $\{1, \ldots, n\}$, where $1 \leq k \leq n$. Hence, by the Exponential Formula, the corresponding exponential generating functions satisfy

$$\bar{b}(x) = e^{\bar{a}(x)}. \tag{12.8}$$

If we know the a_n's, then we can obtain the b_n's and vice versa. However, we do not know the a_n's or the b_n's. This step is where the two-coloring comes in handy.

Let d_n be the number of simple bipartite graphs on n labeled vertices, *provided with a two-coloring*. If c_n is the number of *connected* simple bipartite graphs, provided with a two-coloring, then we have, as above, that $d_n = \sum_{\{S_1,\ldots,S_k\}} c_{|S_1|} \cdots c_{|S_k|}$, where the sum is taken over all proper partitions $\{S_1, \ldots, S_k\}$ of $\{1, \ldots, n\}$, where $1 \leq k \leq n$. Hence, just as before, by the Exponential Formula, the corresponding exponential generating functions satisfy

$$\bar{d}(x) = e^{\bar{c}(x)}. \tag{12.9}$$

So again, if we know the c_n's, then we can obtain the d_n's and vice versa. This time, however, we can actually obtain one of these, namely the d_n's. We are able to do this in the following way.

Given the n labeled vertices, we start by deciding which are the red ones; say we color k of them red, $0 \leq k \leq n$. That can be done in precisely $\binom{n}{k}$ ways; the remaining $n - k$ of them we color blue. Now that we have the two-coloring (and hence also the bipartition!), we know that each edge must connect one of the k red vertices to one of the $n - k$ blue vertices. Thus, the possible set of edges is among the $k(n-k)$ edges forming the complete bipartite graph, connecting each red vertex with each blue one. There are precisely $2^{k(n-k)}$ ways of choosing edges to make up our two-colored graph with k red vertices and $n - k$ blue vertices. Summing up, we have that

$$d_n = \sum_{k=0}^{n} \binom{n}{k} 2^{k(n-k)}. \tag{12.10}$$

Now that we know $\bar{d}(x)$, we also know $\bar{c}(x)$ by (12.9). Is there a way that we can connect one of them with either $\bar{a}(x)$ or $\bar{b}(x)$? Indeed there is, since every simple, connected, bipartite graph has precisely two red/blue vertex colorings. We have that $c_n = 2a_n$, and hence

$$\bar{c}(x) = 2\bar{a}(x). \tag{12.11}$$

By (12.8), (12.9), and (12.11) we obtain that

$$\bar{b}(x) = e^{\bar{a}(x)} = e^{\bar{c}(x)/2} = \sqrt{\bar{d}(x)}.$$

Together with (12.10), we have the following theorem.

Theorem 12.44 If b_n denotes the number of simple bipartite graphs on n labeled vertices, then the corresponding exponential generating function $\bar{b}(x)$ is given by

$$\bar{b}(x) = \sqrt{\sum_{n\geq 0} \left(\sum_{k=0}^{n} \binom{n}{k} 2^{k(n-k)} \right) \frac{x^n}{n!}}.$$

Remark 12.45 Note that by (12.10), we have $d_0 = 1$, and hence taking the formal square root of a power series, using the General Binomial Theorem, is perfectly valid.

The result in Theorem 12.44 can be used to further obtain the generating function $\bar{b}_2(x)$ for $(b_{2:n})_{n\geq 0}$, where $b_{2:n}$ is the number of simple *two-connected* bipartite graphs on n labeled vertices, as is done in Harary and Robinson [25]. There, in addition, a recurrence relation for $b_{2:n}$ is established, as was done for c_n, the number of simple graphs on n vertices, in Corollary 12.41.

In the next section we examine even graphs.

12.7.3 Even Graphs

A graph is even if, and only if, each of its vertices has an even degree. By Theorem 5.7, the Eulerian graphs are precisely the connected even graphs. We start by counting the simple even graphs on n labeled vertices. For this we make use of a direct combinatorial bijection and obtain the following theorem, which might seem a bit surprising at first.

Theorem 12.46 The number ϵ_n of even simple graphs on n labeled vertices is equal to the number of simple graphs on $n-1$ labeled vertices. That is,

$$\epsilon_n = 2^{\binom{n-1}{2}}.$$

Moreover, this equality is given by a constructive bijection between these two classes of simple graphs.

Proof For $n \in \mathbb{N}$, let $\mathcal{G}_{n-1}$ be the set of all simple graphs on the labeled set $\{u_1, \ldots, u_{n-1}\}$ of vertices, and let $\mathcal{E}_n$ be the set of all even graphs on the labeled set $\{u_1, \ldots, u_n\}$ of vertices. It suffices to define a bijection $f : \mathcal{G}_{n-1} \to \mathcal{E}_n$, which we do as now illustrated.

For each $G \in \mathcal{G}_{n-1}$, by Corollary 1.27 (a direct consequence of the Hand-Shaking Theorem 1.26) the number of vertices in G of odd degree is even. We form $f(G) \in \mathcal{E}_n$ by connecting u_n to each vertex of G that has odd degree, as depicted in Figure 12.15, where the $n-1$ labeled vertices $\{u_1, \ldots, u_{n-1}\}$, which constitute the simple graph G, are inside the oval. The map f is clearly both injective and surjective. Hence, we have the theorem. ■

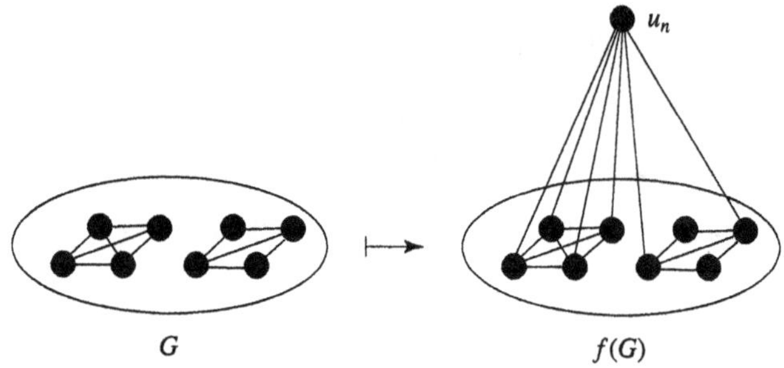

Figure 12.15: The formation of an even simple graph on n labeled vertices from a simple graph on $n-1$ labeled vertices.

By the slick bijection given in the proof of Theorem 12.46, we obtain as simple an answer as we could possibly have hoped for. If we at this point want to count the simple Eulerian graphs on n labeled vertices, can the result of Theorem 12.46 help us? Sure enough—since the Eulerian graphs are the connected even graphs, we can continue as when we counted the connected simple graphs, by using the Exponential Formula.

If E_n denotes the number of simple Eulerian graphs on n labeled vertices, then the number ϵ_n of even graphs on n labeled vertices satisfies

$$\epsilon_n = \sum_{\{S_1,\ldots,S_k\}} E_{|S_1|} \cdots E_{|S_k|},$$

where the sum is taken over all proper partitions $\{S_1, \ldots, S_k\}$ of $\{1, \ldots, n\}$ and where k runs through $\{1, \ldots, n\}$. Hence, just as before, by the Exponential Formula, we obtain the following theorem. The second formula we leave as Exercise 12.51.

Theorem 12.47 For a nonnegative integer n, let E_n be the number of simple Eulerian graphs on n labeled vertices, and let $\bar{E}(x)$ be the corresponding exponential generating function.

1. If $\bar{\epsilon}(x)$ is the exponential generating function of the sequence $(\epsilon_n)_{n\geq 0}$ of the number of simple even graphs on n labeled vertices from Theorem 12.46, then
$$\bar{\epsilon}(x) = e^{\bar{E}(x)}.$$
2. The E_n's satisfy $E_1 = 1$, and $E_2 = 0$, and for $n \geq 3$ we have the following recursion
$$E_n = 2^{\binom{n-1}{2}} - \frac{1}{n}\left(\sum_{i=1}^{n-1}(n-i)\binom{n}{i}2^{\binom{i-1}{2}}E_{n-i}\right). \tag{12.12}$$

Example 12.48 Clearly, as stated in Theorem 12.47, we have $E_1 = 1$ and $E_2 = 0$. Since there is only one simple even and connected graph on three vertices, namely K_3, we have $E_3 = 1$. We can now use (12.12) to compute E_4, and we obtain

$$\begin{aligned} E_4 &= 2^{\binom{3}{2}} - \frac{1}{4}\left(\sum_{i=1}^{3}(4-i)\binom{4}{i}2^{\binom{i-1}{2}}E_{4-i}\right) \\ &= 8 - \frac{1}{4}(12E_3 + 12E_2 + 8E_1) \\ &= 3. \end{aligned}$$

In Figure 12.16 we show all the Eulerian simple graphs on four labeled vertices. Note that they are precisely all the cycles on four labeled vertices, something reflected in the fact that if $n = 4$, then the number of labeled cycles on n vertices is $(n-1)!/2 = 3$ as well.

Remark 12.49 The basic fact that a simple Eulerian graph G on four labeled vertices $\{u_1, u_2, u_3, u_4\}$ is indeed a cycle can be seen as follows. Clearly, $G \subseteq K_4$, and exactly one of the three edges with u_1 as an endvertex is not contained in G, say $\{u_1, u_2\}$. Since G is connected, both u_1 and u_2 have two neighbors in G, namely u_3 and u_4. Since G is even, the edge $\{u_3, u_4\}$ cannot be in G either.

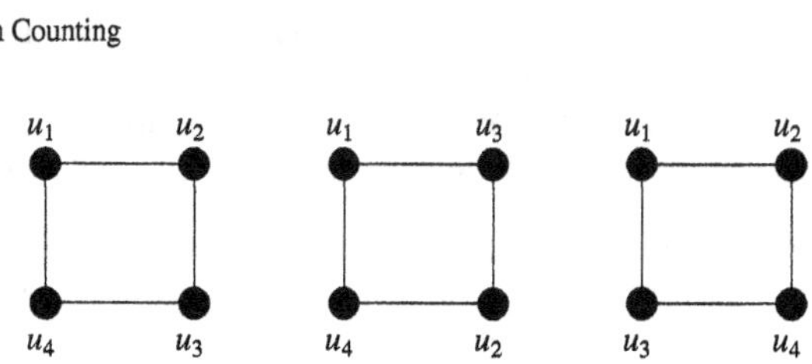

Figure 12.16: All the Eulerian simple graphs on four labeled vertices.

This chapter has provided an introduction to graph counting. For additional general-counting results the reader is referred to the references mentioned throughout this chapter.

12.8 Exercises

12.1. Complete the proof of Observation 12.5 by showing that for any $m, n \in \mathbb{N}$, where $0 \le m \le \binom{n}{2}$, the number $g_n(m)$ of simple graphs on n labeled vertices with m edges is given by

$$g_n(m) = \binom{\binom{n}{2}}{m}.$$

12.2. Use the second part of Observation 12.5 to prove its first part.

12.3. How large must $n \in \mathbb{N}$ be in order for the number of simple labeled graphs on $V = \{u_1, \ldots, u_n\}$ to fill a 300-page book, when written in the usual decimal notation expansion? [*Note!* Assume that each of the 300 pages has 50 lines with each line containing 75 digits.]

12.4. Let $n \in \mathbb{N}$ and $N = \lceil\sqrt{2n \lg n}\rceil + 1$, where lg is the base-2 logarithm. Use Cayley's Theorem to show that the number of labeled simple graphs on N vertices is larger than the number of labeled trees on n vertices.

12.5. Let $k, N \in \mathbb{N} \cup \{0\}$. Prove that

$$\binom{N}{N} + \binom{N+1}{N} + \cdots + \binom{N+k}{N} = \binom{N+k+1}{N+1},$$

and thereby complete the proof of Theorem 12.8. Can you find a combinatorial proof for this equation?

12.6. List all digraphs having three directed edges and no directed loops on three labeled vertices $V = \{u_1, u_2, u_3\}$.

12.7. Let $n, m \in \mathbb{N}$. Derive an expression, as in Theorem 12.8, for the number of digraphs on n labeled vertices and m directed edges and having no directed loops. Use your formula to find an expression for the number of digraphs on n labeled vertices and at most m edges and having no directed loops.

12.8. List all digraphs on three labeled vertices $V = \{u_1, u_2, u_3\}$ having three directed edges.

12.9. Let $n, m \in \mathbb{N}$. Derive an expression, as in Theorem 12.8, for the number of digraphs on n labeled vertices and m directed edges. Use your result to find an expression for the number of digraphs on n labeled vertices and at most m edges.

12.10. Let $m, m_e, m_l \in \mathbb{N} \cup \{0\}$ be such that $m_e + m_l = m$. Show that the number of graphs on n labeled vertices having m_l loops and m_e nonloop edges is given by

$$\binom{m_e + \binom{n}{2} - 1}{m_e}\binom{m_l + n - 1}{m_l}.$$

12.11. Let $k_1, k_2, N \in \mathbb{N} \cup \{0\}$. Prove that

$$\sum_{i=0}^{N}\binom{i + k_1 - 1}{i}\binom{N - i + k_2 - 1}{N - i} = \binom{N + k_1 + k_2 - 1}{N}.$$

Use this equation and Exercise 12.10 to give an alternative proof of Theorem 12.8, part 3.

12.12. Does the transformation duality discussed in Example 12.10 hold for graphs on $n \geq 4$ vertices as well as for graphs on $n = 3$ vertices? If not, can it be modified to hold?

12.13. Compute the generating function $\dot{g}_n(x)$ for the labeled loopless graphs on n vertices, as in Example 12.17.

12.14. Compute the generating function $\ddot{g}_n(x)$ for the labeled graphs on n vertices, as in Example 12.17.

12.15. Let $\vec{g}_n(m)$ denote the number of simple digraphs on n labeled vertices and m directed edges. (Here "simple" means that there is at most one directed edge from u to v.) Compute the corresponding generating function $\vec{g}_n(x)$ for labeled simple digraphs on n vertices.

12.16. Consider a sequence $T_0, T_1, T_2, \ldots$ of binary trees, where both T_0 and T_1 are the trivial single vertex tree and for each $k \geq 2$ the binary tree T_k has T_{k-2} as its left subtree and T_{k-1} as its right subtree. Let b_k denote the number of vertices of T_k. Use generating functions to obtain a formula for b_k. How fast does b_k grow when k tends to infinity?

12.17. For $n \in \mathbb{N}$ show that

$$\binom{1/2}{n} = \frac{2(-1)^{n-1}}{n4^n}\binom{2n-2}{n-1}.$$

Does a similar formula hold for $\binom{1/3}{n}$?

12.18. As in Exercise 3.39 in Chapter 3, let b_n be the number of binary trees on n vertices. Use their recursive property from that exercise,

$$\begin{aligned} b_1 &= 1, \\ b_2 &= 1, \\ b_n &= b_{n-1} + (b_1 b_{n-2} + b_2 b_{n-3} + \cdots + b_{n-2} b_1), \end{aligned}$$

and generating functions to find a closed formula for b_n.

12.19. As for the Catalan numbers, we can let t_n be the number of regular ternary trees on $3n+1$ vertices, as in Exercise 3.38 in Chapter 3. Use their recursive property from that exercise,

$$t_0 = 1,$$
$$t_1 = 1,$$
$$t_n = \sum_{i+j+k=n-1} t_i t_j t_k,$$

and generating functions to find a closed formula for t_n.

12.20. Prove Theorem 12.20. In addition, show that every element of $\mathbb{R}[[x]]$ has the form $a(x) = x^k \alpha(x)$, where $\alpha(x)$ is an invertible element of $\mathbb{R}[[x]]$ and k is some nonnegative integer.

12.21. Let $a(x) \in \mathbb{R}[[x]]$ and $a_0 \neq 0$. If $a(x)^{-1} \in \mathbb{R}[[x]]$ is its multiplicative inverse, and $[x^n]a(x)^{-1} = \alpha_n$, then the α_n's are completely determined by the infinite list of equations

$$\begin{aligned} a_0\alpha_0 &= 1, \\ a_0\alpha_1 + a_1\alpha_0 &= 0, \\ a_0\alpha_2 + a_1\alpha_1 + a_2\alpha_0 &= 0, \\ &\vdots \\ a_0\alpha_n + a_1\alpha_{n-1} + \cdots + a_n\alpha_0 &= 0, \\ &\vdots \end{aligned}$$

Use this to find the inverse of $1 + 2x + 3x^2 + \cdots \in \mathbb{R}[[x]]$ directly.

12.22. If $f(x) = 1 + 2x + 3x^2 + \cdots = \sum_{n\geq 0}(n+1)x^n$, then $F(x) = \sum_{n\geq 0} x^n = 1/(1-x)$ has $f(x)$ as a derivative $F'(x) = \frac{d}{dx}F(x) = f(x)$. Use this fact to find the multiplicative inverse of $f(x)$ in $\mathbb{R}[[x]]$.

12.23. A student claims to have found the generating function of the number of Hamiltonian simple planar graphs on n labeled vertices. The student claims that the generating function is given by

$$f(x) = e^{e^{\sqrt{1+x^2}}}.$$

Explain why it is immediate that the student is wrong.

12.24. The embarrassed student from Exercise 12.23 admits that the first conjectured generating function was not correct and states that the correct one is given by

$$f(x) = e^{e^{\sqrt{1+x^2}-1}}.$$

Is it possible for this to be the generating function for the number of Hamiltonian simple planar graphs on n labeled vertices?

12.25. Find a closed form for the generating function for the number of labeled graphs on n vertices and on two *labeled* edges, where loops are possible.

12.26. Express the generating function for the number of labeled graphs on n vertices and on m *labeled* edges in terms of a well-defined (and finite!) operator applied to a known function.

12.27. For $S = \{1, 2, 3, 4, 5\}$, list all the weak ordered partitions of S into two parts (S_1, S_2). Next, list all the proper partitions $\{S_1, S_2\}$ of S into two nonempty parts.

12.28. Show that the number of ways to write 5 as a sum of three positive integers is only two, namely $5 = 1 + 1 + 3$ and $5 = 1 + 2 + 2$. Use this fact to show that $S(5, 3) = 25$ without listing all the possibilities.

12.29. For $n \in \mathbb{N}$ show that the Stirling numbers of the second kind satisfy $S(n, n) = 1$ and that $S(n, n-1) = \binom{n}{2}$.

12.30. Let $\bar{S}_k(x) = \sum_{n\geq 0} S(n, k)\frac{x^n}{n!}$ be the exponential generating function of the Stirling numbers of the second kind for a fixed $k \in \mathbb{N}$. Use the recursive formula from Theorem 12.26 to show that the formal derivative of $\bar{S}_k(x)$ satisfies

$$\bar{S}'_k(x) = k\bar{S}_k(x) + \bar{S}_{k-1}(x).$$

Then use induction to show that

$$\bar{S}_k(x) = \frac{1}{k!}(e^x - 1)^k.$$

12.31. Use the Binomial Theorem on $(e^x - 1)^k$ from Exercise 12.30 to obtain the following explicit formula for the Stirling numbers of the second kind:

$$S(n, k) = \frac{1}{k!}\sum_{i=1}^{k}(-1)^{k-i}\binom{k}{i}i^n.$$

12.32. In Example 12.28 we established that $B(4) = 13$. List all the 13 unrestricted nonempty partitions of the set $\{1, 2, 3, 4\}$.

12.33. Let $\bar{B}(x) = \sum_{n\geq 0} B(n)\frac{x^n}{n!}$ be the exponential generating function of the Bell numbers. Use (12.5) to show that

$$\bar{B}(x) = e^{e^x - 1}. \tag{12.13}$$

12.34. By taking the derivative of both sides of (12.13) give another proof of the fact that the Bell numbers $B(n)$ satisfy the recursion (12.6) from Theorem 12.29.

12.35. Use the exponential generating function from Example 12.31 to compute the first six terms c_0, c_1, c_2, c_3, c_4, and c_5 of $\bar{c}(x)$.

12.36. Prove Lemma 12.32.

12.37. Use the exponential generating function $\bar{d}(x)$ from Example 12.33 to compute the first four nonzero terms of the sequence $(d_n)_{n\geq 0}$.

12.38. Let c_n be the number of ways that a simple labeled graph on $n \geq 0$ vertices can be formed by the union of a simple cycle and a nontrivial path. Compute the exponential generating function $\bar{c}(x)$ for the sequence $(c_n)_{n\geq 0}$, and use it to compute the first four nonzero terms of this sequence.

12.39. Let c_n be the number of ways that a simple labeled *plane* graph on $n \geq 0$ vertices can be formed by the union of a simple plane cycle and a nontrivial plane path. Compute the exponential generating function $\bar{c}(x)$ for the sequence $(c_n)_{n\geq 0}$, and use it to compute the first four nonzero terms of this sequence.

12.40. Find the exponential generating function of the number of ways that a simple labeled plane graph can be formed by the union of a simple plane cycle and isolated vertices, some of which are located inside the cycle.

12.41. Find the exponential generating function of the number of ways that a simple labeled plane graph can be formed by the union of a simple cycle, a nontrivial path, a clique on four or more vertices, a star on four or more vertices, and some isolated vertices. [*Note!* Why is it necessary to specify both the clique and the star to be "on four or more vertices," when using the General Labeled Counting Lemma 12.32?]

12.42. Show that the exponential generating function $\bar{p}(x)$ of the number of ways that a simple graph on n labeled vertices can be formed by the union of simple paths of *nonnegative* length (that is, a single vertex is considered a path of length zero!) is given by

$$\bar{p}(x) = e^{\frac{x(2-x)}{2(1-x)}}.$$

12.43. Show that the exponential generating function $\bar{c}_{ps}(x)$ of the number of ways that a simple plane graph on n labeled vertices can be formed by the union of simple cycles is given by

$$\bar{c}_{ps}(x) = \frac{e^{-x(\frac{x}{2}+1)}}{1-x}.$$

12.44. Show that the exponential generating function $\bar{c}(x)$ of the number of ways that a simple graph on n labeled vertices can be formed by the union of simple cycles is given by

$$\bar{c}(x) = \sqrt{\frac{e^{-x(\frac{x}{2}+1)}}{1-x}}.$$

12.45. Show that the exponential generating function $\bar{\bar{c}}(x)$ of the number of ways that a graph on n labeled vertices can be formed by the union of cycles, each on one or more vertices, is given by

$$\bar{\bar{c}}(x) = \sqrt{\frac{e^{x(\frac{x}{2}+1)}}{1-x}}.$$

12.46. Show that Theorem 12.38 also holds if $\bar{g}(x)$ is the exponential generating function of the number of graphs (respectively, loopless graphs) on $n \geq 0$ labeled vertices, and $\bar{c}(x)$ is the exponential generating function of the number of connected graphs (respectively, loopless graphs) on $n \geq 1$ labeled vertices, where the total number of edges is bounded by m.

12.47. Let c_n (respectively, $\dot{c}_n$) be the number of connected (respectively, connected loopless) graphs on $n \geq 0$ labeled vertices. Find a recursive formula for c_n (respectively, $\dot{c}_n$) similar to the one given in Corollary 12.41.

12.48. If c_n is the number of simple connected graphs on $n \geq 1$ labeled vertices, then give a combinatorial proof of the recursive formula for c_n given in Corollary 12.41. [*Hint:* (Harary and Palmer) A *rooted* graph G is simply a graph with a distinguished vertex. The number r_n of rooted graphs on n vertices is $r_n = ng_n$, where $g_n = 2^{\binom{n}{2}}$ is the number of simple graphs on n labeled vertices. Also, the root is in exactly one component with $k \in \{1, \ldots, n\}$ vertices, and hence $r_n = k\binom{n}{k}c_k g_{n-k}$.]

12.49. Show that the number $\beta_n(m)$ of bipartite graphs on n labeled vertices and m edges (possible multiple) with a given vertex two-coloring is given by

$$\beta_n(m) = \sum_{k=0}^{n} \binom{n}{k} \binom{m + k(n-k) - 1}{k(n-k) - 1}.$$

12.50. Note that Corollary 1.27 holds also for graphs that are not necessarily simple. In spite of this fact, can the bijection given in the proof of Theorem 12.46 be extended to nonsimple graphs?

12.51. Prove the second part of Theorem 12.47. [*Hint:* Use Lemma 12.40.]

12.52. Use (12.12) to show that the number E_5 of simple Eulerian graphs on five labeled vertices is $E_5 = 38$. Also, show directly that an Eulerian graph on five vertices is of one of the following four types: K_5, $K_5 - E(C_3)$, $K_5 - E(C_4)$, or $K_5 - E(C_5) = C_5$. Use this result to show directly that $E_5 = 38$.

CHAPTER

13

Graph Algorithms

Chapter Goals

Provide a discussion of graph algorithms.

Recap the algorithms already presented in this book.

Discuss algorithm efficiency.

Present a breadth-first search algorithm.

Present a depth-first search algorithm.

Present an algorithm for connected components.

Describe DIJKSTRA'S SHORTEST PATH ALGORITHM.

Provide Java code for the algorithms given here.

13.1 Introduction

Thus far, our text has focused primarily on developing the theory of graphs and their applications. We have introduced many key concepts involving simple graphs, general graphs, and digraphs. Their structures and properties have been analyzed, and many examples have been presented of how graphs can be used to model problems in various settings. Throughout the text, we have sprinkled in an occasional algorithm. In this chapter we focus more on the algorithms themselves. Although our brief coverage of graph algorithms does not do this subject justice, hopefully this exposure will help the reader model problems and develop computerized solutions for them.

The huge field of graph algorithms overlaps both computer science and mathematics. Graph algorithms have been developed to solve many different problems. For example, there are algorithms to search graphs, find cliques and independent sets, color graphs, find shortest paths, determine connected components, and so on. Since many problems involve large graphs, it is important to develop efficient algorithms for solving them. Suppose we want to develop an algorithm to check whether a graph having n vertices contains a Hamiltonian cycle. A brute-force approach could exhaustively consider all $(n-1)!/2$ possible cycles of the vertices, and check whether the appropriate edges were present in the graph for a given ordering. Assuming that each permutation could be evaluated in one nanosecond (that is, evaluated in 10^{-9} seconds), how large a graph could we practically test? Table 13.1 presents values of n and the corresponding time needed to process a graph under

Number of Vertices n	Time in Seconds
10	~ 0
15	44
20	6.082×10^7
25	3.102×10^{14}
100	4.666×10^{146}

Table 13.1: Times required to search for a Hamiltonian circuit in a graph, using a brute-force approach.

our assumptions. The table shows that we can solve an instance of the problem having 10 vertices instantaneously, a problem with 15 vertices in a little over 44 seconds, a problem having 20 vertices in 1.9 years, and so forth. Obviously, this algorithm cannot handle a graph having more than 15 vertices in an efficient manner, if at all. Thus exhaustive search is not a practical approach for solving large problems.

Many problems that we want to solve in computer science, mathematics, economics, and other sciences are modeled by graphs having hundreds or even thousands of vertices. In early 2006, a digraph modeling the World Wide Web had 8,000,000,000 vertices and roughly 80,000,000,000 edges! As our simple analysis involving the brute-force algorithm for Hamiltonian cycles illustrates, we need to be able to develop efficient algorithms to solve such problems. Graph algorithms are a rich subject, and numerous books are available on this topic—for example, Even [14] and Valiente [55]. Many computer science algorithm books also contain chapters on graph algorithms, for example, Brassard and Bratley [6], Cormen et al. [10], Goodrich and Tamassia [20], and Manber [37]. When developing an algorithm for a graph theory problem, it is wise to consult these books or search on the Internet for information about related algorithms. Many resources are available on graph algorithms. We will examine a handful of algorithms in this chapter to get started.

13.2 Recap of Algorithms Already Presented

A number of algorithms have already been presented in this book. It is worth summarizing them. Table 13.2 lists all of the explicit algorithms contained in the text and provides their corresponding page numbers. Many proofs throughout the text also provide implicit algorithms for solving problems. When presenting algorithms, we are always careful to specify their *input* and their *output*. The steps in the algorithm are always specified via a high-level description, so a skilled programmer could take that description and then code the algorithm in a programming language of choice. The syntactic details of programming languages are omitted, and the semantics of the statements that we use are self-explanatory. We abstract and convey the essence of algorithms in an unambiguous series of well-defined steps.

Regarding the input and output conventions, we abstract there too. In previous algorithms, for example, we stated that the input was a simple undirected graph and that the output was a list of edges. These high-level descriptions are easy for us to read and understand. However, the actual codings used to represent a graph and a list of edges would need to be made precise in order to implement these objects. There are choices to be made at many different levels when actually programming an algorithm. For example, one could decide to represent a graph using an adjacency matrix, incidence matrix, or edge listing. Having made this choice, one could use various data structures to represent the graph coding. The program-

Algorithm	Page Number
AUGMENTING PATH	344
BREADTH-FIRST SEARCH	397
CONNECTED COMPONENTS	404
DELTA PLUS ONE COLORING	241
DEPTH-FIRST SEARCH	401
DIJKSTRA'S SHORTEST PATH	408
EULERIAN CIRCUIT	138
GREEDY COLORING	244
KRUSKAL'S	123
KUHN-MUNKRES	350
PRIM'S	125
RESTRICTED HAMILTONIAN CYCLE	144
SIMPLICIAL COLORING	285

Table 13.2: Table of algorithms described in this text.

ming language that one is using will also impact the choice of data structures. Similarly, there will be many options for how the output of the algorithm can be represented. Sometimes, both the input and the output conventions that are required may be dictated to a programmer, since the code developed may need to interface directly with an existing computer program.

We will not delve into the wide variety of issues involved in actually coding an algorithm. The exercises ask the reader to implement a number of algorithms, and this will provide valuable experience. For the algorithms described in the chapter, we present their Java implementations. By carefully studying this code, the reader will learn about many of the important technical issues involved in programming graph algorithms.

13.3 Algorithm Efficiency

Given two algorithms for solving the same problem, how does one know which algorithm is preferable? This is a difficult question to answer. Suppose the input and the output conventions for the algorithms are the same, so neither one has an advantage in that department. What if one algorithm runs faster on graphs having large cliques but the other runs faster on graphs having large independent sets? How do we make a fair comparison? Or what type of comparison can we make? Perhaps one algorithm is easier to code than another. Our focus on algorithm analysis will not be at the implementation level, but rather abstracted to a higher, less-detailed analysis. The high-level algorithm analysis, when done properly, applies to the implementation-level analysis as well. An important aspect of algorithm analysis is asymptotic growth. In the theory of algorithms two models of algorithm analysis are worth mentioning here. They involve *worst-case* execution time and *average-case* execution time. We explain these ideas in what follows.

The field of analysis of algorithms is a broad one, and the interested reader is referred to Knuth [32] or Cormen et al. [10] for more details on the subject. The running times of graph algorithms, and for that matter any algorithms, are usually described in terms of their *input size*. The input size for graph algorithms is usually specified in terms of the number of vertices, always denoted by n, or the number of vertices and edges. The number of edges is universally denoted by m. These conventions allow us to make statements such as GRAPH ALGORITHM A runs in $O(n^2)$ time, $O(n \lg n)$ time, or $O(nm)$ time, among others. The O-notation (also called Big-Oh) is used to hide constant multiplicative factors or constant additive factors. So, we write $O(n^2)$, for example, rather than the more accurate but cumbersome $4.8n^2 + 8n + 12{,}210$. Formally, we can state the following definition.

Definition 13.1 Let $f, g : \mathbb{N} \to \mathbb{R}^+$ be two functions.

1. We say that $g(n) \in \Omega(f(n))$, if there are constants $N \in \mathbb{N}$ and $A \in \mathbb{R}$ such that $g(n) \geq Af(n)$ for all $n \geq N$. (Omega)
2. We say that $g(n) \in O(f(n))$, if there are constants $N \in \mathbb{N}$ and $B \in \mathbb{R}$ such that $g(n) \leq Bf(n)$ for all $n \geq N$. (Big-Oh)
3. We say that $g(n) \in \Theta(f(n))$, if $g(n) \in \Omega(f(n))$ and $g(n) \in O(f(n))$. (Theta)

In the algorithm-analysis literature it is common to write "$g(n) = X$" instead of "$g(n) \in X$," even though X is a set. We follow this convention here. Note that if $g(n) = \Omega(f(n))$ and $g(n) = O(f(n))$, then there are $N_1 \in \mathbb{N}$ and $A \in \mathbb{R}$ such that $g(n) \geq Af(n)$ for all $n \geq N_1$, and there are $N_2 \in \mathbb{N}$ and $B \in \mathbb{R}$ such that $g(n) \leq Bf(n)$, when $n \geq N_2$. If we now let $N = \max(\{N_1, N_2\})$, then we have the following observation.

Observation 13.2 For functions $f, g : \mathbb{N} \to \mathbb{R}^+$, we have that $g(n) = \Theta(f(n))$ if, and only if, there are constants $N \in \mathbb{N}$ and $A, B \in \mathbb{R}$ such that $Af(n) \leq g(n) \leq Bf(n)$ for all $n \geq N$.

The following example briefly explains the essence of all three notations—Big-Oh, Omega, and Theta—from Definition 13.1.

Example 13.3 Suppose we have two functions $f, g : \mathbb{N} \to \mathbb{R}^+$ given by $f(n) = 4.8n^2 + 8n + 12{,}210$ and $g(n) = 3.14n + \lg n$. Here we have $f(n) = \Omega(n^2)$, since $f(n) \geq n^2$ for all $n \in \mathbb{N}$ (that is, we can take $N = 1$ and $A = 1$). Also, we have that $f(n) = O(n^2)$, since $f(n) \leq 20n^2$, when $n \geq 110$ (that is, we can take $N = 110$ and $B = 20$). Therefore, we see that $f(n) = \Theta(n^2)$. Likewise, we have that $g(n) = \Omega(n)$, since $g(n) \geq n$ for all $n \geq 1$ (that is, we can take $N = 1$ and $A = 1$). Further, we have that $g(n) = O(n)$, since $g(n) \leq 10n$ for all $n \geq 1$. Hence, we have here that $g(n) = \Theta(n)$.

Suppose now that we have a function $h : \mathbb{N} \to \mathbb{R}^+$ which is defined by

$$h(n) = \begin{cases} f(n) & \text{if } n \text{ is odd,} \\ g(n) & \text{if } n \text{ is even.} \end{cases}$$

Here we have that $h(n) \geq n$ for all $n \in \mathbb{N}$, and hence $h(n) = \Omega(n)$. Also, we have that $h(n) = O(n^2)$, since $h(n) \leq 20n^2$ for all $n \geq 110$. Note, however, that $h(n)$ is neither a $\Theta(n)$ nor a $\Theta(n^2)$ function.

During preliminary algorithm analyses we are usually concerned with orders of magnitude and not constant factors. The notion of time in this context refers to an algorithmic step—that is, one unit of time is one step. A step involves an "atomic" operation such as an addition, comparison, or assignment.

We should note that the actual input size, in terms of the number of bits specified to the program, is not the same as n. For example, if there are n vertices, then each vertex may actually require $\lg n$ bits to represent. The vertices of a 4-vertex graph could be coded using two bits as 00, 01, 10, and 11. The bit string 01100011 might signify vertices 2, 3, 1, and 4 in this order. This string has length eight, since two bits are required to represent each vertex. Extrapolating using this coding, we see that to write down the vertices of an n-vertex graph would require $n \lg n$ bits. An edge could be expressed as two vertices, thus requiring a total of $2 \lg n$ bits to represent. So, the edges of an m-edge graph could be coded using $2m \lg n$ bits. In this book we are concerned only with representing running times in terms of n, m, or both. We do not actually look at the exact number of bits required to represent an input or an output in our analyses.

Suppose we have a graph algorithm called DEMO. Suppose further that we want to measure DEMO's running time in terms of the number of vertices in its input graph. If DEMO is a general-purpose algorithm, it should be able to handle graphs having 10 vertices as well as those having 1,000 vertices, or even 1,000,000 vertices. As we know, there are many different graphs having, say, 1,000 vertices. ALGORITHM DEMO will inevitably run for different lengths of time, then, even on graphs having the same number of vertices. To cope with this situation, the worst-case time measure is based on the single input of each size that requires the algorithm to run the longest, whereas the average case time simply averages the running times over all inputs of a given size. More formally, we can make the following definition.

Definition 13.4 Let A be an algorithm. For each $n \in \mathbb{N}$, let

$$t_A(n) = \max(\{1, \text{ maximum running time of } A \text{ on inputs of length } n\}).$$

The function $t_A(n)$ is the *worst-case running time* of algorithm A. We write $t(n)$ when A is understood. For each $n \in \mathbb{N}$, let

$$t'_A(n) = \max(\{1, \text{ average running time of } A \text{ on all inputs of length } n\}).$$

The function $t'_A(n)$ is the *average-case running time* of algorithm A. We write $t'(n)$ when A is understood.

Note it is implicit in Definition 13.4 that the algorithm handles all input sizes. The worst-case running time is usually much easier to determine than the average-case running time, and in Exercise 13.9, we ask for an intuitive explanation why. In this book we focus only on worst-case running times.

A great deal of skill is required for the design and analysis of algorithms. Perhaps the best way to begin acquiring that skill is to examine a number of algorithms. In the remainder of this chapter, we do this and explore several well-known and important graph algorithms.

13.4 Breadth-First Search

The first algorithm that we describe is the BREADTH-FIRST SEARCH ALGORITHM. This algorithm is very useful for exploring all of the vertices of a graph. The idea is to start at a designated source vertex and then explore the neighbors of the original vertex. Once all of these have been examined, their neighbors are explored. This process repeats until all vertices of the entire graph have been visited. In this way a spanning forest of the input graph can be constructed. If the original graph is connected, the exploration forms a structure called a *breadth-first search tree*. Pseudocode for the BREADTH-FIRST SEARCH ALGORITHM is provided after a few preliminaries. The output of the algorithm for this variant of breadth-first search is a breadth-first search tree. For simplicity we assume that the input graph is connected.

Prior to presenting the BREADTH-FIRST SEARCH ALGORITHM, we define the *queue* data structure that is useful for expressing the algorithm. A queue can be thought of as a line

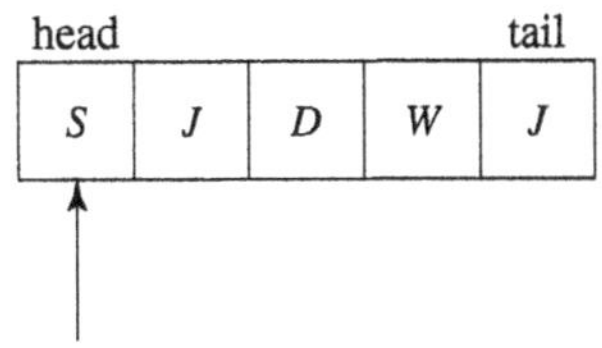

Figure 13.1: A sample queue.

such as a grocery-store checkout line, where entries are removed from the front, often called the *head*, and added at the rear, usually called the *tail*. An example of a queue is shown in Figure 13.1. There are five items in this queue, and we say the queue has *size* five. We see that S was added to the queue first, followed by J, D, W, and J.

Three operations can be performed on a queue. They are ENQUEUE, DEQUEUE, and EMPTY. These correspond, respectively, to adding an item to the queue at the tail, removing an item from the head, and checking to see if there is anything in the queue. More formally, the operation ENQUEUE(Q, u) adds u to the tail of Q, thereby increasing the size of Q by one item. The operation DEQUEUE(Q) removes one item from the head of the nonempty queue Q and returns that element. This reduces the queue size by one. The operation EMPTY(Q) returns a value TRUE if the queue has no items and FALSE otherwise. Armed with the operations on the queue data structure, we can now present the BREADTH-FIRST SEARCH ALGORITHM.

```
BREADTH-FIRST SEARCH ALGORITHM
Input: A simple connected undirected graph G = (V, E) on n vertices and a
designated source vertex s.
Output: A breadth-first search tree T of G.
begin
    T_1 = ({s}, ∅);
    i = 1;
    mark s;
    initialize Q to the empty queue;
    ENQUEUE(Q, s);
    while not EMPTY(Q) do {
        x = DEQUEUE(Q);
        for all edges {x, y} where y is not marked do {
            mark y;
            i = i + 1;
            T_i = T_{i-1} ∪ {x, y};
            ENQUEUE(Q, y);
        }
    }
    output T_n;
end.
```

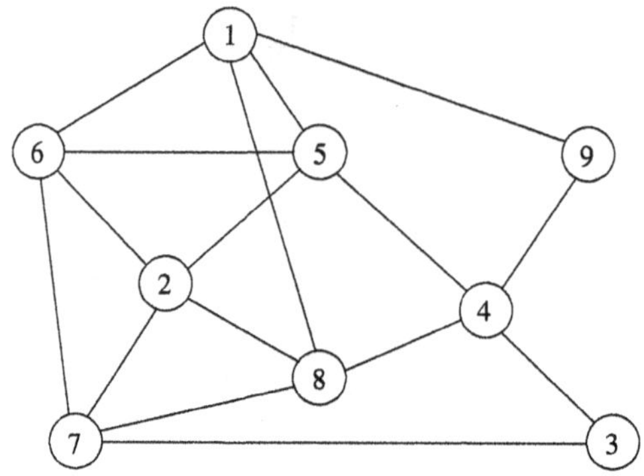

Figure 13.2: Graph to be explored via breadth-first search and depth-first search.

Before analyzing the running time of the algorithm, we present an example.

Example 13.5 We use the graph of Figure 13.2 to illustrate the BREADTH-FIRST SEARCH ALGORITHM. Notice that the input graph is connected and consists of nine vertices. In this example the source vertex s is vertex 1. Figure 13.3 illustrates the search as it progresses from vertex 1. In the graph, vertices 6, 8, 5, and 9 were explored in this order (arbitrarily chosen, as are other orders throughout this example), and the edges between these vertices and vertex 1

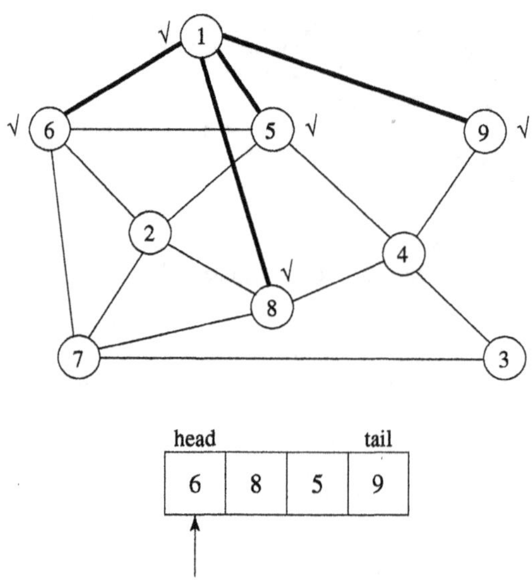

Figure 13.3: The neighbors of vertex 1 have been explored and queued up during the BREADTH-FIRST SEARCH ALGORITHM.

were added to the breadth-first search tree being constructed. Edges added to the tree are shown in thick lines. Vertices 6, 8, 5, and 9 have been queued up, in this order, as shown in the figure. These vertices plus vertex 1 already have been marked. A vertex marking is denoted by a check mark ($\surd$) in the figure.

From this point vertex 6 is dequeued and its neighbors 1, 5, 2, and 7 are explored. Since vertices 1 and 5 are already marked, no new edges are added to the tree based on them. On the other hand, vertices 2 and 7 are unmarked. The edges $\{2, 6\}$ and $\{6, 7\}$ are therefore added to the breadth-first search tree. Vertex 2 and next vertex 7 are marked and added to the queue. The search from vertex 8 results in the edge $\{4, 8\}$ being added to the breadth-first search tree and the vertex 4 being marked and queued. Since all of the edges from vertex 5 lead to marked vertices, no new edges are added during its exploration. Figure 13.4 shows a snapshot of the search status just before vertex 9 is dequeued. Vertices 9 and 2 are then dequeued, and since all of their neighbors are marked, no new edges are added during their exploration. When vertex 7 is explored, we add edge $\{3, 7\}$ to the tree, and mark and queue vertex 3. No additional edges are added to the tree while the queue is being emptied.

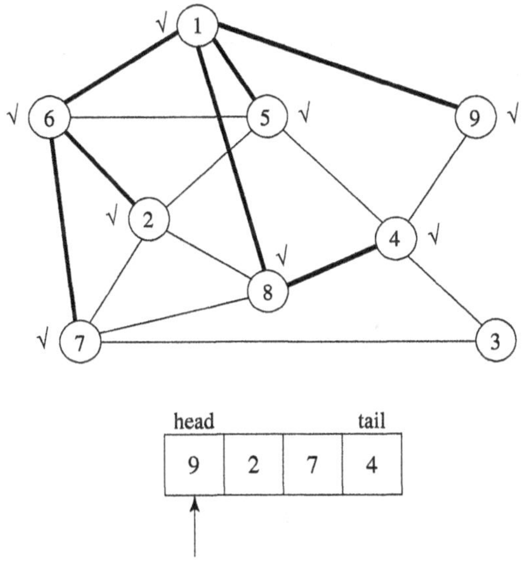

Figure 13.4: The breadth-first search of the graph shown in Figure 13.2 continues. All vertices but one have been marked.

The resulting breadth-first search tree is shown in Figure 13.5. In Exercise 13.17 we ask the reader to investigate other orders for visiting the neighbors of the vertex that one is currently expanding.

We now turn our discussion to determining the worst-case running time of the BREADTH FIRST SEARCH ALGORITHM. Consider the algorithm on input $G = (V, E)$ and a vertex s, where $|V| = n$ and $|E| = m$. Since the graph is connected, it has at least $n - 1$ edges, and so $n = O(m)$. Suppose $V = \{u_1, \ldots, u_n\}$. The algorithm begins with several straightforward

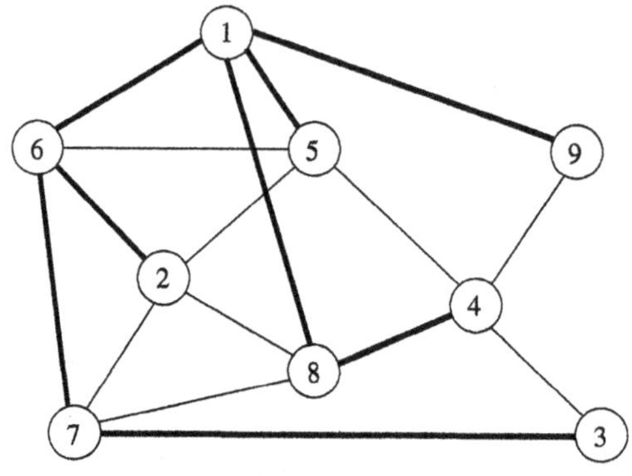

Figure 13.5: The breadth-first search tree constructed from the graph shown in Figure 13.2.

initializations. These require constant time, $O(1)$. The time-consuming parts of the algorithm are the **while** loop and the **for** loop contained inside it. Note that every vertex will at some point be queued, and each operation within the **for** loop can be done in constant time. When a vertex is dequeued, all edges emanating from it are explored. Thus, each edge is explored twice, once from each endvertex. So, the worst-case running time of the algorithm is given by

$$t(n) = O(\sum_{i=1}^{n} d_G(u_i)).$$

Therefore, by the Hand-Shaking Theorem, $t(n) = O(m)$. It is intuitively clear that a breadth-first search of a graph must explore all edges of the graph. Thus, up to constant factors, our BREADTH-FIRST SEARCH ALGORITHM is optimal—that is, it runs as efficiently as possible.

In the next section we investigate another important search algorithm.

13.5 Depth-First Search

Next we describe the DEPTH-FIRST SEARCH ALGORITHM. In contrast to the BREADTH-FIRST SEARCH ALGORITHM, which explored all the neighbors of a given vertex at a time, the DEPTH-FIRST SEARCH ALGORITHM explores only one neighbor at a time, and then tries to go "deeper" into the graph. The idea is to start at a given vertex and then explore a path in the graph as far as possible. When it is no longer possible to go "forward," the algorithm backtracks one level and then tries again to go deeper. This process repeats until all vertices of the entire graph have been visited. In this way a spanning forest of the input graph can be constructed. If the original graph is connected, the exploration forms a structure called a *depth-first search tree*. Pseudocode for the DEPTH-FIRST SEARCH ALGORITHM is provided in what follows. The output of the algorithm for this variant of depth-first search is a depth-first search tree. Note that this algorithm makes use of a recursive procedure.

```
DEPTH-FIRST SEARCH ALGORITHM
Input: A simple, connected, undirected graph G = (V, E).
Output: A depth-first search tree T of G.

procedure DFSSEARCH(H)
// H is an undirected graph.
begin
    for all v ∈ V do
        color[v] = white;
    for all v ∈ V do
        if color[v] = white then DFS(v);
end

procedure DFS(v)
// vertex v has not been visited yet
begin
    color[v] = gray;
    for all w ∈ N(v) do
        if color[w] = white then {
          add edge {v, w} to T;
          DFS(w);
        }
end

main
begin
    DFSSEARCH(G);
end.
```

Before analyzing the running time of the algorithm, we present an example.

Example 13.6 We use the graph of Figure 13.2 to illustrate the DEPTH-FIRST SEARCH ALGORITHM. In this example the search begins from vertex 1. Figure 13.6 shows the search as it progresses. When a choice exists, by convention we assume here in this example that lower-numbered vertices get visited first. In practice, vertices are usually visited based on their order in the input encoding.

The initial call to DFSSEARCH(G) results in all vertices being colored white. This marking signifies that none of the vertices have yet been visited by the search. Suppose vertex 1 is selected initially. This choice results in the call DFS(1). Vertex 1 is colored gray. Vertex 1's lowest-numbered neighbor, vertex 5, is then selected. The edge $\{1, 5\}$ is added to the depth-first search tree, and the call DFS(5) is made. Vertex 5 is colored gray. Vertex 5's lowest-numbered neighbor is vertex 1, which has already been colored gray. Vertex 5's next lowest-numbered neighbor is vertex 2. The edge $\{2, 5\}$ is added to the depth-first search tree, and the call DFS(2) is made. Vertex 2 is then colored gray. Vertex 2's lowest-numbered neighbor is vertex 5. Since vertex 5 has already been colored gray, the algorithm explores the next lowest-numbered neighbor, namely vertex 6. The edge $\{2, 6\}$ is added to the depth-first search tree, and the call DFS(6) is made. Vertex 6 is colored gray. Vertex 6's two lowest-numbered neighbors, vertices 1 and 2, are already colored gray, so vertex 7 is explored. The

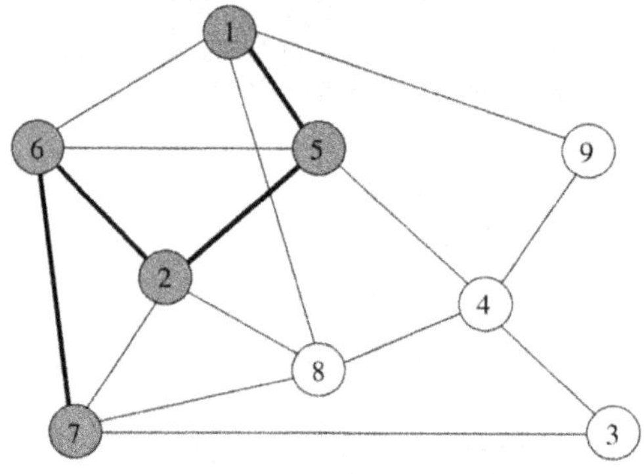

Figure 13.6: A depth-first search proceeding from vertex 1. Vertex 7 has just been visited.

edge {6, 7} is added to the depth-first search tree, and the call DFS(7) is made. Vertex 7 is colored gray. Figure 13.6 depicts the search up to this point.

Note that in the figures for depth-first search we have shown in thick lines the edges "along which" new vertices are visited. These edges comprise the depth-first search tree. For example, Figure 13.6 shows that vertex 7 was explored as a neighbor of vertex 6, not from vertex 2. The search continues from vertex 7, resulting in vertices 3, 4, and 8 being visited and colored gray. Figure 13.7 illustrates this progression of the search, and the additional edges that are added to the depth-first search tree.

From vertex 8 the search attempts to go deeper by trying vertices 1, 2, and 4. However, they are all already colored gray. Now the search backtracks to vertex 4. Earlier the search had tried unsuccessfully to visit vertices 3 and 5 from vertex 4, and actually did successfully visit vertex 8. So, now the search proceeds to vertex 9. The edge {4, 9} is added to the depth-first search tree. Vertex 9 is colored gray. At this point all vertices have been colored gray, although the search does not "know" this yet. So, the search continues trying to explore and

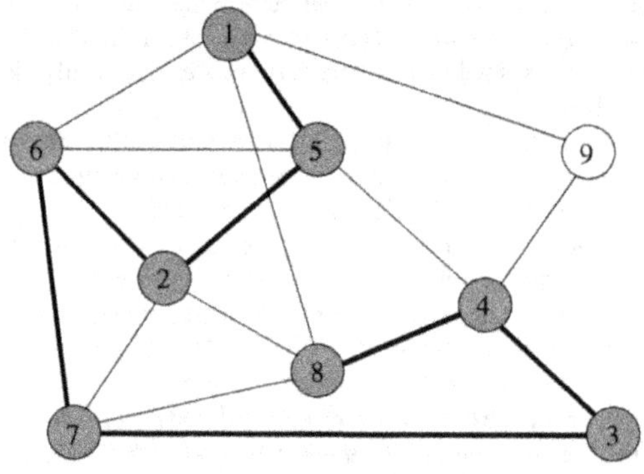

Figure 13.7: A depth-first search that has just visited vertex 8.

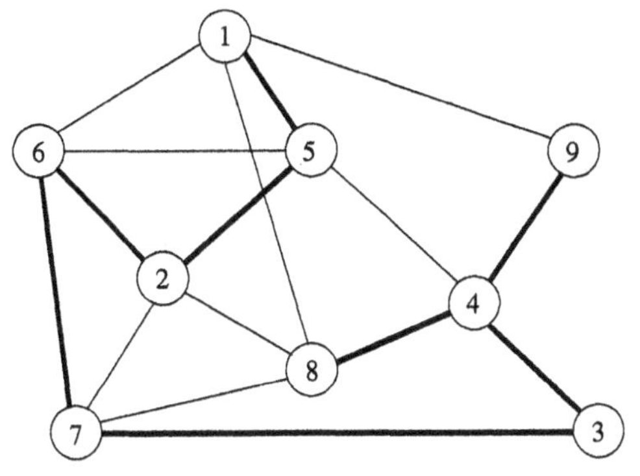

Figure 13.8: The depth-first search tree constructed from the graph shown in Figure 13.2.

backtracks, eventually backtracking all the way to vertex 1, where vertices 6, 8, and 9 are tried. At this point the search discovers that it is finished, since there are no more vertices to try. If there were another connected component, the search would jump into it by choosing an arbitrary vertex to explore. Thus, the algorithm actually works for general graphs and constructs a *depth-first search forest*.

The resulting depth-first search tree for the completed search is shown in Figure 13.8. In Exercise 13.23, we ask the reader to investigate other orders for visiting the neighbors of the vertex that one is currently expanding.

We now discuss determining the worst-case running time of the DEPTH-FIRST SEARCH ALGORITHM. Consider the algorithm on input $G = (V, E)$, where $|V| = n$ and $|E| = m$. Since each vertex is visited exactly once, the procedure DFS($\cdot$) is called n times. Assuming that a data structure such as an adjacency list is used to represent the graph, successive neighbors on the adjacency list can be accessed in one step. Since from each vertex all of its neighbors are checked, by the Hand-Shaking Theorem the total number of checks is simply twice the number of edges of G, or the value $2m$. Thus, the algorithm's worst-case running time is $O(\max(\{n, m\})$. Note that if G is connected, then G has at least $n - 1$ edges, and hence the running time is $O(\max(\{n, m\}) = O(m)$, as was the case for the BREADTH-FIRST SEARCH ALGORITHM. This running time is optimal, because it is no larger than the actual encoding of the graph itself. That is, any algorithm which performs a depth-first search must at least read its entire input.

In the next section we present an important application of the DEPTH-FIRST SEARCH ALGORITHM to determine the connected components of a graph.

13.6 Connected Components

Throughout this book, we have talked about the connected components of a graph. The notion was formally presented in Definition 2.7. In this section we describe the CONNECTED COMPONENTS ALGORITHM for determining the connected components of a graph. The algorithm relies on the techniques developed in the DEPTH-FIRST SEARCH ALGORITHM from §13.5. The idea is to explore all vertices in a given component using depth-first search.

Vertices that are in the first component explored are given a component number of 1. Once all vertices in a given component have been numbered, the algorithm increments the component number and jumps into another component. The vertices there are given the new component number. This process is repeated until all components are explored, and all vertices are numbered.

```
CONNECTED COMPONENTS ALGORITHM
Input: A graph G = (V, E) on n vertices, where the vertices are labeled from
1,...,n.
Output: An array components[·] of size n, where for all vertices i in the jth
connected component of G components[i] = j.
procedure SEARCH(u)
// u is a vertex in V
begin
    color[u] = gray;
    components[u] = component-number;
    for all w ∈ N(u) do
        if color[w] = white then SEARCH(w);
end

main
begin
    component-number = 1;
    for all v ∈ V do
        color[v] = white;
    while color[v] = white for some v ∈ V do {
        color[v] = gray;
        SEARCH(v);
        component-number = component-number + 1;
    }
end.
```

Before analyzing the running time of the algorithm, we present an example.

Example 13.7 We use the graph of Figure 13.9 to illustrate the CONNECTED COMPONENTS ALGORITHM. This graph has four connected components and eleven vertices. The algorithm begins by coloring all vertices white. This color signifies that none of the vertices have yet been classified into their respective components. The **while** loop in the **main** routine continues to execute until all vertices have been colored gray. In this example we assume that the first vertex chosen to search is vertex 1. Vertex 1 is colored gray. The call SEARCH(1) results in components[1] receiving a value of 1. This assignment indicates that vertex 1 is in the first component being searched.

Next the neighbors of vertex 1 are searched. In keeping with our convention of exploring lower-numbered vertices first, the algorithm tries neighbor 7 first. Since vertex 7 is colored white, the call SEARCH(7) results. Vertex 7 is colored gray, and components[7] is assigned the value 1. Figure 13.10 illustrates the situation up until this point of the algorithm's execution. The neighbors of vertex 7 are then explored.

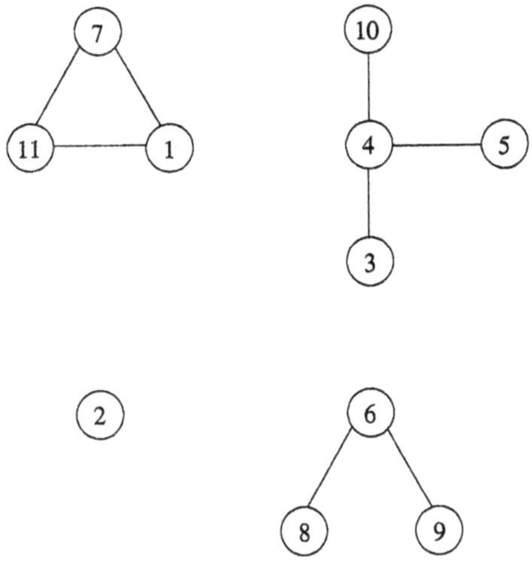

Figure 13.9: Graph used to illustrate the CONNECTED COMPONENTS ALGORITHM.

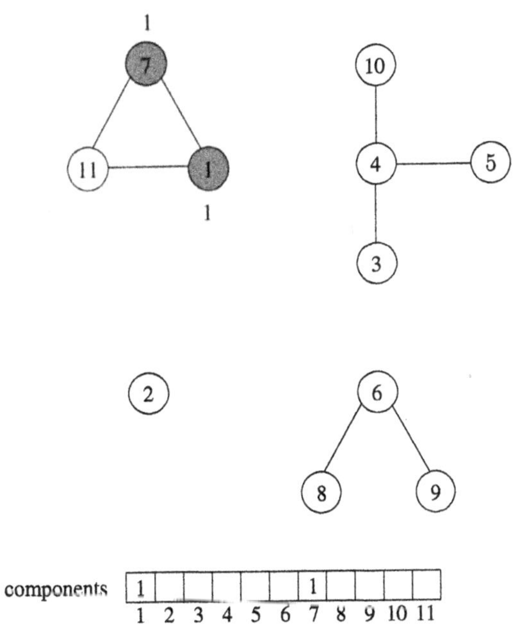

Figure 13.10: A snapshot of the CONNECTED COMPONENTS ALGORITHM right after vertex 7 has been explored.

The algorithm next explores neighbor 1. Since vertex 1 is colored gray, the algorithm moves on to neighbor 11 of vertex 7. This fact produces the call SEARCH(11). The result is that vertex 11 is colored gray, and components[11] is assigned the value 1. The algorithm next explores the neighbors of vertex 11. Since these are both colored gray, the algorithm backtracks to vertex 7. The algorithm has already explored all of vertex 7's neighbors, so the algorithm further backtracks to vertex 1. From vertex 1 only neighbor 7 had been explored, so neighbor 11 is tried next. Since vertex 11 is colored gray, no further calls result, and the algorithm "knows" that its work in this component is finished.

In the routine **main** suppose that vertex 3 is next chosen to be explored. Figure 13.11 illustrates the situation after vertex 3 and its neighbor vertex 4 have been explored. Notice that these vertices are labeled with the value 2, signifying that they are in the second connected component. The figure shows that both vertices 3 and 4 have been colored gray.

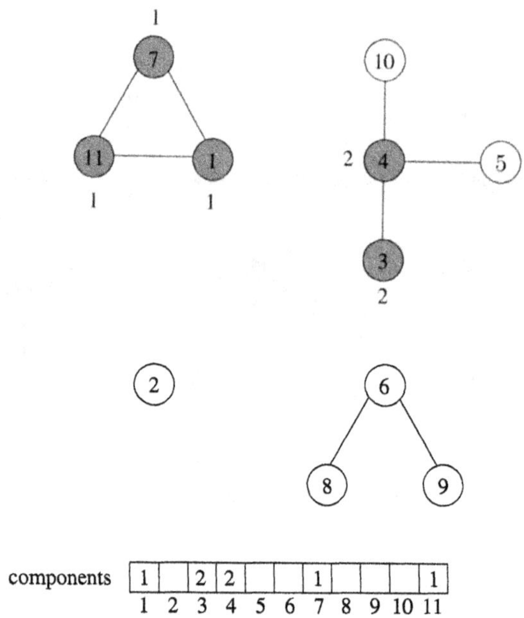

Figure 13.11: A snapshot of the CONNECTED COMPONENTS ALGORITHM after vertex 4 has been explored.

The algorithm finishes exploring the second connected component. Next we assume that the algorithm selects the isolated vertex 2 to search. The algorithm later moves into the fourth connected component and labels all of those vertices with a value 4, coloring them gray in the process. Since no more white vertices remain, the algorithm halts. Figure 13.12 shows the final result of the algorithm. The array components[·] contains the output of the algorithm. This output can be manipulated to utilize the information about the connected components of the input graph.

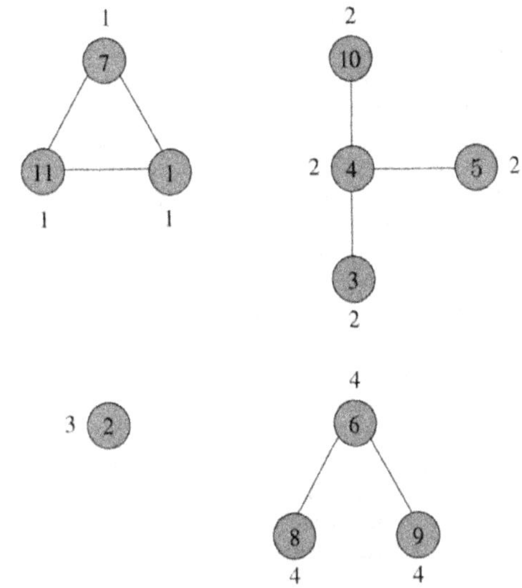

components	1	3	2	2	2	4	1	4	4	2	1
	1	2	3	4	5	6	7	8	9	10	11

Figure 13.12: A complete execution of the CONNECTED COMPONENTS ALGORITHM on the graph shown in Figure 13.9.

We now discuss determining the worst-case running time of the CONNECTED COMPONENTS ALGORITHM. Consider the algorithm on input $G = (V, E)$, where $|V| = n$ and $|E| = m$. Since each vertex is visited exactly once by the **while** loop, the procedure SEARCH(v) is called n times. Assuming that a data structure such as an adjacency list is used to represent the graph, successive neighbors on the adjacency list can be accessed in one step. Since from each vertex all of its neighbors are checked, by the Hand-Shaking Theorem the total number of checks is simply twice the number of edges of G, or the value $2m$. Thus, the algorithm's worst-case running time is $O(\max(\{n, m\})$. This running time is optimal, since each edge must be checked.

13.7 Dijkstra's Shortest Path Algorithm

In this section we cover a classic algorithm for computing the costs of shortest paths between a designated source vertex and all other vertices in a weighted digraph. The algorithm, called DIJKSTRA'S SHORTEST PATH ALGORITHM, is named for its founder the late Edsger Dijkstra—the 1972 ACM Turing Award winner. In this case the *cost* of a path is defined as the sum of the weights of the edges on the path. This algorithm has many applications, including network routing protocols and route-finding problems.

DIJKSTRA'S SHORTEST PATH ALGORITHM takes as input a weighted digraph $G = (V, E)$ with a nonnegative weight function W and a designated source vertex $s \in V$, where $V =$

$\{1, \ldots, n\}$. Without loss of generality, we assume that the source vertex s is vertex 1. In the algorithm the values of the weight function are denoted by $W_{ij} \geq 0$ for $1 \leq i, j \leq n$. Here W_{ij} indicates the cost of the edge from vertex i to vertex j. If no such edge exists, the weight is assumed to be ∞.

DIJKSTRA'S SHORTEST PATH ALGORITHM maintains a set of vertices U for which the cost of a shortest path from the source vertex s to each vertex in U has not yet been computed. (The U stands for "unknown.") An array pathcost[·] is maintained to keep track of the values of the shortest path estimates to all vertices. During each iteration of the main **for** loop, one vertex v is removed from U. Removal of vertex v from U signifies that the algorithm has found the cost of a shortest path to v. We ask the reader to prove this fact in Exercise 13.35. This exercise shows that DIJKSTRA'S SHORTEST PATH ALGORITHM correctly computes the costs of shortest paths from s.

The algorithm next updates the shortest path estimates maintained in the array pathcost[·]. A cost estimate for a vertex u either remains the same or is improved by taking a path through v and then following the edge from v to u. Although it appears as though updating pathcost[u] should involve comparisons with every vertex in $V \setminus U$ as well as v itself, these comparisons were in fact already made for each $t \in V \setminus U$, when t was removed from U. Note that those comparisons are sufficient, since the value pathcost[t] has not been modified. The algorithm is presented next.

DIJKSTRA'S SHORTEST PATH ALGORITHM

Input: A weighted digraph $G = (V, E)$ with a nonnegative weight function W, $V = \{1, \ldots, n\}$, and a designated source vertex $s = 1 \in V$.
Output: The array pathcost[·] of size n, where pathcost[i] contains the cost of a shortest path between the source vertex s and vertex i.

```
begin
   U = {2, ..., n};
   for i = 2 to n do pathcost[i] = W_{1i}; //initialize distances
   for i = 1 to n − 2 do { //main loop
      U_min = {x | x ∈ U and pathcost[x] is a minimum};
      v = an element from U_min;
      U = U \ {v};
      for every u ∈ U do
         pathcost[u] = min({pathcost[u], pathcost[v] + W_{vu}});
   }
end.
```

Before analyzing the running time of the algorithm, we present an example.

Example 13.8 We illustrate DIJKSTRA'S SHORTEST PATH ALGORITHM using the digraph shown in Figure 13.13. The source vertex s is vertex 1. The algorithm begins by setting $U = \{2, 3, 4, 5, 6\}$ and initializing the array pathcost[·]. In this case only vertices 3 and 4 are adjacent to vertex 1. Their corresponding values in the array pathcost[·] are set to 3 and 5, respectively. All other values (except for the source) are set to ∞. In the figures corresponding to this example, we show edges as being thick if they are included in some current shortest-path estimate. In Figure 13.13 the edges $(1, 3)$ and $(1, 4)$ are grayed, indicating that these two edges are involved in shortest path estimates, namely the path from 1 to 3 and the path from

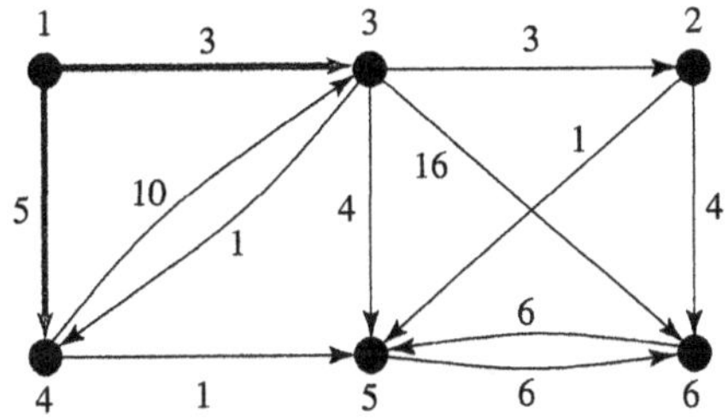

	1	2	3	4	5	6
pathcost		∞	3	5	∞	∞

$U = \{2, 3, 4, 5, 6\}$

Figure 13.13: A weighted digraph used to illustrate DIJKSTRA'S SHORTEST PATH ALGORITHM.

1 to 4. The values of pathcost[·] and U are shown in all the figures corresponding to this example.

Figure 13.14 shows that vertex 3 is selected during the first iteration of the main **for** loop. This choice results in improved estimates for shortest paths to vertices 2, 4, 5, and 6. Since vertex 3 is removed from U during this iteration, we know that the cost of a shortest path to vertex 3 has been computed. In this case the path $1 \rightarrow 3$ is the shortest path to vertex 3, and this path has a cost of 3. The gray edges (1, 3) and (3, 6), for example, illustrate the current shortest path from vertex 1 to vertex 6.

Figure 13.15 shows that vertex 4 is selected during the second iteration of the main **for** loop. This choice results in an improved estimate for a shortest path to vertex 5. Notice

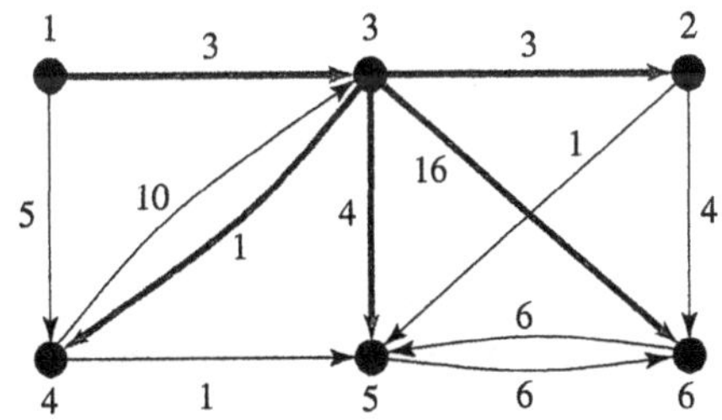

	1	2	3	4	5	6
pathcost		6	3	4	7	19

$U = \{2, 4, 5, 6\}$

Figure 13.14: Vertex 3 is selected for removal from U. Paths to the remaining vertices in U are updated.

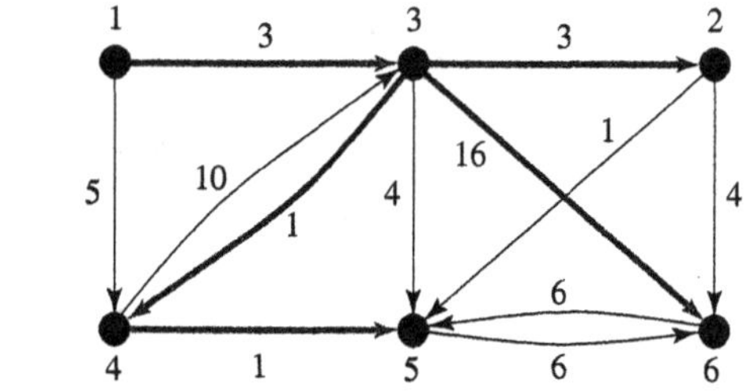

pathcost		6	3	4	5	19
	1	2	3	4	5	6

$U = \{2, 5, 6\}$

Figure 13.15: Vertex 4 is selected for removal from U. Paths to the remaining vertices in U are updated.

that W_{42} and W_{46} are equal to ∞, so no improvements in the known best paths to vertices 2 and 6 are made at this stage. Since vertex 4 is removed from U during this iteration, we know that the cost of a shortest path to vertex 4 has been computed. In this case the path $1 \rightarrow 3 \rightarrow 4$ is the shortest path to vertex 4, and this path has a cost of 4. The gray edges (1, 3) and (3, 4) illustrate the shortest path from vertex 1 to vertex 4. Note that the gray edges (1, 4) and (4, 5) are included to illustrate the shortest path estimate from vertex 1 to vertex 5.

Figure 13.16 shows that vertex 5 is selected during the third iteration of the main **for** loop. This choice results in an improved estimate for a shortest path to vertex 6. Since vertex

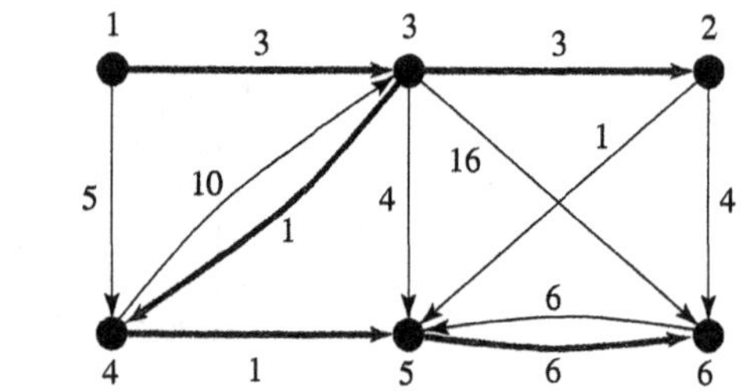

pathcost		6	3	4	5	11
	1	2	3	4	5	6

$U = \{2, 6\}$

Figure 13.16: Vertex 5 is selected for removal from U. Paths to the remaining vertices in U are updated.

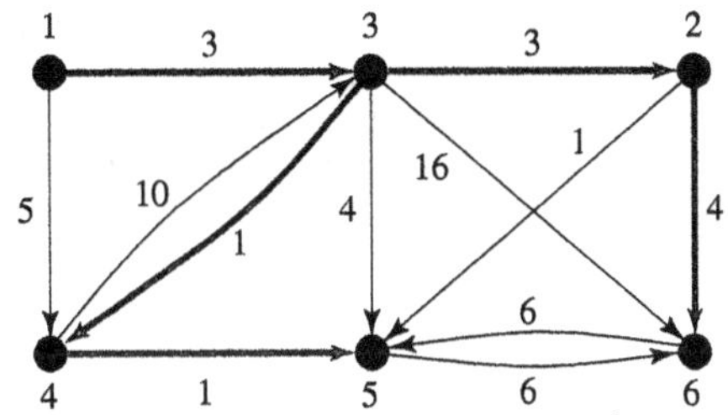

pathcost						
		6	3	4	5	10
	1	2	3	4	5	6

$U = \{6\}$

Figure 13.17: Vertex 2 is selected for removal from U. Paths to the remaining vertices in U are updated.

5 is removed from U during this iteration, we known that the cost of a shortest path to vertex 5 has been computed. In this case the path $1 \to 3 \to 4 \to 5$ is the shortest path to vertex 5, and this path has a cost of 5. The gray edges $(1, 3)$, $(3, 4)$, and $(4, 5)$, for example, illustrate the shortest path from vertex 1 to vertex 5.

Figure 13.17 shows that vertex 2 is selected during the final iteration of the main **for** loop. This choice results in an improved estimate for a shortest path to vertex 6. Since vertex 2 is removed from U during this iteration, we know that the cost of a shortest path to vertex 2 has been computed. In this case the path $1 \to 3 \to 2$ is the shortest path to vertex 2, and this path has a cost of 6. No additional improvements are possible, as only one vertex remains in the set U. So, a shortest path to vertex 6 has a cost of 10. Edges involved in the shortest paths computed by the algorithm are shown in gray. Note that the edge $(3, 4)$, for example, is used in several shortest paths.

Notice that the algorithm does not explicitly output the shortest paths it finds. In Exercise 13.41 we ask the reader to modify the algorithm so that it does compute the paths themselves.

Now we turn to analyzing the running time of DIJKSTRA'S SHORTEST PATH ALGORITHM. Suppose that $G = (V, E)$ has n vertices and that the weight function W is stored in an $n \times n$ matrix. The first two initialization steps require $O(n)$ time. The main **for** loop is executed $O(n)$ times. The inner **for** loop is the most expensive step inside the main **for** loop. This loop can be executed in $O(n)$, since U has $O(n)$ elements. Thus, the algorithm's worst-case running time is $O(n) \times O(n) = O(n^2)$.

13.8 Java Source Code

In this chapter we have presented a number of important algorithms in graph theory, including breadth-first search, depth-first search, connected components, and DIJKSTRA'S SHORTEST PATH ALGORITHM. We have used these algorithms to give a flavor of the types of algorithms available for exploring graphs and computing graph properties. Throughout the text, we have sprinkled algorithms here and there. Additionally, we have provided many references to graph algorithms. However, up until now, we have not presented any specific implementations of algorithms. In the remainder of this section we provide Java source code for the algorithms described in this chapter. It is important to study this code carefully. The code is commented extensively and should be easy to follow when coupled with the descriptions given in this chapter.

CODE LISTING : : **BFS.java**

```java
package algorithms;
import graph.Graph;
import java.util.Vector;
/** This class demonstrates a Breadth-First Search Algorithm. */
public class BFS {
  /**
   * An implementation of the Breadth-First Search Algorithm
   *  @param  ARGS                    input application parameters
   */
  public static void main(String args[]) {
    // create the graph from a file
    Graph graph = new Graph("bfsExample.graph");
    // make the graph undirected
    graph.makeUndirected();
    // print the graph as an adjacency list
    System.out.println("Undirected Graph: ");
    graph.print();
    // ALG: T1 = ({s}, 0);
    Graph bfsTree = new Graph(graph.size());
    // ALG: mark s;
    graph.visited(graph.getRoot());
    // ALG: initialize Q to be empty queue;
    Vector queue = new Vector();
    // ALG: ENQUEUE(Q, s);
    queue.add(new Integer(graph.getRoot()));
    // ALG: while not EMPTY(Q) do
    while (queue.size() > 0) {
      // ALG: x = DEQUEUE(Q);
      int x = ( (Integer) queue.firstElement()).intValue();
      // ALG: DEQUEUE(Q);
      queue.remove(queue.firstElement());
      // ALG: for all edges {x, y}
      for (int i = 0; i < graph.childrenCount(x); i++) {
        int y = graph.getChild(x, i);
        // ALG: where y is not marked do
        if (graph.isVisited(y)) continue;
        // ALG: mark y
        graph.visited(y);
        // ALG: Ti = Ti-1 U {x, y};
        bfsTree.addDirectedEdge(x, y);
        // ALG: ENQUEUE(Q, y)
        queue.add(new Integer(y));
      }
    }
    // ALG: output Tn;
    System.out.println("Breadth-First Search Tree: ");
    bfsTree.print();
    System.exit(0);
  }
```

continue...

continued::**BFS.java**

```
}
```


CODE LISTING::**DFS.java**

```
package algorithms;
import graph.Graph;
/** This class demonstrates a Depth-First Search Algorithm. */
public class DFS {

  /**
   * Traverses a forest using Depth-First Search Algorithm.
   *  @param  GRAPH             Graph the graph that will be traversed
   */
  public static Graph DFSSearch(Graph graph){
    if (graph == null) return null; // check for valid input
    // graph is an undirected graph.
    // ALG: for all v (that belong to) V do
    // ALG:     color[v] = white;
    graph.reset();
    // create an empty graph representing the depth-first search tree
    Graph dfsTree = new Graph(graph.size());
    // ALG: for all v (that belong to) V do
    for (int i = 1; i <= graph.size(); i++) {
      // ALG: if color[v] = white then DFS(v);
      if (!graph.isVisited(i)){
        // execute a depth-first search and
        // store the depth-first search tree in the dfsTree variable
        DFS(i, graph, dfsTree);
      }
    }
    return dfsTree;
  }

  /**
   * Recursive implementation of a Depth-First Search Algorithm. Each
   * node is marked as "Visiting" when it is first visited and marked
   * as "Visited" when the recursive function leaves the node.
   *  @param  NODE        the node that has not been visited yet
   *  @param  GRAPH       the graph
   *  @param  DFSTREE     the dfsTree will be stored in this variable
   */
  public static void DFS(int node, Graph graph, Graph dfsTree) {
    // check if the input is valid
    if (graph == null || dfsTree == null)return;
    int child;
    // mark the node as "Visiting"
    // ALG: color[v] = gray;
    graph.visiting(node);
    // ALG: for all w (that belong to) N(v) do
    // for all children of the visiting node
    for (int i = 0; i < graph.childrenCount(node); i++) {
      child = graph.getChild(node, i);
      // ALG: if color[w] = white then {
      // if the child was not visited yet
      if (!graph.isVisited(child)) {
        // ALG: add edge {v, w} to T;
```

continue...

continued::**DFS.java**

```
        // add the edge to the depth-first search tree
        dfsTree.addDirectedEdge(node, child);
        // ALG: DFS(w);
        // go to the next node
        DFS(child, graph, dfsTree);
      }
    }
    // all children are visited, mark this node as "Visited"
    graph.visited(node);
  }

  /**
   * An implementation of the Depth-First Search Algorithm
   * @param ARGS input application parameters
   */
  public static void main(String[] args) {
    // create the graph from a file
    Graph graph = new Graph("dfsExample.graph");
    // make the graph undirected
    graph.makeUndirected();
    // print the graph as an adjacency list
    System.out.println("Undirected Graph: ");
    graph.print();

    // ALG: DFSSEARCH(G);
    Graph dfsTree = DFSSearch(graph);
    if (dfsTree ≠ null){
      // print entered and exited times for each node
      System.out.println("Enter and Exit times:");
      graph.printTimes();
      // print the depth-first search tree graph as an adjacency list
      System.out.println("Depth-First Search Tree:");
      dfsTree.print();
    }
    System.exit(0);
  }
}
```

java2LaTeX © 2002 rusi@acm.org

CODE LISTING::**Components.java**

```
package algorithms;

import graph.Graph;

/** This class demonstrates the Connected Components Algorithm. */
public class Components {

  /**
   * Search the graph from the selected node and mark all nodes
   * that can be reached with the sequential number for that
   * connected component
   * @param GRAPH Graph undirected graph
   * @param NODE  int vertex that has not been visited yet
   * @param COUNT int a number for the connected component
   */
  public static void search(Graph graph, int node, int count){
```

continue...

continued::**Components.java**

```
    // ALG: color[u] = gray;
    graph.visiting(node);
    // ALG: components[u] = component-number;
    graph.mark(node, count);
    int child;
    // ALG: for all w (that belong to) N(u) do
    for (int i = 0; i < graph.childrenCount(node); i++) {
      child = graph.getChild(node, i);
      // ALG: if color[w] = white then SEARCH(w);
      if (!graph.isVisited(child)) search(graph, child, count);
    }
  }

  /**
   * A Connected Components Algorithm is used to find the
   * number of connected components.
   *  @param   ARGS                input application parameters
   */
  public static void main(String[] args) {
    // create the graph from a file
    Graph graph = new Graph("ccExample.graph");
    // make the graph undirected
    graph.makeUndirected();
    // print the graph as an adjacency list
    System.out.println("Undirected Graph: ");
    graph.print();
    // ALG: component-number = 1;
    int count = 1;
    // ALG: for all v (that belong to) V do
    // ALG:    color[v] = white;
    graph.reset();
    // ALG: while color[v] = white
    //           for some v (that belongs to) V do
    for (int i = 1; i <= graph.size(); i++) {
      if (graph.isVisited(i)) continue;
      // ALG: SEARCH(G, v);
      search(graph, i, count);
      // ALG: component-number = component-number + 1;
      count = count + 1;
    }
    System.out.println("Number of connected components: "
           + (count-1));
    System.out.println("Node marks:");
    graph.printMarks();
    System.exit(0);
  }
}
```

java2LaTeX © 2002 rusi@acm.org

CODE LISTING::**ShortestPath.java**

```
package algorithms;
import graph.Graph;
import graph.Utils;
/** This class demonstrates the Dijkstra's Shortest Path Algorithm. */
public class ShortestPath {
  /**
```

continue...

continued::**ShortestPath.java**

```
   * The Dijkstra's Shortest Path Algorithm is used to find the
   * shortest path between a designated source vertex s = 1 and
   * all other nodes of the graph.
   *  @param  ARGS              input application parameters
   */
  public static void main(String[] args) {
    // create the graph from a file
    Graph graph = new Graph("shortestExample.graph");
    // make the graph undirected
    graph.makeUndirected();
    // print the graph as an adjacency list
    System.out.println("Undirected Graph: ");
    graph.print();
    // ALG: U = {2, ... , n};
    int[] nodes = new int[graph.size()-1];
    for (int i = 0; i < nodes.length; i++) nodes[i] = i+2;
    // set all weights that are unknown to an arbitrary high number
    Utils.matrixSubst(graph.getWeightMatrix(), -1, 99999);
    // print the weights for each path
    System.out.println("Weight matrix:");
    graph.printWeights();
    // initialize distances
    // ALG: for i = 2 to n do pathcost[i] = W_1i;
    for (int i = 2; i <= graph.size(); i++)
      graph.mark(i, graph.getWeight(1, i));
    // ALG: for i = 1 to n - 2 do { // main loop
    for (int i = 1; i <= graph.size()-2; i++){
      // ALG: U_min = { x | x (that belongs to) U
      //          and pathcost[x] is a minimum};
      int w = 0;
      // from all nodes, find the minimum
      for (int j = 1; j < nodes.length; j++)
        if (graph.getMark(nodes[j])
            < graph.getMark(nodes[w])) w = j;
      // ALG: v = an element from U_min;
      int v = nodes[w];
      // ALG: U = U \ {v};
      nodes = Utils.remove(nodes, w);
      // ALG: for every u (that belongs to) U do
      for (int u = 0; u < nodes.length; u++){
        // ALG: pathcost[u] =
        //          min({pathcost[u], pathcost[v] + W_vu});
        graph.mark(nodes[u], Math.min(graph.getMark(nodes[u]),
                        graph.getMark(v) +
                        graph.getWeight(v, nodes[u])));
      }
    }
    System.out.println("Shortest paths from node 1:");
    graph.printMarks();
    System.exit(0);
  }
}
```

java2LaTeX © 2002 rusi@acm.org

13.9 Exercises

13.1. Report three Web sites that actually contain code for a graph algorithm.

13.2. How many books can you find having "Graph Algorithms" in their title?

13.3. Using your favorite search engine, determine which graph algorithm is the most popular on the Web. Popularity should be gauged by the number of search engine hits.

13.4. Construct a table that evaluates the following functions $\lg n$, n, $n \lg n$, n^2, n^3, and 2^n at the values 10^1, 10^2, 10^3, 10^4, and 10^5. If these are the running times in terms of seconds, for which values and which functions is less than a day required, under a year, and over a year?

13.5. Show that for functions $f, g : \mathbb{N} \to \mathbb{R}^+$, we have the following:

$$\begin{aligned} O(f(n)) + O(g(n)) &= O(f(n) + g(n)), \\ O(f(n)) + O(g(n)) &= O(\max(\{f(n), g(n)\})), \\ O(f(n)) \cdot O(g(n)) &= O(f(n)g(n)), \\ f(n) \cdot O(g(n)) &= O(f(n)g(n)). \end{aligned}$$

Similarly, show that the previous equations also hold for Ω or Θ instead of O.

13.6. Show for each $k \in \mathbb{N}$, we have that

$$O(f(n))^k = O(f(n)^k),$$

and use this fact to show that if $p(x)$ is a polynomial, then

$$p(O(f(n))) = O(p(f(n))) = O((p \circ f)(n))).$$

13.7. Show that $\lg(O(f(n))) = O(\lg(f(n)))$.

13.8. Is it always true that $a^{O(f(n))} = O(a^{f(n)})$, where a is a positive real number?

13.9. Intuitively, explain why you would expect it to be easier to determine the worst-case running time rather than the average-case running time of an algorithm.

13.10. Assume an algorithm A has a running time given by $t_{1;A}(n) = \sqrt{n}$ for 25% of all the possible inputs of size n, and that the running time is given by $t_{2:A}(n) = n^{1.2}$ for the other 75% of the inputs of size n. What is the worst-case running time for A as a function of n, and what is the average running time for A as a function of n? Compute the Θ estimates for both cases.

13.11. Suppose an algorithm B has a running time given by $t_{1;B}(n) = \sqrt{n}$ for all but 1,000 of the possible $n \lg n$ inputs of size n, and that the running time is given by $t_{2:B}(n) = n^{1.2}$ for the 1,000 remaining inputs of size n. What is the worst-case running time for A as a function of n, and what is the average running time for A as a function of n? Compute the Θ estimates for both cases.

13.12. How many bits are required to store a graph using edge lists?

13.13. Which graph representation among adjacency matrix, incidence matrix, and edge lists is more efficient in terms of the number of bits required to represent the graph?

13.14. Implement the BREADTH-FIRST SEARCH ALGORITHM using a *stack* as the underlying data structure. [*Hint:* Research the stack data structure on the World Wide Web.]

13.15. Modify the BREADTH-FIRST SEARCH ALGORITHM to handle graphs that are not connected.

13.16. Modify the BREADTH-FIRST SEARCH ALGORITHM so that it outputs the distance from the start vertex to each vertex. These values are called *level numbers*.

13.17. In Example 13.5 will a different breadth-first search tree result if a different order is chosen for exploring the neighbors of the vertex that was just dequeued? If so, provide an example of another breadth-first search tree that the algorithm would construct. Try exploring greedily by always visiting the lower-numbered neighbors first. Does this result in a different breadth-first search tree?

13.18. Write an algorithm that takes as input a graph G and the vertices u and v, and outputs the order in which these vertices are visited in a breadth-first search that uses a stack as an underlying data structure. Write a breadth-first search algorithm that performs the same task but uses a queue as its underlying data structure. For what types of inputs will the answers be the same?

13.19. Implement a DEPTH-FIRST SEARCH ALGORITHM using your favorite programming language other than Java.

13.20. Can you implement a DEPTH-FIRST SEARCH ALGORITHM using a *queue* as an underlying data structure?

13.21. Modify the DEPTH-FIRST SEARCH ALGORITHM to output ordered pairs (i, j), where i denotes a vertex number and j signifies that vertex i was the jth vertex colored gray by the DEPTH-FIRST SEARCH ALGORITHM.

13.22. Modify the DEPTH-FIRST SEARCH ALGORITHM so that it outputs the distance from the start vertex to each vertex. These values are called *level numbers*.

13.23. In Example 13.6 will a different depth-first search tree result if a different order is chosen for exploring the neighbors of the vertex that was just visited? If so, provide an example of another depth-first search tree that the algorithm would construct. Try exploring greedily by always visiting the higher-numbered neighbors first. Does this result in a different depth-first search tree?

13.24. Write an algorithm that takes as input a graph G and the vertices u and v, and outputs the order in which these vertices are visited in a depth-first search.

13.25. For what types of inputs will the breadth-first search and depth-first search trees be the same?

13.26. Choose an appropriate input representation for graphs. Implement the CONNECTED COMPONENTS ALGORITHM using your favorite programming language besides Java. Comment your code.

13.27. Our CONNECTED COMPONENTS ALGORITHM was implemented based on depth-first search. Implement a CONNECTED COMPONENTS ALGORITHM using breadth-first search.

13.28. Code an algorithm in your favorite programming language that outputs the size of the second-largest connected component of a graph.

13.29. Define an algorithm for a "new" search technique that combines features of breadth-first search and depth-first search. For what graphs, if any, does your search result in a breadth-first search? How about a depth-first search?

13.30. In your favorite programming language implement an algorithm to find a cut set with respect to a given pair of vertices u and v.

13.31. Choose an appropriate input representation for graphs. Implement the DELTA PLUS ONE COLORING ALGORITHM using your favorite programming language. Comment your code.

13.32. Choose an appropriate input representation for graphs. Implement the EULERIAN CIRCUIT ALGORITHM using your favorite programming language. Comment your code.

13.33. Choose an appropriate input representation for graphs. Implement PRIM'S ALGORITHM using your favorite programming language. Comment your code.

13.34. Choose an appropriate input representation for graphs. Implement KRUSKAL'S ALGORITHM using your favorite programming language. Comment your code.

13.35. Prove that DIJKSTRA'S SHORTEST PATH ALGORITHM correctly computes shortest paths by showing that once a vertex is removed from the set U, its shortest path from s is known. [*Hint:* Use induction to prove facts about vertices in U and in $V \setminus U$.]

13.36. Explain what can "go wrong" with DIJKSTRA'S SHORTEST PATH ALGORITHM if we allow negative weights.

13.37. Implement DIJKSTRA'S SHORTEST PATH ALGORITHM using your favorite programming language besides Java.

13.38. Write a two-page biography of Edsger Dijkstra. Document your sources.

13.39. Do the magnitudes of the weights impact the running time of DIJKSTRA'S SHORTEST PATH ALGORITHM? Carefully justify your answer.

13.40. Devise your own shortest path algorithm based on breadth-first search. How does its running time compare with that of DIJKSTRA'S SHORTEST PATH ALGORITHM?

13.41. Modify DIJKSTRA'S SHORTEST PATH ALGORITHM so that your version explicitly outputs the actual minimum cost paths themselves.

13.42. Write a one-page paper that discusses three practical applications of DIJKSTRA'S SHORTEST PATH ALGORITHM.

13.43. Suppose DIJKSTRA'S SHORTEST PATH ALGORITHM has just been run on a weighted digraph G with source vertex s. If another vertex v is now added to G, how can we most effectively make use of the information already determined by the algorithm to compute the shortest distance from s to v? Do other path weights need to be updated? If so, how can we do this computation efficiently?

13.44. Modify DIJKSTRA'S SHORTEST PATH ALGORITHM to compute the costs of shortest paths between *all* pairs of vertices (in both directions).

13.45. Modify DIJKSTRA'S SHORTEST PATH ALGORITHM so that it works for weighted undirected graphs.

APPENDIX A

Greek Alphabet

This appendix contains a list of the lower- and upper-case Greek letters, as well as their spelling in English.

Lower Case	Upper Case	Name
α	A	alpha
β	B	beta
γ	Γ	gamma
δ	Δ	delta
ϵ	E	epsilon
ζ	Z	zeta
η	H	eta
θ	Θ	theta
ι	I	iota
κ	K	kappa
λ	Λ	lambda
μ	M	mu
ν	N	nu
ξ	Ξ	xi
o	O	omicron
π	Π	pi
ρ	P	rho
σ	Σ	sigma
τ	T	tau
υ	Υ	upsilon
ϕ	Φ	phi
χ	X	chi
ψ	Ψ	psi
ω	Ω	omega

APPENDIX

B

Notation

The notation used throughout this book is described below. It is divided into the following categories: function, graph, miscellaneous, number, and set. Alphabetizing is done either by the symbols themselves *or* by the meaning of the symbols. Many of the concepts are standard, and a definition is not given here. Instead we simply explain the notation. Items not defined in the book, but that are perhaps less familiar, are defined here. Concepts defined in the book the reader can usually trace through the index.

- **Function**

f, g, and h	Variables used to represent functions.
$f \circ g$	Function composition.
f^{-1}	Inverse of function f.
ι_X	Identity map from X to X.
$O(f(n))$	Set of functions whose growth rate is at most order $f(n)$.
$\Omega(f(n))$	Set of functions whose growth rate is at least order $f(n)$.
$\Theta(f(n))$	Set of functions whose growth rate is both $O(f(n))$ and $\Omega(f(n))$.
$n^{O(1)}$	Set of polynomially bounded functions.

- **Graph**

G'	Abstract dual of G.
$\mathsf{A}(G)$	Adjacency matrix of the graph G.
$\text{Anc}(u)$	Set of u's ancestors in a rooted tree.
$\text{Ave}(T)$	Average level of the leaves in a tree T.

$\mathsf{B}(G)$	Incidence matrix of the graph G.
$B(G)$	Block graph.
$\mathrm{BC}(G)$	Block-cutpoint graph of G.
$\chi(G)$	Chromatic number of G.
$\chi_G(t)$	Chromatic polynomial of G.
$\mathrm{Chi}(u)$	Set of u's children in a rooted tree.
$K_{m_1,\ldots,m_k}$	Complete k-partite graph with partition sizes of $m_1, \ldots, m_k$.
K_m^k	Complete k-partite graph with all partitions of size m.
δ_X	Deficiency of a bipartite graph at the part X.
$\gamma(G)$	Domination number of G.
$\cong$	Graph isomorphism relation.
$\mathrm{Des}(u)$	Set of u's descendants in a rooted tree.
$d_G(u)$	Degree of vertex u in G.
$d(u)$	Degree of vertex u in the graph under consideration.
$\mathrm{diam}(G)$	Diameter of the graph G.
$\partial_G(u, v)$	Distance between u and v in the graph G.
$\partial(u, v)$	Distance between u and v in the graph under consideration.
∂W	Infinitesimal flow.
$\chi_d(G)$	Distance-d chromatic number of G.
$\chi'(G)$	Edge chromatic number of G.
$E(u)$	Eccentricity of u.
$E_G(u)$	Eccentricity of u in the graph G.
$\beta'(G)$	Edge covering number of G.
$\chi(S)$	Euler characteristic of surface S.
G and H	Variables used to represent graphs.
$G(E)$	Set of edges of the graph G.
$G[F]$	Subgraph of the graph $G = (V, E, \phi)$ induced by the set of edge $F \subseteq E$.
$G \cup \{u, v\}$	The simple graph obtained by adding the edge between u and v to G. That is, $G \cup \{u, v\} = (V(G), E(G) \cup \{\{u, v\}\})$.
$G(V)$	Set of vertices of the graph G.
$G[W]$	Subgraph of the graph $G = (V, E, \phi)$ induced by the set of vertices $W \subseteq V$.
(G, W)	Weighted graph having weight function W.
$G' \subseteq G$	G' is a subgraph of G.
$\overline{G}$	Complement of G.
pq	Composite walk, trail, or path.

$G \cdot e$	Contraction of G by edge e.
G/e	Simple contraction of G by edge e.
$\beta(G)$	Vertex covering number of G.
$\nu(G)$	Crossing number of G.
$\vec{G}$	Directed graph G.
ϕ	Edgemap of a graph.
$G - e$	Graph G with edge e removed.
$G - v$	Graph G with vertex v removed.
$\gamma(G)$	Genus of G.
G^*	Geometric dual of G.
$\mathrm{Hd}(u)$	Head of the edge u in the underlying digraph.
ht(G)	Height of the tree T.
$\ell\mathrm{max}$	Height of a rooted tree.
p^{-1}	Inverse walk, trail, or path.
$\kappa(G)$	Vertex connectivity of G.
$\kappa'(G)$	Edge connectivity of G.
$\ell_T(u)$	Level of vertex u in rooted tree T.
$\ell(u)$	Level of vertex u in a rooted tree.
$\delta_G(t)$	Matching defect polynomial of G.
$\alpha'(G)$	Matching number of G.
Δ	Maximum degree of a specific graph.
$\Delta(G)$	Maximum degree of a vertex in G.
$\delta(G)$	Minimum degree of a vertex in G.
$N(u)$	Set of neighbors of vertex u in the graph under consideration.
$N_G(u)$	Set of neighbors of vertex u in the graph G.
$\mathrm{par}(u)$	Parent of u in a rooted tree.
$\mu(G)$	Number of perfect matchings of G.
G^d	The dth power of G.
$\mu_C(t)$	Rook polynomial of the chessboard C.
S and T	Variables used to represent trees.
$S(V_1, V_2)$	Cut set consisting of all edges having one endvertex in V_1 and the other in V_2.
$\tau(G)$	Set of all spanning trees of G.
$\mathrm{Tl}(u)$	Tail of edge u in the underlying digraph.
(T,r)	Tree rooted at r.
$T_k(n)$	Turán graph.

(u, v)	The two-tuple consisting of u and v. If u and v are vertices in a graph, this pair represents the directed edge from u to v.
$\{u, v\}$	Undirected edge between u and v.
$\text{val}(f)$	Value of the flow of f.

- **Miscellaneous**

//	Indicates a comment in an algorithm.
$\det(M)$	Determinant of the square matrix M.
$\dim_F(W)$	Dimension of the vector space W over field F.
i through n	Variables representing natural numbers.
Ω	Ohm.
$\text{Rank}(A)$	Rank of the matrix A.
$B_{-1}(G)$	Reduced incidence matrix of G.
M^t	Transpose of matrix M.
u, v, and w	Variables used to represent vertices.

- **Number**

$B(n)$	Bell numbers.
$\binom{n}{k}$	Value of $\frac{n!}{k!(n-k)!}$.
C_n	The nth Catalan number.
$\lceil x \rceil$	Least integer not less than x.
$n!$	Factorial. Product $1 \cdot 2 \cdots (n-1) \cdot n$. Note that 0! is defined to be 1.
$(t)_n$	Falling factorial, $(t)_n = t(t-1)\cdots(t-n+1)$.
$t^{\underline{n}}$	Falling factorial, $t^{\underline{n}} = t(t-1)\cdots(t-n+1)$.
$\lfloor x \rfloor$	Greatest integer not exceeding x.
$\lg n$	Maximum of 1 and $\lceil \lg_e n \rceil$.
$S(n, k)$	Stirling numbers of the second kind.

- **Set**

$\|S\|$	Cardinality of set S.
$\mathbb{C}$	Set of complex numbers, $\mathbb{C} = \{x + yi : x, y \in \mathbb{R}\}$, where $i = \sqrt{-1}$ denotes the imaginary number satisfying $i^2 = -1$.
$\in$	Element of a set.
$\varnothing$	Empty set. The set containing no elements.
$\mathbb{Z}$	Set of integers, $\{\dots, -2, -1, 0, 1, 2, \dots\}$.
$\max(S)$	Set of all maximal elements of poset $(S, \preceq)$.
$\min(S)$	Set of all minimal elements of poset $(S, \preceq)$.
$\mathbb{N}$	Set of natural numbers, $\{1, 2, \dots\}$.
$(S, \preceq)$	Partially ordered set.
$\mathbf{P}(S)$	Power set of S—the set of all possible subsets of S.
$\mathbb{Q}$	Set of rational numbers, $\mathbb{Q} = \{a/b : a, b \in \mathbb{Z} \text{ and } b \neq 0\}$.
$\mathbb{R}$	Set of real numbers.
$[a; b]$	Real numbers x such that $a \leq x \leq b$.
$[a; b[$	$[a; b] \setminus \{b\}$.
$]a; b]$	$[a; b] \setminus \{a\}$.
$]a; b[$	$[a; b] \setminus \{a, b\}$.
$\sim$	Denotes a relation.
$\setminus$	Set difference.
$\cap$	Set intersection.
$\cup$	Set union.
$\subseteq$	Subset.
Δ	Symmetric difference.

APPENDIX

C

Top Ten Online References

This appendix contains ten uniform resource locators (URLs) relating to graph theory. These Web pages are good starting points for exploring graph theory online. All references were accessed in April 2006. Note that a query of "Graph Theory" to the Google search engine resulted in 5,830,000 hits on April 9, 2006.

Each citation for a Web presentation provides its title (as capitalized on the Web page), where possible an "author" name, and URL. For the sake of readability, we have omitted the leading `http://` on URLs. Sometimes we just include the URL for the "main page" of a presentation, although many of its "subpages" may also be of interest. In this list, URLs are sorted by title.

1. *Combinatorics.*
   `www.combinatorics.org/`
2. *Graph Theory.*
   `en.wikipedia.org/wiki/Graph_theory`
3. Daniel P. Sanders. *Graph Theory Resources.*
   `www1.cs.columbia.edu/~sanders/graphtheory`
4. Daniel P. Sanders. *The Graph Theory White Pages.*
   `www1.cs.columbia.edu/~sanders/graphtheory/people/`
5. Stephen C. Locke. *Graph Theory.*
   `www.math.fau.edu/locke/graphthe.htm`
6. *Graph Theory Open Problems.*
   `dimacs.rutgers.edu/~hochberg/undopen/graphtheory/graphtheory.html`
7. *Graph Theory: Textbooks and Resources.*
   `www.graphtheory.com`
8. Chris K. Caldwell. *Graph Theory Tutorials.*
   `www.utm.edu/departments/math/graph/`
9. *Journal of Graph Theory.*
   `www3.interscience.wiley.com/cgi-bin/jhome/35334`
10. *Encyclopedia of Integer Sequences.*
    `www.research.att.com/~njas/sequences/`

Bibliography

We used the references listed below, many of which contain material not covered in this book. The references are a good place to begin further exploration of more advanced topics. The parenthesized number(s) indicate the page(s) in this book on which the reference is cited.

[1] G. Agnarsson, R. Greenlaw, and M. M. Halldórsson. On powers of chordal graphs and their colorings. *Congressus Numerantium*, 100(144):41–65, 2000. (294, 295)

[2] K. Appel and W. Haken. *Every Map is Four Colorable*. Providence, RI: American Mathematical Society, first edition, 1989. (249)

[3] R. Axelsson. Netafræði. Unpublished, 1989. (xiv)

[4] R. Balakrishnan and P. Paulraja. Powers of chordal graphs. *Journal of the Australian Mathematical Society, Series A*, 35(2):211–217, 1983. (296)

[5] B. Bollobás. *Graph Theory, An Introductory Course*, volume 63 of Graduate Texts in Mathematics. New York: Springer-Verlag, first edition, 1979. (323, 240)

[6] G. Brassard and P. Bratley. *Fundamentals of Algorithmics*. Upper Saddle River, NJ: Prentice Hall, 1996. (393)

[7] A. Cayley. On the theory of analytical forms called trees. *Philosophical Magazine*, 13:172–176, 1857. (101)

[8] N. Chandrasekharan, R. Laskar, and C. Veni Madhavan. Chromatic polynomials of chordal graphs. *Congressus Numerantium*, 61:133–142, 1988. (290)

[9] M. Chudnovsky, N. Robertson, P. Seymour, and R. Thomas. Progress on perfect graphs. *Mathematical Programming B*, 97(1–2):405–422, 2003. (288)

[10] T. H. Cormen, C. E. Leiserson, R. L. Rivest, and C. Stein. *Introduction to Algorithms*. Cambridge, MA: MIT Press, second edition, 2001. (393, 394)

[11] N. Deo. *Graph Theory*. Upper Saddle River, NJ: Prentice Hall, 1979. (xiv, 121, 365)

[12] R. Diestel. *Graph Theory*, volume 173 of Graduate Texts in Mathematics. New York: Springer-Verlag, second edition, 1997; third edition, 2005. (xiv, 323)

[13] G. A. Dirac. Connectivity theorems for graphs. *Quarterly Journal of Math*, 3(2):171–174, 1952. (143)

[14] S. Even. *Graph Algorithms*. New York: Computer Science Press, 1979. (393)

[15] I. Fary. On straight line representation of planar graphs. *Acta Scientiarum Mathematicarum*, 11:229–233, 1948. (199)

[16] R. Fritsch and G. Fritsch. *The Four-Color Theorem*. New York: Springer-Verlag, 1998. (248, 252, 257)

[17] M. Gardner. Mathematical games. *Scientific American*, 226(1):104–107, 1972. (332)

[18] F. Gavril. The intersection graphs of subtrees in trees are exactly the chordal graphs. *Journal of Combinatorial Theory B*, 16:47–56, 1974. (288)

[19] A. Ghouilá-Houri. Une condition suffisante d'existence d'un circuit hamiltonien. *Comptes Rendus de l'Académie des Sciences*, 25:495–497, 1960. (146)

[20] M. Goodrich and R. Tamassia. *Algorithm Design*. New York: John Wiley & Sons, 2002. (393)

[21] R. Gould. *Graph Theory*. San Francisco: Benjamin/Cummings, first edition, 1988. (xiv)

[22] L. R. Graham, D. E. Knuth, and O. Patashnik. *Concrete Mathematics*. Boston: Addison-Wesley, second edition, 1998. (373)

[23] M. J. Greenberg and J. R. Harper. *Algebraic Topology, A First Course*. Boston: Addison-Wesley, 1981. (226)

[24] F. Harary and E. M. Palmer. *Graphical Enumeration*. New York: Academic Press, 1973. (356, 364)

[25] F. Harary and R. W. Robinson. Labeled bipartite blocks. *Canadian Journal of Mathematics (Journal Canadien de Mathématiques)*, 31(1):60–68, 1979. (384)

[26] T. W. Haynes, S. T. Hedetniemi, and P. J. Slater. *Domination in Graphs. Advanced Topics*. Monographs and Textbooks in Pure and Applied Mathematics, 209. New York: Marcel Dekker, 1998. (312)

[27] T. W. Haynes, S. T. Hedetniemi, and P. J. Slater. *Fundamentals of Domination in Graphs*. Monographs and Textbooks in Pure and Applied Mathematics, 208. New York: Marcel Dekker, 1998. (312)

[28] P. Hell and J. Nešetřil. *Graphs and Homomorphisms*, volume 28 of Oxford Lecture Series in Mathematics and Its Applications. New York: Oxford University Press, 2004. (234)

[29] D. A. Huffman. A method for the construction of minimum redundancy codes. *Proceedings of the Institute of Radio Engineers*, 40:1098–1101, 1952. (93)

[30] R. E. Jamison and R. Laskar. Elimination orderings of chordal graphs. *Proceedings of the Seminar on Combinatorics and Applications (Calcutta, 1982)*. Calcutta: Indian Statistical Institute, 1984, pages 192–200. (288)

[31] D. Kőnig. Über Graphen und ihre Anwendung auf Determinantentheorie und Mengenlehre. *Mathematische Annalen*, 77:453–465, 1916. (333)

[32] D. E. Knuth. *Fundamental Algorithms*, volume 1 of *The Art of Computer Programming*. New York: Addison-Wesley, third edition, 1997. (93, 394)

[33] K. Kuratowski. Sur le problème des courbes gauches en topologie. *Fundamenta Mathematicae*, 15:271–283, 1930. (207)

[34] R. Laskar and D. Shier. On powers and centers of chordal graphs. *Discrete Applied Mathematics*, 6(2):139–147, 1983. (296)

[35] C. G. Lekkerkerker and J. C. Boland. Representation of a finite graph by a set of intervals on the real line. *Fundamenta Mathematicae*, 51:45–64, 1962/1963. (283)

[36] L. Lovász and M. D. Plummer. *Matching Theory*, volume 29 of North-Holland Mathematics Studies, 121. Annals of Discrete Mathematics. Amsterdam: North-Holland Publishing Company, 1986. (340, 342)

[37] U. Manber. *Introduction to Algorithms: A Creative Approach*. New York: Addison-Wesley, 1989. (393)

[38] T. McKee. How chordal graphs work. *Bulletin of the Institute of Combinatorics and Its Applications*, 9:27–39, 1993. (289)

[39] R. Merris. *Graph Theory*. Wiley-Interscience Series in Discrete Mathematics and Optimization. New York: Wiley-Interscience, first edition, 2001. (xiv, 280)

[40] J. W. Moon. *Counting Labelled Trees*. From lectures delivered to the Twelfth Biennial Seminar of the Canadian Mathematical Congress (Vancouver, 1969). Canadian Mathematical Monographs, No. 1. Montreal: Canadian Mathematical Congress, 1970. (104)

[41] A. Mowshowitz. The characteristic polynomial of a graph. *Journal of Combinatorial Theory, Series B*, 12:177–193, 1972. (342)

[42] J. R. Munkres. *Topology, A First Course*. Englewood Cliffs, NJ: Prentice-Hall, first edition, 1975. (201)

[43] O. Ore. A problem regarding the tracing of graphs. *Review of Elementary Mathematics*, 6:49–53, 1960. (143)

[44] C. H. Papadimitriou and K. Steiglitz. *Combinatorial Optimization: Algorithms and Complexity*. Mineola, NY: Dover Publications, 1998. (185)

[45] J. Riordan. *Introduction to Combinatorial Analysis*. Wiley Series in Probability and Mathematical Statistics. New York: John Wiley & Sons, 1958. (365)

[46] H. E. Robbins. A theorem on graphs, with an application to a problem in traffic control. *The American Mathematical Monthly*, 46:281–283, 1939. (163)

[47] K. Rosen. *Discrete Mathematics and Its Applications*. New York: McGraw-Hill, fourth edition, 1999. (73)

[48] A. Schrijver. *Theory of Linear and Integer Programming*. Wiley-Interscience Series in Discrete Mathematics and Optimization. New York: Wiley Interscience, 1998. (185)

[49] A. Seidenberg. A simple proof of a theorem of Erdős and Szekeres. *Journal of London Mathematical Society*, 34:352, 1959. (74)

[50] R. P. Stanley. *Enumerative Combinatorics, Volume 1*, number 49 of Cambridge Studies in Advanced Mathematics. New York: Cambridge University Press, 1997. (356, 371)

[51] R. P. Stanley. *Enumerative Combinatorics, Volume 2*, number 62 of Cambridge Studies in Advanced Mathematics. New York: Cambridge University Press, 1999. (87, 356, 371)

[52] C. Thomassen. Resistance and currents in infinite electrical networks. *Journal of Combinatorial Theory*, 49(B):87–102, 1990. (118)

[53] W. P. Thurston. *Geometry and Topology of 3-Manifolds*. Lecture Notes. Princeton, NJ: Princeton University Press, first edition, 1977–1978. (282)

[54] W. P. Thurston. *Three-Dimensional Geometry and Topology*, volume 1 of Princeton Mathematical Series. Princeton, NJ: Princeton University Press, 1997. (283)

[55] G. Valiente. *Algorithms on Trees and Graphs*. New York: Springer-Verlag, 2002. (393)

[56] D. J. A. Welsh and M. B. Powell. An upper bound for the chromatic number of a graph and its applications to timetabling problems. *Computer Journal*, 10:85–87, 1967. (241)

[57] D. B. West. *Introduction to Graph Theory*. Upper Saddle River, NJ: Prentice Hall, second edition, 2001. (xiv, 240, 319, 323)

[58] H. Wilf. *Generatingfunctionology*. New York: Academic Press, second edition, 1994. (356)

[59] R. J. Wilson. *Introduction to Graph Theory*. Upper Saddle River, NJ: Prentice Hall, fourth edition, 1996. (xiv, 231)

Index

C

D

E

F

H

I

J

K

S

T

U

V

W

Y

Z

www.ingramcontent.com/pod-product-compliance
Lightning Source LLC
LaVergne TN
LVHW080525210926
848481LV00022B/2952
* 9 7 8 0 1 3 1 4 2 3 8 4 8 *